BOOK SALE

PROBABILISTIC
MECHANICAL DESIGN

Probabilistic Mechanical Design

EDWARD B. HAUGEN

Associate Professor
Aerospace and Mechanical Engineering
The University of Arizona
Tucson, Arizona

A WILEY-INTERSCIENCE PUBLICATION

JOHN WILEY & SONS   New York · Chichester · Brisbane · Toronto

Copyright © 1980 by John Wiley & Sons, Inc.

All rights reserved. Published simultaneously in Canada.

Reproduction or translation of any part of this work beyond that permitted by Sections 107 or 108 of the 1976 United States Copyright Act without the permission of the copyright owner is unlawful. Requests for permission or further information should be addressed to the Permissions Department, John Wiley & Sons, Inc.

Library of Congress Cataloging in Publication Data:

Haugen, Edward B
    Probabilistic mechanical design.

    "A Wiley-Interscience publication."
    Bibliography: p.
    Includes index.
    1. Engineering design. 2. Probabilities.

I. Title.
TA174.H34   620'.00425   80-13428
ISBN 0-471-05847-5

Printed in the United States of America

10 9 8 7 6 5 4 3 2 1

*To My Wife Mercedes,
and to Johnnie and Cecilia.*

"The human understanding is of its own nature prone to suppose the existence of more order and regularity in the world than it finds. Once adopted such an opinion (either as being the received opinion or as being agreeable to itself) draws all things else to support and agree with it. And though there be greater number and weight of instances to be found on the other side, yet these it either neglects and despises, or else by some distinction sets aside and rejects; in order that by this great and pernicious predetermination the authority of its former conclusions may remain inviolate", Bacon *(The Novum Organum).*

Preface

During the development of the various branches of science and engineering the initial solutions have sometimes seemed, in retrospect, to be special cases of more general problems. Such has been the case with mechanical design and analysis.

Until recently, many of the implications due to the probabilistic nature of engineering phenomena were not fully realized, even though from early in the 20th century, with the announcement of Heisenberg's uncertainty principle, the whole of science (including engineering) was recognized as ultimately based philosophically on the concepts of experimental probability. However, some of the necessary supporting science and technology had not, until recently, sufficiently evolved to provide a base for a design theory and methodology incorporating probabilistic and statistical concepts.

Beginning with a probability-based design theory and eliminating all variability in the design variables, the residual is a special case, corresponding to a theory based on the deterministic assumption. It is essentially this special case into which classical deterministic design theory has evolved in its main features. Over the past several centuries an enormous methodology and supporting literature has been developed around the special case of assumed determinism.

Although the variables have been identified and the forms of their functional relationships correctly described in most classes of engineering problems, the behavior of many individual variables has until recently remained imperfectly understood. Also, probability mathematics, for quantifying functions of random variables, has only recently been developed to support design synthesis and analysis.

The focus of effort in this book is on engineering design involving random variables. Also, certain points of existing theory in which the assumption of determinism has led to untenable positions have been modified. Other points of theory have been developed, whose need under deterministic assumptions was not revealed.

The task undertaken here is to present theory and methodology in sufficient detail for the student and practicing engineer to readily use the approach in design. Numerous examples are given, pursuing a reliability specification, in which both over- and underdesigning is avoided. Variability information is used, not suppressed, and meaningful information is developed to provide a basis for management decisions (since uncertainties that exist can be expressed and resource-use alternatives stated). The reader will find that probabilistic design extends and supplements classical theory.

Probabilistic Mechanical Design is a textbook as well as a reference book for engineers. As a textbook, it is intended for use in a second course in the area of mechanical design. It is the experience of the author that upper-division undergraduate students can be taught this material. For a one-semester design course, material can be selectively presented from Chapters 2, 3, 4, 5, 8, and 9.

Although this book focuses on the design of mechanical and structural components and systems, the concepts discussed apply to other specialities in mechanical engineering (power systems, etc.) and to other engineering disciplines such as nuclear, electrical, and mining. Using their ingenuity, readers from other fields of interest can adapt probabilistic methods to their needs.

This book was in part developed from the author's class notes and unsupported research at the University of Arizona and in part with the support of the U.S. Army Armament Command, Rodman Engineering Laboratory, Rock Island. In particular, thanks are due to Mr. R. Stanley Thompson, chief of the Rodman Engineering Laboratory and Mr. Ernest A. Felsted, former director of quality assurance.

It is impossible to acknowledge all of the many contributors to the development of probabilistic design. However, the references amount to a partial listing. Faculty colleagues such as Dr. Duane Dietrich at the University of Arizona deserve thanks for assisting in their areas of specialization, as does Dr. Larry B. Scott, head of the department, for providing a climate conducive to research in the Department of Aerospace and Mechanical Engineering.

Special thanks are due to Dr. Charles Mischke of Iowa State University for critically reviewing the entire manuscript and offering many valuable suggestions. The manuscript was polished during the author's sabbatical leave at Cranfield Institute of Technology, Cranfield, Bedford, England. Thanks are due to Professor Denis Howe and Dean John L. Stollery of the Cranfield Institute for their generous support and to Professor Lennart Carlsson at Chalmers Tekniska Högskola, Göteborg, Sweden for his support.

EDWARD B. HAUGEN

Tucson, Arizona
August 1980

Contents

PREFACE   vii

1 INTRODUCTION   1

    1.1   Definition of Design, 1
    1.2   Mechanical Engineering Design, 1
    1.3   Design Development, 7
    1.4   Design Theory, 9
    1.5   Classical versus the Probabilistic Approach, 13

2 SUPPORTING MATHEMATICS   16

    2.1   Introduction, 16
    2.2   Properties of Numbers, 17
    2.3   Probability Considerations, 18
    2.4   Expected Value $E(x)$ and Variance $V(x)$, 20
    2.5   Algebra of Expectation, 23
            2.5.1   Mean and Variance of Sum $X+Y$, 26
            2.5.2   Mean and Variance of Difference $X-Y$, 27
            2.5.3   Statistics of Product of Independent Random Variables X and Y, 28
            2.5.4   Mean and Variance of Quotient X/Y, 30
    2.6   Algebra of Normal Distributions, 37
            2.6.1   Introduction, 37
            2.6.2   Statistics of Sum $Z=(X+Y)$, 38
            2.6.3   Statistics of Difference $Z=(X-Y)$, 41
            2.6.4   Statistics of Product $Z=XY$, 43
            2.6.5   Product of a Random Variable and Its Inverse, 49
            2.6.6   Statistics of Quotient $Z=(Y/X)$, 50

      2.6.7   Distribution of a Quotient, 52
      2.6.8   Statistics of Quadratic Form, 54
      2.6.9   Coefficient of Variation, 54
      2.6.10   Laws of Combination, 56
      2.6.11   Characteristics, 56
      2.6.12   Closure, 56
      2.6.13   Convergence, 57
      2.6.14   Normal Distribution as a Model, 57

2.7   Statistics of Arbitrary Functions, 58
      2.7.1   Mean Value, 58
      2.7.2   Variance of Random-valued Functions, 59

2.8   Characteristics of Lognormal Random Variables, 66
      2.8.1   Introduction, 66
      2.8.2   Lognormal Density Function, 67
      2.8.3   Transformation of Parameters, 68
      2.8.4   Interpretation of Statistics, 69
      2.8.5   Measures of Central Tendency, 71

2.9   Monte Carlo Simulation, 72
      2.9.1   Monte Carlo Concept, 72
      2.9.2   Random-number Generation, 73
      2.9.3   Sample Size and Variational Bands, 73

Addenda, 74

3 LOADING RANDOM VARIABLES   77

3.1   Introduction, 77

3.2   Experimental Loads Measurement, 80
      3.2.1   Data Reduction, 80
      3.2.2   Goodness-of-fit Test, 84
      3.2.3   Graphical Estimates, 85

3.3   Sampling Requirements, 86
      3.3.1   Central Limit Theorem, 86
      3.3.2   Estimation of Scatter Based on Range, 91

3.4   Force and Moment as Random Variables, 92
      3.4.1   Avoidance of Overload, 98

3.5   Correlated Engineering Variables, 99

3.6   Correlated and Independent Loads, 101

Contents   xi

    3.7   Constant- and Random-amplitude Loads, 106

           3.7.1   Constant-amplitude Dynamic Loads (Zero Means), 107
           3.7.2   Narrow-band Random Loads (Zero Means), 109
           3.7.3   Simultaneous Action of Static and Dynamic Loads, 109
           3.7.4   Examples of Dynamic Loads, 110
           3.7.5   Loading Distributional Forms, 112

    3.8   Parameters Modifying Loading, 114

           3.8.1   Introduction, 114
           3.8.2   Force Amplitude, 114
           3.8.3   Escalation Rates, 115
           3.8.4   Frequency, 116

    3.9   Unifying the Loads Model, 116
    3.10   Complete Profile of Loading, 117
    3.11   Load Record Ensembles, 120
    3.12   Random Variables and Random Processes, 120
    3.13   Stationary and Ergodic Processes, 122

4   DESCRIBING COMPONENT GEOMETRY WITH RANDOM VARIABLES   125

    4.1   Introduction, 125
    4.2   Distributional Characteristics, 127
    4.3   Dimensional Determination, 128

           4.3.1   Geometric Tolerance and Tolerance Limits, 130
           4.3.2   Tolerances on Finished Metal Products, 133

    4.4   Linear Dimensional Combinations, 139

           4.4.1   Confidence Limits on Sum and Difference Random Variables, 152

    4.5   Nonlinear Dimensional Combinations, 152
    4.6   Tolerancing by Computer (Monte Carlo Simulation), 153
    4.7   Geometry versus Strength, 157

5   THE STRESS RANDOM VARIABLE   160

    5.1   Introduction, 160
    5.2   Analysis of Stress, 161
    5.3   Variation of Stress with Orientation of Cross Section, 164

           5.3.1   Confidence Limits on Stress, 167

- 5.4 Sum of Random Normal Stresses, 167
- 5.5 Static Biaxial Tension or Compression, 170
- 5.6 Invariance of Sums of Combined Stresses, 173
- 5.7 Principal Stresses as Random Phenomena, 174
- 5.8 Triaxial Stresses, 176
- 5.9 Torsional Stresses, 176
- 5.10 Bending Stresses, 182
- 5.11 Shearing Stresses, 183
- 5.12 Superposition of Random Variables, 185
- 5.13 Unit Strain, 187
- 5.14 The Quality of Elasticity, 189
    - 5.14.1 Modulus of Elasticity, 189
    - 5.14.2 Poisson Ratio, 192
- 5.15 Stress–Strain Relations, 193
    - 5.15.1 Uniaxial Stress, 193
    - 5.15.2 Static Biaxial Stress, 193
- 5.16 Random Fatigue Stresses, 194
    - 5.16.1 Dynamic Stress Models, 196
- 5.17 Multiaxial Dynamic Stresses, 198
- 5.18 Estimating Stress Statistics with Finite Elements, 202
- 5.19 Lewis Model for Dynamic Stresses, 208
    - 5.19.1 Bending Stresses, 209
    - 5.19.2 Modified Lewis Model, 209
- 5.20 Surface Stress, 214
    - 5.20.1 Spur Gear Stresses, 214
    - 5.20.2 Ball-bearing Stresses, 217

6 RANDOM-DEFLECTION VARIABLES 220

- 6.1 Introduction, 220
- 6.2 Deflection of Beams, 226
- 6.3 Deflection of a Simply Supported Beam (Concentrated Load), 229
- 6.4 Deflection With Load Superposition, 231
- 6.5 Indeterminate Problems, 234
- 6.6 Strain Energy in Tension and Compression, 236

Contents xiii

   6.7   General Expression for Strain Energy, 242
   6.8   Castigliano's Theorem, 245
   6.9   Random Column Behavior, 249
   6.10   Mechanical Vibration, 260
        6.10.1   Vibration and Damping, 263
        6.10.2   Steady-state Vibration, 264
        6.10.3   Natural Frequencies, 266

7 GEOMETRIC STRESS CONCENTRATION   273

   7.1   Introduction, 273
   7.2   Stress Concentration and Static Loads, 279
   7.3   Determining Theoretical K_t, 280
   7.4   Probabilistic Approach, 281
        7.4.1   Estimating K_t Statistics by Monte Carlo Simulation, 281
   7.5   Statistics of K_t Envelopes, 286

8 DESCRIBING MATERIALS BEHAVIOR WITH RANDOM VARIABLES   287

   8.1   Introduction, 287
   8.2   Definitions, 289
   8.3   Tension Testing, 292
   8.4   Proportional Limit, 292
   8.5   Yield Point, 293
   8.6   Tensile Yield Strength, 293
   8.7   Ultimate Tensile Strength, 293
   8.8   Shear Strength, 296
   8.9   Torsional Strength, 302
   8.10   Tensile Ductility, 304
   8.11   Fracture Toughness, 307
   8.12   Resilience, 308
   8.13   Hardness, 312
        8.13.1   Introduction, 312
        8.13.2   Hardness Measuring Systems, 312
        8.13.3   Hardness versus Mechanical Properties, 313
        8.13.4   Statistical Aspects of Hardness, 316

8.14 Fatigue, 316
- 8.14.1 Endurance Limit and Fatigue Strength, 319
- 8.14.2 Fatigue Strength under Fluctuating Stresses, 332
- 8.14.3 Materials Behavior Subject to Narrow-band Random Loading (Zero Mean), 338
- 8.14.4 Materials behavior Subject to Narrow-band Random Loading (Nonzero Mean), 342

8.15 Strength, a Random Process, 350
- 8.15.1 Random Processes in Design, 355

9 DESIGN AND ANALYSIS 356

9.1 Introduction, 356
9.2 Classical Approach to Design, 357
9.3 Probabilistic Approach to Design, 357
- 9.3.1 General Reliability Model, 360
- 9.3.2 Normal Stress and Strength, 363
- 9.3.3 Design for Random Static Axial Loads, 369
- 9.3.4 Design for Random Static Bending Loads, 376
- 9.3.5 Design for Nonnormal Stress, 386
- 9.3.6 Additional Stress–Strength Combinations (Distributions), 395
- 9.3.7 Consideration of Cycles to Failure, 397
- 9.3.8 Design for Endurance, 399
- 9.3.9 Upper Bound on Requirements for a Specified R, 412
- 9.3.10 Gears and Bearings, 424
- 9.3.11 Reliabilty Assurance, 428

Addendum, 433

10 STRENGTH OF MECHANICAL COMPONENTS 436

10.1 Introduction, 436
10.2 Maximum Normal Stress Theory, 438
10.3 Maximum Shear Stress Theory, 443
10.4 Distortion Energy Theory, 446
10.5 Significant Stress–Strength Models, 454
- 10.5.1 Significant Strength Models, 454
- 10.5.2 Significant Stress Models, 455

Contents

- **10.6** Strength and Stress Modifiers, 456
  - 10.6.1 Strength Modifiers, 456
  - 10.6.2 Stress Modifiers, 457
- **10.7** Design to Avoid Fatigue Failure, 467
  - 10.7.1 Constant-amplitude Stresses (Zero Means), 469
  - 10.7.2 Bearing Selection, 478
  - 10.7.3 Fatigue Due to Narrow-band Random Loading (Zero Mean), 479
  - 10.7.4 Design for Narrow-band Random Loading (Zero Mean), 482
- **10.8** Fluctuating Stress (Nonzero Mean), 485
  - 10.8.1 Constant-amplitude Dynamic Stresses, 485
  - 10.8.2 Random Dynamic Stresses (Nonzero Mean), 493
- **10.9** Design for Multiaxial Dynamic Loading, 494
- **10.10** Multiple-member Systems, 495
  - 10.10.1 Pin-jointed Systems, 495
- **10.11** Probabilistic Fracture Mechanics, 501
  - 10.11.1 Monotonic Loading, 501
  - 10.11.2 Design to Avoid Running Cracks (Monotonic Loading), 504
  - 10.11.3 Design to Avoid Fracture (Dynamic Loading), 510

11 MECHANICAL ELEMENT OPTIMIZATION   515

- **11.1** Introduction, 515
  - 11.1.1 Optimum Design of Mechanical Components, 519
- **11.2** Optimum Design: R Specified, Stress Gradient Zero, 522
- **11.3** Optimum Design: R Specified, Surface Stress Gradient, 524
- **11.4** Optimum Design: R Specified, Cross-section Stress Gradient, 532
- **11.5** Analysis with Opposing Failure Modes, 540

12 RELIABILITY CONFIDENCE INTERVALS   546

- **12.1** Introduction, 546
  - 12.1.1 Confidence Interval of a Normal Random Variable of Known Standard Deviation, 546
  - 12.1.2 Confidence Interval of a Normal Random Variable of Unknown Standard Deviation, 547

12.1.3 Confidence Interval of the Standard Deviation of a Normally Distributed Random Variable, 548

12.2 Confidence Intervals on Differences of Random Variables, 550

12.2.1 Confidence Interval on $\mu_S - \mu_s$, σ_S and σ_s Known, 550

12.2.2 Confidence Interval for Difference between Mean Values of Two Normal Random Variables of Unknown but Equal Standard Deviations [15, p. 221], 553

12.2.3 The General Case of Confidence, 556

12.3 Confidence Interval on Reliability, 558

12.3.1 Confidence Interval on Reliability, σ_S and σ_s Known, 558

12.3.2 Confidence Interval on Reliability, $\sigma_S = \sigma_s$ but Unknown, 559

12.3.3 Confidence Interval on Reliability, General Case, 561

APPENDIXES   563

REFERENCES   618

INDEX   623

Notation

Notation in this book has posed a problem because of the coupling of probability theory and statistics with mechanical engineering. As much as possible, the familiar meanings attached to various symbols have been retained. However, a symbol frequently had one meaning in probability theory and statistics and a different one in engineering language. The following is an attempt to provide an easily understood system of notation.

In this book μ and σ denote the universe mean value and standard deviation of engineering variables, respectively, and subscripts are utilized to identify particular random variables; that is, μ_E and σ_E are the mean value and standard deviation of the elastic modulus, and μ_S and σ_S are the mean value and standard deviation of strength. The sample means and standard deviations of the elastic modulus and strength are denoted by $\bar{E}$ and s_E and by $\bar{S}$ and s_S.

Consistent with engineering practice, uppercase S denotes strength, and subscripts identify specific kinds, such as ultimate (S_u), yield (S_y), and endurance (S_e). Since statistics of engineering variables are estimated from samples, ultimate strength is indicated as $(\bar{S}_u, s_{S_u})$, yield strength by $(\bar{S}_y, s_{S_y})$, and so on.

Applied stress is denoted by lowercase s and the sample statistics by $(\bar{s}, s_s)$. A sample standard deviation is denoted by script s, with subscript lowercase s indicating that the standard deviation is that of stress.

In the following table, random variables are described by couples of sample mean values and sample standard deviations. Constants are represented by single symbols.

a	Real-valued constant
$(\bar{a}, s_a)$	$\frac{1}{2}$ Crack length
$(\bar{a}_{cr}, s_{a_{cr}})$	Critical crack length
$(\bar{A}, s_A)$	Area
b	Real-valued constant
$(\bar{B}, s_B)$	Sheet or plate thickness
c	Constant or damping coefficient

$(\bar{c}, s_c)$	Distance to neutral axis
$(\bar{c}_c, s_{c_c})$	Critical damping coefficient
C	Bearing radial load
C_x and C_y	Coefficients of variation of X and Y, respectively
C_D and C_L	Drag and lift coefficients, respectively
$(\bar{d}, s_d)$	Diameter
$(\bar{D}, s_D)$	Diameter
$(\bar{e}, s_e)$	Eccentricity
$(\bar{E}, s_E)$	Modulus of elasticity
$E(x)$	Expected value of $x = \mu_x$
$f(x)$	Probability density function of X
$F(x)$	Cumulative distribution function
$(\bar{F}, s_F)$	Force or load
g	Gravitational constant
$(\bar{G}, s_G)$	Shear modulus of elasticity
$(\bar{h}, s_h)$	Height
$(\overline{Hp}, s_{Hp})$	Horsepower
i	Index
$(\bar{I}, s_I)$	Moment of inertia
$(\overline{ID}, s_{ID})$	Inner diameter
$(\bar{J}, s_J)$	Polar moment of inertia
k	Numerical constant
$(\bar{k}, s_k)$	Spring constant
$(\bar{K}, s_K)$	Stress intensity factor
$(\bar{K}_a, s_{K_a})$	Surface factor
$(\bar{k}_b, s_{k_b})$	Size factor
$(\bar{K}_c, s_{K_c})$	Plane stress fracture toughness
$(\bar{K}_{1c}, s_{K_{1c}})$	Plane strain fracture toughness
K_α	One-sided tolerance factor
$(\bar{K}_f, s_{K_f})$	Fatigue stress concentration factor
(K_t, s_{K_t})	Geometric stress concentration
$\Delta_{K_{th}}$	Threshold fracture toughness
$(\bar{l}, s_l)$	Length, $\pm \Delta l$ — tolerance range
$(\bar{L}, s_L)$	Load
$L(t)$	Random process
$(\bar{M}, s_M)$	Moment
N	Number of gear teeth
n	Sample size or cycles to failure
$(\overline{OD}, s_{OD})$	Outer diameter

Notation

p	Probability or pitch diameter
$(\bar{P}, s_P)$	Force of load or pressure
P_f	Probability of failure ($p_f = 1 - R$)
P_c	Diametral pitch
$(\bar{P}_{cr}, s_{P_{cr}})$	Critical load
Q	Notch sensitivity factor
$(\bar{r}, s_r)$	Radius, range, sample correlation coefficient
$(\bar{r}_c, s_{r_c})$	Radius of curvature
$(\bar{r}_g, s_{r_g})$	Radius of gyration
R	Reliability
$(\bar{R}_1, s_{R_1})$ and $(\bar{R}_2, s_{R_2})$	Reactions at supports
RA	Area reduction
$(\bar{s}, s_s)$	Applied stress, static or dynamic
$(\bar{s}_n, s_{s_n})$	Normal stress
$(\bar{s}_1, s_{s_1})$ and $(\bar{s}_2, s_{s_2})$	Principal stresses
$(\bar{s}_{eq}, s_{s_{eq}})$	Equivalent stress, static or dynamic
$(\bar{s}_c, s_{s_c})$	Surface or contact stress
$(s_{cr}, s_{s_{cr}})$	Critical stress
s_R	Stress radius vector
s_x	Sample standard deviation of x
$(\bar{S}, s_S)$	Allowable stress or strength
$(\bar{S}_u, s_{S_u})$	Tensile ultimate strength
$(\bar{S}_y, s_{S_y})$	Tensile yield strength
$(S_{su}, s_{S_{su}})$	Shear strength
$(\bar{S}_e, s_{S_e})$	Endurance limit
$(\bar{S}_f, s_{S_f})$	Fatigue limit
S_R	Strength radius vector
$(\bar{S}_{se}, s_{S_{se}})$	Shear endurance or elastic limit
$(\bar{S}_{sy}, s_{S_{sy}})$	Shear yield strength
$(\bar{t}, s_t)$	Thickness or time
$(\bar{T}, s_T)$	Torque
$(\bar{U}, s_U)$	Strain energy
$(\bar{v}, s_v)$	Velocity or shearing force
$V(x)$	Variance of x
$(\bar{w}, s_w)$	Width
$(\bar{w}_t, s_{w_t})$	Transmitted gear load
X	Random variable

x	Realization of X
$\bar{x}$	Sample mean of X
ΔX	Tolerance on X
$(\bar{y}, s_y)$	Deflection
α	Crack configuration coefficient or allowable fraction defective
$(\bar{\delta}, s_\delta)$	Strain, total elongation
$(\bar{\varepsilon}, s_\varepsilon)$	Unit strain or error
ν	Confidence coefficient
$(\bar{\gamma}, s_\gamma)$	Shearing strain
$(\bar{\nu}, s_\nu)$	Poisson ratio
μ_x	Mean value of X
σ_x	Standard deviation of X
$(\bar{\tau}, s_\tau)$	Shear or torsional stress
$(\bar{\omega}, s_\omega)$	Frequency
$(\bar{\omega}_n, s_{\omega_n})$	Natural frequency
ρ	Correlation coefficient or density
ϕ	Phase angle or starting condition

Engineering Units

There will be a displacement in the near future of United States customary units by the usage of SI (international metric system) units in the field of mechanical engineering. It does seem likely, however, that the use of both United States customary and SI systems of units, modules, sizes, ratings, and so on will continue.

**Commonly Used Conversion Factors:
United Stated Customary to SI Units**[a]

Quantity	Conversion	Factor
Plane angle	deg to rad	$1.745 \cdot E - 02$
Length	in. to m	$2.54 \cdot E - 02$
	ft to m	$3.048 \cdot E - 01$
	mi to m	$1.609 \cdot E + 03$
Area	in^2 to m^2	$6.452 \cdot E - 04$
	ft^2 to m^2	$9.290 \cdot E - 02$
Volume	ft^3 to m^3	$2.832 \cdot E - 02$
	in^3 to m^3	$1.639 \cdot E - 05$
	liter to m^3	$1.000 \cdot E - 03$
Velocity	ft/min to m/sec	$5.08 \cdot E - 03$
	ft/sec to m/sec	$3.048 \cdot E - 01$
	km/hr to m/sec	$2.778 \cdot E - 01$
	mi/hr to m/sec	$4.470 \cdot E - 01$
	mi/hr to km/hr	$1.609 \cdot E + 00$
Mass	oz (avoirdupois) to kg	$2.832 \cdot E - 02$
	lb (avoirdupois) to kg	$4.536 \cdot E - 01$
	slugs to kg	$1.459 \cdot E + 01$
Acceleration	ft/sec^2 to m/sec^2	$3.048 \cdot E - 01$
	standard gravity to m/sec^2	$9.807 \cdot E + 00$
Force	kgf to N	$9.806 \cdot E + 00$
	lbf to N	$4.448 \cdot E + 00$
	poundal to N	$1.382 \cdot E - 01$
Bending torque	kgf·m to N·m	$9.806 \cdot E + 00$
	lbf·m to N·m	$1.429 \cdot E - 01$
	lbf·ft to N·m	$1.356 \cdot E + 00$
Pressure, stress	kgf/m^2 to Pa[a]	$9.806 \cdot E + 00$
	$poundal/ft^2$ to Pa	$1.488 \cdot E + 00$

**Commonly Used Conversion Factors:
United Stated Customary to SI Units**[a]

Quantity	Conversion	Factor
	lbf/ft² to Pa	$4.788 \cdot E+01$
	lbf/in² to Pa	$6.892 \cdot E+03$
Energy, work	Btu (IT) to J	$1.055 \cdot E+03$
	caloric (IT) to J	$4.487 \cdot E+00$
	ft/lbf to J	$1.356 \cdot E+00$
Power	hp [550 ft (lbf)/sec)] to W	$7.457 \cdot E+02$

Source: ASME Orientation and Guide for Use of SI Units.
[a] $Pa = pascal = newtons/meter^2 = N/m^2$.

PROBABILISTIC
MECHANICAL DESIGN

CHAPTER 1

Introduction

1.1 DEFINITION OF DESIGN

"To design is to formulate a plan for the fulfillment of a human need," [1]. Krick [2] considers design as a "decision-making process." Initially, the need for design may be well defined; however, the problem may often be somewhat nebulous. There is now a choice of philosophies available for carrying out mechanical design: (1) design based on theory of probability and (2) design based on deterministic assumptions.

Design problems, in contrast to analysis problems, rarely have unique solutions. Hence attempts are made to optimize design solutions to achieve important objectives. The satisfactory solution of today, however, may become unsatisfactory later because of new knowledge, developments in theory, changes in priorities, and so on.

In engineering, a design problem has an authentic purpose, namely, the development of an end result by taking definite action or by producing an article that has physical reality and utility. Regardless of the words used to describe the design function, to engineers it remains a process requiring the tools of engineering.

1.2 MECHANICAL ENGINEERING DESIGN

Mechanical engineering design covers a broad spectrum of activity, including heating and ventilating, fluid machinery, power plants, weapons systems, turbines, production machinery, internal and external combustion engines, and household appliances. The field is so broad that a single person is rarely concerned with more than a portion of it.

One classification is presented in Figure 1.1. Because the activities of most mechanical engineers to some extent touch on design, this category is centrally located in Figure 1.1. The design process may require that behavior of materials, methods of production, geometric dimensions, and so on be precisely defined or that groups of components be assembled into a device or system, as indicated in Figure 1.2.

Knowledge from mathematics, physics, English, mechanics of materials, dynamics, graphics, production processes, fluid mechanics, heat transfer, and

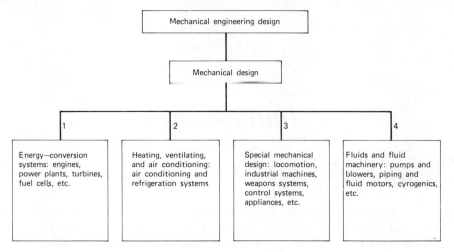

Figure 1.1 Mechanical design [1, p. 5].

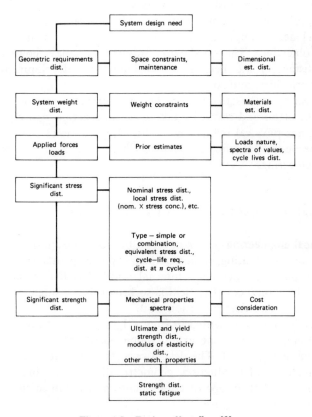

Figure 1.2 Design effort flow [3].

probability theory and statistics is utilized in design. The engineer integrates concepts from several disciplines. He could not estimate the probability of adequate performance if he were ignorant of failure theories, stress analysis, or dynamic behavior of materials.

In modern society we become increasingly dependent on the proper functioning of increasingly complicated mechanical devices and systems, many of very recent origin. However, in many respects engineering methodology until recently has been slow to change in response to new demands. Theory and methodology sufficient to satisfy former (simpler) requirements have experienced difficulty with some current problems.

Recent concern with the reliability of designed mechanical products has led to a reevaluation of the foundations of design. The response of engineers has

Table 1.1 Diameters of Test Sections of Grooved Reserve Bending Specimens[a]

Diameter (in.)	Diameter (in.)
0.1799	0.1803
0.1808	0.1807
0.1804	0.1804
0.1803	0.1805
0.1805	0.1796
0.1807	0.1805
0.1805	0.1801
0.1804	0.1802
0.1806	0.1804
0.1805	0.1805
0.1816	0.1802
0.1808	0.1795
0.1800	0.1810
0.1808	0.1801
0.1806	0.1804
0.1802	0.1804
0.1805	0.1805

[a]Nominal radius, $r = 0.250$ in.; nominal diameter, $d = 0.180$ in.; number of specimens, 34.

$\hat{\mu}_x = 0.18042$ Note: $\mu_x = \bar{x}$
$\hat{\sigma}_x = 0.00038043$ $\hat{\sigma}_x = \sigma_x$

(These specimens were used in fatigue tests at the calculated stress level of 60,300 psi) [4].

Table 1.2 Rockwell C Hardness Data of 41 R. R. Moore Machine Specimens [4]

Hardness Rockwell C Measurements of Same Specimen	Average Rockwell C Specimen Hardness
22, 22, 21, 22, 22	21.8
22, 22, 22, 22, 21	21.8
21, 22, 21, 21, 20	21.0
21, 22, 22, 21, 22	21.6
22, 21, 21, 20, 20	20.8
22, 22, 22, 21, 22	21.8
22, 22, 22, 22, 21	21.8
21, 22, 22, 20, 21	21.2
20, 22, 22, 22, 22	21.6
21, 18, 18, 21, 17	19.0
21, 21, 19, 21, 21	20.6
22, 21, 21, 22, 21	21.4
20, 19, 22, 21, 18	20.0
22, 22, 22, 22, 23	22.2
22, 21, 21, 20, 21	21.0
21, 19, 22, 22, 20	20.8
22, 21, 21, 20, 22	21.2
21, 21, 20, 22, 22	21.2
21, 20, 22, 21, 21	21.0
22, 20, 22, 22, 21	21.4
21, 21, 21, 21, 20	20.8
22, 22, 20, 22, 20	21.2
22, 22, 18, 22, 18	20.4
22, 21, 21, 22, 21	21.4
20, 20, 19, 20, 18	19.4
24, 22, 20, 21, 21	21.6
21, 20, 20, 20, 21	20.4
20, 19, 20, 20, 21	20.0
21, 19, 19, 21, 20	20.0
20, 21, 22, 20, 20	20.6
21, 20, 19, 23, 21	20.8
21, 21, 22, 22, 21	21.4
21, 22, 19, 20, 22	20.8
21, 21, 20, 20, 21	20.6
21, 21, 20, 21, 21	20.8
21, 20, 21, 19, 21	20.4
21, 20, 18, 22, 22	20.6
21, 21, 19, 20, 21	20.4
20, 21, 18, 20, 20	19.8
20, 20, 20, 21, 22	20.6
20, 20, 20, 20, 20	20.0

been intense research activity over the past several decades, resulting in:

1. The development of finite-element analysis (definition of stress).
2. Fatigue research (material behavior and physics of failure).
3. The development of fracture mechanics (crack propagation and failure).
4. The development of probabilistic design (closer correspondence between predicted and actual performance of mechanical components).
5. Optimization (improvement in efficiency and economy).

Reevaluation of the design process involves a close scrutiny of the various classes of primary design variables, such as loads, materials behavior, and geometry.

Tables 1.1 through 1.3 and Figures 1.3 through 1.8 indicate the usually observed design variable behavior. Table 1.1 and Figure 1.3 illustrate typical

Table 1.3 Ultimate Tensile-Strength Data of 40 Wire[a] Specimens of SAE 4340 Cold-drawn and Annealed Steel Wires Obtained by Using Riehle Testing Machine [4]

Ultimate Tensile Strength (psi)	
103,779	102,325
102,906	105,377
104,796	104,796
103,197	106,831
100,872	105,087
97,383	103,633
101,162	107,848
98,110	99,563
104,651	105,813
102,906	108,430
103,633	106,540
102,616	101,744
106,831	103,799
102,325	100,145
104,651	103,799
105,087	103,488
106,395	102,906
100,872	101,017
104,360	104,651
101,453	103,197
103,779	105,337
101,162	101,744
102,470	106,104
105,232	100,726
103,924	101,598

[a]Nominal wire diameter = 0.0937 ± 0.0005 in.

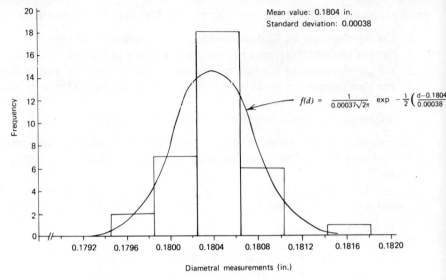

Figure 1.3 Reduced diametral measurements.

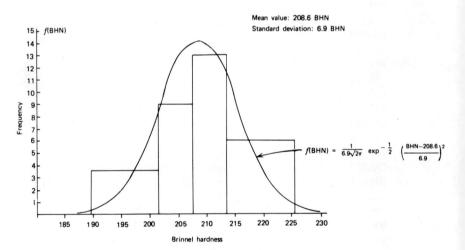

Figure 1.4 Distribution of Brinnel hardness.

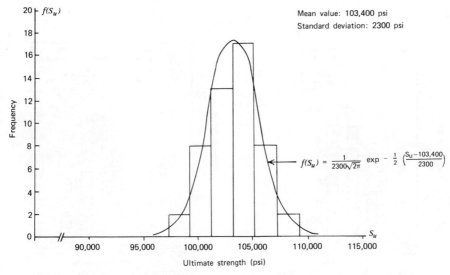

Figure 1.5 Static ultimate strength.

variation in element cross-section geometry (due to machining), which contributes to statistical variations in moment of inertia among a group of components. Table 1.2 and Figure 1.4 illustrate typical variations in Rockwell hardness, and hence in tensile and endurance strength, within a group of SAE 4340 steel test specimens. Table 1.3 and Figure 1.5 illustrate the variation found in the mechanical properties of titanium alloys. Figures 1.7 and 1.8 illustrate spectra of acceleration intensities in aerospace and automotive design situations.

To achieve designs in which performance is predictable, it may be postulated that

1. Models must preserve the essential characteristics of individual engineering variables, that is, indicate that variables are multivalued random phenomena.

2. Models of functions of engineering variables must preserve essential characteristics, that is, indicate that functions are (also) multivalued random phenomena.

1.3 DESIGN DEVELOPMENT

The engineering task of developing a device from concept to an operational system has been conceived as occurring in the following four stages [4]: (1) design, (2) development, (3) production, and (4) service. These stages may be interrelated to some extent. For example, service failures require possible

Ti-4Al-3Mo-1V

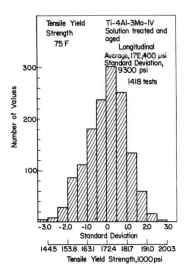
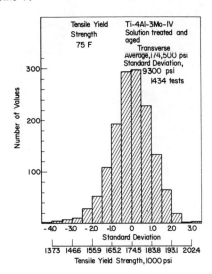
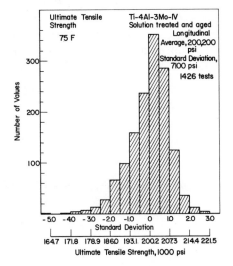
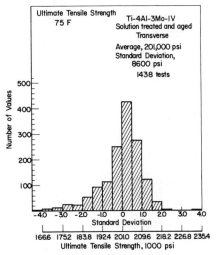

Figure 1.6 Typical strength histograms.

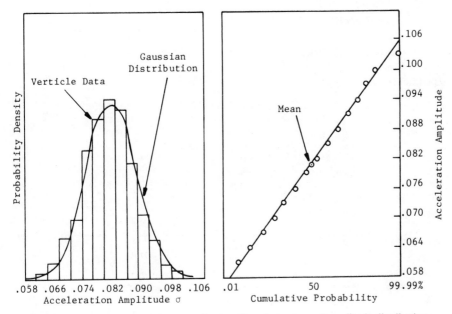

Figure 1.7 Experimental evidence confirming Gaussian nature of amplitude distribution.

reevaluation of the design, the developmental procedure, and/or the production process to explain the cause and to suggest correction. Corrective procedures tend to be minimized when problablistic methodology is employed since the impact of design parameter variability and problablistic parameter interactions have already been taken into account.

The first action taken in the design of a new device is to establish a priori quantitative design requirements from service requirements of the projected system. Design objectives are established from this set of requirements. The mechanical device or system created must have the required strength, stiffness, and intrinsic reliability while remaining cost competitive.

1.4 DESIGN THEORY

Three bodies of knowledge, with some overlap, are utilized in designing reliable products [8]. Such knowledge is couched in (1) the classical sciences (e.g., physics and mathematics), (2) engineering science, and (3) practical engineering theory (engineering mechanics, strength of materials, material behavioral science, etc.).

Engineering analysis theory identifies variables, material properties, physical and mathematical constants and describes functional relationships involved in a state or process. Engineering design theory identifies the number of decisions under the designer's control, makes some suggestions as to which

Percent of Maximum (Positive) Symmetrical Limit Load Factor (Column 3 of Table I of Specification MIL-A-8861)	Number of Times Per Thousand Hours That Load Factor is Experienced			
	Flight Maneuver Load Spectrum			
	A	*B*	*C*	*D*
35	17,000	25,000		
45	9,500	10,000	10,000	
55	6,500	3,500	3,000	1,000
65	4,500	1,000	1,000	150
75	2,500	500	300	20
85	1,500	200	100	3
95	300	75	30	0.5
105	150	25	10	0.05
115	40	10	3	
125	16	3	2	

Percent of Minimum (Negative) Symmetrical Limit Load Factor (Column 4 of Table I of Specification MIL-A-8861)	*A*	*B*	*C*	*D*
0	500	5	0.7	
10	200	2	0.5	
20	100	1	0.25	
30	60	0.25		
40	35			
50	30			
60	25			
70	20			
80	15			
90	10			
100	5			
110	3			

Figure 1.8 Airplane strength and rigidity reliability requirements, represented loads and fatigue [7].

decisions should be the design variables, and outlines the optimization process needed to establish the "best" decision set (i.e., the optimal design). It permits estimation of performance for specific combinations of initial, boundary, and operating conditions from general considerations, predicting the specific from the general. Prior to a developed probablistic theory, the capability of engineered systems could not be estimated with precision. Optimization was difficult because the design methods were not being geared to entertain generally random situations.

Increasing demands for performance, resulting often in operation near limit conditions, has placed increasing emphasis on precision and realism. Probablistic considerations and rational performance measures are required to overcome the logical and practical problems that have plagued engineers [9].

Realistic description of loads, strength properties, and design modifying factors relate magnitude values with expected frequencies. Recognizing the uncertainty in strength and stress intensities, the performance of products must be considered variable, that is, have a probability of failure associated with them (consistent with experience).

The objective in design is to create systems that satisfy operational and economic requirements with a specified measure of reliability. This requires the following steps:

1. A preliminary design concept must be invented that will be in accordance with anticipated usage and subject to change as economic and operational considerations are clarified.

2. The loads to be imposed on or resisted by the product must be quantitatively estimated. A priori loading estimation must define the range of magnitudes together with the relative frequency of each.

3. From an analysis of the preliminary system, quantitative descriptions of loads in the individual components must be estimated.

4. The material and thermomechanical treatment for each element must be selected, on the basis of mechanical and/or physical properties, purpose of the design, and limitations imposed by economics and feasibility.

5. Statistical descriptions of the significant strength characteristics of the materials must be obtained.

6. A quantitative description of the strength or resistance properties and the failure characteristics of each individual element must be estimated.

7. A description of the collective strength (resistance) and failure characteristics of the system of elements must be estimated.

The system thus described may be optimized in relation to operational demands and economic constraints.

Figure 1.9 shows two possible treatments of a stress function. If the real conditions were such that each variable $X_1, X_2, \ldots, X_n$ were single valued and the relation of these variables to the variable s were described by the function $F(X_1, X_2, \ldots, X_n)$, the transformation from $X_1, X_2, \ldots, X_n$ to s would be a single value. If, however, the variables $X_1, X_2, \ldots, X_n$ were each multivalued, the transformation from $X_1, X_2, \ldots, X_n$ to s would be a distribution of probable values of s. A similar argument applies to the strength function.

In past practice the basic parameters and functions utilized to estimate allowable stress (S) and applied stress (s) were modeled as indicated in Figure 1.10. Actually, the variables and factors relevant in design, analysis, and production of systems usually behave more nearly as indicated in Figure 1.11. Classically, primary design variables and factors appear in strength and/or stress functions and single-valued influences on the designed device or system are calculated. Possible errors of omission (or commission) are presumably taken care of by a safety factor.

The accepted criterion for an adequate strength-limited design has been

$$S > s \cdot (SF), \tag{1.1}$$

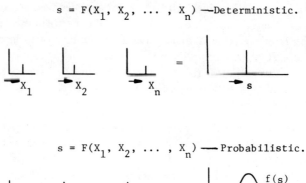

Figure 1.9 Deterministic versus probabilistic representation.

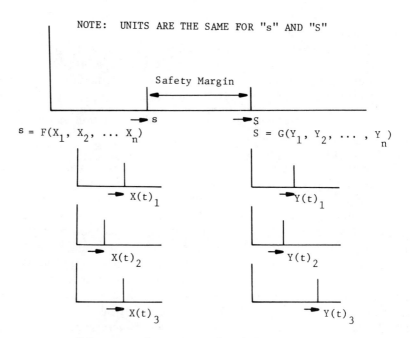

Figure 1.10 Design–deterministic approach.

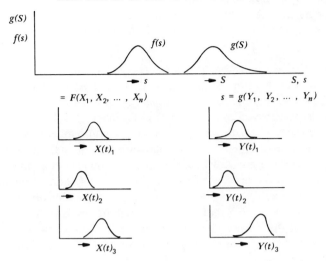

Note: Symbol t refers to service life.

Figure 1.11 Design–probabilistic approach.

where S denotes deterministic component or system strength, SF denotes safety factor, and s denotes single-valued applied load stress. Now, increasingly, design variables and factors appear in allowable stress and/or applied stress functions, and their multivalued influences on the designed system are exploited to yield the desired reliability.

1.5 CLASSICAL VERSUS THE PROBABILISTIC APPROACH

The intent in the design synthesis of a mechanical component is to assure that, for a specified mission

$$(S > s) = (S - s > 0) \qquad (a)$$

Taking probabilities in equation (a), we obtain

$$P(S > s) = P(S - s > 0) = R \qquad (1.2)$$

where $S \equiv (\bar{S}, \sigma_S) \equiv$ strength and $s \equiv (\bar{s}, \sigma_s) \equiv$ applied stress are random variables and R is a probability measure. Equation (1.2) is a probabilistic design criterion. In other words, this criterion states that the probability that strength S, exceeds applied stress s equals or exceeds a stated probability R.

If it is assumed that $\sigma_S, \sigma_s \to 0$ in equation (1.2), then $R \to 1$, because S and s become single valued. The expression

$$P(S - s > 0) = 1$$

describes the case of design assuming constant values $S > s$ with probability 1. Strengthening the inequality $S > s$ by introducing a constant $k > 1$,

$$S \geqslant k \cdot s$$

is equivalent to the classical design criterion [equation (1.1)] when $k = SF$. Such a simple shift in design criteria as that from equation (1.1) to equation (1.2) has far-reaching effects in the design process itself.

Probabilistic theory admits a (small) probability of failure (consistent with experience), treats strength and stress as multivalued phenomena (consistent with test and measurement results), and depends on statistical numbers for modeling variables and statistical algebra for constructing functions. Functions of design variables are treated as functional relationships among random variables, each having a defining probability density function. The model of the function is a multivariate probability density function.

Until present the values of design variables have been viewed as invariant. Actually, it is the probability density functions of design variables that are invariant, in the limit.

In addition, consider that the moment that engineering phenomena are treated as multivalued and random (a fact in the real world), we are dealing with nonorderable sets, with members no longer as simple and well behaved as the real-number models utilized thus far in engineering. Considerations regarding which theory in the past was silent are clearly important. In addition to the spectra of values that variables may exhibit, the likelihood of each value impacts on performance. In such sets of variables, we cannot say of the members that one is larger or smaller than another. The result is that the concept of principal stresses, which rests on ordering, must be reexamined and restated. The same applies to the familiar failure theories. Another unique feature of the random variables discussed previously is that they may be either statistically independent or correlated. The consequence is that differing solutions to design problems involving numerically identical variables becomes a real possibility. Numerous other unique features, attributable to random phenomena and having an impact in design, are treated in the text.

The reasons for adopting probabilistic approaches are summarized as follows:

1. Theoretical validity

   a. Phenomena encountered in engineering consideration are inherently probabilistic, (i.e., multivalued).
   b. Reliability R and failure probability p_f relate directly to performance.
   c. Close correspondence between predicted and actual performance requires consideration of variability.
   d. Consistent reliability among the components of a system is needed.
   e. Utilization of strength and stress models, which have admissible ranges of values $(0, \infty)$, takes cognizance of abnormally large stresses and abnormally low strengths of rare occurrence.

2. Reliability
   a. Reliability goals should be incorporated in the design methodology, not achieved by costly redesign, after test, often impractical as a major route to reliable design.
   b. Known levels of performance are often needed.
   c. Known levels of safety are often needed.
3. Economic considerations
   a. Improved design will minimize costly testing programs of total devices and systems.
   b. The time from design inception to completion of performance-adequate systems will be reduced.
   c. It is necessary to minimize the inertia of moving parts and weight to be lifted or moved (i.e., optimize); this is especially essential in view of energy considerations and use of nonrenewable resources.
   d. There is a need for rational approaches to costing and to establishing policy regarding replacements and warranties.
   e. Overprecision and ability to meet performance requirements, while larger manufacturing tolerances results in economies.

This book focuses primarily on the problem of incorporating real-world randomness into the engineering design rationale.

CHAPTER 2

Supporting Mathematics

Since engineering involves functional relationships, this chapter treats functions of one, two, and n random variables, providing formulas for estimating the statistics, that is, the mean values and standard deviations or variances, of functions in terms of the statistics of its variables.

2.1 INTRODUCTION

In engineering basic design variables are described by parameters that must be modeled quantitatively. For this purpose, numbers are used to express magnitude in a concise way. However, as indicated in Chapter 1, engineering variables are inherently multivalued, assuming various values at random. Generally a variable that eludes predictability by exhibiting different magnitudes is called a *random variable* (synonym *variate*).

A random variable may be defined as "a real finite-valued function defined on a sample space, on whose events a probability function has been defined." The manipulation of random variables and their functions requires special algebras. The basic axioms defining the properties of numbers within a given system, together with the rules defining the relationships, constitute an algebra. There are many applications in engineering in which new versatile numbers are needed, because not only effects of mean values but the entire spectrum of values each variable may exhibit must be accounted for in the evaluation of a function.

Random variables can be treated as specialized numbers. The sum of two random variables is a random variable, the product of two random variables is a random variable, and so on. Only the mean values* and variances or standard deviations[†] of functions are estimated here, for the following reasons:

1. The mean value and variance of any specified two-parameter distribution [11, p. 37] uniquely define it, and two-parameter distributions are employed extensively in this book.

---
*One measure of central tendency.
[†]Measure of scatter.

2. The statistics (mean value and variance) of random-valued functions can be accurately estimated regardless of functional complexity.
3. Methods for developing the distributions of functions of random variables, such as transformation of variables, become very complicated except for the simplest functions.

This two-statistic approach has found important applications in statistical communications and control theory and in time-series analysis.

2.2 PROPERTIES OF NUMBERS

All properties within a number system follow by logical deduction from a few basic axioms, which form the working basis of the particular system [12]. The axioms as usually given are five in number. It is understood that the number is identical to the random variable in this discussion. The *axioms of the basic operations* are as follows:

1. Addition
    a. Any two numbers, x and y, have a unique sum $x+y$.
    b. The associative law holds; that is, if x, y, and z are any numbers, $x+(y+z)=(x+y)+z$.
    c. There exists an identity element for addition, number 0, such that $x+0=x$ for any number x.
    d. There is a uniary minus $(-)$. Corresponding to any number x there exists a number $-x$ such that $x+(-x)=0$, the negative of x.
    e. The commutative law holds; that is, if x and y are any numbers, $(x+y)=(y+x)$. The difference between x and y is defined as $(x-y)=x+(-y)$, called *subtraction*.

2. Multiplication
    a. Any two numbers x and y have a unique product, xy.
    b. The associative law holds; that is, if x, y, and z are any numbers, $x \cdot (yz) = (xy) \cdot z$.
    c. There exists an identity element for multiplication, number $1 \neq 0$, such that $(x \cdot 1) = x$ for any number x; this is called *unity*.
    d. Corresponding to any number $x \neq 0$ there exists a number x^{-1} (multiplicative inverse), such that $x \cdot x^{-1} = 1$, called the *reciprocal*.
    e. The commutative law holds; that is, if x and y are any numbers, $xy = yx$. The quotient of x and y ($y \neq 0$) is defined as $(x/y) = x \cdot y^{-1}$, called *division*.

3. Addition and multiplication
    a. The distributive law holds; that is, if x, y, and z are any numbers, $x \cdot (y+z) = (xy + xz)$.

In addition to random variables useful in engineering, these postulates hold for the rational, real, and complex numbers.

Engineering variables are found to be either statistically independent, by which it is inferred that they are in no way interdependent, or correlated when the random value taken by one (or more) variable(s) in a function is (are) related to the random value taken by another variable(s) in the same function [13, p. 235]. In engineering problems the variables are usually found to be unrelated, with correlation coefficient $\rho=0$, or linearly related with correlation coefficient $\rho=\pm 1$. For instance, a dimensional variable is probably not statistically related to a material strength. A common instance of linear correlation occurs when a variable is squared ($\rho=1$).

2.3 PROBABILITY CONSIDERATIONS

In applications of probability theory and statistics in engineering design the class of random variables of greatest utility have finite mean values and variances and may, in principal, assume any value within ranges of definition; that is, they are continuous. However, with engineering measurements it is necessary to digitize because of limitation on the resolution of values. Hence an engineering definition of a continuous random variable is that a random variable "is continuous, in an engineering sense, if it may assume values regardless of the fineness of our ability to discriminate or digitize measurements over the range of definition."

Referring to Figure 2.1, the *normal* is the most widely used of all distributions. Furthermore, empirical evidence indicates that the normal distribution provides a good representation for many engineering variables. The normal distribution has the further advantage for many problems that it is tractable mathematically [13, p. 72]. It is seen subsequently that numerous functions of normal random variables are normally distributed or approximately normally distributed.

The *gamma* (γ) distribution is used as a model for random variables bounded at one end. It is not used as a model in this book and is mentioned here for completeness.

The *exponential* distribution is sometimes a reasonable model for dynamic loading and applied stress. The exponential is a special case of the γ distribution.

The *lognormal* distribution is used as a model for the distribution of cycle life of engineering materials. This distribution is sometimes of interest because products and quotients of lognormally distributed random variables remain lognormally distributed.

The two-parameter *Weibull* distribution is used as a model for bearing life and for cycles or time to failure. Also, the normal (and other distributions) can be shown to be special cases of the Weibull distribution.

Distribution Name	Parameters	Probability Density Function	Expected Value	Variance
Normal	$-\infty < \mu < \infty,$ $\sigma > 0$	$f(x) = \dfrac{1}{\sigma\sqrt{2\pi}} \exp\left[-\dfrac{1}{2}\left(\dfrac{x-\mu}{\sigma}\right)^2\right]$ $-\infty < x < \infty$	μ	σ^2
Gamma	$\lambda > 0,$ $\eta > 0$	$f(x) = \begin{cases} \dfrac{\lambda^\eta}{\Gamma(\eta)} x^{\eta-1} e^{-\lambda x}, & x \geq 0 \\ 0 & \text{elsewhere} \end{cases}$	$\dfrac{\eta}{\lambda}$	$\dfrac{\eta}{\lambda^2}$
Exponential	$\lambda > 0$	$f(x) = \begin{cases} \lambda e^{-\lambda x}, & x \geq 0 \\ 0 & \text{elsewhere} \end{cases}$	$\dfrac{1}{\lambda}$	$\dfrac{1}{\lambda^2}$
Rayleigh	$\sigma > 0$	$f(x) = \begin{cases} \left(\dfrac{x}{\sigma^2}\right)\exp\left(\dfrac{-x^2}{2\sigma^2}\right), & x \geq 0 \\ 0 & \text{elsewhere} \end{cases}$	$\dfrac{(\sigma^2\pi)^{1/2}}{\sqrt{2}}$	$0.429\sigma^2$
Weibull	$\eta > 0, \sigma > 0$	$f(x) = \begin{cases} \dfrac{\eta}{\sigma}\left(\dfrac{x}{\sigma}\right)^{\eta-1} \exp\left[-\left(\dfrac{x}{\sigma}\right)^\eta\right], & x \geq 0 \\ 0 & \text{elsewhere} \end{cases}$	$\sigma\Gamma\left(\dfrac{1}{\eta}+1\right)$	$\sigma^2\left\{\Gamma\left(\dfrac{2}{\eta}+1\right) - \left[\Gamma\left(\dfrac{1}{\eta}+1\right)\right]^2\right\}$
Lognormal	$-\infty < \mu < \infty;$ $\sigma > 0$	$f(x) = \dfrac{1}{\sigma x\sqrt{2\pi}} \exp\left[-\dfrac{1}{2\sigma^2}(\log x - \mu)^2\right],$ $x \geq 0$	$e^{\mu + 1/2\sigma^2}$	$e^{2\mu+\sigma^2}(e^{\sigma^2}-1)$

Figure 2.1 Continuous distributions [13].

The *Rayleigh* distribution is sometimes a useful model for dynamic loading and applied stress when the instantaneous intensities are normally distributed. The Rayleigh distribution is further discussed and utilized later.

The exponential and Rayleigh distributions are one-parameter families. The normal, γ, lognormal, and two-parameter Weibull are two-parameter families.

With increasing numbers of parameters, models are more flexible as to the distributions of values that they may represent. The three-parameter Weibull and lognormal distributions are very adaptable in modeling test data. The price of flexibility comes in the difficulty of mathematical manipulation of such distributions. The three-parameter Weibull distribution is intractable mathematically, and methods for operating with functional combinations of such random variables, except by simulation, are as yet largely nonexistent.

The two-parameter distribution models used in this book provide a great deal of flexibility, and the mathematics of functional combinations remains relatively tractable. Referring to Figure 2.1, the probability density function $f(x)$ of a continuous random variable X is the function of the real variable x with the property that for every pair of real numbers a, b, with $a < b$:

$$P(a < X < b) = \int_a^b f(x)\,dx; \qquad F(x) = P(X < x) = \int_{-\infty}^x f(t)\,dt$$
(see Appendix 1)

$$f(x) = F'(x); \qquad \int_{-\infty}^\infty f(x)\,dx = 1 \quad (-\infty \leqslant x \leqslant \infty)$$

2.4 EXPECTED VALUE $E(x)$* AND VARIANCE $V(x)$†

Assume a random sample of n values (x_i) (see Figure 2.2) from the population of a random variable X having a known or unknown distribution

$$x_1, x_2, x_3, \ldots, x_n$$

The mean value $(\bar{x})$ and standard deviation (s_x) of the sample are

$$\bar{x} = \frac{x_1 + x_2 + \cdots + x_n}{n}; \qquad s_x = \sqrt{\frac{1}{n-1} \sum_{i=1}^n (x_i - \bar{x})^2}$$

where $\bar{x}$ is an efficient estimator of the population mean μ_x and is unbiased [15, p. 214]. Also, s_x is an efficient and unbiased estimator of σ_x.

---
*Theoretical measure of central tendency.
†Theoretical measure of scatter.

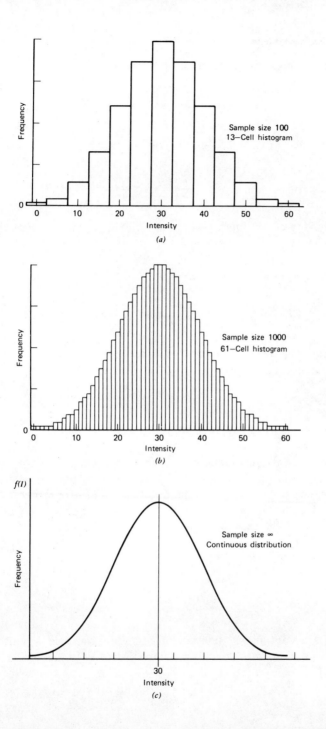

Figure 2.2 Relationships among histograms, (section 3.2.2). Based on finite samples and a theoretical distribution. Standard deviation $\sigma = 10$ (in all cases). (a) A sample consists of a small collection from some larger aggregate about which information is needed [57, p. 1]. (b) Samples are examined and facts learned from which inferences about aggregate or population are made.

22   Supporting Mathematics

If $f(x)$ is the probability function of X, then

$$\mu_x = E(x) = \sum_{i=1}^{\infty} x_i \cdot f(x_i) \qquad (2.1)$$

and the variance $V(x)$ of X is given by [9, p. 59]

$$V(x) = \sum_{i=1}^{\infty} [x_i - E(x)]^2 \cdot p_i \qquad (2.2)$$

where $p_i = f(x_i)$.

Similarly, for a continuous random variable X with probability density function $f(x)$,

$$E(x) = \int_{-\infty}^{\infty} x \cdot f(x) \cdot dx = \mu_x \qquad (2.3)$$

$$V(x) = \int_{-\infty}^{\infty} [x - E(x)]^2 \cdot f(x) \cdot dx = E(x - \mu_x)^2 = \sigma_x^2 \qquad (2.4)$$

If X is a random variable having a variance $V(x)$, the standard deviation σ_x of the population is

$$\sigma_x = \sqrt{V(x)} \; {}^* \qquad (2.5)$$

The following formulas [11, p. 84] on expectation were utilized in deriving expressions for the expected values (means or averages) and variances (or standard deviations) for functions of one variable [equation (2.6)] and for two independent variables [equation (2.7)]. If $\phi(x) = x$, the expression for the mean of x (i.e., μ_x) is obtained [equation (2.3)], and if $\phi(x) = [x - E(x)]^2$, the expression for the variance of x, [i.e., $V(x)$] is obtained [equation (2.4)]. Corresponding to the expected value and variance just mentioned, the estimators $\bar{x}$ and s_x are†

$$\lim_{n \to \infty} \bar{x} \approx \mu_x$$

$$\lim_{n \to \infty} s_x \approx \sigma_x = \sqrt{V(x)}$$

If $\phi(x)$ is a continuous function of a random variable X and if X is a continuous random variable with probability density function $f(x)$, then

$$E[\phi(x)] = \int_{-\infty}^{\infty} \phi(x) \cdot f(x) \, dx \qquad (2.6)$$

---

*The standard deviation σ_x is widely used in engineering because its units are the same as those of the mean, μ_x.
†It should be noted that in engineering applications, only estimators based on samples are available (see Figure 2.2).

If $\phi(x,y)$ is a continuous function of the real variables x and y and if (X, Y) is a continuous combined random variable with joint density function $f(x,y)$, then

$$E[\phi(x,y)] = \int_{-\infty}^{\infty}\int_{-\infty}^{\infty} \phi(x,y) \cdot f(x,y)\, dx\, dy \tag{2.7}$$

2.5 ALGEBRA OF EXPECTATION

Means and standard deviations of functions of one and two continuous independent random variables are discussed in this section (distributions are not specified). The population is that of mutually independent continuous random variables that have finite mean values and standard deviations and one- or two-parameter probability density functions.

The formulas for estimating mean values and standard deviations of functions [equations (2.8) through (2.35)] constitute what is referred to as "the algebra of expectation" in the literature.

STATISTICS OF FUNCTION OF ONE RANDOM VARIABLE

If a is a constant and X is a random variable with expectation $E(x)$, consider the function ax. Let $\phi(x) = ax$; then by equation (2.6), we obtain

$$E[\phi(x)] = E(ax) = \int_{-\infty}^{\infty} \phi(x) \cdot f(x)\, dx = \int_{-\infty}^{\infty} ax \cdot f(x)\, dx$$

$$E(ax) = a \int_{-\infty}^{\infty} x \cdot f(x) \cdot dx = aE(x) = a\mu_x \tag{2.8}$$

where [from equation (2.3)]:

$$\int_{-\infty}^{\infty} x \cdot f(x) \cdot dx = E(x)$$

If X is a random variable with expectation $E(x)$, then

$$E(1 \cdot x) = 1 \cdot E(x) = E(x) \tag{2.9}$$

If a is a constant and X a random variable with variance $V(x)$, then $V(ax) = a^2 V(x)$. Equation (2.4) yields

$$V(x) = \int_{-\infty}^{\infty} [x - E(x)]^2 f(x)\, dx = E(x - \mu_x)^2 = \sigma_x^2$$

when $\phi(x) = ax$. Then [9, p. 97]

$$V(ax) = E[ax - E(ax)]^2$$
$$= E[a(x - E(x))]^2$$
$$= E(a)[E[x - E(x)]] E(a) E[x - E(x)]$$
$$V(ax) = a^2 E[x - E(x)]^2 = a^2 V(x) \quad (2.10)$$

Example A

The mean value and standard deviation of the horizontal force component, F_h, acting as shown in Figure 2.3, is needed. Determining the statistics of the product of a random variable (μ_F, σ_F) and a constant, $\cos 30° = 0.866$, require equations (2.8) and (2.10).

$$\mu_{F_h} = \mu_F \cos 30° = 0.866 \cdot \mu_F$$
$$\sigma_{F_h} = \sigma_F \cos 30° = 0.866 \cdot \sigma_F$$

Thus
$$(\mu_{F_h}, \sigma_{F_h}) = (0.866 \mu_F, 0.866 \sigma_F) \text{ lb.}$$

If a is a real number (constant) and if X is a random variable with expectation $E(x)$, consider the function $x + a$. Note that

$$\int_{-\infty}^{\infty} f(x) \, dx = 1$$

Let
$$\phi(x) = x + a; \quad \therefore E[\phi(x)] = E[x + a].$$
$$E(x + a) = \int_{-\infty}^{\infty} f(x)(x + a) \, dx$$
$$= \int_{-\infty}^{\infty} f(x) x \, dx + \int_{-\infty}^{\infty} f(x) a \, dx$$
$$E(x + a) = E(x) + a \int_{-\infty}^{\infty} f(x) \, dx = E(x) + a \quad (2.11)$$

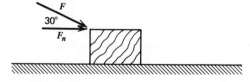

Figure 2.3 Statistics of a force component.

If a is a constant and X is a random variable with variance $V(x)$, then $V(x+a) = V(x)$, as shown by

$$V(x+a) = E[x+a-E(x+a)]^2$$
$$= E[x+a-E(x)-a]^2$$

and by equation (2.4)

$$V(x+a) = V(x) \qquad (2.12)$$

If a is a real number and X is a random variable with expectation $E(x) = \mu_x$, then letting $\phi(x) = a$ and utilizing equation (2.6):

$$\mu_a = E(a) = a \qquad (2.13)$$

If a is a constant and X is a random variable with variance $V(x)$:

$$V(a) = E[a-E(a)]^2 = E(a-a)^2 = 0 \qquad (2.14)$$

and $\sigma_a = 0$. The zero and unit elements are degenerate random variables represented by point $(0,0)$ and by point $(1,0)$. The expression for the variance of the random variable X in terms of the mean (μ_x) and the mean square $E(x^2)$ is developed from the basic definition of variance:

$$V(x) = E[x-E(x)]^2 = E\left[x^2 - 2xE(x) + [E(x)]^2\right]$$
$$= E(x^2) - E[2x \cdot E(x)] + E[E(x)]^2$$
$$= E(x^2) - 2E(x) \cdot E(x) + [E(x)]^2$$
$$= E(x^2) - 2[E(x)]^2 + [E(x)]^2$$
$$V(x) = E(x^2) - [E(x)]^2 \qquad (2.15)$$

[*Note:* In random vibration $E(x^2)$ is called the *mean square* (*MS*) value, and when $E(x) = 0$, the $MS = V(x)$. Also, the root mean square (RMS) = $\sqrt{V(x)}$ = σ_x.] Equation (2.15) directly yields a very useful result:

$$E(x^2) = V(x) + [E(x)]^2$$

and

$$E(x^2) = \sigma_x^2 + (\mu_x)^2 \qquad (2.16)$$

The term $E(x^2)$ is the expected value of the random variable X squared.

It is shown in equation (2.36h) that

$$V(x^2) = 4\mu_x^2\sigma_x^2 + 2\sigma_x^4,$$

$$\sigma_{x^2} = \sqrt{4\mu_x^2\sigma_x^2 + 2\sigma_x^4} \qquad (2.17)$$

$$\sigma_{x^2} \approx 2\mu_x\sigma_x \qquad (2.18)$$

Derivation of the statistics of the inverse of the random variable X is deferred [see equations (2.33) and (2.35)].

STATISTICS OF FUNCTIONS OF TWO INDEPENDENT RANDOM VARIABLES

2.5.1 Mean and Variance of Sum, $X + Y$

If X and Y are random variables with mean values μ_x and μ_y, then we let

$$\phi(x,y) = x + y$$

and apply equation (2.7):

$$E[\phi(x,y)] = E(x+y) = \int_{-\infty}^{\infty}\int_{-\infty}^{\infty}(x+y)g(x,y)\,dx\,dy$$

$$= \int_{-\infty}^{\infty} x\,dx \int_{-\infty}^{\infty} g(x,y)\,dy + \int_{-\infty}^{\infty} y\,dy \int_{-\infty}^{\infty} g(x,y)\,dx \qquad (a)$$

From elementary probability we obtain [9, p. 66]:

$$\int_{-\infty}^{\infty} g(x,y)\,dy = f(x) \qquad (b)$$

$$\int_{-\infty}^{\infty} g(x,y)\,dx = f(y)* \qquad (c)$$

Substituting equations (b) and (c) into equation (a) yields

$$E[x+y] = \int_{-\infty}^{\infty} xf(x)\,dx + \int_{-\infty}^{\infty} yf(y)\,dy$$

$$\mu_{x+y} = E[x+y] = E(x) + E(y) = \mu_x + \mu_y \qquad (2.19)$$

(*Note:* Finite sums and products of consistent estimators are consistent estimators of the corresponding parameters, and the mean is a consistent estimator [15, p. 136].)

---

*The term $g(x,y)$ is the joint probability density of X and Y, and $f(y)$ is called the *marginal density of Y*.

If X and Y are independent random variables with variances $V(x)$ and $V(y)$, the variance of the sum is

$$V(x+y) = V(x) + V(y)$$

developed from the basic definition of variance

$$V(x+y) = E[(x+y) - E(x+y)]^2$$
$$= E\{[x - E(x)] + [y - E(y)]\}^2$$
$$= E\{[x - E(x)]^2 + 2[x - E(x)][y - E(y)] + [y - E(y)]^2\}$$
$$= E[x - E(x)]^2 + 2E\{[x - E(x)][y - E(y)]\} + E[y - E(y)]^2.$$

Now, $x - E(x)$ is a function of random variable x minus a constant, as is $y - E(y)$. Note that $x - E(x)$ and $y - E(y)$ are independent since x and y are independent.

$$E\{[x - E(x)][y - E(y)]\} = E[x - E(x)]E[y - E(y)]$$

and since the expected value of a constant is the constant $E[E(x)] = E(x)$, and

$$[E(x) - E(x)][E(y) - E(y)] = 0$$

Thus

$$V(x+y) = E[x - E(x)]^2 + E[y - E(y)]^2$$
$$V(x+y) = V(x) + V(y) \tag{2.20}$$

$$\sigma_{x+y} = \sqrt{\sigma_x^2 + \sigma_y^2} \tag{2.21}$$

By induction, the variance of a finite sum of independent random variables is the sum of the variances of those variables.

2.5.2 Mean and Variance of Difference $X - Y$

If X and Y are random variables with mean values μ_x and μ_y, then*

$$\mu_{x-y} = E(x-y) = \mu(x) - \mu(y) \tag{2.22}$$

If X and Y are independent random variables, with variances $V(x)$ and $V(y)$, then*

$$V(x-y) = V(x) + V(y) \tag{2.23}$$

$$\sigma_{x-y} = \sqrt{\sigma_x^2 + \sigma_y^2} \tag{2.24}$$

*The developments are similar to those for sums.

Example B

This example illustrates a need to estimate the statistics of the difference of two independent random variables. Equations (2.22) and (2.24) provide formulae for estimating the statistics of a difference in terms of the statistics of the components.

A particular spur gear tooth is subject to a peak force F, inducing a tensile stress of (34,000; 2600) psi at each revolution. Production processing has introduced a compressive residual stress of (11,600; 1500) psi at the base of the tooth. Estimate the effective applied stress statistics, $\bar{s}$ and s_s.

Effective stress is the difference between the applied tensile and residual compressive stresses; thus

$$\bar{s} = 34{,}000 - 11{,}600 = 22{,}400 \text{ psi}$$

$$s_s = \sqrt{(2{,}600)^2 + (1{,}500)^2} = 3000 \text{ psi}$$

2.5.3 Statistics of Product of Independent Random Variables X and Y

If X and Y are independent random variables with expectations $E(x)$ and $E(y)$, then $E(xy) = E(x) \cdot E(y)$, shown as follows. Let

$$\phi(x,y) = xy$$

Equation (2.7) yields

$$E(xy) = \int_{-\infty}^{\infty} \int_{-\infty}^{\infty} (xy) \cdot f(x,y) \cdot dx\, dy$$

Since X and Y are independent* $f_1(x) \cdot f_2(y) = f(x,y)$,

$$E(xy) = \int_{-\infty}^{\infty} \int_{-\infty}^{\infty} (xy) \cdot f_1(x) \cdot f_2(y)\, dx\, dy$$

$$E(xy) = \int_{-\infty}^{\infty} x f_1(x)\, dx \int_{-\infty}^{\infty} y f_2(y)\, dy$$

$$E(xy) = E(x) \cdot E(y) \tag{2.25}$$

*Continuous random variables X and Y are independent if and only if their joint probability density function is the product of their respective marginal density functions [15, p. 79]:

$$\lambda(x,y) = f(x) \cdot g(y)$$

If X and Y are independent random variables with variance $V(x)$ and $V(y)$ and expected values μ_x and μ_y, then the variance $V(xy)$ is

$$V(xy) = E(xy - \mu_x\mu_y)^2 = E\left[x^2y^2 - 2xy\mu_x\mu_y + \mu_x^2\mu_y^2\right]$$

$$= E(x^2y^2) - 2\mu_{xy}E(xy) + \mu_{xy}^2$$

$$V(xy) = E(x^2y^2) - 2\mu_{xy}\mu_x\mu_y + \mu_{xy}^2 = E(x^2y^2) - \mu_{xy}^2$$

$$= E(x^2)E(y^2) - \mu_{xy}^2$$

$$V(xy) = (\sigma_x^2 + \mu_x^2)(\sigma_y^2 + \mu_y^2) - \mu_{xy}^2$$

$$V(xy) = \sigma_x^2 \cdot \mu_y^2 + \sigma_y^2 \cdot \mu_x^2 + \sigma_x^2 \cdot \sigma_y^2 \tag{2.26}$$

$$\sigma_{xy} = \sqrt{\mu_x^2\sigma_y^2 + \mu_y^2\sigma_x^2 + \sigma_x^2\sigma_y^2} \tag{2.27}$$

Example C

This example illustrates the estimation of the statistics of a product of two independent random variables. Equations (2.25) and (2.27) provide formulas for estimating the statistics of a product in terms of the statistics of the components.

If a bending equation were to be applied to make a rough estimation of the stress in a gear tooth, the moment at the weakest section (Figure 2.4) would be given by $M = F_b l$, for utilization in the subsequent stress estimation. Given

$$\left(\overline{F}_b, s_{F_b}\right) = (3800; 460) \text{ lb}$$

$$\left(\overline{l}, s_l\right) = (0.730, 0.016) \text{ in.}$$

The statistics of M are

$$\overline{M} = 3800 \cdot 0.730 = 2770 \text{ in. lb}$$

$$s_M = \left[(3700)^2 \cdot (0.016)^2 + (0.730)^2 \cdot (460)^2 + *(460)^2(0.016)^2\right]^{1/2}$$

$$= 340 \text{ in. lb}$$

*Note that the product of the variances, $(460)^2 \cdot (0.016)^2$, may be eliminated with negligible error, in this case.

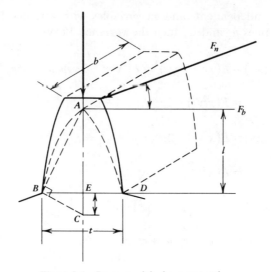

Figure 2.4 Beam model of a gear tooth.

The statistics of the ratio of two independent random variables (x/y) can be estimated by (1) integrating equation (2.7.) and (2) a Taylor series expansion (see Section 2.7). A review of the literature indicates that the method of partial derivatives has been used in the past without knowledge as to whether the statistics of such functions existed and whether the series converge.

Sanders, Dietrich, and Haugen developed existence theorems for negative moments of the probability density function of the ratio of two arbitrary random variables that may, in principle, not exist. However, it has also been shown that if the denominator probability density function takes on positive values only, that is,

$$a < y < b$$

for $a > 0$ and $b > 0$, the statistics (moments) do exist. Expected error resulting from use of the preceding estimation methods is considered in Section 2.6.

2.5.4 Mean and Variance of Quotient X/Y

If X and Y are independent random variables with expectations $E(x)$ and $E(y)$ and if the probability density function of X is $f(x)$ and $f(y)$ is the probability density function of Y, then let

$$\phi(x,y) = \frac{x}{y} \tag{a}$$

Substitution of equation (a) into equation (2.7) yields

$$E[\phi(x,y)] = E\frac{x}{y} = \int_{-\infty}^{\infty}\int_{-\infty}^{\infty} \frac{x}{y} f(x,y)\,dx\,dy$$

$$E\left(\frac{x}{y}\right) \approx \frac{E(x)}{E(y)} = \frac{\mu_x}{\mu_y} \qquad (2.28)$$

If X and Y are independent random variables with variances $V(x)$ and $V(y)$ and expectations $E(x)$ and $E(y)$, then

$$V[\phi(x,y)] = E\left(\frac{x}{y} - \frac{\mu_x}{\mu_y}\right)^2$$

$$= \int_{-\infty}^{\infty}\int_{-\infty}^{\infty}\left[\left(\frac{x}{y}\right) - \frac{\mu_x}{\mu_y}\right]^2 f(x,y)\,dx\,dy$$

Thus

$$V\frac{x}{y} \approx \frac{\mu_x^2 \sigma_y^2 + \mu_y^2 \sigma_x^2}{\mu_y^4} \qquad (2.29)$$

$$\sigma_{x/y} \approx \sqrt{\frac{\mu_x^2 \sigma_y^2 + \mu_y^2 \sigma_x^2}{\mu_y^4}} \qquad (2.30)$$

(*Note:* Division by the zero random variable, $(0,0)$, is not defined.)

Example D

This example illustrates the estimation of the statistics of a quotient of two independent random variables. Equations (2.28) and (2.30) provide formulas for estimating the statistics of a quotient in terms of the statistics of the numerator and the denominator.

$(\bar{s}, s_s) = (15{,}000; 1{,}600)$ psi are the statistics of stress in the material of a component having a modulus of elasticity $(\bar{E}, s_E) = (29 \cdot 10^6, 3.3 \cdot 10^5)$ psi. If the model for elastic strain is $\varepsilon = (s/E)$ estimate the statistics of ε.

$$\bar{\varepsilon} = \frac{\bar{s}}{\bar{E}} = \frac{15{,}000}{29 \cdot 10^6} = 5.2 \cdot 10^{-4} \qquad (2.31)$$

$$s_\varepsilon = \sqrt{\frac{\bar{s}^2 s_E^2 + \bar{E}^2 s_s^2}{\bar{E}^4}} \qquad (2.32)$$

$$s_\varepsilon = \sqrt{\frac{(1.5 \cdot 10^4)^2 (3.3 \cdot 10^5)^2 + (29 \cdot 10^6)^2 (1.6 \cdot 10^3)^2}{(29 \cdot 10^6)^4}}$$

$$= 5.6 \cdot 10^{-5}$$

The mean and variance of an inverse

$$z = \frac{1}{x}$$

Then in couple notation we obtain

$$(\mu_z, \sigma_z) = \frac{(1,0)}{(\mu_x, \sigma_x)} \quad (a)$$

Equations (2.28) and (2.33a) yields

$$\mu_z \approx \frac{1}{\mu_x} \quad (2.33)$$

The expression for the variance of $1/x$ is obtained by using equations (2.29) and (a):

$$V\left(\frac{1}{x}\right) \approx \frac{\sigma_x^2}{\mu_x^4} \quad (2.34)$$

$$\sigma_{1/x} \approx \frac{\sigma_x}{\mu_x^2} \quad (2.35)$$

The algebra of expectation for arbitrary independent two-parameter random variables has the following elements of structure in common with the algebra of the real numbers [9 and 101]:

1. Addition
   a. Uniqueness of sum.
   b. Associative law in addition.
   c. Existence of identity element for addition.
   d. Unary minus operation leads to subtraction.
   e. Commutative law in addition.

2. Multiplication
   a. Uniqueness of product.
   b. Associative law in multiplication.
   c. Existence of element for multiplicative inverse leads to division.
   d. Commutative law for multiplication.

The distributive law holds in combinations of addition and multiplication. Thus far in this section we have considered simple functions of one and two mutually independent random variables of unspecified distribution, each of which is uniquely described by (one or) two finite-valued parameters. Formulas for estimating the means and standard deviations of the functions in terms of the statistics of the component random variables have been developed.

In Section 2.6 the random variable set is (1) restricted to normally distributed random variables having finite-valued means and standard deviations and (2) expanded to include correlated in addition to independent normal random variables. The consideration of possible correlation between random variables in a function requires the introduction of additional techniques, specifically, moment-generating functions, which include coefficients ρ that account quantatively for correlation effects.

Moment-generating Functions. For those distributions whose moments are finite, it is possible to write moment generating functions. Such a function is by definition $E(e^{xt})$, where X is a random variable and t is a continuous variable. The expectation is a function of t, which may be written [9, p. 70], utilizing equation (2.6),

$$m(t) = E[e^{xt}] = \int_{-\infty}^{\infty} e^{xt} f(x) \, dx \tag{a}$$

where $f(x)$ is the probability density function of the random variable X. Differentiating equation (a) r times with respect to t yields

$$\frac{d^r m(t)}{dt^r} = \int_{-\infty}^{\infty} x^r e^{xt} f(x) \, dx$$

$$\frac{d(e^{xt})}{dt} = x e^{xt}$$

and setting $t = 0$,

$$\frac{d^r m(0)}{dt^r} = E[x^r] = \mu'_r$$

The term on the left is the rth derivative of $m(t)$ at $t = 0$. Consequently, the moments of a distribution may be formally obtained by differentiation.

The moments of a function of a random variable X must often be evaluated, for example, the moments of a function $h(x)$, where x has the probability density function $f(x)$. The rth moment of $h(x)$ is given by equation (b) [see equation (2.6)],

$$E[h(x)]^r = \int_{-\infty}^{\infty} [h(x)]^r f(x) \, dx \tag{b}$$

The moment-generating function of $h(x)$ is

$$m(t) = E[(e^{t \cdot h(x)})] = \int_{-\infty}^{\infty} e^{t \cdot h(x)} f(x) \, dx \tag{c}$$

The previous results may be extended to multivariate distributions. Consider three random variables X, Y, and Z with a joint density function

$f(x,y,z)$. The rth moment of z is, for example,

$$E[z^r] = \int\int\int_{-\infty}^{\infty} z^r f(x,y,z)\,dz\,dy\,dx$$

Also

$$E[x^q y^r z^s] = \int\int\int_{-\infty}^{\infty} x^q y^r z^s f(x,y,z)\,dz\,dy\,dx$$

where q, r, and s are any positive integers, including zero.

The joint moment-generating function must also be defined. For example, for three random variables X, Y and Z,

$$m(t_1, t_2, t_3) = E[e^{t_1 x} e^{t_2 y} e^{t_3 z}]$$
$$= E[\exp(t_1 x + t_2 y + t_3 z)]$$

The rth moment of the random variable Z is obtained by differentiating the moment-generating function r times with respect to t_3 and then setting $t_1 = t_2 = t_3 = 0$. Similarly, a joint moment $E[x^q y^r z^s]$ is obtained by differentiating q times with respect to t_1, r times with respect to t_2, and s times with respect to t_3 and then setting $t_1 = t_2 = t_3 = 0$.

Univariate Normal Moment-generating Function. Of interest in probabilistic design are the moments of functions of normally distributed random variables X that have probability density functions $f(x)$:

$$f(x) = \frac{1}{\sigma_x \sqrt{2\pi}} \exp\left[-\frac{1}{2}\left(\frac{x - \mu_x}{\sigma_x}\right)^2\right]$$

The moment-generating function of X is

$$m(t) = E(e^{tx}) = e^{t\mu} E(e^{t(x - \mu_x)})$$

and

$$m(t) = \exp\left[t\mu_x + \frac{\sigma_x^2 t^2}{2}\right] \qquad (d)$$

Example. Illustrate the application of equation (d) by deriving expressions for the mean and standard deviation of $\phi(x) = x^2$. Assume that X is normally distributed; thus

$$E(x^2) = \mu_{x^2} = \frac{d^2(m)}{dt^2}\bigg|_{t=0}$$

$$m(t) = \exp\left(t\mu + \frac{\sigma^2 t^2}{2}\right) = e^R$$

The first derivative of $m(t)$ with respect to t is

$$\frac{\partial m(t)}{\partial t} = (\mu_x + \sigma_x^2 t)e^R$$

The second derivative of $m(t)$ with respect to t is

$$\frac{\partial^2 m(t)}{\partial t^2} = (\sigma_x^2)e^R + (\mu_x + \sigma_x^2 t)^2 e^R$$

Thus $E(x^2) = \mu_{x^2}$ is

$$E(x)^2 = \frac{\partial^2 m(t)}{\partial t^2}\bigg]_{t=0} = \sigma_x^2 + \mu_x^2 \tag{e}$$

To derive an expression for the variance of $\phi(x) = x^2$, we utilize the basic definition for $V[\phi(x)]$,

$$V(x^2) = E[x^2 - E(x^2)]^2$$

$$= E(x^4) - 2x^2 E(x^2) + E[E(x^2)]^2$$

$$V(x^2) = E(x^4) - E2x^2(\sigma_x^2 + \mu_x^2) + (\sigma_x^2 + \mu_x^2)^2 \tag{f}$$

In the expression for $V(x^2)$ it is seen that the expressions for $E(x^4)$ and $E(x^2)$ are required. Thus the third and fourth derivatives of $m(t)$ with respect to t must be obtained and evaluated at $t = 0$:

$$\frac{\partial^3 m(t)}{\partial t^3} = (\sigma_x^2 + \mu_x^2)e^R + (2\mu_x \sigma_x^2 t)e^R + (\sigma_x^4 t^2)e^R$$

$$= (\sigma_x^2 + \mu_x^2)(\mu_x + \sigma_x^2 t)e^R + (2\mu_x \sigma_x^2)e^R + 2\mu_x \sigma_x^2 t(\mu_x + \sigma_x^2 t)e^R$$

$$+ (2\sigma_x^4 t)e^R + \sigma_x^4 t^2(\mu_x + \sigma_x^2 t)e^R$$

$$\frac{\partial^3 m(t)}{\partial t^3} = \{\mu_x^3 + \mu\sigma_x^2 + \mu_x^2 \sigma_x^2 t + \sigma_x^4 t + 2\mu_x \sigma_x^2 + 2\mu_x^2 \sigma_x^2 t$$

$$+ 2\mu_x \sigma_x^4 t^2 + 2\sigma_x^4 t + \mu_x \sigma_x^4 t^2 + \sigma_x^6 t^3\} e^R$$

$$E(x^3) = \mu_x^3 + \mu_x \sigma_x^2 + 2\mu_x \sigma_x^2 = \mu_x^3 + 3\mu_x \sigma_x^2 \tag{g}$$

$$\frac{\partial^4 m(t)}{\partial t^4} = \{\mu_x^2 \sigma_x^2 + \sigma_x^4 + 2\mu_x^2 \sigma_x^2 + 2\sigma_x^4\} e^R$$

$$+ (\mu_x + \sigma_x^2 t)e^R(\mu_x^3 + \mu_x \sigma_x^2 + 2\mu_x \sigma_x^2)$$

$$\frac{\partial^4 m(t)}{\partial t^4} = \{\mu_x^2 \sigma_x^2 + \sigma_x^4 + 2\mu_x^2 \sigma_x^2 + 2\sigma_x^4\} e^R$$

$$+ (\mu_x^4 + 3\mu_x^3 \sigma_x^2)e^R$$

$$E(x^4) = \mu_x^4 + 6\mu_x^2 \sigma_x^2 + 3\sigma_x^4 \tag{h}$$

By substituting equations (e) and (h) into (f), the expression for $V(x^2)$ is obtained:

$$V(x^2) = E(x^4) - 2x^2(\sigma_x^2 + \mu_x^2) + (\sigma_x^2 + \mu_x^2)^2$$

$$= \mu_x^4 + 6\mu_x^2\sigma_x^2 + 3\sigma_x^4 - \sigma_x^4 - 2\mu_x^2\sigma_x^2 - \mu_x^4$$

$$V(x^2) = 4\mu_x^2\sigma_x^2 + 2\sigma_x^4$$

$$\sigma_{x^2} = \sqrt{4\mu_x^2\sigma_x^2 + 2\sigma_x^4} \qquad \text{(i)}$$

Bivariate Normal Moment-generating Function. Of interest in probabilistic design are the moments of functions of normally distributed random variables X and Y, which have bivariate probability density functions $f(x,y)$ [9, p. 72]:

$$f(x,y) = \frac{1}{2\pi\sigma_x\sigma_y(1-\rho^2)^{1/2}}$$

$$\times \exp\left[-\frac{1}{2(1-\rho^2)}\left[\left(\frac{x-\mu_x}{\sigma_x}\right)^2 - 2\rho\frac{(x-\mu_x)(y-\mu_y)}{\sigma_x\sigma_y} + \left(\frac{y-\mu_y}{\sigma_y}\right)^2\right]\right]$$

(j)

The bivariate moment-generating function in random variables X and Y is

$$m(t_1, t_2) = E[\exp(t_1 x + t_2 y)] = \int\int_{-\infty}^{\infty} e(t_1 x + t_2 y) f(x,y) \, dy \, dx$$

and

$$m(t_1, t_2) = \exp\left[t_1\mu_x + t_2\mu_y + \frac{1}{2}(t_1^2\sigma_x^2 + 2\rho t_1 t_2 \sigma_x\sigma_y + t_2^2\sigma_y^2)\right] = e^R \qquad \text{(k)}$$

An example employing equation (k) is as follows. Let

$$\phi(x,y) = xy = z$$

Since the random variables X and Y may be either statistically independent or correlated, the bivariate normal moment-generating function is utilized to derive the expected value of $E(xy)$:

$$E[z] = E[xy] = \left.\frac{\partial^2 m}{\partial t_1 \partial t_2}\right|_{t_1 = t_2 = 0}$$

The first partial derivative of equation (k) with respect to t_1 is

$$\frac{\partial m}{\partial t_1} = \left(\mu_x + t_1 \sigma_x^2 + 2\rho t_2 \sigma_x \sigma_y\right) e^R$$

and the second partial derivative with respect to t_2 is

$$\frac{\partial^2 m}{\partial t_1 \partial t_2} = (\rho \sigma_x \sigma_y) \cdot e^R + \left(\mu_y + \sigma t_1 \sigma_x \sigma_y + t_2 \sigma_y^2\right)\left(\mu_x + \rho t_2 \sigma_x \sigma_y + t_1 \sigma_x^2\right) \cdot e^R$$

Thus the expected value $E(xy)$ is

$$E[z] = E(xy) = \left. \frac{\partial^2 m}{\partial t_1 \partial t_2} \right|_{t_1 = t_2 = 0} = \mu_x \mu_y + \rho \cdot \sigma_x \sigma_y \qquad (1)$$

STATISTICS OF FUNCTIONS OF TWO RANDOM VARIABLES (INDEPENDENT OR CORRELATED)

2.6 ALGEBRA OF NORMAL DISTRIBUTIONS [9]

Means and standard deviations of functions of one and two independent or correlated normal random variables are estimated by utilizing the statistics of the component random variables and ρ.

Consider the population of both mutually independent and correlated normally distributed random variables that have finite mean values μ and standard deviations σ [16].

2.6.1 Introduction

The formulas for mean values and standard deviations of sums, differences, products, quotients, and so on (equations 2.36 to 2.62) constitute what is referred to as the "algebra of normal functions".

The expressions for mean values and variances of functions of a single random variable (Section 2.5) are valid in the algebra of normal functions. The formulas for the statistics of sums, differences, products, and quotients in Section 2.5 are valid for the special case of independent normally distributed random-variable pairs.

2.6.2 Statistics of Sum $Z = (X + Y)$

The expected value (mean) is

$$E[z] = \mu_z = E\phi(x,y) = E(x+y)$$

Equation (2.7) yields

$$E[\phi(x,y)] = E(x+y) = \int_{-\infty}^{\infty}\int_{-\infty}^{\infty} (x+y)g(x,y)\,dx\,dy$$

$$\mu_{x+y} = E[x+y] = E(x) + E(y) = \mu_x + \mu_y \tag{2.36}$$

Variance is expressed by $V(x+y) = V(z)$. Employing the basic definition of variance, we obtain

$$V[\phi(x,y)] = E[\phi(x,y) - E(\phi(x,y))]^2$$

$$V(x+y) = E[(x+y) - (\mu_x + \mu_y)]^2$$

and expanding the preceding expression on the right,

$$V(x+y) = E\big[x^2 + 2xy - 2x\mu_x - 2x\mu_y + y^2 - 2y\mu_x - 2y\mu_y + \mu_x^2 + 2\mu_x\mu_y + \mu_y^2\big]$$

Equation (e) yields

$$E(x^2) = \mu_x^2 + \sigma_x^2$$
$$E(y^2) = \mu_y^2 + \sigma_y^2$$

Equation (l) yields

$$E(2xy) = 2(\mu_x\mu_y + \rho\sigma_x\sigma_y)$$

from which

$$V(x+y) = \sigma_x^2 + 2\rho\sigma_x\sigma_y + \sigma_y^2 \tag{2.37}$$

If $\rho = 0$, and X and Y are statistically independent, then

$$V(z) = \sigma_x^2 + \sigma_y^2$$

If $\rho = +1$ and X and Y are identical, then

$$\sigma_z = \sigma_{x+y} = \sqrt{\sigma_x^2 + 2\sigma_x\sigma_y + \sigma_y^2} = \sigma_x + \sigma_y = 2\sigma_x = 2\sigma_y.$$

If $\rho = -1$ and X and Y are identical, then

$$\sigma_z = \sigma_{x+y} = \sqrt{\sigma_x^2 - 2\sigma_x\sigma_y + \sigma_y^2} = 0$$

that is, the additive inverse exists in the algebra of normal functions.

The sum of any two independent normally distributed random variables is closed (i.e., is a normal random variable).

Example E Sum of Independent Random Variables

The following example illustrates the estimation of the statistics of the sum of two independent normally distributed random variables. Equations (2.36) and (2.37) provide formulas for estimating the statistics of a sum in terms of the statistics of the components $\rho = 0$.

Given acceleration due to gravity $(g, 0)$ and a base acceleration $(1.65g, 0.15g)$, two independent normal random variables, and from equations (2.36) and (2.37), the effective acceleration random variable (a) is

$$(\bar{a}, s_a) = (g, 0) + (1.65g, 0.15g) = (2.65g, 0.15g)$$

where g is assumed to be a degenerate random variable (i.e., $\sigma_g = 0$).

Example F Sum of Correlated Random Variables

See example B in Chapter 5, where correlated normal stress random variables are summed.

The mean and variance estimation of a linear combination [9] is

$$W = aX + bY + cZ + \cdots +$$

where X, Y, and Z are independent normal random variables and a, b, and c are arbitrary constants. The mean estimator of the linear combination is

$$\bar{w} = a\bar{x} + b\bar{y} + c\bar{z} + \cdots + \qquad (2.38)$$

The variance estimator of the linear combination is

$$s_w^2 = a^2 s_x^2 + b^2 s_y^2 + c^2 s_z^2 + \cdots + \qquad (2.39)$$

Example G

This example illustrates a common problem involving the statistics of the sum of several independent normally distributed random variables. Equations (2.38) and (2.39) provide formulas for estimating the statistics of the sum in terms of the statistics of the components.

The overall tolerance of a stack must be controlled. The (stack) assembly consists of a crank (C), two washers (W_1 and W_2), and two side bars (B_1 and B_2). The nominal dimensions (averages) are

$$\bar{c} = 0.750 \text{ in.}$$

$$\bar{w}_1 = 0.060 \text{ in.}$$

$$\bar{w}_2 = 0.060 \text{ in.}$$

$$\bar{b}_1 = 0.450 \text{ in.}$$

$$\bar{b}_2 = 0.450 \text{ in.}$$

The stack average $= \bar{s} = 1.770$ in.

Assume that the dimensional requirements for the stack are 1.770 in., with a stack tolerance of ± 0.025 in. All components have the same variance, σ^2. No more than the fraction 0.0027 of the assemblies can exceed tolerance.

The variance of the stack is [equation (2.39)]:

$$V(s) = \sigma_s^2 = 5\sigma^2$$

A fraction 0.0027 outside tolerance corresponds to the density under a normal probability density function curve beyond ± 3.0 standard deviations from the mean (Appendix 1). Thus the stack tolerance, 0.025, amounts to

$$\frac{0.025}{\sigma_s} \approx 3.0$$

and the stack standard deviation is

$$\sigma_s = 0.0083 \text{ in.}$$

The variance of the stack is $(\sigma_s)^2$; thus

$$\sigma_s^2 = 0.000069$$

The maximum variance of each component is

$$\sigma^2 = \frac{0.000069}{5} = 0.000014$$

and the maximum component standard deviation σ is

$$\sigma = \sqrt{0.000014} = 3.7 \cdot 10^{-3} \text{ in.}$$

A maximum component tolerance is $\pm \Delta \approx \pm 3\sigma = \pm 0.011$ in.

2.6.3 Statistics of Difference $Z = (X - Y)$

The mean, $E(z) = \mu_z$, is expressed as

$$\mu_z = E[x - y] = E(x) - E(y) = \mu_x - \mu_y \qquad (2.40)$$

The variance, $V(z) = \sigma_z^2$, is expressed as

$$V(z) = E[z - \mu_z^2] = \sigma_z^2 = \sigma_x^2 - 2\rho \cdot \sigma_x \sigma_y + \sigma_y^2 \qquad (2.41)$$

If $\rho = 0$ and X and Y are statistically independent, then

$$\sigma_z^2 = \sigma_x^2 + \sigma_y^2$$

If $\rho = +1$ and X and Y are identical, then

$$\sigma_z = \sqrt{\sigma_x^2 - 2\sigma_x \sigma_y + \sigma_y^2} = \sigma_x - \sigma_y = 0$$

If $\rho = -1$ and X and Y are identical, then

$$\sigma_z = \sqrt{\sigma_x^2 + 2\sigma_x \sigma_y + \sigma_y^2} = \sigma_x + \sigma_y = 2\sigma_x = 2\sigma_y$$

Example H Difference of Independent Random Variables

This example illustrates the estimation of the statistics of the difference of two independent normally distributed random variables. Equations (2.40) and (2.41) provide the formulas for estimating the statistics of the difference in terms of the statistics of the components, illustrated in Figure 2.5.

Assume that two normally distributed independent load components X_1 and X_2 are aligned but opposed to one another:

$$(\bar{x}_1, s_1) = (200, 45) \text{ lb}$$

$$(\bar{x}_2, s_2) = (100, 15) \text{ lb}$$

Compute the statistics of the difference random variable $X = (X_1 - X_2)$:

$$\bar{x} = \bar{x}_1 - \bar{x}_2 = 200 - 100 = 100 \text{ lb}$$

Noting that $\rho = 0$:

$$s_x = \sqrt{s_1^2 + s_2^2} = \sqrt{(45)^2 + (15)^2} = 47 \text{ lb}$$

Thus

$$(\bar{x}, s_x) = (\bar{x}_1, s_1) - (\bar{x}_2, s_2) = (100, 47) \text{ lb}$$

The difference random variable X is normally distributed.

42 Supporting Mathematics

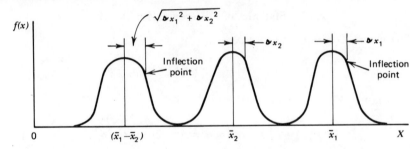

Figure 2.5 Difference of two independent normal random variables.

Example I Difference of Correlated Random Variables

This example results from an inference based on differencing considerations of normally distributed random variables. Equations (2.40) and (2.41) are required. Classically, a necessary condition for equilibrium is that (assumed) deterministic forces acting in any direction must sum to zero:

$$\Sigma F = 0 \tag{a}$$

The probabilistic analogue of equation (a) is

$$\Sigma \left(\bar{F}, \sigma_F \right) = (0, 0) \tag{b}$$

State the necessary and sufficient conditions for equilibrium that satisfy equation (b) for a pair of opposed normally distributed force random variables, at a given instant. The required conditions are

1. $\bar{F}_1 - \bar{F}_2 = 0$.
2. $\sigma_{F_1} = \sigma_{F_2}$.
3. F_1 and F_2 correlated, with $\rho = +1$.

If F_1 and F_2 are dynamic loads, they must be in phase.

Example J

In mechanical design, a common problem is the random mating of bearings and shafts (Figure 2.6). This problem involves the difference between two statistically independent normal random variables. Given the outer diameter (OD) statistics of the shaft, we obtain

$$\left(\bar{D}_s, \sigma_s \right) = (1.048, 0.0013) \text{ in.}$$

and the inner diameter (ID) statistics of the bearing

$$\left(\bar{D}_B, \sigma_B \right) = (1.059, 0.0017) \text{ in.}$$

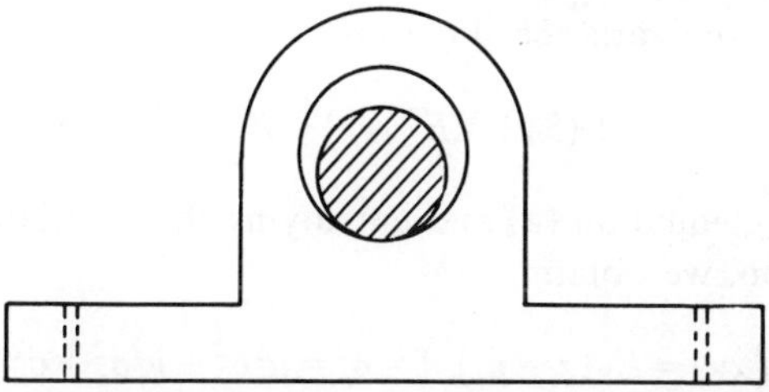

Figure 2.6 Mating of shaft and bearing.

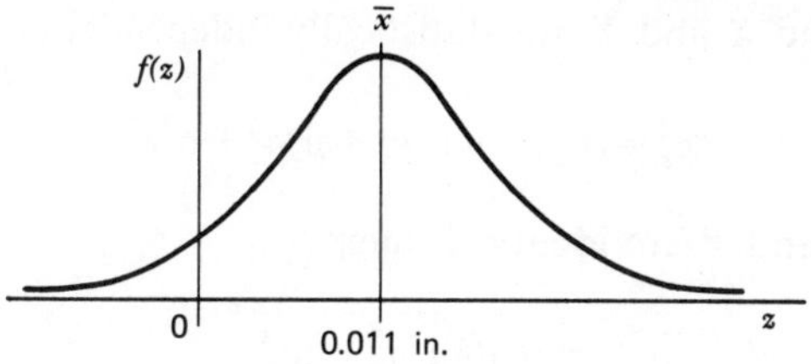

Figure 2.7 Clearance probability density function.

What are the clearance statistics (Z) on random pairing of such bearings and shafts?
Equations (2.40) and (2.41) provide formulas for estimating the statistics of the difference random variable,

$$Z = D_B - D_S$$

and

$$\bar{Z} = \bar{D}_B - \bar{D}_S = 0.011 \text{ in.} = (1.059 - 1.048) \text{ in.}$$

and with $\rho = 0$:

$$s_z = \sqrt{(0.0013)^2 + (0.0017)^2} = 0.0021 \text{ in.}$$

Thus $(\bar{z}, s_z) = (0.011, 0.0021)$ in. (see Figure 2.7).

2.6.4 Statistics of Product $Z = XY$

The expected value (mean), μ_z, is derived as follows. Let $\phi(x,y) = xy$ and $E(xy) = E(z)$. Then, from the derivation on p. 39 yielding equation (l), we obtain

$$E[z] = \mu_{xy} = \mu_x \mu_y + \rho \cdot \sigma_x \sigma_y \tag{2.42}$$

When $\rho = 0$, $\mu_z = \mu_x \mu_y$, and $\rho = +1$ and X and Y are identical,

$$\mu_z = \mu_x^2 + \sigma_x^2 = \mu_y^2 + \sigma_y^2$$

44 Supporting Mathematics

Given $\phi(x,y) = xy$, the variance, $V(xy)$ is

$$V(xy) = E[(xy) - E(z)]^2 \tag{a}$$

Thus by expanding equation (a) and employing the bivariate normal moment generating function, we obtain

$$V(xy) = E[(z - \mu_z)^2] = \sigma_z^2 = \mu_x^2 \sigma_y^2 + \mu_y^2 \sigma_x^2 + \sigma_x^2 \sigma_y^2$$
$$+ 2\rho \cdot \mu_x \mu_y \cdot \sigma_x \sigma_y + \rho^2 \cdot \sigma_x^2 \sigma_y^2 \tag{2.43}$$

and when $\rho = 0$ and X and Y are statistically independent:

$$\sigma_z^2 = (\sigma_{xy})^2 = \mu_x^2 \sigma_y^2 + \mu_y^2 \sigma_x^2 + \sigma_x^2 \sigma_y^2 \tag{2.44}$$

If $\rho = +1$ and X and Y are identical, then

$$\sigma_z = \sqrt{4\mu_x^2 \sigma_x^2 + 2\sigma_x^4}\,^*$$

[see equation (i), p. 38].

If $\rho = -1$ and X and Y are identical, then

$$\sigma_z = \sqrt{2} \cdot \sigma_x^2 = \sqrt{2} \cdot \sigma_y^2$$

Example K Product of Independent Random Variables

What is the allowable load variate P when the normally distributed parameters are A (cross-sectional area) and S_y (tensile yield strength), given that $P = AS_y$.

Equations (2.42) and (2.44) are required to estimate the statistics of P from the statistics of S_y and A.

Given

$$(\overline{A}, s_A) = (10, 0.4) \text{ in.}^2 \text{ and } (\overline{S}_y, s_{S_y}) = (5 \cdot 10^4, 3 \cdot 10^3) \text{ lb/in.}^2$$
$$\overline{P} = \overline{A}\,\overline{S}_y = 5 \cdot 10^5 \text{ lb.}$$

and since P and A are independent with $\rho = 0$,

$$s_P = \sqrt{(0.4)^2 \cdot (5 \cdot 10^4)^2 + (3 \cdot 10^3)^2 \cdot (10)^2 + (0.4)^2 \cdot (3 \cdot 10^3)^2} = 3.6 \cdot 10^4 \text{ lb}$$

Thus the statistics of P are

$$(\overline{P}, s_P) = (5 \cdot 10^5, 3.6 \cdot 10^4) \text{ lb.}$$

---

*In the usual case in engineering $2\sigma_x^4$ can be ignored as inconsequential.

Example L   Products of Correlated Random Variables

Consider the equation $U=(F\delta/2)$ for strain energy, where δ is a linear function of F (i.e., correlated $\rho=+1$) within the elastic limit.

DISTRIBUTION OF A PRODUCT

The product of two normally distributed random variables is not (strictly) normally distributed (i.e., *closed*). However, under certain conditions the products closely approximate normality, that is, are *robust*. Robustness is defined as follows. The errors of assuming normality depend on the use to which this assumption is put. Many statistical methods derived under this assumption remain valid under moderate deviations and are thus said to be robust [13, p. 74]. Note that analysis of variance is a method that is robust under deviations from normality as are physical variables with averages many standard derivations removed from the value zero.

Product Probability Density Function.   Heckman and Dietrich [17] considered numerous parameter combinations in studying the probability density function of the product $Y_1=(X_1 \cdot X_2)$, where X_1 and X_2 are normal random variables.

The shape of product probability density functions is sensitive not to the value of the dimension parameter, but to values of σ_{x_1}/μ_{x_1} and σ_{x_2}/μ_{x_2} (see Tables 2.1 through 2.4 and Figures 2.8 and 2.9). It can be concluded that:

1. For $0.1 \leq (\mu_{x_1}/\mu_{x_2}) \leq 10$, $0.005 \leq (\sigma_{x_1}/\mu_{x_1}) \leq 0.20$ and $0.005 \leq (\sigma_{x_2}/\mu_{x_2}) \leq 0.20$; the product probability density function is unimodal.

Table 2.1.   Product $Y_1 = X_1 \cdot X_2$

Mean Ratio	Variate X_1	Variate X_2	Coefficient of Variation (%) X_1	Coefficient of Variation (%) X_2	Robust Normal	γ	Lognormal
1	(1,000, 50)	(1,000, 50)	5	5	Yes	Yes	Yes
1	(1,000, 200)	(1,000, 50)	20	5	Yes	No	No
1	(1,000, 200)	(1,000, 200)	20	20	No	No	No
0.1	(1,000, 50)	(10,000, 50)	5	0.5	Yes	Yes	Yes
0.1	(1,000, 50)	(10,000, 200)	5	2	Yes	Yes	Yes
0.1	(1,000, 200)	(10,000, 50)	20	0.5	Yes	No	No
0.1	(1,000, 200)	(10,000, 200)	20	2	Yes	No	No
1	(10,000, 50)	(10,000, 50)	0.5	0.5	Yes	Yes	Yes
1	(10,000, 50)	(10,000, 200)	0.5	2	Yes	Yes	Yes
1	(10,000, 200)	(10,000, 200)	2	2	Yes	Yes	Yes

Table 2.2. Product $Y_1 = X_1 \cdot X_2$

Mean Ratio μ_{X_1}/μ_{X_2}	Variate X_1		Variate X_2		Coefficient of Variation (%) X_1	X_2	Robust Normal	γ	Lognormal
0.1	(2,000,	100)	(20,000,	100)	5	0.5	Yes	Yes	Yes
1	(3,000,	150)	(3,000,	150)	5	5	Yes	Yes	Yes
10	(5,000,	100)	(500,	25)	2	5	Yes	Yes	Yes
1	(20,000,	4000)	(20,000,	1,000)	20	5	Yes	Yes	No
10	(20,000,	100)	(2,000,	100)	0.5	5	Yes	Yes	Yes

Table 2.3. Product $Y_1 = X_1 \cdot X_2$

Mean Ratio μ_{X_1}/μ_{X_2}	Variate X_1		Variate X_2		Coefficient of Variation (%) X_1	X_2	Robust Normal
1	(1,000,	200)	(1,000,	60)	20	6	Yes
1	(1,000,	50)	(1,000,	60)	5	6	Yes
1	(1,000,	60)	(1,000,	60)	6	6	Yes
10	(10,000,	200)	(1,000,	60)	2	6	Yes
10	(10,000,	50)	(1,000,	60)	0.5	6	Yes
1	(1,000,	50)	(1,000,	75)	5	7.5	Yes
10	(10,000,	50)	(1,000,	75)	0.5	7.5	Yes
0.1	(1,000,	20)	(1,000,	750)	2	7.5	Yes
0.1	(1,000,	75)	(1,000,	75)	7.5	7.5	Marginal

Table 2.4. Product $Y_1 = X_1 \cdot X_2$

Mean Ratio μ_{X_1}/μ_{X_2}	Variate X_1		Variate X_2		Coefficient of Variation (%) X_1	X_2	γ	Robust Lognormal
1	(1,000,	50)	(1,000,	75)	5	7.5	Yes	Yes
1	(1,000,	75)	(1,000,	75)	7.5	7.5	Yes	Yes
10	(10,000,	50)	(1,000,	75)	0.5	7.5	Yes	Yes

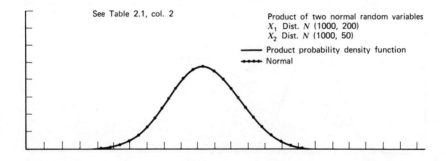

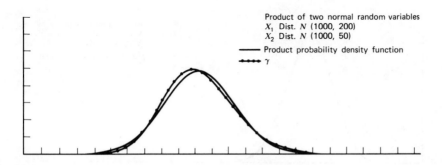

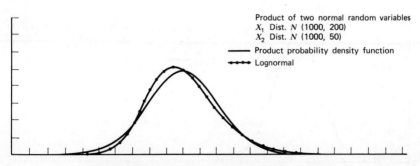

Figure 2.8 Product of two normal random variables.

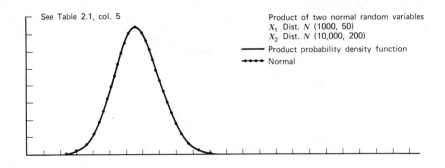

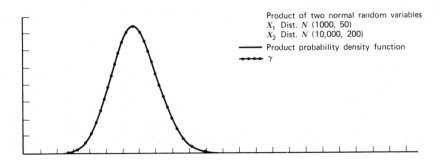

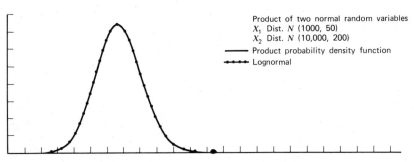

Figure 2.9 Product of two normal random variables.

2. For $0.1 \leqslant (\mu_{x_1}/\mu_{x_2}) \leqslant 10$, $0.005 \leqslant \sigma_{x_1}/\mu_{x_1} \leqslant 0.20$ and $0.005 \leqslant (\sigma_{x_2}/\mu_{x_2}) \leqslant 0.20$, a sufficient condition for a *robust normal approximation* by the product probability density function is that either (σ_{x_1}/μ_{x_1}) or $(\sigma_{x_2}/\mu_{x_2}) \leqslant 0.075$.

2.6.5 Product of a Random Variable and its Inverse

Assume a normally distributed random variable X, with statistics

$$(\mu_x, \sigma_x)$$

and inverse $z = (1/x)$. According to the algebra of expectation:

$$\mu_z, \sigma_z = \mu_{1/x}, \sigma_{1/x} = \frac{1}{\mu_x} \; ; \; \frac{\sigma_x}{\mu_x^2}$$

Consider the product $(\mu_x, \sigma_x)(\mu_z, \sigma_z)$. If it is assumed that the distribution of Z is robust as indicated by Figure 2.10, the mean of the product may be approximated as follows:

$$\mu_{xz} \approx \mu_{x(1/\mu_x)} = 1$$

The correlation of x and $1/x$ is negative linear. Thus for $\rho = -1$ and applying equation (2.43):

$$\sigma^2_{x(z)} \approx \mu_x^2 \cdot \frac{\sigma_x^2}{\mu_x^4} + \frac{1}{\mu_x^2}\sigma_x^2 + \sigma_x^2 \cdot \left(\frac{\sigma_x^2}{\mu_x^4}\right) - 2\mu_x \frac{\sigma_x^2}{\mu_x^3} + \frac{\sigma_x^4}{\mu_x^4} \approx \frac{2\sigma_x^4}{\mu_x^4} \qquad (a)$$

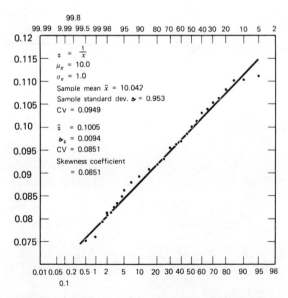

Figure 2.10 Fit of random variable $z = (1/x)$ to Gaussian probability density function model when $N(\mu_x, \sigma_x) = (10, 1)$ (Monte Carlo Simulation, $n = 200$) (CV = coefficient of variation).

Example M

If $(\sigma_x/\mu_x)=0.10$, then $\sigma^2_{x(z)}\approx 0.0002$, and equation (a) is found to yield a satisfactory approximation to zero for many purposes in engineering.

2.6.6 Statistics of Quotient $Z=(Y/X)$

The expected value (mean) is $E(z)=\mu_z$. If $\phi(x,y)=(y/z)$, then by utilizing equation (2.7), we obtain

$$E[\phi(x,y)] = E\left[\frac{y}{x}\right] = E[z] = \int_{-\infty}^{\infty}\int_{-\infty}^{\infty} \left(\frac{y}{x}\right) f(x,y)\,dx\,dy$$

$$\mu_z = \mu_{y/x} \approx \frac{\mu_y}{\mu_x}\left[1+\frac{\sigma_x}{\mu_x}\left(\frac{\sigma_x}{\mu_x}-\rho\frac{\sigma_y}{\mu_y}\right)\left(1+3\frac{\sigma_x^2}{\mu_x^2}+\cdots+\right)\right]^*$$

If the second and higher powers in the last term on the right are negligible relative to the first term, then

$$\mu_z = \mu_{y/x} \approx \frac{\mu_y}{\mu_x}\left[1+\frac{\sigma_x}{\mu_x}\cdot\left(\frac{\sigma_x}{\mu_x}-\rho\frac{\sigma_y}{\mu_y}\right)\right] \qquad (2.45)$$

For the case where $\rho=0$ and X and Y are statistically independent,

$$\mu_z = \mu_{y/x} \approx \frac{\mu_y}{\mu_x}\cdot\left[1+\left(\frac{\sigma_x}{\mu_x}\right)^2\right]$$

When $\sigma_x^2\mu_y \ll \mu_x^3$,

$$\mu_z = \mu_{y/x} = \frac{\mu_y}{\mu_x} \qquad (2.46)$$

The variance is $V(y/z)=\sigma^2_{y/x}$. If $\phi(x,y)=(y/x)$, then by the definition of variance

$$V\frac{y}{x} = E\left(\frac{y}{x}-E\frac{y}{x}\right)^2. \qquad (a)$$

If equation (a) is expanded bivariate normal moment-generating functions are utilized, then (by a Taylor series approximation)

$$\sigma_z^2 = \frac{\mu_y^2}{\mu_x^2}\left[\frac{\sigma_x^2}{\mu_x^2}+\frac{\sigma_y^2}{\mu_y^2}-2\rho\frac{\sigma_x\sigma_y}{\mu_x\mu_y}\right] \qquad (2.47)$$

*See Haugen [9, p. 122].

and (by Merrill's approximation)

$$\sigma_z^2 \approx \frac{\mu_y^2}{\mu_x^2}\left[\frac{\sigma_x^2}{\mu_x^2} + \frac{\sigma_y^2}{\mu_y^2} - 2\rho\frac{\sigma_x\sigma_y}{\mu_x\mu_y} + 8\frac{\sigma_x^4}{\mu_x^4} + \cdots + \right] \quad (2.48)$$

Equations (2.47) and (2.48) differ only in the fourth and higher-order terms. For the case where $\rho=0$ and X and Y are statistically independent,

$$\sigma_z^2 = \sigma_{y/x}^2 \approx \frac{\sigma_x^2\mu_y^2 + \sigma_y^2\mu_x^2}{\mu_x^4} \quad (2.49)$$

This result is in agreement with that by the method given in Section 2.7.

For the case where $\rho = +1$ and X and Y are identical, by equation (2.47) or (2.48), neglecting fourth and higher-order terms,

$$\sigma_{y/x} \approx 0; \quad \mu_z \approx \frac{\mu_y}{\mu_x} = 1, \quad \text{(b)}$$

In other words, equations (b) approximate the unit random variable, $(1,0)$.

Example N   Quotient of Independent Random Variables

Many problems in engineering design require the estimation of the statistics of the quotient of two independent normally distributed random variables. Equations (2.45) and (2.47) or equations (2.28) and (2.30) provide formulas for estimating the statistics of a quotient in terms of the statistics of the components (for an example, see Example D, Section 2.5).

Example O   Quotient of Correlated Random Variables

This example illustrates the estimation of the statistics of the quotient of two correlated normally distributed random variables. Equations (2.45) and (2.47) provide the formulas for estimating the statistics of the quotient in terms of the statistics of the components.

The interaction between loading and deformation is important in mechanical design. Loading intensity, in pounds per square inch (i.e., stress, s) is a random variable $(\bar{s}, \mathcal{I}_s)$. Also, deformation, in inches per linear inch (i.e., strain, ε) is a random variable $(\bar{\varepsilon}, \mathcal{I}_\varepsilon)$.

A unique feature of truly elastic behavior is that deformation (strain) is a linear function of load intensity, hence a linear function of stress, and they are correlated $\rho = +1$.

The ratio of two correlated random variables, $E=(s/\varepsilon)$, is called Hooke's law:

$$\bar{E} \approx \frac{\bar{s}}{\bar{\varepsilon}} \cdot \left[1+\left(\frac{\partial_\varepsilon}{\bar{\varepsilon}}\right)\cdot\left(\frac{\partial_\varepsilon}{\bar{\varepsilon}}-\rho\frac{\partial_s}{\bar{s}}\right)\right]$$

$$\partial_E \approx \frac{\bar{s}}{\bar{\varepsilon}} \cdot \left[\frac{\partial_\varepsilon^2}{\bar{\varepsilon}^2}+\frac{\partial_s^2}{\bar{s}^2}-2\rho\frac{\partial_\varepsilon \partial_s}{\bar{\varepsilon}\bar{s}}\right]^{1/2}$$

The modulus of elasticity, E, is the quotient of two correlated random variables.

2.6.7 Distribution of a Quotient

The quotient of two normally distributed random variables is not (strictly) normally distributed (unless the denominator is a constant). However, under certain conditions quotients closely approximate normality, or are robust.

The *robustness of quotients* of normal random variables has been studied by utilizing numerical methods for integration of the probability density functions [16]. Normal, γ, and lognormal probability density functions have been tested for fit with the empirical probability density functions of quotients (see Table 2.5 and Figure 2.11). It can be concluded that:

Table 2.5. Quotient $Y=(X_1/X_2)$

Mean Ratio	Variate				Coefficient of Variation (%)		Robust		
	X_1		X_2		X_1	X_2	Normal	γ	Lognormal
1	(1,000,	50)	(1,000,	50)	5	5	No	Yes	Yes
1	(1,000,	50)	(1,000,	200)	5	20	No	No	No
1	(1,000,	200)	(1,000,	50)	20	5	Yes	No	No
1	(1,000,	200)	(1,000,	200)	20	20	No	No	No
0.1	(1,000,	50)	(10,000,	50)	5	0.5	Yes	Yes	Yes
0.1	(1,000,	50)	(10,000,	200)	5	2	Yes	Yes	Yes
0.1	(1,000,	200)	(10,000,	50)	20	0.5	Yes	No	No
0.1	(1,000,	200)	(10,000,	200)	20	2	Yes	No	No
10	(10,000,	50)	(1,000,	50)	0.5	5	No	Yes	Yes
10	(10,000,	200)	(1,000,	50)	2	5	No	Yes	Yes
10	(10,000,	200)	(1,000,	200)	2	20	No	No	No
1	(10,000,	50)	(10,000,	50)	0.5	0.5	Yes	Yes	Yes
1	(10,000,	50)	(10,000,	200)	0.5	2	Yes	Yes	Yes
1	(10,000,	200)	(10,000,	50)	2	0.5	Yes	Yes	Yes
1	(10,000,	200)	(10,000,	200)	2	2	Yes	Yes	Yes

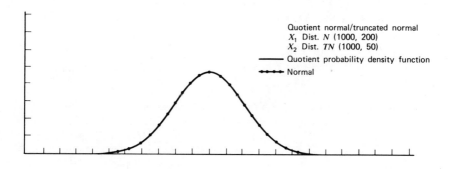

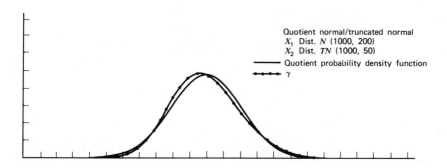

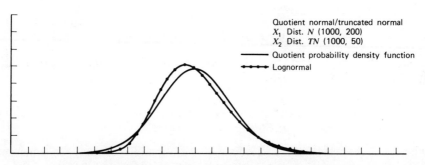

Figure 2.11 Quotient of normal: truncated normal distributions.

1. For $0.1 \leqslant (\mu_{x_1}/\mu_{x_2}) \leqslant 10$, $0.005 \leqslant (\sigma_{x_1}/\mu_{x_1}) \leqslant 0.20$ and $0.005 \leqslant (\sigma_{x_2}/\mu_{x_2}) \leqslant 0.20$; the quotient probability density function is unimodal.
2. For $0.1 \leqslant \mu_{x_1}/\mu_{x_2} \leqslant 10$, $0.005 \leqslant (\sigma_{x_1}/\mu_{x_1}) \leqslant 0.20$ and $0.005 \leqslant (\sigma_{x_2}/\mu_{x_2}) \leqslant 0.20$; a sufficient condition for a robust approximation of normality is that $(\sigma_{r_2}/\mu_{y_2}) \leqslant 0.05$.

The statistics of a root, $Z = X^{1/2}$, are expressed as

$$\mu\sqrt{x} \approx \left[\frac{1}{2}\left(\mu_x + \sqrt{\mu_x^2 - \sigma_x^2}\right)\right]^{1/2} \qquad (2.50)$$

$$\sigma\sqrt{x} \approx \frac{1}{2}\cdot\left(\frac{\sigma_x}{\mu_x}\right) \qquad (2.51)$$

2.6.8 Statistics of Quadratic Form

Let

$$Z = h(x) = aX^2 + bX + c$$

where X is a normally distributed random variable. Then

$$E(z) = a(\mu_x^2 + \sigma_x^2) + b\mu_x + c = \mu_z \qquad (2.52)$$

and

$$\sigma_z^2 = \sigma_{h(x)}^2 = \sigma_x^2(2a\mu_x + b)^2 + 2a^2\sigma_x^4 \qquad (2.53)$$

2.6.9 Coefficient of Variation [9]

The coefficient of variation of a random variable X that has mean μ_x and standard deviation σ_x is defined as

$$CV_x = \frac{\sigma_x}{\mu_x} = C_x. \qquad (2.54)$$

where CV represents the coefficient of variation. Recall that for sums and differences of random variables, $Z = (X \pm Y)$, the variance of Z is given by

$$\sigma_z^2 = \sigma_x^2 + \sigma_y^2$$

Consider the product $Z = XY$, with X and Y normal and $\rho = 0$. It can be shown that the coefficient of variation, C_z^2, is approximately equal to $C_x^2 + C_y^2$. Equation (2.61) yields

$$\sigma_z^2 = \mu_x^2\mu_y^2\cdot\left[\frac{\sigma_x^2}{\mu_x^2} + \frac{\sigma_y^2}{\mu_y^2} + \frac{\sigma_x^2\sigma_y^2}{\mu_x^2\mu_y^2}\right]$$

Dividing by $\mu_x^2\mu_y^2$ yields

$$C_z^2 = C_x^2 + C_y^2 + C_x^2 C_y^2$$

For C_x and C_y small,

$$C_x^2 C_y^2 \ll C_x^2 + C_y^2$$

Thus

$$C_z^2 \approx C_x^2 + C_y^2 \tag{2.55}$$

Next, consider the quotient $Z = (X/Y)$, with X and Y statistically independent; by using equation (2.30), we obtain

$$\sigma_z^2 = \frac{\sigma_x^2 \mu_y^2 + \sigma_y^2 \mu_x^2}{\mu_y^4}$$

If a derivation similar to that for products of variances is used, then

$$C_z^2 = \frac{C_y^2 + C_x^2}{1 + C_y^2}$$

and, for C_y small, $C_y^2 \ll 1$ and

$$C_z^2 \approx C_x^2 + C_y^2 \tag{2.56}$$

It is notable that the expressions for variance σ_z^2 of sums and differences of normally distributed random variables may be formally similar to the coefficients of variation, C_z^2, of products and quotients.

2.6.10 Laws of Combination

The normally distributed random variables for which the laws of combination are generally operable are those for which closure (see Section 2.6.12) constraints are not exceeded.

The algebra of normal functions has the following elements of structure in common with the algebra of the real numbers [9]:

1. Addition
    a. Uniqueness of sum.
    b. Associative law in addition.
    c. Existence of identity element for addition.
    d. Unary minus operation leads to subtraction.
    e. Commutative law in addition.

2. Multiplication
   a. Uniqueness of product.
   b. Associative law in multiplication.
   c. Existence of element for multiplicative inverse leads to division.
   d. Commutative law for multiplication. In combinations of addition and multiplication the distributative law holds.

2.6.11 Characteristics

Equality of normal random variables is expressed as

$$(\mu_a, \sigma_a) = (\mu_b, \sigma_b)$$

if and only if $\mu_a = \mu_b$ and $\sigma_a = \sigma_b$.

Order relations among normal random variables is expressed as

$$(\mu_a, \sigma_a) \neq (\mu_b, \sigma_b) \tag{a}$$

$$(\mu_a, \sigma_a) > (\mu_b, \sigma_b) \quad \text{or} \quad (\mu_a, \sigma_a) < (\mu_b, \sigma_b) \tag{b}$$

Equation (a) does not imply equations (b). If the mean value μ_a of a normal variate is plotted on the abscissa and standard deviation σ_a on the ordinate, (μ_a, σ_a) represents a point in a plance (see Figure 2.12), analogous to the complex plane.

In the sense of ordering, the expressions "greater than" or "less than" are meaningless in relations of (normal or other) random variables, except for degenerate variables.

2.6.12 Closure

Given any two normal variates (μ_x, σ_x) and (μ_y, σ_y), from the population, we may associate with them other elements, such as $(\mu_{x+y}, \sigma_{x+y})$ and (μ_{xy}, σ_{xy}). There is closure under a binary operation, and the operation is uniquely defined if for all elements (in this case normal variates) X and Y in

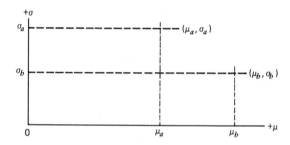

Figure 2.12 Normal variate representation in a plane.

the population, the element $(X + Y)$ or (XY) is a unique element in the same population; thus:

1. The population is closed under addition.
2. Closure is approximated under the operation of multiplication, provided that the variances are moderate.

2.6.13 Convergence

If the dispersion in each member of the population of normal random variables converges to zero as a limit, the population converges to the real numbers; that is, the real numbers are imbedded in the population of normal random variables.

2.6.14 Normal Distribution as a Model [13, p. 72]

Material behavioral properties, dimensional values, and external loads take on positive values. Thus it has been argued that the normal distribution has limited value as a model of engineering random variables, as its range is $(-\infty, \infty)$.

However, in the usual case of an engineering random variable (Figure 2.13) $\mu_x \gg \sigma_x$ and $(\sigma_x/\mu_x) \ll 1$.
Since it is the negative density of X that is questioned, consider Table 2.6 (see Appendix 1).

Example P

Consider a normal model in which $\mu_x = 5\sigma_x$, $C_x = \frac{1}{5}$ [i.e., $100 \cdot (\sigma_x/\mu_x) = 20\%$]. For an engineering variable, $C_x = 0.20$ represents a random variable that has large scatter. However, the negative fraction of the density in this case is negligible.

$$\frac{1}{\sigma_x\sqrt{2\pi}} \int_{-\infty}^{\infty} \exp\left[-\frac{1}{2}\left(\frac{x-\mu_x}{\sigma_x}\right)^2\right] dx$$

$$-\frac{1}{\sigma_x\sqrt{2\pi}} \int_{0}^{\infty} \exp\left[-\frac{1}{2}\left(\frac{x-\mu_x}{\sigma_x}\right)^2\right] dx = 0.00000024$$

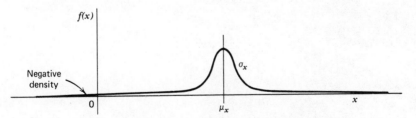

Figure 2.13 Typical engineering variable (normal model).

Table 2.6. Negative Density of Normal Truncated Random Variables

(μ_x/σ_x)	Negative Density	Normalizing Constant for Truncation at Zero
3	0.001350	0.998650
4	0.00003167	0.99996833
5	0.00000024	0.99999976
6	0.000000001	0.999999999
7	$0.0_{10}\,13$	$0.9_{10}\,87$

If the distribution were truncated at zero, $F(0 < x < \infty) = 0.99999976$, the probability density function of the truncated normal distribution would be

$$f(x) = K \frac{1}{\sigma_x \sqrt{2\pi}} \exp\left[-\frac{1}{2}\left(\frac{x - \mu_x}{\sigma_x}\right)^2\right] \quad (0 \leqslant x < \infty);$$

here

$$K = \frac{1}{0.99999976} \approx 1$$

STATISTICS OF FUNCTIONS OF n VARIABLES

2.7 STATISTICS OF ARBITRARY FUNCTIONS*

The mean values and variances of arbitrary functions of n random variables can be estimated by Taylor series expansions [9]. This procedure, which involves partial derivatives, is very useful, as is indicated by the following examples.

2.7.1 Mean Value

Consider the function $y = f(X_1, X_2, \ldots, X_n)$, where $X_1, X_2, \ldots, X_n$ are independent random variables that have expected values μ_{x_i} and variances $V(x_i)$, $[i = 1, 2, \ldots, n]$. To estimate the mean value and standard deviation of y, expand y in a Taylor series about $\mu_{x_1}, \mu_{x_2}, \ldots, \mu_{x_n}$:

$$E(y) = f[E(x_1), E(x_2), \ldots, E(x_n)]$$
$$+ \frac{1}{2} \sum_{i=1}^{n} \frac{\partial^2 y}{\partial x_i^2} \cdot E[x_i - E(x_i)]^2 \quad (2.57)$$

$$E(y) = f[E(x_1), E(x_2), \ldots, E(x_n)]$$
$$+ \frac{1}{2} \sum_{i=1}^{n} \frac{\partial^2 y}{\partial x_i^2} \cdot V(x_i) \quad (2.58)$$

*For a numerical approach, see Ref. 18 and 19.

If the value of the second term on the right is inconsequential compared to the value of the first term, then

$$E(y) = f[E(x_1), E(x_2), \ldots, E(x_n)] \tag{2.59}$$

2.7.2 Variance of Random-valued Functions

To estimate the variance by a Taylor series expansion, the results in equation (2.58) are used with equation (2.16),

$$V(y) = E(y^2) - [E(y)]^2$$

Squaring equation (2.58) yields

$$[E(y)]^2 = [f[E(x_1), E(x_2), \ldots, E(x_n)]]^2$$

$$+ [f[E(x_1), E(x_2), \ldots, E(x_n)]] \left[\sum_{i=1}^{n} \frac{\partial^2 y}{\partial x_i^2} V_{(x_i)} \right]$$

$$+ \left[\frac{1}{2} \sum_{i=1}^{n} \frac{\partial^2 y}{\partial x_i^2} V_{(x_i)} \right]^2 \tag{2.60}$$

To obtain $E(y^2)$, square equation (2.58) and take the expected values term by term,

$$E(y^2) = [f[E(x_1), E(x_2), \ldots, E(x_n)]]^2 + \sum_{i=1}^{n} \left(\frac{\partial y}{\partial x_i} \right)^2 V_i$$

$$+ \left[f[E(x_1), E(x_2), \ldots, E(x_n)] \sum_{i=1}^{n} \frac{\partial^2 y}{\partial x_i^2} V_i \right]$$

$$+ \sum_{i=1}^{n} \left(\frac{\partial^2 y}{\partial x_i} \right) \left(\frac{\partial^2 y}{\partial x_i^2} \right) E[x_i - E(x_i)]^2 \tag{2.61}$$

The square root of the difference between equations (2.60) and (2.61) yields the standard deviation of the function y. If the second and higher terms can be ignored without consequential error, then

$$\sigma_y \approx \left[\sum_{i=1}^{n} \left(\frac{\partial y}{\partial x_i} \right)^2 \cdot \sigma_{x_i}^2 \right]^{1/2} \tag{2.62}$$

Equations (2.59) and (2.62) provide estimators of the mean values and standard deviations of functions of independent random variables. If the variables are correlated, similar expressions for the mean and standard deviation [13], involving the correlation coefficients, can be derived (see Section 2.6). For a function whose first derivative (in the range of interest) is very small, the higher terms in the Taylor expansion cannot be ignored. If the variables are normal, the function is approximately normal [15].

Example Q

Given that an expression for moment of inertia of a solid cylindrical bar is

$$I \approx \frac{\pi d^4}{64} \quad \text{(a)}$$

Estimate the statistics of I by utilizing equations (2.59) and (2.62).

$$\bar{I} \approx f[E(x_1), E(x_2), \ldots, E(x_n)] = \frac{\pi \bar{d}^4}{64} \quad \text{(a')}$$

Since

$$f(x_1) = f(d) = \frac{\pi}{64} \cdot d^4,$$

$$\frac{\partial I}{\partial d} = \frac{\pi}{16} d^3$$

Thus

$$\mathbin{\raise.3ex\hbox{$\scriptstyle s$}}_I \approx \left[\left(\frac{\partial I}{\partial d} \right)^2 \cdot \mathbin{\raise.3ex\hbox{$\scriptstyle s$}}_d^2 \right]^{1/2} = \left(\frac{\pi}{16} \right) \bar{d}^3 \cdot \mathbin{\raise.3ex\hbox{$\scriptstyle s$}}_d \quad \text{(a'')}$$

Example R

Given the circuit as shown, and that R_1 and R_2 are statistically independent resistors, each normally distributed:

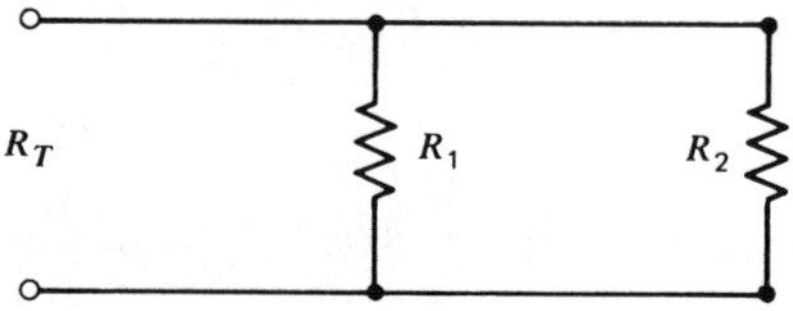

$$(\bar{R}, \mathbin{\raise.3ex\hbox{$\scriptstyle s$}}_{R_1}) = (10{,}000,\ 300)\ \Omega.$$

$$(\bar{R}, \mathbin{\raise.3ex\hbox{$\scriptstyle s$}}_{R_2}) = (20{,}000,\ 600)\ \Omega. \quad \text{(a)}$$

Estimate the resistance statistics $(\bar{R}_T, \mathbin{\raise.3ex\hbox{$\scriptstyle s$}}_{R_T})$. The solution is as follows:

$$\frac{1}{R_T} = \frac{1}{R_1} + \frac{1}{R_2} \quad \text{(b)}$$

and

$$R_T = \frac{R_1 R_2}{R_1 + R_2}$$

Estimate the mean of R_T [see equation (2.59)] by substituting values (mean) from equation (a) into equation (b):

$$\bar{R}_T = \frac{\bar{R}_1 \bar{R}_2}{\bar{R}_1 + \bar{R}_2} \approx \frac{(10{,}000)(20{,}000)}{10{,}000 + 20{,}000} = 6.66 \cdot 10^3 \, \Omega \qquad (b')$$

The partial derivatives required by equation (2.62) for estimating s_{R_T} are given by equation (c) and evaluated at values [equation (a)]:

$$\frac{\partial R_T}{\partial R_1} = \frac{(\bar{R}_2)^2}{(\bar{R}_1 + \bar{R}_2)^2} = 0.444$$

$$\frac{\partial R_T}{\partial R_2} = \frac{(\bar{R}_1)^2}{(\bar{R}_1 + \bar{R}_2)^2} = 0.111 \qquad (c)$$

$$s_{R_T} \approx \left[\left(\frac{\partial R_T}{\partial R_1}\right)^2 \cdot s_{R_1}^2 + \left(\frac{\partial R_T}{\partial R_2}\right)^2 \cdot s_{R_2}^2 \right]^{1/2} \qquad (b'')$$

Substituting from equations (a) and (c) into equation (b''):

$$s_{R_T} \approx \left[(0.444)^2 \cdot (300)^2 + (0.111)^2 \cdot (600)^2 \right]^{1/2} = 149 \, \Omega$$

Thus

$$(\bar{R}_T, s_{R_T}) \approx (6.66 \cdot 10^3, 149) \, \Omega$$

Note that for a parallel system, dispersion reflected by the standard deviation is decreasing at a faster rate than is the mean value.

The Probabilistic Pythagorean Theorem. Heckman and Dietrich [17] investigated the Pythagorean random variable (see Figure 2.14). Given

$$Y = \sqrt{X_1^2 + X_2^2}^{\,*} \qquad (a)$$

with X_1 and X_2 mutually independent normal random variables that have statistics (μ_1, σ_1) and (μ_2, σ_2), the estimator of the mean from equation (2.59) is

$$\mu_y \approx \sqrt{\mu_1^2 + \mu_2^2} \qquad (2.63)$$

*For the special case of $\mu_1 = \mu_2 = 0$, the random variable Y is Rayleigh distributed.

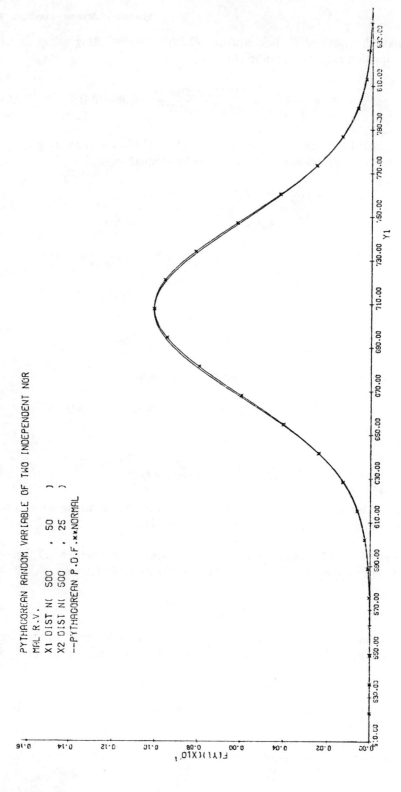
PYTHAGOREAN RANDOM VARIABLE OF TWO INDEPENDENT NOR
MAL R.V.
X1 DIST N(500 , 50)
X2 DIST N(500 , 25)
--PYTHAGOREAN P.D.F.**NORMAL
F(Y1)(X10^-1)
Y1
0.00
0.02
0.04
0.06
0.08
0.10
0.12
0.14
0.16
510.00
530.00
550.00
570.00
590.00
610.00
630.00
650.00
670.00
690.00
710.00
730.00
750.00
770.00
790.00
810.00
830.00

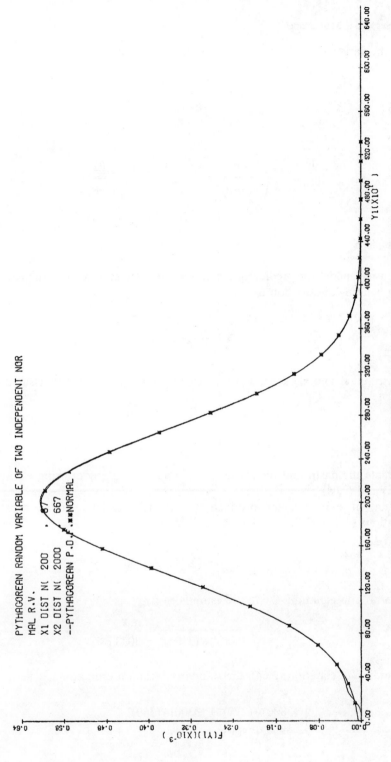

Figure 2.14 Comparison of Pythagorean density and normal approximation.

64 Supporting Mathematics

and the variance of Y by equation (2.62) is

$$\sigma_y^2 \approx \frac{\mu_1^2 \cdot \sigma_1^2 + \mu_2^2 \cdot \sigma_2^2}{\mu_1^2 + \mu_2^2} \tag{2.64}$$

where

$$\frac{\partial y}{\partial x_1} = -\frac{\mu_1}{\sqrt{\mu_1^2 + \mu_2^2}}; \quad \frac{\partial y}{\partial x_2} = -\frac{\mu_2}{\sqrt{\mu_1^2 + \mu_2^2}}$$

Example S

The usual model for predicting maximum shear stress in a shaft subjected to both bending and torsion is

$$\tau_{max} = \sqrt{\left(\frac{s_x^2}{2}\right) + (\tau_{xy})^2} \text{ psi} \tag{a}$$

Given that s_x and τ_{xy} are independent and normally distributed, assume that

$$(\bar{s}_x, s_{s_x}) = (7600, 580) \text{ psi}$$

$$(\bar{\tau}_{xy}, s_{\tau_{xy}}) = (3300, 360) \text{ psi}$$

Estimate the mean and the standard deviation of τ_{max} by utilizing equations (2.63) and (2.64).

First, since $s_x/2$ is the product of a random variable and a constant, equations (2.8) and (2.10) yield

$$\left(\frac{\bar{s}_x}{2}, s_{s_x/2}\right) = (3800, 290) \text{ psi}$$

According to equation (2.63), the estimator of $\bar{\tau}_{max}$ is

$$\bar{\tau}_{max} = \sqrt{(3800)^2 + (3300)^2} \approx 5033 \text{ psi}$$

According to equation (2.64), the standard deviation estimator $s_{\tau_{max}}$ is

$$s_{\tau_{max}} = \left[\frac{(3800)^2 \cdot (290)^2 + (3300)^2 \cdot (360)^2}{(3800)^2 + (3300)^2}\right]^{1/2} \approx 320 \text{ psi}$$

Sensitivity Analysis. Equation (2.62) provides a valuable tool with which to draw sensitivity inferences, because the terms $(\partial y/\partial x_i)^2 \cdot s_{x_i}^2$ disclose the contributions of each x_i in equation (2.62) to the magnitude of σ_y. For instance, it is desirable to minimize the variability in applied stress, s. An examination of the expression for σ_s will disclose which variables can most profitably be closely controlled. Further, the tolerances on some variables (producing negligible effects) may sometimes be relaxed.

Example T

Examine the sensitivity of the standard deviation on stress to the contributions due to applied force and the relevant dimensional random variables. A cantilever beam formula (rough approximation) sometimes used to predict stress (s) in a spur gear tooth is

$$s = \frac{6P_t \cdot l}{Ft^2} \text{ lb/in.}^2 \qquad (a)$$

Given

$$P_t = \text{Force} = (\overline{P}_t; s_{P_t}) = (5400, 360) \text{ lb}$$

$$l = \text{Tooth height} = (\overline{l}; s_l) = (1.30, 0.040) \text{ in.}$$

$$F = \text{Face width} = (\overline{F}, s_F) = (0.625, 0.025) \text{ in.}$$

$$t = \text{Root thickness} = (\bar{t}, s_t) = (0.75, 0.035) \text{ in.} \qquad (b)$$

The partial derivatives of s with respect to P_t, l, F, and t (evaluated at the mean values) are required in equation (2.62) to estimate s_s:

$$\frac{\partial s}{\partial P_t} = \frac{6\bar{l}}{\overline{F}\bar{t}^2} = 22$$

$$\frac{\partial s}{\partial l} = \frac{6\overline{P}_t}{\overline{F}\bar{t}^2} = 92{,}200$$

$$\frac{\partial s}{\partial F} = \frac{-6\overline{P}_t \bar{l}}{\overline{F}^2 \bar{t}^2} = -191{,}700$$

$$\frac{\partial s}{\partial t} = \frac{-12\overline{P}_t \bar{l}}{\overline{F}\bar{t}^3} = -319{,}500 \qquad (c)$$

Substitution of the algebraic expressions [equation (c)] into equation (2.62) yields

$$s_s = \left[\left(\frac{\partial s}{\partial P_t}\right)^2 \cdot s_{P_t}^2 + \left(\frac{\partial s}{\partial l}\right)^2 \cdot s_l^2 + \left(\frac{\partial s}{\partial F}\right)^2 \cdot s_F^2 + \left(\frac{\partial s}{\partial t}\right)^2 \cdot s_t^2\right]^{1/2}$$

$$s_s = \left[\left[\frac{6\bar{l}}{Ft^2}\right]^2 \cdot s_{P_t}^2 + \left[\frac{6\bar{P}_t}{Ft^2}\right]^2 \cdot s_l^2 + \left(\frac{-6\bar{P}_t\bar{l}}{F^2t^2}\right)^2 \cdot s_F^2 + \left[\frac{-12\bar{P}_t\bar{l}}{Ft^3}\right]^2 \cdot s_t^2\right]^{1/2} \quad \text{(a)}$$

Substitution of the numerical values from equations (b) and (c) into equation (a') yields

$$s_s = [6.39 \cdot 10^7 + 1.36 \cdot 10^7 + 2.30 \cdot 10^7 + 12.5 \cdot 10^7]^{1/2} = 15{,}000 \text{ psi} \quad \text{(d)}$$

Equation (2.59) yields

$$\bar{s} \approx \frac{6\bar{P}_t \cdot \bar{l}}{Ft^2} = 119{,}800 \text{ psi} \quad \text{(a'')}$$

Equation (d) indicates that the variability in t is the largest contributor to variability in s. Elimination of variability in t would result in a reduction in s_s to approximately 10,000 psi.

2.8 CHARACTERISTICS OF LOGNORMAL RANDOM VARIABLES

2.8.1 Introduction [20]

The importance of the normal distribution as a statistical model is undeniable, especially considering the central limit theorems for sums of random variables. A theorem that applies to engineering states that the sum of a large number of independent random variables is most likely to be approximately normally distributed [9].

There are also processes in which the elementary variables are "multiplicatively related." The logarithms of the variables are additive, and, by a central limit theorem [21], the logarithm of the product has been shown to be approximately normally distributed. If the random variable Y is normally distributed and X is defined as

$$Y = \ln X, \quad (2.65)$$

then the random variable X is said to be lognormally distributed.

According to the "additive" central limit theorem, $\ln X_n$ is normally distributed; and by definition, X_n (the product of many variables) is lognormally distributed.

In another form the "additive" central limit theorem asserts that the arithmetic mean of n observations from the same random variable has an approximate normal distribution for large n [13]. The analogue "multiplicative" form asserts that the geometric mean has an asymptotic lognormal distribution [21].

2.8.2 Lognormal Density Function

Because the lognormal can be easily transformed to the normal form, it is a skew distribution simple to utilize with certain functional combinations of random variables. For example, let Y be normally distributed:

$$Y \approx N(\mu_Y, \sigma_Y)$$

The density function of Y is (see Figure 2.15)

$$f_Y(y) = \frac{1}{\sigma\sqrt{2\pi}} \exp\left\{-\frac{1}{2\sigma^2}(y-\mu_y)^2\right\} \qquad (2.66)$$

Letting X be defined as in equation (2.65) the density function of X is

$$Y = \ln X,$$

$$f_X(x) = \frac{1}{x} f_Y(\ln x)$$

$$f_X(x) = \frac{1}{x\sigma_Y\sqrt{2\pi}} \exp\left[-\frac{1}{2c_y^2}(\ln x - \mu_y)^2\right] \quad (-\infty < \mu_y < +\infty,\ \sigma_y > 0,\ x \geq 0)$$

$$(2.67)$$

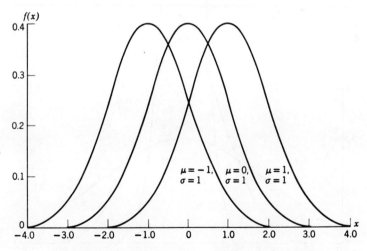

Figure 2.15 Normal curves: σ constant, μ varied [13].

2.8.3 Transformation of Parameters [22]

The parameters μ_y and σ_y in the normal density function (Figs. 2.15 and 2.16) are associated with the normal variate $Y = \ln X$. The transformations to parameters of the lognormal random variable, X (called "measured" parameters), are

$$\mu_x = \exp\left(\mu_y + \tfrac{1}{2}\sigma_y^2\right) \tag{2.68}$$

$$\sigma_x^2 = \exp(2\mu_y + 2\sigma_y^2) - \exp(2\mu_y + \sigma_y^2) \tag{2.69}$$

Expressions for μ_y and σ_y are sometimes needed in terms of the measured parameters μ_X and σ_X; these are

$$\sigma_y^2 = \ln(\mu_x^2 + \sigma_x^2) - \ln(\mu_x^2) \tag{2.70}$$

$$\mu_y = \ln(\mu_x) - \tfrac{1}{2}\left[\ln(\mu_x^2 + \sigma_x^2) - \ln(\mu_x^2)\right] \tag{2.71}$$

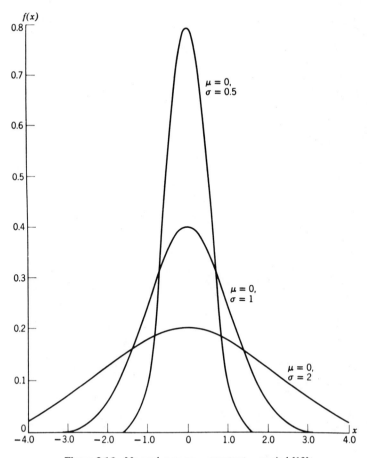

Figure 2.16 Normal curves: μ constant, σ varied [13].

Example U

The logarithms of a sample of observations x_i are normally distributed with sample mean $\bar{y}=2.0$ and variance $s_y^2=0.04$. What are the estimators of the mean and variance?

Given

$$Y \sim N(\bar{y}, s_y^2) = N(2, 0.4)$$

According to equation (2.68), the mean of x is

$$\bar{x} = \exp\left(\bar{y} + \tfrac{1}{2}s_y^2\right) = \exp\left[2 + \tfrac{1}{2}(0.04)\right] = 7.5$$

According to equation (2.69), the variance of x is

$$s_x^2 = \exp\left(2\bar{y} + 2s_y^2\right) - \exp\left(2\bar{y} + s_y^2\right)$$

$$= \exp[2(2) + 2(0.041)] - \exp[2(2) + (0.04)] = 2.3$$

2.8.4 Interpretation of Statistics

Statistics of continuous distributions describe the location, the scale, or the shape of a distribution [23]. In the normal distribution μ is a location statistic and σ a scale statistic. There is no shape statistic.

Lognormal distribution statistics, μ and σ, require a unique interpretation. Consider the three-parameter lognormal distribution

$$f(x) = \frac{1}{(x-\lambda)\sigma\sqrt{2\pi}} \exp\left\{\frac{-1}{2\sigma^2}[\ln(x-\lambda) - \mu]^2\right\}$$

$$(-\infty < \mu < +\infty, -\infty < \lambda < +\infty, x \geqslant \lambda, \sigma > 0) \qquad (2.72)$$

where λ affects the location of the distribution. For $\lambda = 0$, the distribution is located entirely in the nonnegative region, that is, the familiar two-parameter form. Thus λ can be regarded as a lognormal location statistic.

Figures 2.17 and 2.18 show that the scale of the distribution is sensitive to changes in μ whereas the shape is affected by changes in σ. For constant μ and decreasing values of σ, the distribution approaches symmetry. For the coefficient of variation (σ/μ) in the neighborhood of 0.05, the lognormal distribution closely approximates the normal [2].

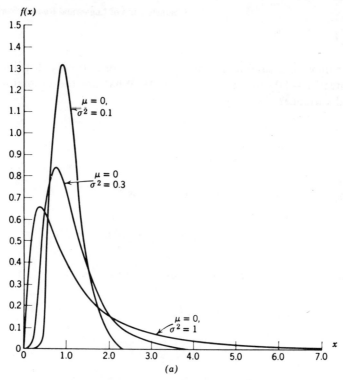

Figure 2.17 Lognormal curves: μ constant, σ^2 varied [13].

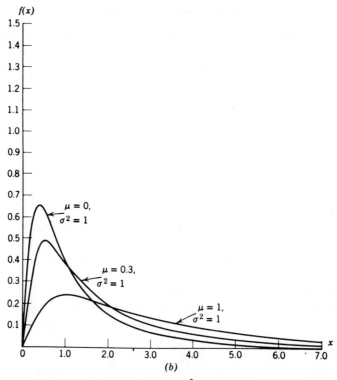

Figure 2.18 Lognormal curves: σ^2 constant, μ varied [13].

2.8.5 Measures of Central Tendency

The statistics used to describe distributions of random variables include the expected value, $E(x)$, the median, $\tilde{x}$, and the mode, $\hat{x}$. The relative magnitude of these statistics provides information concerning the symmetry of the distribution. For unimodel, continuous distributions (see Figure 2.19),

1. $\mu_x = \tilde{x} = \hat{x}$, distribution symmetric.
2. $\hat{x} < \tilde{x} < \mu_x$, distribution skewed positive.
3. $\mu_x < \tilde{x} < \hat{x}$, distribution skewed negative.

The expected value (mean) of the lognormally distributed random variable X was [equation (2.68)]

$$E(x) = \exp\left(\mu_y + \tfrac{1}{2}\sigma_y^2\right) \quad (2.73)$$

The median and the mode are given by [23]

$$\tilde{x} = \exp(\mu_y) \quad (2.74)$$

$$\hat{x} = \exp(\mu_y - \sigma_y^2) \quad (2.75)$$

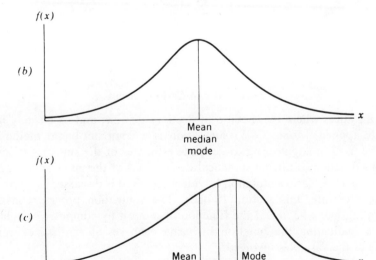

Figure 2.19 Relationship between mean, median, and mode [13].

72 Supporting Mathematics

Table 2.7 Characteristics of Lognormal Distribution

Standard Deviation	Coefficient of Variation	Coefficient of Skewness	Ratio of Mean to Median
0.0	0.	0.	1.0000
0.05	0.0500	0.1502	1.0013
0.10	0.1003	0.3018	1.0050
0.15	0.1508	0.4560	1.0113
0.2	0.2020	0.6143	1.0202
0.25	0.2540	0.7783	1.0317
0.3	0.3069	0.9495	1.0460
0.35	0.3610	1.1300	1.0632
0.4	0.4165	1.3219	1.0833
0.45	0.4738	1.5277	1.1066
0.5	0.5329	1.7502	1.1331

The geometric mean (GM) of the lognormal variate x is

$$GM_x \equiv (x_1 \cdot x_2 \cdot \ \cdots \ \cdot x_n)^{1/n}$$

The characteristics of lognormal distribution are presented in Table 2.7. Then

$$\ln GM_x = \frac{1}{n} \ln \left(\prod_{i=1}^{n} x_i \right) = \frac{1}{n} (\ln x_1 + \ln x_2 + \cdots + \ln x_n)$$

The logarithm of the geometric mean equals the arithmetic mean of the log values

$$GM_x = \exp\left(\frac{1}{n} \sum_{i=1}^{n} \ln x_i \right) = \exp(\mu_Y) \equiv \tilde{X}$$

2.9 MONTE CARLO SIMULATION [9, 13]

2.9.1 Monte Carlo Concept

In instances where the method of Sections 2.5, 2.6, and so on may not be readily applied, Monte Carlo simulation is a computer-based method that may be used in engineering to generate estimates of the statistics describing various functions, utilizing statistical descriptions of the involved variables.

Since Monte Carlo simulation is widely used and is discussed in great detail in the literature, this section is brief. The simulation process consists of generating many values of the function of interest by computer calculations, that is, evaluating the function by using each set of synthesized random values of the variables involved.

Consider the function y, described by the equation

$$y = x_1 + x_2 + x_3 - x_4 \tag{a}$$

in example I, Section 4.6 (in Chapter 4). Assume further that for each of x_1, x_2, x_3, and x_4 we have samples of 1000 measurements or test values. Since the structure of the function y is given by equation (a), y can be estimated from the known values of x_1, x_2, x_3, and x_4. Furthermore, if it is possible to obtain synthetic values of the x_i values instead of each x_i by drawing 1000 random values from each distribution, these sets of random values can be used to estimate 1000 random values of y. Such a procedure is indicated in Figures 4.17 and 7.11. The 1000 values of y can be used to estimate the statistics of y (mean value and standard deviation). High-speed computers that quickly and economically estimate the statistics of complex functions (e.g., dynamic stress and strength functions) have led to wide usage of Monte Carlo methods in engineering.

2.9.2 Random-number Generation

Sets of random values for the distributions of the variables in a function are required in the Monte Carlo process. Random values for a uniform distribution of interval $(0,1)$ and for a normal distribution $(\mu,\sigma)=(0,1)$ can be found in M. G. Kendall and B. Babington-Smith, *Tracts for Computers*, no. 24, Cambridge University Press, Cambridge, 1939.

Although uniform and normal variates are available on punched cards, it is often more convenient to obtain needed random values directly from already developed programs. Such values are called *pseudorandom* because they are determined by using deterministic equations.* In program Monti (given in Figure 7.10) generators for normally distributed random numbers are included as subroutines. Statistical tests are usually applied to the generated values of y to indicate the most likely type of distribution that models the synthesized values (see Chapter 3).

To obtain a random value $Y_N(\mu,\sigma)$ from a normal distribution described by statistics μ and σ, Y_N is given by

$$Y_N(\mu,\sigma) = \sigma Z_N + \mu$$

where Z_N is a random value from a standardized normal distribution described by statistics $\mu=0$ and $\sigma=1$. Procedures involving other distributions are found in the literature [13].

2.9.3 Sample Size and Variational Bands

Since in engineering usage, Monte Carlo simulation involves random values, the estimation, of a functional value involves statistical variations. Hence with estimates will be associated variational bands (see Section 3.3.1). The larger the sample in a simulation, the narrower the variational band in the determination. Variation can be controlled by the number of replications. In engineering the permissible magnitude of variation is often specified, that

---

*See K. D. Tocher, *The Art of Simulation*, Van Nostrand, Princeton, N. J., 1963.

is, the confidence interval (e.g., $\bar{y}+\varepsilon$) within which μ_y will be found with a probability $1-\alpha$.

Thus to determine, a priori, the required numbers of determinations of y, we must specify [13]

ε = Maximum variation in estimate of μ_y

$1-\alpha$ = Desired confidence that $\bar{y}$ does not differ from μ_y by more than $\pm\varepsilon$

s_y = Initial estimate of the standard deviation of y

The approximate number n of Monte Carlo replications is

$$n \simeq \left[\frac{(Z_{1-\alpha/2})s_y}{\varepsilon} \right]^2$$

where $Z_{1-\alpha/2}$ is the $(1-\frac{\alpha}{2})$ 100 percent point of the standard normal distribution.

It is notable that the statistical methods for obtaining confidence bounds on estimates of statistics and estimating the required number n of determinations to obtain required precision, as discussed in Sections 3.3.1, are applicable to Monte Carlo processes.

ADDENDA

Chebysheff and Camp Meidell Inequalities. It is often desirable to make probability statements regarding time to failure, cycles to failure, and so on. Also, although estimators of the means and standard deviations may be available, little may be known of the distributional form of the relevant random variables. In such situations the Chebysheff or Camp–Meidell inequality can be of assistance in making management and/or engineering probability statements and in avoiding gross errors [9].

With only the mean values and standard deviations estimators of cycle life, Table 2.8 may be useful in estimating an upper bound on failure probability. If the distribution of the variable can be assumed to be continuous and unimodal, the Camp–Meidell inequality can be utilized to produce a probability statement.

When the range of values that a variable is likely to assume can be estimated, the method of Section 3.3.2 can be applied to approximate the standard deviation, and again, if it is known that the distribution is likely to be unimodal and continuous, the Camp–Meidell inequality can be applied.

Limits for Engineering Random Variables. In engineering problems the use of a probability density function with infinite tails as a distribution model is not literally justified. For example, the normal probability density function is

Table 2.8. Bounds on Failure Probability

Random Variable Characteristics	Inequality Model	Total Tail Probability at t Standard Deviations			
		$t=2$	$t=3$	$t=4$	$t=5$
Normally distributed	Normal Distribution $P(\|x-\mu_x\| \geqslant t\sigma_x) \leqslant 1 - \int_{-t}^{t} \frac{e^{-u^2/2}}{\sqrt{2\pi}} du$	0.0456	0.0074	0.0001	0.0000011
Distribution continuous and unimodal	Camp–Meidell Inequality $P\{\|x-\mu_x\| \geqslant t\sigma_x\} \leqslant \frac{1}{2.25 t^2}$ $t>1;\ P \leqslant 1$	0.1111	0.0493	0.0277	0.0178
Little known of distribution	Chebysheff's Inequality $P\{\|x-\mu_x\| \geqslant t\sigma_x\} \leqslant \frac{1}{t^2}$ $t>1;\ P \leqslant 1$	0.2500	0.1111	0.0625	0.0400

often used as a model (see Section 8.3) for tensile strengths of steel. However, infinite positive or negative strengths do not exist. A finite distributional model might be defined on the interval $0<x<b$, where b is a finite number.

Regarding the choice of limits for engineering random variables, Benjamin and Cornell [24] state, "To include an event such as ∞ in the sample space does not necessarily mean the engineer thinks it is a possible outcome of the experiment. The choice of ∞ as an upper limit is simply a convenience; it avoids choosing a specific, arbitrarily large number to limit the sample space."

Accuracy of Approximate Calculations. Since engineering representations usually involve estimators known only to a limited number of significant figures, the accuracy of calculations based thereon is an important consideration. The following is a summary of rules for estimating accuracy.

1. *Products and quotients.* If k_1 and k_2 are the first significant figures of two numbers that are each correct to n significant figures, and if neither number is of the form $k(1000\ldots)\times 10^p$, their product or quotient is correct to:

    $n-1$ significant figures if $k_1 \geqslant 2$ and $k_2 \geqslant 2$

    $n-2$ significant figures if either $k_1 = 1$ or $k_2 = 1$

2. *Powers and roots.* If k is the first significant figure of a number that is correct to n significant figures, and if this number contains more than one digit different from zero, then its pth power is correct to

    $n-1$ significant figures if $p \leqslant k$

    $n-2$ significant figures if $p \leqslant 10k$

and its rth root is correct to

n significant figures if $rk \geqslant 10$

$n-1$ significant figures if $rk < 10$

3. *Logs and antilogs.* If k is the first significant figure of a number that is correct to n significant figures, and if this number contains more than one digit different from zero, then for the absolute error in its common logarithm, we have

$$E_a < \frac{1}{4k \times 10^{n-1}}$$

If a logarithm (to the base 10) is not in error by more than two units in the mth decimal place, the antilog is certainly correct to $m-1$ significant figures.

When computing with rounded or approximate numbers of unequal accuracy, a sound and safe rule is to retain from the beginning one more significant figure in the more accurate numbers than are contained in the least accurate number and then round off the result to the same number significiant figures as the least accurate number. Further rules are:

1. In the case of addition, retain in the more accurate numbers one more decimal digit than is contained in the least accurate number.

2. By retaining one more digit in the more accurate numbers, we reduce to zero the errors of those terms and thus reduce the error of the final result.

3. In the case of subtraction or addition of only two numbers, round off the more accurate number to the same number of decimal places as the less accurate number to the same number of decimal places as the less accurate one before subtracting or adding.

A comprehensive discussion of accuracy of approximate calculations is found in Scarborough [25].

CHAPTER 3

Loading Random Variables

The following topics are considered in this chapter:

Experimental loads measurement and data reduction.
Sample size to achieve required precision.
Estimation of variables in the absence of data.
Static loads—forces and moments.
Dispersion of loading in a system.
Effects of multiple loads on a component.
Independent force combinations.
Correlated force combinations.
Dynamic loading—constant amplitude and random.
Dynamic superimposed on static loads.
Dynamic amplitude, escalation rates, and frequency.
Unifying the loads model.
The generalized model—a random process.

3.1 INTRODUCTION

"Loading in mechanical design" refers to surface traction and body forces that influence behavior of mechanical systems, thereby inducing reactive strains. This chapter contains a general discussion of static and dynamic loads.*

Classification of loads as either static or dynamic is based on the effects of the loads on mechanical components and systems. The effects referred to are strains and deflections. The effects induced by static loads usually remain

---

*Detailed consideration of particular loading problems can be found in books such as the *Fatigue Design Handbook* [3], *Metal Fatigue* [26], and *Stress, Strain and Strength* [27, 28]. Published papers provide reports of loads studies, for example, "Random Load Fatigue Testing, A State of the Art Survey" [29], "Random Loads Spectrum Test to Determine Durability of Structural Components of Automotive Vehicles" [30], "Structural Design Criteria for Boost Vehicles by Statistical Methods" [31], and "Development of a Random Load Life Prediction Model" [32].

78 Loading Random Variables

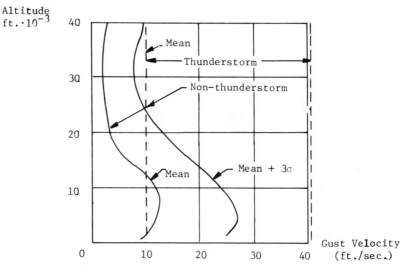

Figure 3.1 Statistics of atmospheric gust velocities.

essentially constant, if elastic limits are not exceeded, whereas dynamic effects fluctuate in time. As is seen later, failure modes associated with static loads are very different from those resulting from dynamic loading.

It is often advantageous to refrain from initially specifying the nature of the loading variables during model development and analysis. Subsequently, multivalued and random aspects are recognized. Loading determination involves the identification and measurement of relevant parameters and the use of this information to estimate total forces on the system. Figure 3.1 illustrates a variable loading parameter of importance in the design of aircraft structures.

Loading definitions utilized in classical design [33] are:

Limit load: maximum load anticipated in service.

Ultimate load: maximum load that a part or structure must be capable of supporting.

Proof load: product of limit load and proof factor (<1.0).

Proof and ultimate factors of safety: design factors to provide for possibility of loads greater than those expected in normal conditions of operation, uncertainties in design, and variations in structural strength.

In probablistic design the single-valued loads defined in the preceding list are replaced by load statistics ($\bar{L}, s_L$) estimated from magnetic tape records of actual histories, for example, or by sample probability density functions based on samples of load measurements that describe the range and the probability of load values for particular environments. The prior estimation of both static and dynamic load spectra for design usage is often an involved

Table 3.1 Loading Regimes [35]

Strain rate (sec^{-1})		Usual method of loading	Dynamic considerations in testing	
10^7, 10^6, 10^5, 10^4	Hypervelocity impact	Light gas gun or explosively driven projectile	Shock–wave propagation	Inertia forces important
10^3, 10^2	Impact	Mechanical or explosive impact	Elastic–plastic wave propagation	Inertia forces important
10^1, 10^0, 10^{-1}	Dynamic	Fast–acting hydraulic or pneumatic machine	Mechanical resonance in specimen and machine	Inertia forces important
10^{-2}, 10^{-3}, 10^{-4}, 10^{-5}	"Static"	Hydraulic or screw machine	Constant–strain rate test	Inertia forces neglected
10^{-6}, 10^{-7}, 10^{-8}, 0	Creep	Constant–load or stress machine	Strain versus time or creep rate recorded	Inertia forces neglected

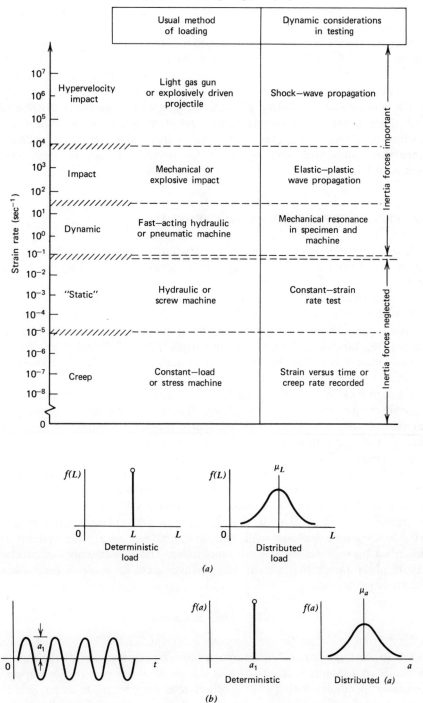

Figure 3.2 Comparison of load models. (*a*) Static. (*b*) Dynamic.

task, as suggested by Table 3.1. However, statistical modeling permits each data point to provide some information concerning the phenomenon studied (see Figure 3.2) [34].

For engineering purposes, the principal task is to estimate random-variable characteristics from data on physical phenomena or experiments. Statistics present a compact method of summarizing all the possible values of a physical phenomenon assumed to be a continuous random variable.

The phenomena accounting for static loads are often normally distributed, and those generating dynamic loads are often normally or lognormally distributed. However, the exact distributional loads model may not be critical since the asymptotical limiting model for stress is usually lognormal (see Section 9.3.9).

3.2 EXPERIMENTAL LOADS MEASUREMENT

In experimental loads study the measurable variables are actual surface strain, acceleration, pressure, weight, time, temperature, and other parameters.

Experimental methods are available for measuring strain reactions to load and load combinations. Brittle coatings provide one whole field method for locating highly strained areas and indicating the general directions of surface strains [36]. Electrical resistance strain gauges (Figure 3.4) and other types of strain follower can thus be located at critical points for measurements of spectra of strain magnitudes [37].

Force intensities are estimated from strain, acceleration, weight, and/or pressure measurements. The central limit theorem is invoked in estimating the number of sample measurements needed to provide estimators of the statistics, $\bar{F}$ and s_F, within required limits with a desired probability, that is, confidence.

The distribution of load peak amplitudes is the useful model for both static and dynamic loads, as peaks are the meaningful values in the load history of a component. For example, accelerations experienced by vehicle components in transit over a typical road, gust loading on an airplane control surface, wind loading on a building wall, and seismic effects on existing systems are described by expressions of peak amplitudes. The consequences of product failure often dictate that a considerable investment be made in establishing the statistics of loading.

3.2.1 Data Reduction

Data from loads studies are processed for further use by plotting as bar charts (or histograms) and reduction to summarizing statistics. If the number of values in a sample is large [38] and the specie of the parent distribution is unknown, goodness-of-fit (χ^2 and other) tests may be made to suggest the most likely distribution type. Figure 1.7b shows acceleration amplitude data

Mean Value: 339,470.6 lbs.
St. Dev.: 130,009.1 lbs.

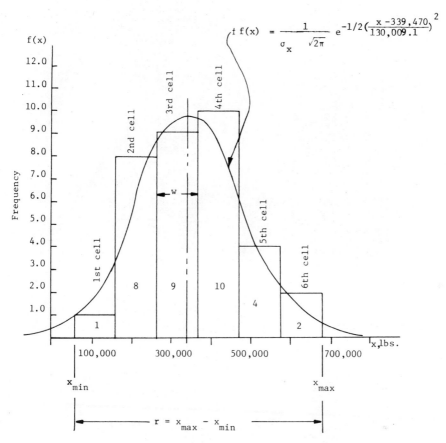

Figure 3.3 Histogram and distribution estimate of anticipated loads.

plotted on probability paper, to test for possible normality. [13]. The distribution model together with estimators of the statistics define a sample function.

Example A   Loads Histogram

The histogram in Figure 3.3 was developed by using data developed in a static loads study (Table 3.2) by the procedure that follows:

1. An appropriate number of cells (k) in the histogram is estimated by using Sturge's rule:*

$$k = 1 + 3.3 \log_{10} n$$

---

*H. A. Sturges, "The Choice of a Class Interval," *ASA*, **21**, 65–66 (1926).

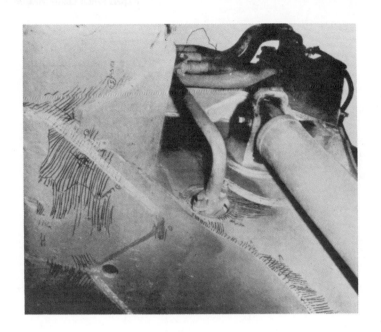

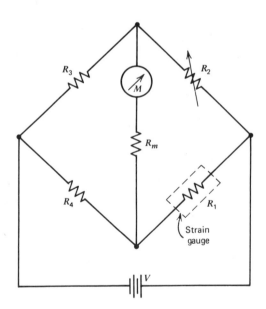

Figure 3.4 Balanced-bridge circuit for static strains.

82

Table 3.2. Static Load Test Values (Hypothetical)[a]

	Test Number	Force (lb)
x_1	3	403,000
x_1	3	403,000
x_2	27	364,000
x_3	33	306,000
etc.	29	214,000
	34	622,000
	18	548,000
	9	391,000
	21	262,000
	32	476,000
	27	395,000
	28	318,000
	4	394,000
	14	238,000
	2	471,000
	16	375,000
	13	492,000
	15	306,000
	8	225,000
	3	250,000
	34	259,000
	20	220,000
	12	175,000
	20	426,000
	17	281,000
x_{max}	18	656,000
	28	283,000
	12	269,000
	8	367,000
x_{min}	21	57,000
	26	173,000
	14	234,000
	15	455,000
x_{33}	10	364,000
x_{34}	31	273,000

[a] Total 34 test measurements.

where n is sample size and

$$k = 1 + 3.3 \log_{10} 34 \approx 6.$$

2. The range (r) of values in the sample is determined:

$$r = x_{max} - x_{min} = (656,000 - 57,000) \text{ lb}$$
$$= 599,000 \text{ lb}$$

3. The width (w) of each cell in the histogram is calculated:

$$w = \frac{r}{k} = \frac{599{,}000}{6} \approx 100{,}000 \text{ lb}$$

4. The cells are defined such that any value x can belong to one and only one cell. Thus cell 1 is

$$(x_{min} \leq x < x_{min} + w)$$

cell 2 is

$$(x_{min} + w \leq x < x_{min} + 2w)$$

and cell 3 is

$$(x_{min} + 2w \leq x < x_{min} + 3w)$$

and so on. It is sometimes advantageous to treat the outlying cells as unbounded on one side.

5. The number of values belonging to each cell is determined from Table 3.2.

On the basis of the mean value and standard deviation estimated from the data, the most likely normal model curve is plotted over the histogram as indicated in Figure 3.3.

A population used to simulate a continuous random variable (often assumed normal) departs from the model in two ways: (1) the sample is limited in size and range, and (2) it is a discrete random variable, not continuous as implied in the theory. The effects of these departures, which apply to loads and other random variables, will scarcely be noticed because they are usually small versus sampling variation [14].

3.2.2 Goodness-of-fit Test

Relative cumulative frequencies may be obtained by utilizing data assigned to histogram cells and then plotted against the cell value on normal probability paper. If the plotted points approximate a straight line, normality can be inferred.* If a straight line is not obtained, the data might better fit another distribution. A convex plot indicates a left-skewed distribution and might suggest extreme value probability paper. A concave plot denotes right skewness and suggests that lognormal paper could be fruitful, with the γ distribution as a possibility. The Weibull paper is useful for left or right skewness [13].

Table 3.3. Spring-deflection data (lb/in.)

3.7	5.4	4.4	4.4	3.4	4.8	4.9	5.1	3.5	4.1
4.3	3.7	1.6	2.7	7.4	2.9	3.9	0.6	2.7	4.7
4.6	4.1	3.3	6.5	3.0	3.1	5.2	3.7	1.7	5.0
2.3	3.7	4.2	3.6	3.4	4.0	2.7	3.8	4.1	2.6
2.9	1.9	3.1	4.7	4.5	5.9	3.0	4.1	4.3	5.3

Example B

Utilizing the spring-calibration data in Table 3.3, determine class intervals, frequencies, and relative cumulative frequencies. To plot these values, see Figure 3.5. It appears likely that this sample is from a normal population.

The observations in Table 3.3 were made on 50 springs selected at random from a population and are in pounds of force required to produce unit deflection [39]. To convert from pounds to newtons, multiply by 4.448.

Bayes Decision Procedures. We may adopt the Bayesian view because the normal model selected in Table 3.3 describes our belief, called a *prior probability distribution*. By using the information given by the prior distribution, the engineer is able to establish a framework from which to make design calculations and research decisions.

Our belief may be supported by probability paper plots, histograms, and other evidence.

3.2.3 Graphical Estimates

If the fit of data using normal probability paper is close, rough estimates of the mean μ and standard deviation σ (e.g., of the spring rate) can be obtained

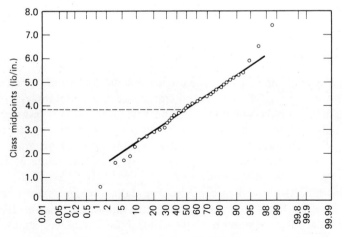

Figure 3.5 Fit of data to normal model [39]. Data from Table 3.2.

Classification of Data From Table 3.3

Class Interval (lb)	Frequency	Relative Cumulative Frequency (%)	Class Midpoint
0.51–1.51	1	2	1.0
1.51–2.51	4	10	2.0
2.51–3.51	14	38	3.0
3.51–4.51	18	74	4.0
4.51–5.51	10	94	5.0
5.51–6.51	2	98	6.0
6.51–7.51	1	100	7.0

from a graph such as that in Figure 3.5. The mean is estimated by the value at which the best fit line crosses the 50% line. The standard deviation is estimated by subtracting the 6.7% value from the 93.3% value and then dividing the difference by 3. Alternatively, the absolute difference between the mean and either the 84% or 16% point gives an estimate of s directly (see Appendix 1).

By using the preceding data above, the estimates can be calculated from the fitted line in Figure 3.5, as follows: 6.7% value $\approx$ 2.0 lb, 50.0% value $\approx$ 3.8 lb, 93.3% value $\approx$ 5.6 lb. The graphical estimate of the mean is 3.8 lb and of the standard deviation, $(5.6 - 2.0)/3 \approx 1.2$ lb.

The number of test values required to define mean force intensity in a load function within acceptable bounds at a specified confidence is discussed next.

3.3 SAMPLING REQUIREMENTS

It is often necessary to estimate the magnitudes of loads (and other design variables, design factors, and empirical measures) from observations based on tests and from measurements made during an experimental development (see Figure 3.9).

3.3.1 Central Limit Theorem

A frequent question concerns the degree of confidence with which it can be stated that the estimate of a parameter made from a sample is near the corresponding population value. More precisely, if an estimator, $\hat{\theta} = \hat{\theta}(x_1, x_2, \ldots, x_n)$ of the parameter θ is used, what is the probability that the error $\hat{\theta} - \theta$ will not be more than some number? For instance, let $\hat{\theta} = \bar{x}$ and $\theta = \mu_x$, where $\bar{x} = [(x_1 + x_2 + \cdots + x_n)/n]$ is the estimator of the mean value.

For the problem of estimating the population mean using a sample of values, the central limit theorem [9, p. 53] gives approximately the distribution of the sample mean. In essence, the central limit theorem states that the sample mean has an approximate Gaussian distribution with mean μ_x (population) and standard deviation $\sigma/\sqrt{n}$, where σ is the population standard deviation.

If $\bar{x}_n$ is the mean value of a sample of size n from a population having a mean μ_x and standard deviation σ_x, it follows that:

$$P\{a<\bar{x}_n<b\}\to 1 \quad \text{as} \quad n\to\infty \quad \text{if} \quad a<\mu_x<b,$$

$$P\{a<\bar{x}_n<b\}\to 0 \quad \text{as} \quad n\to\infty \quad \text{if} \quad \mu_x<a \quad \text{or} \quad \mu_x>b$$

It may be noted that

$$E(\bar{x}_n)=\mu_x \quad \text{and} \quad V(\bar{x}_n)=\frac{\sigma_x^2}{n}$$

Therefore,

$$z=\frac{\bar{x}_n-\mu_x}{\sigma_x/\sqrt{n}}$$

is a random variable with zero mean and standard deviation of 1.

According to the central limit theorem, for each interval (a,b) the probability (P) that this reduced random variable (z) will assume a value between a and b tends, as n approaches infinity, to the probability that a random variable which is $N(0,1)$ will assume a value between a and b (see Appendix 1).

If $\bar{x}_n$ is a sample mean of a sample size n from an infinite population with mean μ_x and standard deviation σ_x, then for any real numbers a and b with $a<b$,

$$P\left[a<\frac{\bar{x}_n-\mu_x}{\sigma_x/\sqrt{n}}<b\right]\to \frac{1}{\sqrt{2\pi}}\int_a^b e^{-z^2/2}dz \quad \text{as} \quad n\to\infty \quad \text{(a)}$$

or multiplying by $\sigma/\sqrt{n}$ and then adding μ_x in the inequality

$$P\left[\mu_x+\frac{a\sigma_x}{\sqrt{n}}<\bar{x}_n<\mu_x+\frac{b\sigma_x}{\sqrt{n}}\right]\to \frac{1}{\sqrt{2\pi}}\int_a^b e^{-z^2/2}dz \quad \text{as} \quad n\to\infty \quad \text{(b)}$$

Questions pertaining to loads estimation and data requirements, often occurring in engineering, are illustrated in Example C.

Example C

A description of the likely load statistics is needed for a proposed design. The definition should be quite precise; thus 36 test measurements are to be made from a population known to have a variance ($\sigma^2 = 10$ klb^2).

88 Loading Random Variables

1. What is the a priori probability that a sample mean will be obtained for which $\bar{L}$ will differ from μ_L by less than 1 klb? To be determined is

$$P\{|\bar{L}-\mu_L|<1\} = P\{-1<\bar{L}-\mu_L<1\} \qquad \text{(c)}$$

To use the central limit theorem, rewrite equation (c), dividing each term in the inequality by the standard deviation

$$\sigma_{\bar{L}} = \frac{\sigma}{\sqrt{n}} = \frac{\sqrt{10}}{\sqrt{36}} = \frac{\sqrt{10}}{6}$$

Then

$$p\{|L-\mu_L|<1\} = p\left[-\frac{6}{\sqrt{10}} < \frac{\bar{L}-\mu_L}{\sigma/\sqrt{n}} < \frac{6}{\sqrt{10}}\right]$$

and according to equation (b):

$$p\left[-\frac{6}{\sqrt{10}} < \frac{\bar{L}-\mu_L}{\sigma/\sqrt{n}} < \frac{6}{\sqrt{10}}\right] \rightarrow \frac{1}{\sqrt{2\pi}} \cdot \int_{-z}^{z} e^{-z^2/2} dz$$

where $z = (6/\sqrt{10}) \approx 1.90$. The area under the standard normal probability curve (Figure 3.6) between -1.90 and 1.90 gives the required probability. From standard normal area tables (Appendix 1):

$$p\{|\bar{L}-\mu_L|<1\} \approx 0.943$$

Thus the a priori probability of $|\bar{L}-\mu_L|<1$ is 0.943. Experience indicates that a sample of 30 or more is enough for the central limit theorem to yield a reasonable approximation.

2. The probability is 0.95 that the error will be less than what number, ε? That is, what is the number ε such that the a priori probability that a sample

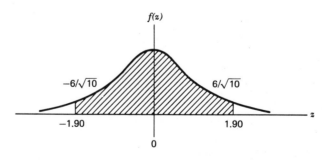

Figure 3.6 Standard normal density curve.

taken for which the sample mean will differ from the population mean by less than ε is 0.95? The error (ε) must be determined such that

$$p\{|\bar{L}-\mu_L|<\varepsilon\} = 0.95 = p\{-\varepsilon<\bar{L}-\mu_L<\varepsilon\}$$

Rewrite the preceding inequality, dividing through by $\sigma_{\bar{L}} = (\sigma/\sqrt{n}) = \sqrt{10}/6$. Thus

$$p\left[-\frac{6\varepsilon}{\sqrt{10}} < \frac{\bar{L}-\mu_L}{\sigma_{\bar{L}}} < \frac{6\varepsilon}{\sqrt{10}}\right] = 0.95$$

According to equation (b), the left-hand member in the preceding expression approximates

$$\frac{1}{\sqrt{2\pi}} \cdot \int_{-\frac{6\varepsilon}{\sqrt{10}}}^{\frac{6\varepsilon}{\sqrt{10}}} e^{-z^2/2}\,dz = \frac{2}{\sqrt{2\pi}} \int_{0}^{\frac{6\varepsilon}{\sqrt{10}}} e^{-z^2/2}\,dz \quad (d)$$

From standard normal area tables, it is found that the area under Figure 3.7 from -1.96 to 1.96 is 0.95 (Appendix 1):

$$\frac{1}{\sqrt{2\pi}} \int_{-1.96}^{1.96} e^{-z^2/2}\,dz = 0.95 \quad (e)$$

By equating upper limits of integration in equations (d) and (e), we obtain

$$\frac{6\varepsilon}{\sqrt{10}} = 1.96$$

$$\varepsilon = 1.03 \text{ klb}$$

Thus the probability is 0.05 that after 36· measurements are made, $\bar{L}$ will differ from μ_L by more than 1.03 klb.

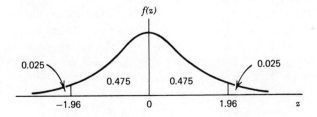

Figure 3.7 Standard normal 95% interval.

3. How large a sample is required to assure that the a priori probability is only 0.05 that $\bar{L}$ will differ from μ_L by more than 1 klb? The number n must be determined such that

$$p\{|\bar{L} - \mu_L| > 1\} = 0.05$$

or

$$p\{|\bar{L} - \mu_L| < 1\} = 0.95 \tag{f}$$

Dividing through the inequality by $\sigma/\sqrt{n}$ and restating equation (f) yields

$$p\left[-\frac{\sqrt{n}}{\sigma} < \frac{\bar{L} - \mu_L}{\sigma_{\bar{x}}} < \frac{\sqrt{n}}{\sigma}\right] = 0.95.$$

Since $\sigma = \sqrt{10}$, it follows that

$$p\left[-\frac{\sqrt{n}}{\sqrt{10}} < \frac{\bar{L} - \mu_L}{\sigma_{\bar{L}}} < \frac{\sqrt{n}}{\sqrt{10}}\right] = 0.95$$

By utilizing equation (b), we obtain:

$$\frac{1}{\sqrt{2\pi}} \int_{-\frac{\sqrt{n}}{\sqrt{10}}}^{\frac{\sqrt{n}}{\sqrt{10}}} e^{-z^2/2} dz = \frac{1}{\sqrt{2\pi}} \int_{-1.96}^{1.96} e^{-z^2/2} dz$$

Equating upper limits yields

$$\frac{\sqrt{n}}{\sqrt{10}} = 1.96; \qquad n = 10(1.96)^2 \approx 39$$

A sample of at least 39 is required to assure that the a priori probability is only 0.05 that $|\bar{L} - \mu_L| \geq 1$.

For further considerations of confidence intervals on the mean and standard deviation of a normal random variable, see Chapter 12.

The discussion in Section 3.3.1 may suggest that test data can be obtained in sufficient quantity. The engineer may, however, be confronted with situations where no data are available. If the range of likely values or a tolerance interval of a variable can be estimated, it is still possible to roughly estimate a needed standard deviation.

3.3.2 Estimation of Scatter Based on Range

As a measure of dispersion in small samples, the range, as a substitute for the sample standard deviation s, has been found to give remarkably high efficiency, as compared to s. This method is highly efficient for samples from normal and moderately skewed populations. For the usual run of experimental data, Table 3.4 can be utilized to obtain estimations of s [14].

The midpoint of the range provides a reasonable estimate of the mean value, based on the ratio of mean to median given in Table 2.8.

If the only available information relating to variability of a design variable X is the tolerance range $\pm \Delta_x$, this may often be utilized to estimate the standard deviation. When the anticipated processing is capable of extensive production within $\bar{x} \pm \Delta_x$ limits, these limits may be considered to define the range of a large sample; and as a first approximation, utilizing Table 3.4:

$$s_x \approx \frac{(\bar{x}+\Delta_x)-(\bar{x}-\Delta_x)}{6} = \frac{\Delta_x}{3}$$

Example D

Assume for this example (see Section 3.2.3) that the sample size is 10. Thus from Table 3.4 the $\pm 1.5\ s_k$ range for the spring constant is approximately $(5.6-2.0) = 3.6$ lb. Consequently:

$$s_k \simeq \frac{3.6}{3} = 1.2 \text{ lb}$$

Example E

In Table 4.4 the tolerance on thickness t of cold rolled carbon steel sheet in widths of 40 to 48 in. and of nominal thickness 0.1874 to 0.1800 in. is ± 0.010 in. Thus the approximate standard deviation on sheet thickness t is

$$s_t \approx \frac{\Delta_t}{3} = \frac{0.010}{3} = 0.0033 \text{ in.}$$

Table 3.4. Estimations of s

If n is Near This Number	s is Roughly Estimated by Dividing Range by
5	2
10	3
25	4
100	5
700	6

Example F

Kent's *Mechanical Engineers Handbook* (pp. 2–57) gives the elastic modulus E of ordinary steel as $29,000,000 \pm 1,000,000$ psi. Thus as a first approximation, $\backprime_E$ is

$$\backprime_E \approx \Delta_{E/3} = 1,000,000/3 = 333,000 \text{ psi}$$

STATIC LOADS

3.4 FORCE AND MOMENT AS RANDOM VARIABLES

Static loads (Figure 3.9) induce reactions in mechanical components, and equilibrium usually develops. Equilibrium may persist over time, with the internal forces essentially constant.

Where simple models can be justified, the principles of statics may be applied to estimate force magnitudes in static load-carrying components, such as frames, hydraulic devices, supports of various kinds, constant-velocity rotating parts, and machine members under actions that do not result in acceleration of the member. Consider a simply supported beam required to carry only its own weight. There may be several very different design considerations:

1. The individual weight of a beam produces reactive forces. The load is deterministic.
2. If strength varies statistically among a population of beams, each with the same weight, Figure 3.8a is the model.
3. If a beam is selected at random from among many equally strong beams, weight may be a random variable and Figure 3.8b is the model.
4. If both beam strengths and weights have independent patterns of variability, Figure 3.8c is the model.

A beam may be required to support a concentrated weight in addition to its own. If beam weight is negligible compared with the concentrated weight, it follows that:

1. If the same weight is applied to each of a random population of beams, Figure 3.8a is the model.
2. If selection were at random from a population of nominally identical weights is applied to a given beam, Figure 3.8b is the model.
3. If a weight selected at random is applied to a beam selected at random, Figure 3.8c is the model.

When total static loading on a mechanical system arises from more than one source, a functional model combining the loads may be written and the

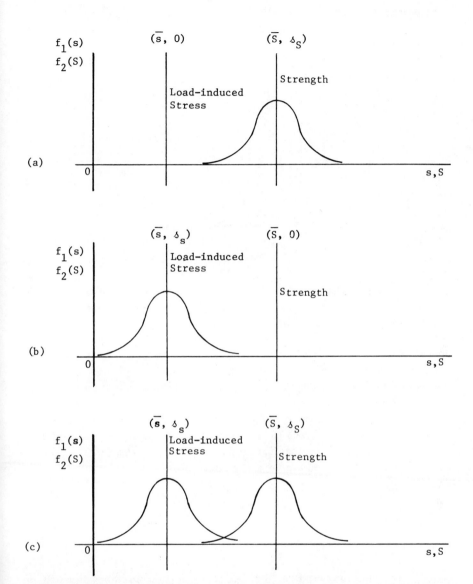

Figure 3.8 Strength versus load-induced stress models. The probability density curves are often nonsymmetrical.

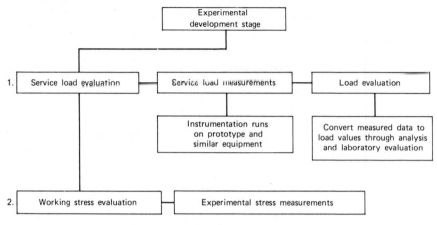

Figure 3.9 Developing loads model [3].

resultant loading statistics calculated therefrom using the methods described in Chapter 2. Examples G and H illustrate the application of the algebra of expectation in estimating loading statistics.

Example G

This example illustrates dispersion of loading from a single source into two load-carrying members.

A rigid ring is supported by two radial wires AB and CD. A load L is applied as shown in Figure 3.10. Determine the mean values and standard deviations of the reactive forces T_1 and T_2 in the support wires.

First, the equilibrium equations are written

$$\sum F_x = 0 = T_2 \sin 45° - T_1 \sin 30° \tag{a}$$

$$\sum F_y = 0 = T_2 \cos 45° + T_1 \cos 30° - L \tag{b}$$

Since L is a random variable, the forces T_1 and T_2 in the wires will be random variables. The angles 30° and 45° are assumed to be deterministic. Solving equations (a) and (b) simultaneously leads to

$$T_1 = 0.732 L \tag{c}$$

$$T_2 = \frac{0.732(0.5)L}{0.707} = 0.516 L \tag{d}$$

$$(\bar{L}, s_L) = (100; 12) \text{ lb} \tag{e}$$

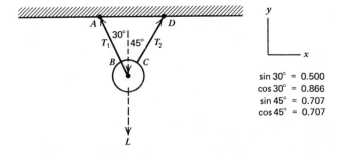

Figure 3.10 Random static loading.

Given. The values μ_L and σ_L are not known, but unbiased estimators $\bar{L}$ and s_L are available from sample information. According to equation (2.8), the mean of a product of a random variable and a constant [equation (3.9)] is $\mu_{ax} = a \cdot \mu_x$; thus

$$\bar{T}_1 = 0.732(100) = 73.2 \text{ lb}$$

and according to equation (2.10), s_{T_1} is

$$s_{T_1} = \sqrt{12^2(0.732)^2} = 8.8 \text{ lb}$$

Similarly, use of equations (3.10), (2.8), and (2.10) yields

$$\bar{T}_2 = 0.516(100) = 51.6 \text{ lb}$$

$$s_{T_2} = \sqrt{12^2(0.516)^2} = 6.2 \text{ lb}$$

Since T_1 and T_2 are linear functions of L their distributions are of the same general type as the distribution of L (see Figure 3.14). In Section 3.6 it is shown that L, T_1, and T_2 are statistically correlated.

Confidence limits on T_1 and T_2 are determined as follows. If the statistics of L are determined from a sample of n test values L_i ($i = 1, 2, \ldots, n$), then confidence limits on these statistics can be estimated, as discussed in Section 3.3.1 and in Chapter 12. Since $(T_1)_i = L_i(0.732)$ and $(T_2)_i = L_i(0.518)$ for each i, the statistics of T_1 and T_2 can be treated as being estimated from samples of size n. Thus confidence limits on T_1 and on T_2 may be calculated in a manner similar to those of L.

Example H

Given statistically independent force vectors P and Q as shown in Figure 3.11, estimate the statistics of M_T about the origin. The problem is to determine the moment generated by two forces with respect to a point.

96 Loading Random Variables

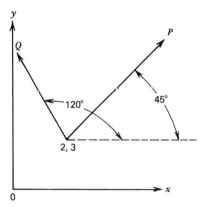

Figure 3.11 Moment due to force vectors.

Assume that the dimensional and angular values in this problem are essentially constant. The statistics of M_T must be estimated from the model

$$M_T = 2F_y - 3F_x$$

where

$$F_x = P_x + Q_x$$
$$F_y = P_y + Q_y$$

Given

$$(\bar{P}, s_p) = (50, 6.5) \text{ lb}$$
$$(\bar{Q}, s_Q) = (30, 2.8) \text{ lb}$$

vectors P and Q are random variables. However, to calculate the statistics of M, it is not necessary to know the distributions of P and Q. The model for the x component of force at the point 2, 3 is

$$F_x = P_x + Q_x \tag{a}$$

Therefore,

$$P_x = P(\cos 45°); \quad Q_x = -Q\cos 60° \tag{b}$$

The model for the y component of force at the point 2, 3 is

$$F_y = P_y + Q_y, \tag{c}$$

thus
$$P_y = P\sin 45°; \qquad Q_y = Q\sin 60°. \tag{d}$$

From equations (2.8) and (2.10), the product of a random variable and a constant [equation (b)] is

$$\bar{P}_x = 50(0.707) = 35.4 \text{ lb}; \qquad s_{P_x} = 6.5(0.707) = 4.6 \text{ lb}$$

$$\bar{Q}_x = -30(0.5) = -15 \text{ lb}; \qquad s_{Q_x} = 2.8(0.5) = 1.4 \text{ lb}$$

Also, according to equation (d):

$$\bar{P}_y = 50(0.707) = 35.4 \text{ lb} \quad \text{and} \quad s_{P_y} = 6.5(0.707) = 4.6 \text{ lb}$$

$$\bar{Q}_y = 30(0.866) = 26 \text{ lb} \quad \text{and} \quad s_{Q_y} = 2.8(0.866) = 2.42 \text{ lb}$$

From equation (2.17), the mean of $(P_x + Q_x) = F_x$ is

$$\bar{F}_x = 35.4 - 15 = 20.4 \text{ lb}$$

and the mean of $(P_y + Q_y) = F_y$ is

$$\bar{F}_y = 35.4 + 26 = 61.5 \text{ lb}$$

According to equation (2.21), the standard deviations of F_x and F_y are

$$s_{F_x} = \sqrt{(1.4)^2 + (4.6)^2} = 4.81 \text{ lb}$$

$$s_{F_y} = \sqrt{(4.6)^2 + (2.42)^2} = 5.20 \text{ lb}$$

Thus the statistics of random variables F_x and F_y are

$$F_x \approx (20.4; 4.81) \text{ lb}$$

$$F_y \approx (61.4; 5.2) \text{ lb} \tag{e}$$

The statistics of F_x and F_y, with the $M_T = 2F_y - 3F_x$ and equations (2.8) and (2.22), yield

$$\bar{M}_T = 2(61.4) - (3) \cdot (20.4) = 61.6 \text{ ft-lb}$$

and application of equations (2.10) and (2.24) yields

$$s_{M_T} \approx \sqrt{[(2) \cdot (5.2)]^2 + [(3) \cdot (4.81)]^2} = 17.8 \text{ ft-lb}$$

98 Loading Random Variables

The statistics of M_T are

$$\left(\overline{M}_T, s_{M_T}\right) \approx (61.6, 17.8) \text{ ft-lb}$$

3.4.1 Avoidance of Overload

To protect critical or costly parts (or meet safety requirements), it is frequently necessary to design with a high probability that the safety element in a system fails (or reacts) at load intensities lower than the capability of more critical components (see Figure 3.12). For example, gear teeth may need protection from an overload, or piping, tubes, fittings, and pressure vessels in a pneumatic or hydraulic system may require protection from excessive pressure.

Overload can be avoided by the following measures:

1. Protect a system by limiting the strength of an expendable component (e.g., shear pins, fuses) (Figure 3.13).
2. Provide devices that directly modify the distribution of applied loads (e.g., governors).

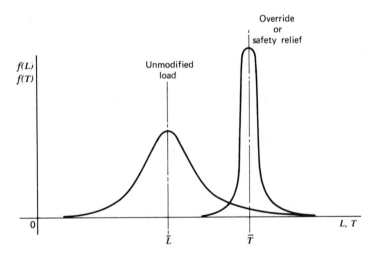

Figure 3.12 Restricting a load envelope.

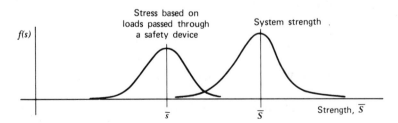

Figure 3.13 Stress based on loads passed through a safety device.

3. Protect a system by limiting the load intensity experienced (e.g., mechanical overrides, safety throw-out devices, shutdown due to excessive deflection triggered by limit switches).

In limiting load intensity, what is the distribution of loads passed through a safety device, where T is the tripping force and L is the unmodified load?

The distribution of loading L' passed through without tripping the safety device is

$$L' = T - L > 0$$

For design purposes, it is necessary to calculate the statistics of applied stress s associated with L' that reaches the system. The design calculations are then carried out such that $P(S > s) \geqslant R$.

CORRELATION

3.5 CORRELATED ENGINEERING VARIABLES [13]

In Chapter 2 the concept of correlation was important in the development of the algebra of normal functions. In reality a statistical algebra would not exist without correlation. Correlated effects are observed in both static and dynamic systems.

An example of loading correlation is observed when the random load resisted by a component in a system is linearly related to another random load resisted by the same component.

In Example G of Section 3.4 a random value taken by L is linearly related to the random values observed in components T_1 and T_2 at the same instant. The correlation coefficient in this case is $\rho = +1$, and the distributions of T_1 and T_2 are easily shown to be of the same general kind as the distribution of L. Similarly, loads F_1 and F_2 may be correlated when they arise from pneumatic pressure pulses in a dynamic system. The correlation coefficient ρ is by definition

$$\rho = E\left[\frac{x - \mu_x}{\sigma_x} \cdot \frac{y - \mu_y}{\sigma_y}\right] = \frac{E(xy) - \mu_x \mu_y}{\sigma_x \sigma_y} *$$

The sample correlation coefficient r is [14, p. 64]

$$r = \frac{\sum_{i=1}^{n}(x_i - \bar{x})(y_i - \bar{y})}{\left[\sum_{i=1}^{n}(x_i - \bar{x})^2 \cdot \sum_{i=1}^{n}(y_i - \bar{y})^2\right]^{1/2}}$$

*The numerator in the equation for ρ is called the *covariance of X and Y*.

The sample correlation coefficient r is estimated from n pairs of observations of the random variables X and Y:

$$(x_1,y_1),(x_2,y_2),\ldots,(x_n,y_n)$$

Noting Figure 3.14, the n pairs of observations will be

$$[l_1,(T_1)_1],[l_2,(T_1)_2],\ldots,[l_n,(T_1)_n]$$

and $\bar{L}$ and $\bar{T}_1$ are

$$\bar{L} = \frac{l_1 + l_2 + \cdots + l_n}{n}$$

$$\bar{T}_1 = \frac{(T_1)_1 + (T_1)_2 + \cdots + (T_1)_n}{n}$$

Since $(T_1)_i = l_i \cdot \tan\delta$,

$$\bar{T}_1 = \bar{L}\tan\delta$$

Furthermore, the variances are linearly related:

$$V(T_1) = (\tan\delta)^2 V(L)$$

The estimator r is

$$r = \frac{\sum_{i=1}^{n}(l_i - \bar{L})\cdot\left[(T_1)_i - \bar{T}_1\right]}{\left[\sum_{i=1}^{n}(l_i - \bar{L})^2 \sum_{i=1}^{n}(T_1)_i - \bar{T})^2\right]^{1/2}}$$

$$r = \frac{\sum_{i=1}^{n}(l_i - \bar{L})\cdot(l_i\tan\delta - \bar{L}\tan\delta)}{\left[\sum_{i=1}^{n}(l_i - \bar{L})^2 \sum_{i=1}^{n}(l_i\tan\delta - \bar{L}\tan\delta)^2\right]^{1/2}}$$

$$r = \frac{\tan\delta \sum_{i=1}^{n}(l_i - \bar{L})^2}{\tan\delta\left[\sum_{i=1}^{n}(l_i - \bar{L})^2 \sum_{i=1}^{n}(l_i - \bar{L})^2\right]^{1/2}} \approx 1$$

The random variables L and T_1 are correlated, with $r = +1$. The mean values and variances are linearly related, and the distributions are of the same general kind.

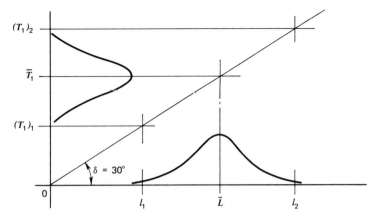

Figure 3.14 Correlation of T_1 and L.

3.6 CORRELATED AND INDEPENDENT LOADS

A single element structure may be loaded by a force, F, acting along the centroidal axis or at an angle such that a bending moment and/or an axial load is generated.

If a structural component is loaded simultaneously by two statistically independent collinear forces, the total effect is estimated by assuming superposition, and the standard deviation is

$$s_F = \sqrt{s_{F_1}^2 + s_{F_2}^2}$$

If the forces are correlated, the standard deviation is

$$s_F = \sqrt{s_{F_1}^2 + s_{F_2}^2 + 2\rho s_{F_1} s_{F_2}}$$

if F_1 and F_2 are normal. The two different effects are illustrated quantitatively in Example I.

Example I

Static loads (F_1) and (F_2) combine additively to produce total tensile loading of a support. Assume they are normally distributed and are described by the following statistics:

$$\left(\bar{F}_1, s_{F_1}\right) = (2700; 410) \text{ lb} \qquad (a)$$

$$\left(\bar{F}_2, s_{F_2}\right) = (3600; 460) \text{ lb} \qquad (b)$$

Let

$$F = F_1 + F_2 \qquad (c)$$

Estimate the effect of independence, $\rho=0$, versus positive correlation, $\rho=+1$, and negative correlations, $\rho=-1$.

1. If F_1 and F_2 are statistically independent; from equations (2.36), (2.37), and (c) with $\rho=0$:

$$\bar{F} = \bar{F}_1 + \bar{F}_2 = 2700 + 3600 = 6300 \text{ lb}$$

$$s_F = \sqrt{s_{F_1}^2 + s_{F_2}^2 + 2\rho s_{F_1} \cdot s_{F_2}} = \sqrt{(410)^2 + (460)^2} = 615 \text{ lb}$$

$$(\bar{F}, s_F) = (6300; 615) \text{ lb}$$

2. If F_1 and F_2 were correlated, $\rho=+1$; from equations (2.36), (2.37), and (c):

$$s_F = \sqrt{(410)^2 + (460)^2 + (2) \cdot (410)(460)} = 870 \text{ lb}$$

$$(\bar{F}, s_F) = (6300, 870) \text{ lb}$$

3. If F_1 and F_2 were correlated, $\rho=-1$; from equations (2.36), (2.37), and (c):

$$s_F = \sqrt{(410)^2 + (460)^2 + (-2)(410)(460)} = 50 \text{ lb}$$

$$(\bar{F}, s_F) = (6300; 50) \text{ lb}$$

Load components $F_1, F_2, \ldots, F_n$ originating from a common source F may each produce an effect on the same mechanical element (such as axial and normal components of a force vector acting on a beam). Such force components are related to F and to each other (see Figures 3.14 and 3.17); F may be either a static or a dynamic random variable.

Example J

If the reactions R_1 and R_2 at the supports of a link (modeled as a simple beam) are to be estimated, considerable insight may be gained by considering correlation. In Figure 3.15 if $c=(a/l)$, then, $R_1=cF$ and $R_2=(1-c)F$. Also, if $\bar{c} \gg s_c$ and $(s_c/\bar{c}) \ll 1$, c may be treated as an approximate constant ($\bar{c}$, ~0). The reactions R_1 and R_2 can be considered correlated with F, with correlation coefficient $\rho \approx 1$. Further, R_1 and R_2 will have distributions of approximately the same general type as F since the transformations from the distribution of F to the distributions of R_1 and R_2 would be nearly linear.

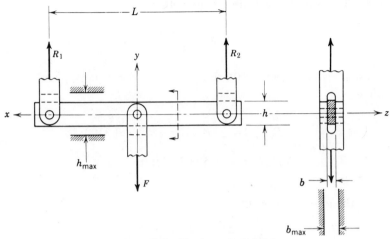

Figure 3.15 Simply supported link.

Example K

Consider a two-member pin jointed motor support bracket (Figure 3.16) assumed to be dynamically loaded by $(\bar{P}, s_P)$, constrained to act vertically and imposing fully reversed loads on members x and y.

Member x, pinned at points A and C, is loaded in tension ($+$) when P acts downward and in compression ($-$) when P acts upward. Estimate the force peak statistics in AC and BC. Since strains in the members are relatively small, angular relationships are considered invariant.

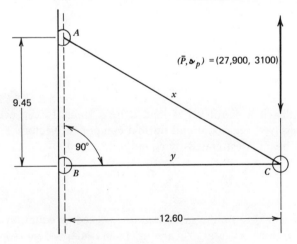

Figure 3.16 Motor support bracket.

Solution. The peak force F_x in member AC is

$$F_x = \frac{5}{3}P$$

From equations (2.8) and (2.10), the product of a constant $\frac{5}{3}$ and a random variable p is

$$F_{AC} \sim \begin{array}{l} (\overline{F}_x, {}^s\!F_x) = \frac{5}{3}(\overline{P}, {}^s\!p) = \frac{5}{3}(27{,}900, 3100) \text{ lb} \\ (\overline{F}_x, {}^s\!F_x) = (46{,}500, 5{,}170) \text{ lb} \end{array}$$

Similarly, the peak force F_y in member y is

$$F_y = \frac{4}{5}F_x = \frac{4}{5} \cdot \frac{5}{3}P = \frac{4}{3}P$$

The value F_y is correlated with P, $\rho = +1$, and F_y is out of phase with F_x (but correlated $\rho = +1$), as it is tensile when F_x is compressive and vice versa. Thus

$$F_{BC} \sim (\overline{F}_y, {}^s\!F_y) = \frac{4}{3}(27{,}900, 3100) = (37{,}200, 4133) \text{ lb}$$

In a given multiaxial loading problem there may be one of two possible situations. Numerically identical load random variables in formally identical functions may generate one distribution if they are correlated and a different distribution if statistically independent.

Example L

Consider a spur gear tooth from an ore crusher, modeled as a cantilever beam (Figure 3.17). The coordinates X, Y, and Z are force vectors, and

$$Z = \sqrt{X^2 + Y^2}$$

Conventionally, if a cantilever is loaded by a single force vector Z (Figure 3.17), it is resolved into axial and normal component vectors X_1 and Y_1. The stress in the beam is estimated by equation (5.23):

$$s = \frac{P}{A} + \frac{Mc}{I}$$

where s is assumed to be the same as in the preceding equation if the beam is loaded by vectors X_2 and Y_2 arising from independent sources, and of magnitudes 12,920 lb and 37,000 lb, respectively. Possibilities relating to

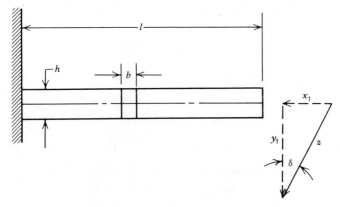

Figure 3.17 Loads X_1 and Y_1 correlated, $\rho = +1$.

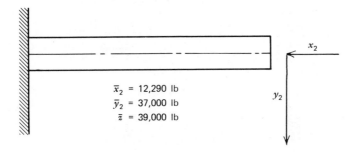

$\bar{x}_2 = 12{,}290$ lb
$\bar{y}_2 = 37{,}000$ lb
$\bar{z} = 39{,}000$ lb

Figure 3.18 Loads X_2 and Y_2 independent, $\rho = 0$.

statistical considerations are as follows:

1. Consider, a cantilever beam (Figure 3.18) loaded by statistically independent random-variable force vectors X_2 and Y_2, each having, for example, a coefficient of variation:

$$\frac{\sigma_{x_2}}{\bar{x}_2} = \frac{\sigma_{y_2}}{\bar{y}_2} = 0.20$$

From the values in Figure 3.18 and the coefficients of variation:

$$(\bar{x}_2, \sigma_{x_2}) = (12{,}290, 2460) \text{ lb}$$

$$(\bar{y}_2, \sigma_{y_2}) = (37{,}000, 7400) \text{ lb} \tag{a}$$

2. Next, calculate the mean value and standard deviation of Z as if the random variable vectors [equation (3.17)] were statistically independent:

$$z = \sqrt{x^2 + y^2}$$

106 Loading Random Variables

According to equation (2.59) or (2.63):

$$\bar{z} \approx \sqrt{\bar{x}_2^2 + \bar{y}_2^2} = 39{,}000 \text{ lb}$$

and equation (2.62) or (2.64):

$$s_z \approx \left[\frac{\bar{x}_2^2 s_x^2 + \bar{y}_2^2 s_y^2}{\bar{x}_2^2 + \bar{y}_2^2} \right]^{1/2}$$

$$s_z \approx \left[\frac{(1.23)^2 \cdot 10^8 \cdot (2.46)^2 \cdot 10^6 + (3.70)^2 \cdot 10^8 \cdot (7.40)^2 \cdot 10^6}{(1.23)^2 \cdot 10^8 + (3.70)^2 \cdot 10^8} \right]^{1/2} = 7070 \text{ lb}$$

Thus

$$(\bar{z}, s_z) \approx (39{,}000, 7070) \text{ lb} \tag{b}$$

3. Last, the beam is loaded by a single random variable Z as described in equation (b). In resolving vector Z, the component random variables X_1 and Y_1 are correlated, $\rho = +1$. Thus

$$\sin \delta = 0.315 = \frac{12.29}{39.00}$$

and

$$\cos \delta = 0.949 = \frac{37.00}{39.00}$$

The correlated random variable components are, employing equations (2.8) and (2.10),

$$(\bar{x}_1, s_{x_1}) = 0.315 \cdot (\bar{z}, s_z) = (12{,}290, 2230) \text{ lb}$$

$$(\bar{y}_1, s_{y_1}) = 0.949 \cdot (\bar{z}, s_z) = (37{,}000, 6700) \text{ lb} \tag{c}$$

The loading in equations (a) and (c) can lead to different requirements structurally, in particular where large variabilities are involved.

DYNAMIC LOADING

3.7 CONSTANT- AND RANDOM-AMPLITUDE LOADS

Mechanical components are often required to resist repetitive or cyclic loads, the peak levels of which may be either constant or random in amplitude. Associated with these may be either zero or nonzero mean or static loads due, for instance, to the weight of a structure.

The behavior of dynamic loaded systems is characterized by changing magnitudes of strains or deflections* in load-resisting components over time. The possibility of fatigue of materials (Section 8.14) becomes a consideration. Figure 3.19 illustrates frequently observed patterns of dynamic loading.

3.7.1 Constant-amplitude Dynamic Loads (Zero Means)

The case of fully reversed sinusoidal loading (with zero mean) (Figure 3.19c) is frequently modeled as

$$f(t) = p \cdot \sin(wt)$$

where p indicates maximum load amplitude; p may be considered a mean value with an element of variability often added to take account of random startup, shutdown, and transient loads; and $f(t)$ may refer to axial, bending, or torsional sinusoidal fully reversed loads.

Example M

Familiar instances of constant-amplitude dynamic loading are the constant-amplitude bending loads resisted by specimens in an R. R. Moore rotating beam fatigue machine and the constant amplitude axial loads imposed on specimens in a Sontag testing machine.

Two distributional models of dynamic fully reversed loadings are of particular interest in design because they yield closed solutions when associated with normally distributed dynamic-strength models (see Chapter 9 and 10):

1. Cases where the instantaneous levels of fully reversed loading are normally distributed. The distributions of load peaks L are Rayleigh distributed, with a probability density function of (see Section 9.3.6)

$$f(L) = \frac{L}{\sigma^2} e^{-L^2/\sigma^2} \quad (0, \infty)$$

where the estimator σ^2 is

$$\sigma^2 = \frac{L_1^2 + L_2^2 + \cdots + L_n^2}{2n}$$

$$\bar{L} = \frac{(\sigma^2 \pi)^{1/2}}{\sqrt{2}}$$

---

*A mechanical system is said to be vibrating when its parts undergo motions that fluctuate in time.

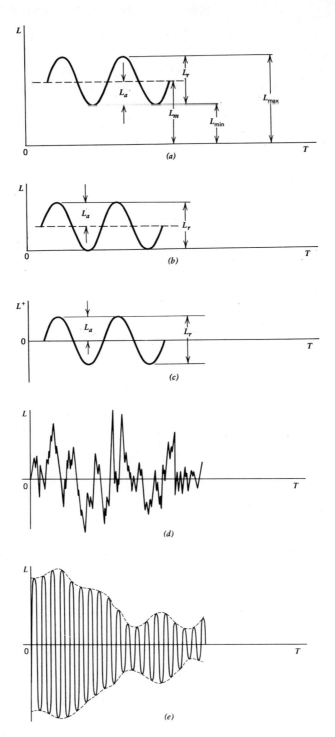

Figure 3.19 Typical fatigue loading. (*a*) Fluctuating loading. (*b*) Repeated loading. (*c*) Reversed loading. (*d*) Random broad band. (*e*) Random narrow band.

2. Cases where the load peaks L can be considered normally distributed or approximately normally distributed (as with numerous natural phenomena). The probability density function is

$$f(L) = \frac{1}{\sqrt{2\pi} \cdot s_L} \cdot \exp\left[-\frac{1}{2}\left[\frac{L-\bar{L}}{s_L}\right]^2\right] \quad (-\infty, \infty), \quad \text{see [40]}$$

where the estimator of $\bar{L}$ is

$$\bar{L} = \frac{L_1 + L_2 + \cdots + L_n}{n}$$

$$s_L = \left[\frac{1}{n}\sum_{i=1}^{n}(L_i - \bar{L})^2\right]^{\frac{1}{2}}$$

3.7.2 Narrow-band Random Loads (Zero Means)

Although constant-amplitude sinusoidal loading has received close attention in the past, it must be considered a special case not too often seen in actual engineering problems. Dynamic loading is usually random, with narrow-band random loading of particular interest. A narrow-band condition implies a band of significant frequencies small compared with the central frequencies (Figure 3.19e).

The usual measure of narrow-band random loading is its root mean square (RMS) value (carrying units of force). Where the mean load is zero, the RMS value amounts to the standard deviation [see equation (2.15)]. In couple notation this is

$$(\bar{L}, s_L) = (0, RMS_L) \text{ lb}$$

Frequency is also often an important consideration in design since it may effect response (see Section 5.16).

The RMS_L will subsequently be used in writing narrow-band stress expressions, see Section 5.16. For the loadings considered in Section 3.7.1 and 3.7.2, the stress at failure to be avoided will be that estimated from appropriate s–n envelopes (considered in Sections 8.14.1 and 8.14.3).

3.7.3 Simultaneous Action of Static and Dynamic Loads

The simultaneous action of static and dynamic loads is observed where a rotating transmission shaft also supports a static load or where ground vehicles are in transit over typical roadways or where aircraft are subjected to random gust loads* in addition to those generated in constant-velocity level flight (exemplified in Figure 3.20).

---
*The exponential distributional model often appears reasonable for gust phenomena.

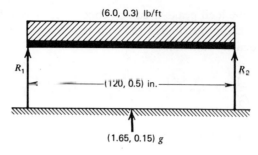

Figure 3.20 Typical dimensions and loading of a transverse support.

Where dynamic loads are acting on a component in addition to static forces, additive superposition is not applicable. This statement follows when mechanical behavior of a material subjected to static and dynamic loading is considered.

The load-effect statistics must be estimated by vector addition of the static- and dynamic-stress vector components acting on a structure (as discussed in Section 5.16). The static and dynamic vectors are usually statistically independent.

The combinations of dynamic loading with a static load (acting simultaneously on a structure) that will require consideration in design are (1) fully reversed sinusoidal loading with a static load, (2) narrow-band random loading with a static load, and (3) broad-band random loading with a static load. (*Note.* Figures 8.39 and 8.40 suggest that for a given RMS_L, the loading, if narrow-band random, provides a somewhat more hostile environment than would a broad-band random loading environment.)

The general model of loading must include the effects of a mean load (static) with an associated dynamic element. This is probably the most frequent loading situation with mechanical systems. A present difficulty in the direct solution of random loading problems is the absence of adequate materials behavior data. Since the behavior of materials in resisting a given specie of loading is very specific, the development of behavioral data from mean loads associated with random dynamic loads should be given high priority [41].

If the static and dynamic load vector components are approximately normally distributed, which is often the case with static loads and with natural dynamic loading phenomena, the effective load vector will very likely be approximately normally distributed (see Section 2.7).

3.7.4 Example of Dynamic Loads

Loading Due to Acceleration. With pistons in reciprocating engines either gas loading or inertia loading due to acceleration may be critical; however, both types of loading are rarely in combination. Gas loading causes high compressive load-induced stresses in the deck of a piston and the area between the deck and pin bore. With inertial loading, high-tensile loads are generated in the area from pin bore to the skirt. With the locking lugs in a weapon, gas loading is critical and inertia loading may be minor.

Example N illustrates inertial loading due to accelerated motion of a rigid airplane (random repeated loading; Figure 3.19b).

Example N

Figure 3.21 illustrates an airplane being trapped on a carrier flight deck and stopped by an arresting hook engaging a cable (with pull T). If the total weight of this type airplane is $(\bar{w}, s_w) = (12{,}000, 600)$ lb and it is decelerated at $(\bar{a}, s_a) = (3.5, 0.25)$ g or $(112.7, 8.05)$ ft/sec², what is the pull T in the arresting hook that is transferred to the airplane structures? Also estimate wheel reaction R. Landing velocity is $(60, 4)$ mph.

Solution. On contact with the arresting cable, the airplane is decelerated horizontally in pure translation, generating a force F;

$$(\bar{F}, s_F) = \frac{(\bar{w}, s_w)}{(g, 0)} (\bar{a}, s_a) \qquad g = (12{,}000, 600)(3.5, 0.25) \text{ lb}$$

From equations (2.25) and (2.27), the product of w and a that describe F is

$$(\bar{F}, s_F) = (42{,}000, 3660) \text{ lb}$$

The forces T and R can now be calculated from equations of equilibrium. Estimate T as follows:

$$\sum F_x = (42{,}000, 3660) + (\bar{T}, s_T)\cos 10° = 0$$

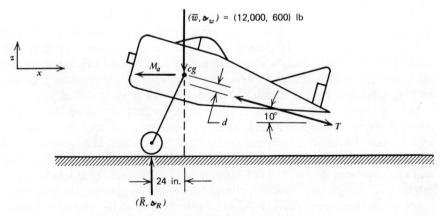

Figure 3.21 Inertial loading due to acceleration.

112 Loading Random Variables

By equations (2.8) and (2.10), the product of a random value (F) and a constant ($1/\cos 10°$) is

$$(\overline{T}, s_T) = \left(\frac{42{,}000, 3660}{0.9848}\right) = (42{,}700, 3720) \text{ lb}$$

Estimate R as follows:

$$\sum F_z = -(12{,}000, 600) + (\overline{R}, s_R) - (42{,}700, 3710) \sin 10° = 0$$

and

$$(42{,}700, 3710) \cdot \sin 10° = (7{,}300, 640) \text{ lb}$$

From equations (2.19) and (2.21), the sum of the two random values describing wheel reaction R is

$$(\overline{R}, s_R) = (12{,}000, 600) + (7300, 640) = (19{,}300, 880) \text{ lb}$$

After the statistics of T and R have been estimated and it is known that these loads will be repeated an estimated n times during the service life of the airplane, the statistics may be used to design the arresting hook and its attaching hardware and the landing-gear structure.

3.7.5 Loading Distributional Forms

Although the exact distributional form of a loading variable is seldom known, the asymptotic limiting form of the distribution of a function of load variables is known. For instance, if total system loading results from a number of summed component loads, the limiting distributional form is likely to be normal. Multiplicative load functions, on the other hand, tend toward a lognormal limiting distributional form.*

Many investigators have observed that the normal distribution often gives a reasonable description of road surface irregularities, hence accelerations, and that natural phenomena frequently appear to be normal or near normally distributed. Normally distributed random vibration filtered through a spring-mass system, which has its resonance within the frequency range of the exciting vibration, retains its normal distribution of peak amplitudes.

Two or more actual service random-load records are never the same, although representative statistics, such as variances (or root mean square levels), may be identical. Computerized servohydraulic test systems with random programmers are capable of reproducing actual dynamic service load random sequences or synthesized random-load sequences. In a test series to validate a design or establish basic materials properties, nonidentical random-load sequences (having the same statistics) must be applied in turn to

*See Bayes decision procedures in Section 3.2.2.

each different specimen or structural component in a test series, to accomplish realistic simulations.

Useful records of vibration phenomena under field conditions may be made with use of transducers such as (1) accelelerometers, (2) variable-resistance strain gauges, (3) pressure gauges, or (4) deformeters to quantitatively measure peaks of dynamic loading. Furthermore, operational-load-generated variable-strain and deflection records can often be utilized in studying response histories (either strain or inferred stress) of existing components in an analysis. If strain histories are those of components operating in normal use environments, asymptotic strain responses may provide a basis from which to nondestructively predict component life.

Reversals represent the key in the fatigue process, and recent research has shown that the absolute values (i.e., the correct mean and the range of the reversals) provide the primary loading information for fatigue predictions [41].

Mechanical systems and their components are typically exposed to loads of many modes, frequencies, and amplitudes. Amplitudes may be constant or random, and frequencies may vary widely, that is, from broad band to narrow band [29]. The existence of combinations of modes of loading may be more damaging than individual modes considered separately [43]. An actual field-stress history generated by dynamic random loading is illustrated in Figure 3.22.

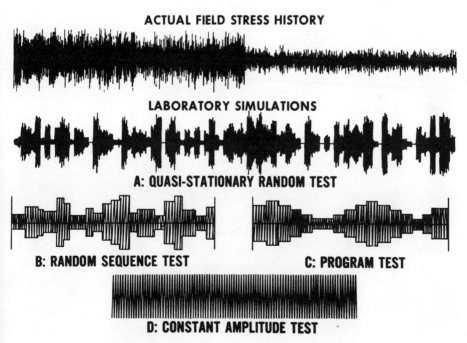

Figure 3.22 Laboratory simulation for service loading.

114 Loading Random Variables

3.8 PARAMETERS MODIFYING LOADING

3.8.1 Introduction

Characteristics of external loads must be determined early in the design process because anticipated internal responses in mechanical elements must be inferred from the action of external applied forces. Figures 3.23 and 3.24 illustrate typical random-load data reduced to statistics.

After determining whether loading is (1) static, dynamic, or a static–dynamic combination and (2) constant or random amplitude, it is necessary to complete the loads model by considering the possible effects of modifiers, in particular, the rates of escalation and frequency.

3.8.2 Force Amplitude

Dynamic-force amplitude is the basic loading parameter to be described. The strain (ε) and internal restoring forces in a member are functions of the applied force. According to Swanson [41], merely determining the history of load reversals or straining does not provide enough information. The coupling of reversals to form ranges of loading and the mean value of adjacent reversals (the mean load) work together to create the reversal grouping. Force intensity (with other considerations) determines whether yield or fracture of a

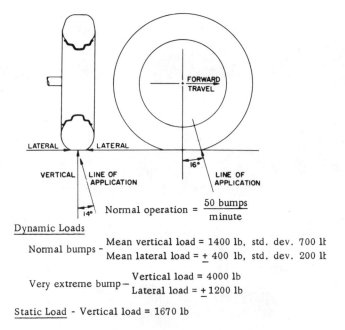

Normal operation = $\dfrac{50 \text{ bumps}}{\text{minute}}$

Dynamic Loads

Normal bumps – Mean vertical load = 1400 lb, std. dev. 700 lb
Mean lateral load = ± 400 lb, std. dev. 200 lb

Very extreme bump – Vertical load = 4000 lb
Lateral load = ± 1200 lb

Static Load - Vertical load = 1670 lb

Figure 3.23 Loads on front wheels of a windrower.

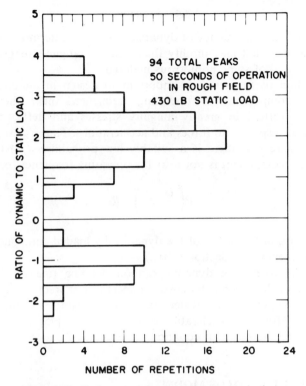

Figure 3.24 Peak vertical drawbar loads in a grain dryer.

component will occur during a single monotonic loading and whether fatigue failure will occur as a result of dynamic loading.

3.8.3 Escalation Rates

Monotonic load escalation rates are a logical part of load definition because the effects modify materials behavior (Chapter 8). Figure 8.6 indicates that both yield (S_y) and ultimate tensile (S_u) strengths are influenced by monotonic strain rate. Thus monotonic load definition may be incomplete without a specification of the escalation rate.

Mean endurance limit $\bar{S}_e$ of ferrous alloys, the measure of resistance to constant amplitude sinusoidal loading, possesses an approximate empirical relationship with mean tensile ultimate strength $\bar{S}_u$, the measure of resistance to monotonic loading. If the empirical relationship $\bar{S}_e \simeq 0.5 \bar{S}_u$ holds in general, the endurance limit may be strain-rate sensitive.

Shock or impulse load peaks are often part of the mix in a dynamic loads spectrum (both constant and random amplitude). It may be argued that shock loads should be part of the makeup of simulated random loads spectra (and included in the estimation of random-load statistics).

3.8.4 Frequency

In Figure 8.35*a, b* frequency of dynamic loading is deterministically shown to be a parameter that may modify the (random) fatigue strength of ferrous materials [44] and of some nonferrous materials.

Consideration of frequency is important to design in another way. The response of components to operating frequencies near their resonant frequencies is (often) to greatly magnify stresses and deflections beyond expected linear responses to forces and accelerations. Thus a design consideration is to assure that there is a suitably low probability (R) that natural frequency of a component is not near the operating frequency, or that

$$P\left(\frac{\omega}{\omega_n} > 3\right) \geq R \qquad (a)$$

where ω_n is natural frequency of the dynamically loaded component and ω is operating frequency (see Section 6.10.2).

In problems where the dynamic random loading (measured as forces, accelerations, or displacements) covers a spectrum of frequencies (narrow or broad band) the power spectral density, for zero mean load functions, yields the variance of force or acceleration or displacement, [45]. In this book it is assumed in the examples that equation (a) is satisfied.

3.9 UNIFYING THE LOADS MODEL

Static, dynamic, and shock phenomena are not unrelated. Actually, monotonic loading is a special case of dynamic loading, as implied in Table 3.1, and the shock effect may in some cases be treated as part of a dynamic model. Many effects due to external loads can be described in terms of peak-force amplitude, escalation rate, and frequency. The entire spectrum of loading conditions may be considered as an integrated family of phenomena, with probabilistic considerations needed to complete the picture.

Static loading has been viewed as $\frac{1}{4}$ cycle of dynamic loading [47], with zero repetitions. The integration of static with dynamic loading has been tacitly implied in the conventional S–N diagram where static stress at failure S_u anchors the curve on the ordinate. In the modified Goodman diagram or

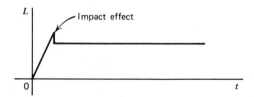

Figure 3.25 Loading constant in time.

Complete Profile of Loading 117

the Gerber parabola, static stress at failure S_u anchors the curve on the abscissa [46]. Both representations imply dynamic loading environments in establishing limits on (load) stress in design.

The shock or impulse manifestation depends only on a sufficiently rapid rate of force buildup. It need not always be considered a specie of loading, since the effect may be associated with static, fully reversed sinusoidal, repeated (and the special case of a single impulse), fluctuating, or random loading (Figure 3.25).

RANDOM PROCESSES

3.10 COMPLETE PROFILE OF LOADING

Until recently, load modeling has been highly idealized. With the development of probabilistic design theory, loading has come to be described by random variables, and to this point in the text loading variability has been discussed in terms of statistics developed from a single sample of data. Such modeling has involved unstated assumptions. In dynamic loading it has been

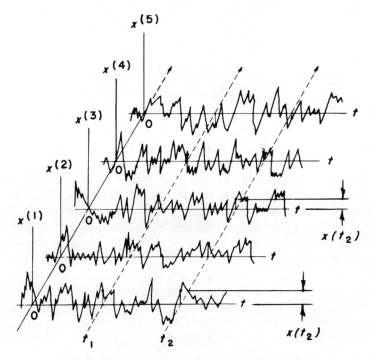

Figure 3.26 Load histories from a vehicle fleet [41].

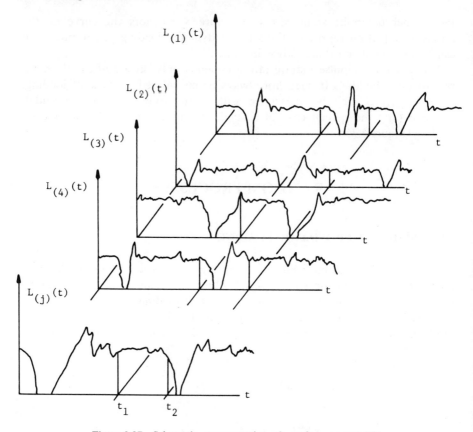

Figure 3.27 Schematic representation of random process $L(t)$.

tacitly assumed that a function developed from a single sample was representative of all possible sample functions. With static loading it has been assumed that the probability density functions at any times t_1 and t_2 (Figure 3.26) were the same (called *stationary*).

The notion of a random process $X(t)$ provides the generalizing concept in loads modeling. It requires the consideration of many possible load-cycle records or samples, continuous (Figure 3.27) or discrete (Figure 3.28). Each sample of loading values is used to generate a sample function, described by a probability density function of $f_i(x)$. With certain random processes (see Section 3.13), a single-sample function does completely define the loading process.

Loading being in general a random process, the models (which are probability density functions) provide complete profiles at any time (t) or cycle (c) [38, Section 5.4.3]. There is a large amount of activity at present directed toward the practical application of random processes in situations where life predictions must be backed by sound experimentation.

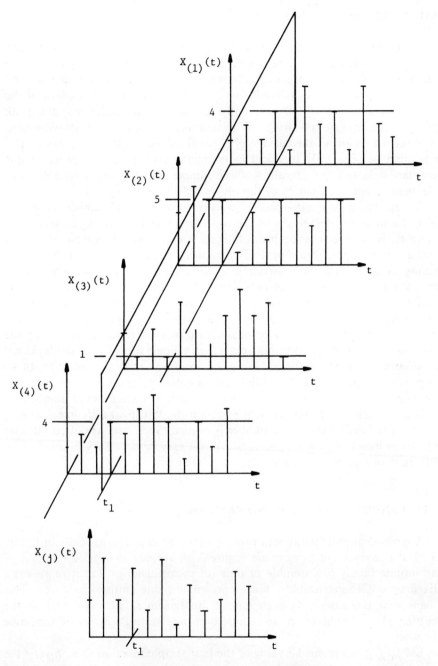

Figure 3.28 Discrete random process.

3.11 LOAD RECORD ENSEMBLES

The concept of randomness includes the notion that in addition to a given record or the results from a test or measurement series, the totality of possible samples of results that might equally well have been produced under the test conditions should be considered. For example, a record could consist of the tensile yield values of 50 nominally identical steel specimens, the peak pressure readings of 100 firings of an automatic weapon, or a 20-sec strain-time record taken on the surface of a panel subject to jet motor noise. The repetition of nominally identical experiments, such as those earlier in this chapter, yield samples of values whose sample statistics are invariably nonidentical; hence multiplicity of sample records is advantageous.

An approximately *deterministic process* (producing a number of nearly identical load samples) might be the constant-amplitude loading of specimens in an R. R. Moore rotating beam fatigue machine [47]. A process of narrow scatter might result from repeatedly measuring the energy released by a drive spring actuating the bolt carrier in a semiautomatic weapon. The energy released by a spring in each of a number of cycles involves small random variations. The statistics of two or more series would, however, probably be very similar. When all conditions under control of the experimenter are maintained the same and the records continually differ from each other, the process is said to be random. In such cases a single record (or sample) is not as meaningful as the statistical variation from one record to another. In an engineering application an individual sample or function (from a given ensemble) may reflect time, cycles, test number, or measurement number.

If an ensemble is to reflect with precision the statistical differences among the sample functions, a large number is needed. Such data are often difficult to obtain in actual design situations. Thus it may be necessary to proceed on the basis of a priori inferences.

3.12 RANDOM VARIABLES AND RANDOM PROCESSES

A typical continuous random process $L(t)$ is shown schematically in Figure 3.29. It is convenient to examine Figure 3.29, keeping in mind the notion of an infinite family or ensemble of time (or cycle) histories. Let t_1 represent a fixed time. Geometrically, t_1 may represent a plane cutting the t axis. This plane simultaneously cuts each sample function at t_1. In loads analysis the cutting plane would often be required to pass through peaks of the same number.

Let $L_{(1)}(t_1)$ represent the value of the first sample function at t_1, $L_{(2)}(t_1)$ the value of the second sample function at t_1, and in general $L_{(j)}(t_1)$ the value of the jth sample function at t_1. This process is continued until the entire ensemble has been evaluated at t_1. Let $L(t_1)$ represent the collection of all possible values obtained from evaluating the entire ensemble at t_1. The value $L(t_1)$ may be considered a random variable.

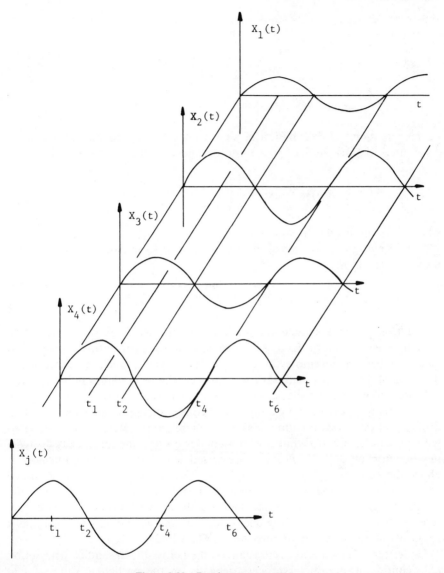

Figure 3.29 Random process $X(t)$.

Loading behavior may be defined from an infinite ensemble of sample functions (finite in the example), and a probability value may be assigned to each possible outcome. This procedure defines a random variable $X(t_1)$. The same procedure could be carried out for any other t, defining a second random variable $X(t_2)$, and so on.

If the distribution of instantaneous load values at any arbitrary time t is normal for a given random process, the distribution of peak values has been found to be Rayleigh distributed, typical of the loading on many automotive and aeronautical structures; thus $X(t) \sim$ Gaussian [i.e., $N(0, \sigma_x^2)$]. The probability density function is

$$f(x) = \frac{1}{\sqrt{2\pi}\,\sigma_x} \exp\left[-\frac{1}{2}\left(\frac{x}{\sigma_x}\right)^2\right] \quad (-\infty \leq x \leq \infty)$$

$\sigma_x = RMS$ value

In this situation the distribution of peak values Y is approximately Rayleigh distributed (defined in Section 2.3 and discussed in the literature [45]).

$$f(y) = \frac{y}{\sigma_y^2} \exp\left[-\frac{1}{2}\left(\frac{y^2}{\sigma_y^2}\right)\right] \quad (0 \leq y \leq \infty)$$

where σ_y^2 is the Rayleigh parameter.

If the Rayleigh parameter σ_y^2 must be estimated from data (e.g., $l_1, l_2, \ldots, l_n$), the estimator σ_l^2 is

$$\sigma_l^2 = \frac{l_1^2 + l_2^2 + \cdots + l_n^2}{2n}$$

where $\sigma_l^2 \approx \sigma_y^2$. For a sample application involving Rayleigh distributed peak loading, see Chapter 10, Example N. In general, the distribution of peak load values across an ensemble (at corresponding cycles) must be determined. Swanson [41] discusses methods for measuring the peak values in a sequence. Schemes for counting cycles are discussed in Chapter 8 [48].

A fluctuating force is an example of a continuous-parameter-load random process, where an alternating tension–compression increment is associated with a mean tensile load. Fatigue strength decays as a monotonically decreasing function of cyclic exposure to external loads, if the stress intensity is above the endurance limit. A multiplicity of such experimentally determined functions will describe this nonstationary random process. In many nominally identical components of the same material, the strength random process over time tends to be monotonically decaying, continuous, and nonstationary, as a result of the random rate of corrosion (Figure 8.45).

Under the most general circumstances, the across-the-ensemble distribution of a continuous-parameter process depends on the particular value of t at which the random variable is defined. For example, the across-the-ensemble average force exerted by new springs for a given deflection is usually higher than the average force exerted by the same springs after several years of service.

3.13 STATIONARY AND ERGODIC PROCESSES

Analysis is simplified if it can be assumed that a random process is stationary, analogous to the assumption of steady-state forced vibration. This assumption, although often not strictly true, may provide useful engineering approximations when suitably interpreted.

A random process is *stationary* if its across-the-ensemble probability distributions are invariant during a shift along the t or the c axis (Figure 3.30). The probability density function at $\tau = (0+\varepsilon)$ also applies 10 min (or cycles) later or 300 min (or cycles) later. This assumption implies that for a given process, the probability density $p(L)$ is universally independent of time, cycles, or other measures. For example, the distribution of static ultimate strength is essentially independent of time. As a consequence, all statistical

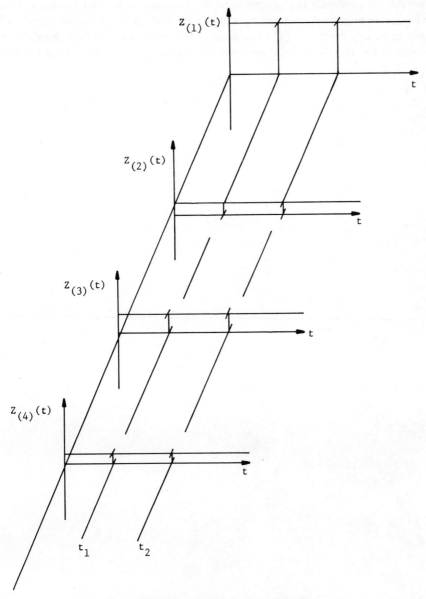

Figure 3.30 Random process $Z(t) = Z$.

parameters based on $p(L)$ are constants independent of time, that is, the mean $E(L)$ and the variance $V(L)$.

In the strict sense, if a process is stationary, it can have no beginning and no end. Each sample must extend from $t = -\infty$ to $t = \infty$ [34]. Real processes in engineering do start and stop. Nonstationary effects associated with starting and stopping may often be neglected if the time of stationary operation is long compared with the starting and stopping intervals. If changes in the statistical properties vary slowly with time, it is sometimes possible to subdivide the process into subprocesses of shorter duration. Each subprocess may be considered approximately stationary. A random process is said to be *ergodic* if it is (1) a stationary random process and (2) the statistics across the ensemble and the statistics of all sample functions are identical, in the limit.

CHAPTER 4

Describing Component Geometry with Random Variables

The following topics are considered in this chapter:

Impact of geometric variability on performance.
Geometry of a random process.
Characteristics of geometry.
Determination of dimensions, tolerances, and limits.
Tolerances on finished products.
Linear and nonlinear dimensional combinations.
Tolerancing by computer (Monte Carlo simulation).
Geometry versus strength.

4.1 INTRODUCTION

Component and system dimensions such as lengths; widths; thicknesses; hole diameters; distances between hole and pin centers; and distances between faces, depths, and other physical features comprise one class of variables of prime importance in mechanical design. Dimensionally described geometry and especially random geometric variations (illustrated in Figure 4.1) of comparable features from component to component, impose sizable direct influences on (1) failure frequency in a component population and (2) probability of failure of a component selected at random from among many.

Statistical dimensional descriptions provide an opportunity to account for machining and processing variability in the design process itself and estimation of the impact of dimensional variability on the performance of a population of like mechanical components has now become a reality. Parameters to account quantitatively for these effects appear in probabilistic design equations. Dimensions are modeled by statistics, and the variability measure

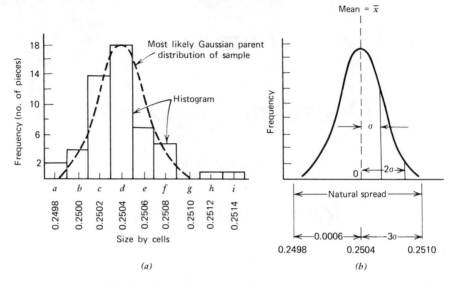

Figure 4.1 Distribution of size of $\tfrac{1}{4}$-in. reamed holes [65].

of each is treated as a design parameter; thus impact of dimensional variability on the failure probability is reflected in design calculations.

Geometric considerations appear as requirements for the design of producible mechanical components because of [35, 49] (1) relative size relationships necessary for the fit of mating parts and for the assembly of systems of components and (2) size requirements bearing on the discharge of functions dependent on strength or stiffness. These conditions must be simultaneously realized through the production processes utilized in manufacture.
It is to be noted that:

1. The fraction of components that will be strengthwise unacceptable may increase with tolerance. The failure probability of a population of mechanical components is a function of the magnitude of geometric scatter and hence of geometric tolerance.

2. Geometric random variables, each with a unique distribution, affect the distributional characteristics of load-induced stress.

Traditionally, given a dimensional design specification (e.g., 2.56 in. ± 0.025 in.),* the view has been that all of the components produced should fall inside the limits. However, dimensional value is a multivalued random phenomenon for which bounds can be established such that no more than a given fraction of components, in the long run, should fall outside the upper and lower specification limits.

---

*To convert from inches to centimeters, multiply by 2.54.

DIMENSIONAL MAGNITUDE—A RANDOM PROCESS

Random processes were discussed in Section 3.12, and it can be argued that the concepts apply for mechanical geometry. If a given geometric feature of a typical mechanical product is studied quantitatively, it is found that any sample of the product provides a random collection of dimensional magnitudes. A large number of samples will result in sample records and statistical parameters, all of which differ in a random manner from each other. Geometric random processes of two kinds can be identified: (1) discrete, when the same feature from one component to another within a sample is considered, and (2) continuous, when for instance thickness over the surface of a metal sheet or along a metal bar must be modeled.

If, as in Chapter 3, the sample statistics of a dimension for a number of samples are identical and are also the same as across-the-ensemble statistics at any sample number, the random process is ergodic, such as casting dimensions. If all across-the-ensemble distributions are identical at any sample number, the random process is stationary. (See Figure 3.30.)

Production processes that employ cutting tools (e.g., as in drilling, milling, turning) are subject to change over time. The same is true of rolling and forging processes. Any dimension such as distance between hole centers, distance between two parallel faces, thickness, and length modified by tool wear results in a geometric random process that is nonstationary. If corrections for tool wear are made periodically, the dimensional values retain the properties of random variables, but the time trend is minimized and it may be possible to consider the random process as approximately ergodic.

Models of geometric random processes provide representative descriptions of random variables to be employed in design and suggest how extensive a testing or measurement program is needed in a particular case to provide the data needed.

4.2 DISTRIBUTIONAL CHARACTERISTICS

1. The determination of the probability density function and required statistic estimators of geometric variables is accomplished by appropriate quantitative studies of the manufacturing process capability and from actual measurement of typical products. If samples are large, goodness-of-fit tests to various distribution models may be carried out.

2. In the absence of directly applicable data, approximations based on previous related studies (or ranges of values to be expected) are used. In design, the problem is sometimes that of estimating, a priori, the geometric variability parameters from process engineering data and tables, as illustrated in Tables 4.13 through 4.16.

128 Describing Component Geometry with Random Variables

Although there are manufacturing processes that produce distributions that are decidedly nonnormal, many are approximately normal.* Many process operations naturally generate normal distributions if they are controlled. Conversely, operations where operators work to the high side of tolerances, for instance, result in distribution that are decidedly skewed. Manually operated processes generate distributions different from those obtained from automatic processes. It is difficult to assure a normal distribution with a mean equal to the nominal specification value when fixed tool setups are involved. A part may be rough machined, milled, and then hardened, in which case heat-treatment effects will cause a shift in the mean dimension and an increase in variability. It is often necessary to machine a part such that it is out of tolerance and then to depend on the heat treatment operation to pull the part within limits.

Production processes never have the capability of exactly reproducing a specified dimension (see Tables 4.13 and 4.14). In a sample some parts will be undersized, many near the average, and still others will be oversized. Dimensional populations possess statistical regularity, however, and as n increases without bound, a sample approaches an exact representation of the underlying random variable; due to the law of large numbers.

4.3 DIMENSIONAL DETERMINATION

If a single item is selected at random from a stream of production and a specified dimension measured, the measured value is comprised of a geometric element and an error element. Production processing, including human influences, accounts for the geometric value. The error element arises from random measuring variability, including a variable human factor illustrated in Figure 4.2. Measuring variability should be determined and removed from the description of dimensional value (see Figure 4.5).

If many items are measured for a specific dimension d, the result is a collection of values:

$$d_1, d_2, d_3, \ldots, d_n$$

from which

$$\bar{d} \approx \frac{d_1 + d_2 + \cdots + d_n}{n}$$

$$s_d \approx \left[\frac{\sum_{i=1}^{n}(d_i - \bar{d})^2}{n} \right]^{1/2}$$

---

*See Bayes decision procedures in Section 3.2.2.

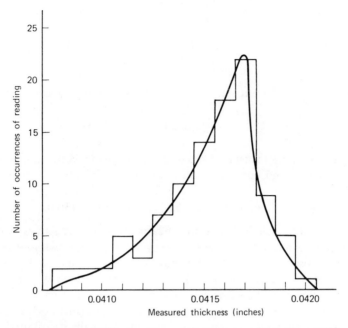

Figure 4.2 Measurement variability (see Figure 4.5). Frequency distribution of 100 micrometer measurements of same dimension made by same experienced inspector. (Reproduced from *Dimensional Control*, published by the Sheffield Corporation.)

Since observed dimensional values include the inherent variability of the item and the variability in the method of measurement, they represent sums of random variables, very likely statistically independent.

According to equations (2.19) and (2.21), the mean of observed values is

$$\overline{\text{observed values}} = \mu^* + \overline{\text{measurement error}}$$

The standard deviation of observed values is

$$s_{\text{observed values}} = \sqrt{s^{2*} + s^2_{\text{measurement error}}}$$

Example A

A determination of the precision of a measurement procedure indicated an error standard deviation of 3.5 min. (milliinches). Observations of the process output indicated that the dimensional standard deviation was 9.5 min. Thus

---
*This is the most representative value.

130 Describing Component Geometry with Random Variables

use of equation (2.21) yields

$$9.5 = \sqrt{s_*^2 + (3.5)^2}$$

The estimator of inherent scatter (with measurement error eliminated) is

$$s_* = 0.0089 \text{ in.}$$

Failure to eliminate measuring circuit error results in overestimation of the standard deviation and hence of natural process variability.

4.3.1 Geometric Tolerance and Tolerance Limits

One must distinguish between specification limits and natural tolerance limits. Specification limits may be set somewhat arbitrarily, without regard to what the process can achieve. Natural tolerance limits state the actual capabilities of a process and are the limits within which all but a given allowable fraction α of the items will (in the long run) fall. Figures 1.3 and 4.3 illustrate values that may be approximately normally distributed. If a dimension in a given design situation is approximately normally distributed, a good design will have equal mean and nominal values and a standard deviation such that only a small allowable fraction α of the parts produced fall outside the specification limits. The natural tolerances should equal or fall within the design specifications. Thus the specification limits should on the average include at least the fraction $1 - \alpha$ of the parts produced.

Example B

If the tolerable number of parts falling outside limits in production is 27 in 10,000 (Appendix 1) and the specifications are 2.560 ± 0.025 in., a process that produces a dimension normally distributed with mean 2.560 and standard

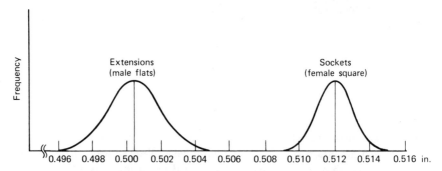

Figure 4.3 Dimensional pdf's probability density functions.

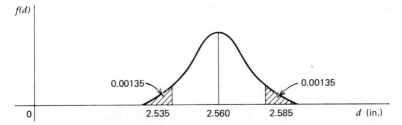

Figure 4.4 Normally distributed dimensional density.

deviation $s \simeq (0.025/3) = 0.0083$ in. will have natural tolerances that coincide with specification limits (see Figure 4.4).

Given symmetrical specifications and knowing the fraction of defectives that can be tolerated (α), the maximum standard deviation that permits the natural tolerances to fall within the specification limits is determined by calculating the number K of standard deviations corresponding to the upper limit

$$K_{0.00135} = \frac{2.585 - 2.560}{0.0083} \approx 3$$

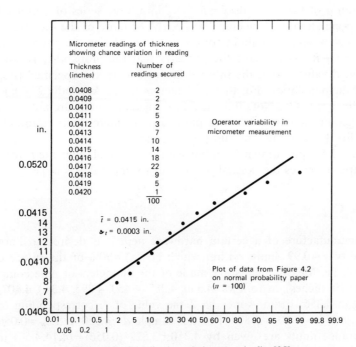

Figure 4.5 Distribution of clearance in fits [35].

where the upper limit is $(2.560+0.025)=2.585$ in. Thus in general:

$$K = \frac{x-\bar{x}}{s_x} \approx \frac{x-\mu_x}{\sigma_x}$$

where μ_x is the population mean, $\bar{x}$ is the mean-value estimator, σ_x is the population standard deviation, and s_x is the standard deviation estimator of the random variable X.

Thus far the discussion has centered around processes assumed to produce normally or approximately normally distributed mechanical component dimensions. Tolerance limits have been defined as those limits within which $100(1-\alpha)\%$ of the product falls. These limits are calculated by adding and subtracting from μ the quantity $K_{\alpha/2} \cdot \sigma$. However, the values μ_x and σ_x are usually unknown, and thus the estimators $\bar{x}$ and s_x must be used.

Where it could be stated that 95% of the time a certain dimension of a manufactured product would lie within $\mu_x \pm 1.96\sigma_x$ (see Appendix 1), such cannot be said of the interval $\bar{x} \pm 1.96 s_x$, as $\bar{x}$ and s_x are themselves random variables. Thus the limits depend on the sample size.

The fraction of components included within $\bar{x} \pm K_{\alpha/2} \cdot s_x$ will not always be $1-\alpha$. However, it is possible to obtain a constant K such that a fixed proportion α of the intervals $\bar{x} \pm K_{\alpha/2} \cdot s_x$ in a large series of samples from a normal population, will include $100(1-\alpha)\%$ or more of the distribution [17]. Thus statistical tolerance limits for a normal population are given by $[L = \bar{x} - K s_x; U = \bar{x} + K s_x]$ and have the property that the probability is equal to a preassigned value α that the interval includes at least a specified proportion $1-\alpha$ of the population. For tolerance factors K, see Appendix 2, $\bar{x} \pm K s_x$ and Appendix 3, $\bar{x} + K s_x$. Note that statistical tolerance limits are not the same as confidence limits, within which a parameter of interest will be found with a probability $1-\alpha$ in the long run (see Appendix 7).

Means of samples drawn from normal parent populations are t-distributed, which approximates the normal model when sample size is large.

Example C

In the manufacture of a certain hardware item it is desired to know with probability $\gamma = 0.99$, limits within which $(1-\alpha) = 90\%$ of the future components will lie. Measurements are made of the diameters of nine components diameter (in inches), as follows: 4.330, 4.287, 4.450, 4.295, 4.340, 4.407, 4.295, 4.388, and 4.356. From Appendix 2 the value of K corresponding to $n=9$, $\gamma = 0.99$, and $\alpha = 0.10$ is 3.822. For this sample $\bar{d} = 4.350$ and $s_d = 0.056$. Thus the tolerance limits are given by $4.350 \pm 3.822 \cdot (0.056) = (4.14, 4.57)$ in. such that with $\nu = 0.99$ no more than 10% will fall outside limits.

It is occasionally useful to specify a single limit $\bar{x} - K_{\hat{s}_x}$ such that a fixed proportion will exceed this limit, or a single limit $\bar{x} + K_{\hat{s}_x}$ such that a fixed proportion will be less than this limit, in which cases Appendix 3 is used.

Example D

According to specification, a manufacturer was required to produce cylinders having a bore with lower limit on diameter that can be assured, with probability $\gamma = 95\%$, that $(1 - \alpha) = 99\%$ of his production be above this limit. A sample of 30 units was taken, and the sample mean and sample standard deviations were $\bar{d} = 9.87$ mm and $\hat{s}_d = 0.059$ mm, respectively. A value of $K = 3.064$ corresponding to $n = 30$, $\gamma = 0.95$, and $\alpha = 0.01$ was obtained from Appendix 3. The required lower tolerance limit was

$$\bar{x} - K_{\hat{s}_x} = 9.87 - (3.064)(0.059) = 9.69 \text{ mm}$$

In addition to the measurable geometric dimensions whose natural tolerances and distributional characteristics arise from special processing, there are many manufactured products that are purchased for use without further processing. Examples are metal sheets and plates, tubes, metal bars, and structural shapes. Dimensional variability in such "vendor-produced" items must often be accepted.

It is frequently necessary to estimate dimensional variability parameters from published tables of manufacturer's tolerances. The nominal dimension is utilized as a mean value estimator, and the ± tolerance range is assumed to approximate a $\pm 3\hat{s}$ interval.

4.3.2 Tolerances on Finished Metal Products

The products described in this section are in forms such as sheets, plates, bars, and tubes. Tables 4.1 through 4.13 provide examples of published tolerance data in the literature.

Table 4.1 Cold-Finished Wire, Rod, and Bar–Aluminum Alloy Tolerances for Rounds, Squares, Hexagons, and Rectangles up to 1.5 in. Thickness or 4 in. Width (Plus or Minus)

Size	Rounds	Squares and Hex.	Rectangles
≤ 0.035	0.0005		0.001
0.036–0.004	0.001		0.0015
0.065–0.500	0.0015	0.002	0.002
0.501–1.000	0.002	0.0025	0.0025
1.001–1.500	0.0035	0.003	0.003
1.501–2.000	0.004	0.005	0.005
1.001–3.000	0.004		0.005
1.001–4.000			0.005

Table 4.2 Rolled Structural Shapes

Dimensions[a]	Tolerance
Thickness of section	Plus or minus $2\frac{1}{2}\%$ of nominal thickness–minimum tolerance, ± 0.010 in.
Overall dimensions length of leg of angles or zees	Plus or minus $2\frac{1}{2}\%$ of nominal–minimum tolerance, $\pm \frac{1}{10}$ in.

Length
  ≤ 20 ft not incl.
  20–30 ft incl.
  > 30 ft
	Minus 0, plus 1/4 in.
	Minus 0, plus 3/8 in.
	Minus 0, plus 1/2 in.

Channels, depth	Plus 3/32 in. minus 1/16 in.
Channels, width of flange	Plus or minus 4% of nominal width
Weight of a lot or shipment of sizes 3 in. or larger	Plus or minus $2\frac{1}{2}\%$ of nominal weight

[a] For sizes under 3 inc. dimension tolerances only apply.

Table 4.3 Rolled-bar Aluminum Alloys: Tolerances for Squares, Hexagons,[a] and Rectangles (Plus or Minus)

Across Flats (in.)	Tolerance	Width of Rectangles (in.)	Tolerance
≤ 0.500	0.006	≤ 1.500	1/64
0.501–0.750	0.003	1.501–4.000	1/32
0.751–1.000	0.012	4.001–6.000	3/64
1.001–2.000	0.016	6.001–10.000	1/10
2.001–4.000	0.020		

[a] Rolled hexagons available only in sizes larger than 1.5 in.; smaller sized cold finished.

Table 4.4 Extruded Rod, Bar, and Shapes; Straightness Tolerances

Circumscribed Circle Diameter (in.)	Thickness (in.)	Allowable Deviation from Straight per Foot of Length
< 1.5	≤ 0.094	0.050
< 1.5	Over 0.004	0.0125
≥ 1.5		0.0125

Table 4.5 Steel Sheet, Carbon, Cold Rolled—Tolerances Thickness[a] Tolerances (Plus or Minus)

Width (in.)	0.2299–0.1875	0.1874–0.1800	0.1799–0.1420	0.1419–0.0972	0.0971–0.0822	0.0821–0.0710	0.0709–0.0568	0.0567–0.0509	0.0508–0.0389	0.0388–0.0344	0.0343–0.0314	0.0313–0.0255	0.0254–0.0195	0.0194–0.0142	0.0141 & less
≤ 3½ incl.												0.003	0.003	0.002	0.002
> 3½ – 6 incl.									0.004	0.004	0.004	0.003	0.003	0.002	0.002
" 6 – 12 "							0.005	0.005	0.005	0.004	0.004	0.003	0.003	0.002	0.002
"12 – 15 "	0.008	0.007	0.007	0.007	0.006	0.006	0.006	0.005	0.005	0.004	0.004	0.003	0.003	0.002	0.002
"15 – 20 "	0.008	0.008	0.008	0.008	0.007	0.007	0.006	0.005	0.005	0.004	0.004	0.003	0.003	0.002	
"20 – 32 "	0.009	0.009	0.009	0.008	0.007	0.007	0.006	0.006	0.005	0.004	0.004	0.003	0.003	0.002	
"32 – 40 "	0.009	0.009	0.009	0.009	0.008	0.007	0.006	0.006	0.005	0.004	0.004	0.003	0.003	0.002	
"40 – 48 "	0.010	0.010	0.010	0.010	0.008	0.007	0.006	0.006	0.005	0.004	0.004	0.003	0.003	0.002	
"48 – 60 "			0.010	0.010	0.008	0.007	0.007	0.006	0.005	0.004	0.004	0.003	0.003	0.002	
"60 – 70 "			0.011	0.011	0.009	0.008	0.007	0.007	0.006	0.005	0.005				
"70 – 80 "			0.012	0.012	0.009	0.008									
"80 – 90 "			0.012	0.012	0.010										
"90			0.012	0.012											

[a] Thickness is measured at any point on the sheet not less than 3/8 in. from an edge.

Table 4.5a Aluminum and Aluminum Alloy Sheet and Plate—Tolerances

Thickness Tolerances (Plus or Minus)
Grades: Alclad 14S; 24S, Alclad 24S, 52S, 61S, 75S and Alclad 75S

Thickness (in.)	≤18	>18–36	>36–48	>48–54	>54–60	>60–66	>66–72	>72–78	>78–84	>84–90	>90–96	>96–120
0.007–0.010	0.001	0.0015										
0.011–0.017	0.0015	0.0015										
0.018–0.028	0.002	0.002	0.0025									
0.029–0.036	0.002	0.002	0.0026									
0.037–0.045	0.0025	0.0025	0.003	0.004								
0.046–0.068	0.003	0.003	0.004	0.005	0.005							
0.069–0.076	0.003	0.003	0.004	0.005	0.006	0.006						
0.077–0.096	0.0035	0.0035	0.005	0.005	0.006	0.006	0.007					
0.097–0.108	0.004	0.004	0.005	0.005	0.007	0.008	0.008	0.009				
0.109–0.140	0.0045	0.0045	0.005	0.006	0.007	0.010	0.010	0.011	0.012			
0.141–0.172	0.006	0.006	0.008	0.008	0.009	0.012	0.013	0.014	0.014	0.016	0.018	
0.173–0.203	0.007	0.007	0.010	0.009	0.011	0.014	0.015	0.016	0.016	0.017	0.018	
0.204–0.249	0.009	0.009	0.011	0.011	0.013	0.016	0.017	0.017	0.017	0.018	0.019	0.020
0.250–0.320	0.013	0.013	0.013	0.013	0.015	0.018	0.018	0.018	0.018	0.019	0.022	0.020
0.321–0.438	0.019	0.019	0.019	0.019	0.020	0.020	0.020	0.020	0.020	0.022	0.024	0.023
0.439–0.625	0.025	0.025	0.025	0.025	0.025	0.023	0.023	0.025	0.025	0.025	0.026	0.026
0.626–0.875	0.030	0.030	0.030	0.030	0.030	0.030	0.030	0.030	0.030	0.035	0.035	0.035
0.876–1.125	0.035	0.035	0.035	0.035	0.035	0.035	0.037	0.037	0.037	0.045	0.045	0.045
1.126–1.375	0.040	0.040	0.040	0.040	0.040	0.040	0.045	0.045	0.052	0.052	0.055	0.055
1.376–1.625	0.045	0.045	0.045	0.045	0.045	0.045	0.052	0.052	0.060	0.060	0.065	0.065
1.626–1.875	0.052	0.052	0.052	0.052	0.052	0.052	0.060	0.060	0.070	0.070	0.075	0.075
1.876–2.250	0.060	0.060	0.060	0.060	0.060	0.060	0.075	0.080	0.080	0.080	0.088	0.088
2.251–2.750	0.075	0.075	0.075	0.075	0.075	0.075	0.090	0.100	0.100	0.100	0.100	0.100
2.751–3.000	0.090	0.090	0.090	0.090	0.090	0.090	0.120	0.120	0.120			

Thickness Tolerances (Plus or Minus)
Grades: 2S and 3S

Thickness (in.)	≤18	>18–36	>36–54	>54–72	>72–90	>90–102	>102–132
0.005–0.007	0.001	0.001	0.002				
0.008–0.010	0.001	0.0015	0.0025				
0.011–0.017	0.0015	0.0015	0.0025	0.0035			
0.018–0.028	0.0015	0.002	0.003	0.004			
0.029–0.036	0.002	0.002	0.0025	0.004	0.005		
0.037–0.045	0.0025	0.0025	0.003	0.005	0.006		
0.046–0.068	0.003	0.003	0.004	0.005	0.006	0.007	
0.069–0.076	0.003	0.003	0.004	0.006	0.007	0.008	
0.077–0.096	0.003	0.004	0.005	0.006	0.008	0.009	
0.097–0.108	0.004	0.005	0.005	0.007	0.009	0.010	
0.109–0.140	0.0045	0.004	0.005	0.007	0.009	0.010	
0.141–0.172	0.006	0.006	0.008	0.009	0.011	0.012	
0.173–0.203	0.007	0.007	0.009	0.011	0.013	0.015	
0.204–0.249	0.009	0.009	0.011	0.013	0.015	0.017	
0.250–0.320	0.013	0.013	0.013	0.015	0.017	0.020	
0.321–0.438	0.019	0.019	0.019	0.019	0.023	0.026	0.026
0.439–0.625	0.025	0.025	0.025	0.025	0.030	0.035	0.035
0.626–0.875	0.030	0.030	0.030	0.030	0.037	0.045	0.045
0.876–1.125	0.035	0.035	0.035	0.035	0.045	0.055	0.055
1.126–1.375	0.040	0.040	0.040	0.040	0.052	0.065	0.065
1.376–1.625	0.045	0.045	0.045	0.045	0.060	0.075	0.075
1.626–1.875	0.052	0.052	0.052	0.052	0.070	0.088	0.088
1.876–2.250	0.060	0.060	0.060	0.060	0.080	0.100	0.100
2.251–2.750	0.075	0.075	0.075	0.075	0.100	0.120	
2.751–3.000	0.090	0.090	0.090	0.090	0.120	0.150	

Table 4.6 Aluminum and Aluminum Alloy Tubing—Tolerances

Round Drawn Tubing
Diameter Tolerances (Plus or Minus)

Nominal Diameter (in.)	Mean Diameter[a] 3S, 4S, 24S, 52S, 61S	Individual Measurement of Diameter (Out of Roundness) Except Soft O, or Thin Wall Tubes[b]	
		3S, 4S, 52S	24S, 61S
0.125– 0.500	0.003	0.003	0.006
0.501– 1.00	0.004	0.004	0.008
1.01 – 2.00	0.005	0.005	0.010
2.01 – 3.00	0.006	0.006	0.012
3.01 – 5.00	0.008	0.008	0.016
5.01 – 6.00	0.010	0.010	0.020
6.01 – 8.00	0.015	0.015	0.030
8.01 –10.00	0.020	0.020	0.040
10.01 –12.00	0.025	0.025	0.050

Round Drawn Tubing
Wall Thickness Tolerances (Plus or Minus)

Nominal Wall Thickness (T) (in.)	Mean Wall[c] Thickness 24S, 61S	Individual Measurements of Wall Thickness	
		24S, 61S	3S, 4S, 52S
0.010–0.035	0.002	10% of T	0.002
0.036–0.049	0.003	10% of T	0.003
0.050–0.120	0.004	10% of T	0.004
0.121–0.203	0.005	10% of T	0.005
0.204–0.300	0.008	10% of T	0.008
0.301–0.375	0.012	10% of T	0.012
0.376–0.500	0.032	10% of T	0.032

[a] Mean dia. is average of any two measurements of dia. taken at right angles to each other at any point along the length of tube.

[b] Thin wall tubes, that is, tubes having a wall thickness of $<2.5\%$ of diameter or <0.020 in. and tubes in the soft (0) temper shall be commercially round. The deviations of individual measurements from the nominal will vary with the alloy and the ratio of wall thickness to diameter.

[c] Mean wall thickness is average of thickness two measurements taken at 120° from each other.

Table 4.6 Continued
Round Tubing and Pipe
Straightness Tolerances

Outside Diameter (in.)	Tolerance
0.375–12.00	0.1 in. in 10 ft or 1 part in 1200 parts of length. Tubing in the soft temper or in diameters less than 3/8 inch is supplied commercially straight, substantially free from kinks and short bends

Pipe Diameter Tolerances

For nominal dia. of $\frac{1}{8}$–$1\frac{1}{2}$ in. incl., the tolerance on the OD is $+1/64$ inch and $-1/32$ in.
For nominal dia. of ≥ 2 in., the tolerance on the OD $\pm 1\%$.

Drawn Tubing and Pipe
Length Tolerances (Plus Only)

Nominal Diameter (in.)	Lengths ≤ 2 ft.	Lengths >2–20 ft.	Lengths >20–30 ft.	Lengths >30 ft.	Coiled Tubing
≤ 0.250	1/8	1/4	3/8	1/2	3%
0.25– 2.00	1/16	1/8	3/16	3/8	2%
2.01– 3.00	1/8	3/16	1/4	5/16	
3.01–10.00	3/16	1/4	5/16	3/8	
10.01–12.00	1/4	5/16	3/8		

A tolerance of 1/64 per inch of OD or fraction thereof will apply on the squareness of all saw cuts.

Table 4.7 Rolling Tolerances
W F Shapes

Rolling Tolerances

Scales of Permissible Tolerances exaggerated for clarity.

Nominal Depth	Depth, d		Width of Flange, b		Out of Square or Parallel, $m - n$	Maximum Depth at any Point, m	Web off Center
	+	−	+	−			
≤ 12 in. incl.	$\frac{1}{8}$ in.	$\frac{1}{8}$ in.	$\frac{1}{4}$ in.	$\frac{3}{16}$ in.	Not more than $\frac{3}{16}$ in.	Not more than $\frac{1}{4}$ in. over theoretical	Not more than $\frac{3}{16}$ in.
>12 in.	$\frac{1}{8}$ in.	$\frac{1}{8}$ in.	$\frac{1}{4}$ in.	$\frac{3}{16}$ in.	Not more than $\frac{1}{4}$ in.	Not more than $\frac{1}{4}$ in. theoretical	Not more than $\frac{3}{16}$ in.

Shapes may have an allowable variation in weight of $\pm 2\frac{1}{2}\%$ from nominal weight.

Source: [58].

Table 4.7 Continued

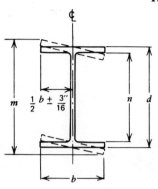 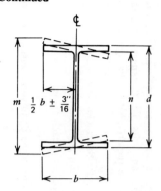

Additional tolerance data are given in Tables 4.14 and 4.15.

4.4. LINEAR DIMENSIONAL COMBINATIONS

Very often mechanical components or component groups must satisfy one of the following dimensional constraints, or some combination of these constraints:

1. Size:

$$y = x_1 + x_2 + \cdots + x_n$$

2. Location:

$$y = x_1 \pm x_2 \pm \cdots \pm x_n$$

3. Clearance:

$$y = x_1 - x_2$$

where $x_1, x_2, \ldots, x_n$ are dimensional random variables.

If the component dimensional distributions of an assembly (which is a linear function) are all normal, the assembly distribution is normal. If the component dimensional distributions are nonnormal and their distribution ranges are approximately homogeneous, then as the number n (of components) increases, the dimensional distribution of linear functions rapidly approaches a normal form (by central limit theorem for sums) [35].

For example, assume that component dimensional characteristics are uniformly distributed. Figure 4.6 shows assembly distributions for $n = 2, 3,$ and 4 rectangular distributions of component dimensions. If n is large, nonhomogeneous components as well as nonnormal component distributions may result

Table 4.8 Permissible Variations in Weight and Thickness of Rolled-steel Structural Shapes and Plates (American Society for Testing Materials: A 7-46)

Permissible Variations

a. One cubic inch of rolled steel is assumed to weigh 0.2833 lb. The cross sectional area or weight of each structural-size shape shall not vary more than 2.5% from the theoretical or specified amounts. The thickness or weights of rectangular sheared mill plates and of universal mill plates shall conform to the requirements of Paragraphs b, c, and d.

b. Plates, when ordered to thickness. No plate shall vary more than 0.01 in. under the thickness specified. The overweight in each lot^a of plates in each shipment shall not exceed the amount given in Table I.

c. Plates, when ordered to weight per square foot. The weight of each lot^a of plates in each shipment shall not vary from the weight ordered more than the amounts given in Table II.

d. Plates over 2 in. in thickness. Each plate over 2 in. in thickness shall conform to the permissible variations over ordered thickness given in Table III.

Permissible Overweights of Plates ≤2 in. in Thickness When Ordered to Thickness

Permissible Excess in Average Weight of Lots for Widths Given, in inches, Expressed in Percentage of Nominal Weights

Specified Thickness (in.)	≤48	Over 48–60, excl.	60–72, excl.	72–84, excl.	84–96, excl.	96–108, excl.	108–120, excl.	120–132, excl.	132–144, excl.	144–168, excl.	>168
$\frac{1}{8}-\frac{1}{4}$, excl.	—	8	9	10	12	14	16	18	19		
$\frac{1}{4}-\frac{5}{16}$, "	6	7	8	9	10	12	14	16	17	18	
$\frac{5}{16}-\frac{3}{8}$, "	5	6	7	8	9	10	12	14	15	16	18
$\frac{3}{8}-\frac{7}{16}$, "	4.5	5	6	7	8	9	10	12	13	14	16
$\frac{7}{16}-\frac{1}{2}$, "	4	4.5	5	6	7	8	9	10	11	12	14
$\frac{1}{2}-\frac{5}{8}$, "	4	4	4.5	5	6	7	8	9	9	10	12
$\frac{5}{8}-\frac{3}{4}$, "	4	4	4	4.5	5	6	7	8	8	9	11
$\frac{3}{4}-1$, "	3.5	4	4	4	4.5	5	6	7	7	8	9
1–2, incl.	3.5	3.5	4	4	4	4.5	5	6	6	7	8

Table 4.9 Cold-finished Carbon Steel Bars or Turned and Polished Rounds, Size Tolerance[a]

Diameter, (in.)	Maximum of Carbon Range ⩽ 0.28%	Maximum of Carbon Range > 0.28–0.55% incl.	Maximum of Carbon Range to 0.55%, incl. Stress Relieved or Annealed after Cold Finishing	Maximum of Carbon Range > 0.55% or All Grades Quenched and Tempered or Normalized and Tempered before Cold Finishing
	All tolerances are in inches and are minus			
⩽ $1\frac{1}{2}$ incl.	0.002	0.	0.004	0.005
> $1\frac{1}{2}$–$2\frac{1}{2}$ incl.	0.003	0.004	0.005	0.006
> $2\frac{1}{2}$–4, incl.	0.004	0.005	0.006	0.007
> 4–6, incl.	0.005	0.006	0.007	0.008
> 6–8, incl.	0.006	0.007	0.008	0.009
> 8–9, incl.	0.007	0.008	0.009	0.010
> 9	0.008	0.009	0.010	0.011

[a]This table includes tolerances for bars that have been annealed, spheroidize annealed, normalized, normalized and tempered, or quenched and tempered before cold finishing. This table does not include tolerances for bars that are spheroidize annealed, normalized, normalized and tempered, or quenched and tempered after cold finishing; the producer should be consulted for tolerances for such bars.

Table 4.9a Cold-finished Carbon Steel Bars Cold Drawn, Ground, and Polished Rounds Size Tolerances (All Tolerances Are in Inches and Are Minus)

Diameter (in.)	Size Tolerance
⩽ $1\frac{1}{2}$, incl.	0.001
> $1\frac{1}{2}$–$2\frac{1}{2}$, excl.	0.0015
$2\frac{1}{2}$–3, incl.	0.002
> 3–4, incl.	0.003

Table 4.9b Cold-finished Carbon Steel Bars Turned Ground, and Polished Rounds Size Tolerances (All Tolerances are in Inches and Are Minus)

Diameter, (in.)	Size Tolerance
⩽ $1\frac{1}{2}$, incl.	0.001
> $1\frac{1}{2}$–$2\frac{1}{2}$, excl.	0.0015
$2\frac{1}{2}$–3, incl.	0.002
> 3–4, incl.	0.003
> 4–6, incl.	0.004[a]
> 6	0.005[a]

[a]For nonresulfurized steels (steels specified to maximum sulfur limits under 0.08%) or for steels thermally treated, the tolerance is increased by 0.001 in.

Table 4.10 Permissible Variations in Diameters and Wall Thicknesses of Round Cold-drawn Seamless Steel Mechanical Tubing

Group	Size OD (in.)	Permissible Variations From					
		OD (in.) >	OD (in.) <	ID (in.) >	ID (in.) <	Wall Thickness (%) >	Wall Thickness (%) <
1	$\frac{3}{16}-\frac{1}{2}$ excl. b	0.004	0	—	—	15	15
2	$\frac{1}{2}-1\frac{1}{2}$ excl. a, b, c	0.005	0	0	0.005	10	10
3	$1\frac{1}{2}-3\frac{1}{2}$ excl. a, b, c	0.010	0	0	0.010	10	10
4	$3\frac{1}{2}-5\frac{1}{2}$ excl. a, c	0.015	0	0.005	0.015	10	10
5	$5\frac{1}{2}-8$ excl. c when wall is <5% of OD	0.030	0.030	0.035	0.035	10	10
6	$5\frac{1}{2}-8$ excl. when wall is 5–7.5% of OD	0.020	0.020	0.025	0.025	10	10
7	$5\frac{1}{2}-8$ excl. a when wall is over 7.5% of OD	0.030	0	0.015	0.030	10	10
8	$8-10\frac{3}{4}$ incl. c when wall is <5% of OD	0.045	0.045	0.050	0.050	10	10
9	$8-10\frac{3}{4}$ incl. when wall is 5–7.5% of OD	0.035	0.035	0.040	0.040	10	10
10	$8-10\frac{3}{4}$ incl. a when wall is over 7.5% of OD	0.045	0	0.015	0.040	10	10

Table 4.11 Permissible Variations in Outside Diameters, Inside Diameters, and Wall Thicknesses of Round Hot-finished Steel Mechanical Tubing

Group	Specified Size, OD (in.)	Ratio of Wall Thickness to OD	Permissible Variations From Diameter and Wall Thickness					
			OD (in.) >	OD (in.) <	ID (in.) >	ID (in.) <	Wall Thickness (%) >	Wall Thickness (%) <
1	<3	All wall thicknesses	0.023	0.023	0.031	0.031	12.5	12.5
2	$3-5\frac{1}{2}$ excl.	All wall thicknesses	0.031	0.031	0.030	0.030	12.5	12.5
3	$5\frac{1}{2}-8$ excl.	All wall thicknesses	0.047	0.047	0.047	0.047	12.5	12.5
4	$8-10\frac{3}{4}$ incl.	≥5% and over	0.047	0.047	0.003	12.5	12.5	
5	$8-10\frac{3}{4}$ incl.	<5%	0.063	0.063	Work	to wall	12.5	12.5

Source: [51].

Table 4.12 Tolerances on Spring Wire Diameter (Inches), Plus and Minus

Diameter, in.	Music Wire	Diameter, in.	Hard-Drawn, Oil-Tempered	Diameter, in.	Alloy Steel, Valve Spring
Up to 0.026	0.0003	Up to 0.075	0.001	Up to 0.148	0.001
0.027 to 0.063	0.0005	0.076 to 0.375	0.002	0.149 to 0.177	0.0015
0.064 to 0.250	0.0010	0.376 and up	0.003	0.178 to 0.375	0.002
				0.376 and up	0.003

Source: *Manual on Design of Helical and Spiral Springs*, Society of Automotive Engineers, Report No. J795, New York, July 1962, Suppl. 9.

Table 4.13 Tolerances on Coil Diameter, Free Length, Load and Load Rate, Cold-Wound Compression, or Extension Springs.

Coil Diameter Tolerance, in., ±			Free-Length Tolerance, in., ±			Load and Load Rate Tolerance		
Mean Coil Diameter, in.	$D/d=$ 3–7.9	$D/d=$ 8–15	Free Length, in.	$D/d=$ 3–7.9	$D/d=$ 8–15	No. Active Coils	Load Tol. %, ±	Rate Tol. %, ±
1/8 or less	0.003	0.004	1/2 or less	0.025	0.040	3 or less	15	10
Over 1/8 to 1/4	0.004	0.006	Over 1/2 to 1	0.035	0.060	Over 3 to 9	10	8
Over 1/4 to 1/2	0.006	0.010	Over 1 to 2	0.050	0.080	Over 9 to 15	8	6
Over 1/2 to 1	0.010	0.016	Over 2 to 4	0.080	0.12	Over 15	7	5
Over 1 to 2	0.016	0.025	Over 4 to 8	0.12	0.19			
Over 2 to 4	0.025	0.042	Over 8 to 16	0.22	0.30			
Over 4 to 8	0.042	0.063	Over 16 to 32	0.35	0.45			

in near-normal assembly distributions. Simulation may be utilized to study the nature of assembly dimensional distributions.

A dimension of a machine assembly or of a single component, is often the sum of several random variables. The probability density functions of the dimensions may or may not be known; however, the statistics of y, where $y = (X_1 + X_2 + \cdots + X_n)$, can be estimated. An additional consideration, that of cost versus tolerance, is illustrated in Figure 4.7.

Assume that the mean-value estimators $\bar{x}_1, \bar{x}_2, \ldots, \bar{x}_n$ and variances $s^2_{x_1}, \ldots, s^2_{x_n}$, respectively, are known. Then, regardless of the distribution of each X_1, using equation (2.19):

$$\bar{y} = \bar{x}_1 + \bar{x}_2 + \cdots + \bar{x}_n$$

and using equation (2.21):

$$s^2_y = s^2_{x_1} + s^2_{x_2} + \cdots + s^2_{x_n}$$

Example E

The size constraint is illustrated by Example G in Chapter 2.

Table 4.14 Practical Tolerances and Surface Finishes Obtained by Common Manufacturing Processes

Process	Surface Finish (rms)	Tolerance Good	Tolerance Best obtainable	Remarks
1. Flame cutting	250–2000	±0.060	±0.020	Distortion
2. Sawing	125–1000	±0.020	±0.005	
3. Shaping	63–1000	±0.010	±0.001	
4. Broaching	32–250	±0.005	±0.0005	
5. Milling	32–500	±0.005	±0.001	
6. Turning	32–500	±0.005	±0.001	
7. Drilling (dia.) (loc.)	63–500	$(+0.010-000)L$ ±0.015	$(+0.002-000)S$ ±0.002 (with jig)	($L \geq 1$ in. dia. ($S \leq \frac{1}{2}$ in. dia.
8. Reaming (dia.)	16–125	±0.002	±0.0005	
9. Hobbing	32–250	±0.005	±0.001	
10. Grinding	8–125	±0.001	±0.0002	
11. Lapping	0–16	±0.0002	±0.000,050	50μ
12. Forming (brake) (roll)	Same as rolling Same as rolling	±0.060 ±0.010	±0.015 ±0.005	
13. Stamping	Same as rolling	±0.010	±0.001	
14. Drawing	Same as rolling	±0.010	±0.002	

Table 4.14 Continued

Process	Surface Finish (rms)	Tolerance		Remarks
		Good	Best obtainable	
15. Forging	125–1000	±0.060 (in./in.)	±0.030 (in./in.)	
16. Rolling (cold)	8–32	±0.010	±0.001	
17. Extrusion (hot)	63–250	±0.020	±0.005	
(cold)	8–63	±0.005	±0.001	
18. Sand casting	250–2000	$\pm \frac{1}{8}$ in./ft.	$\pm \frac{1}{32}$ in./ft.	
19. Permanent mold	40–125	±(0.030 + .002D)	±(0.010 + .002D)	
20. Die casting	40–100	±0.010	±0.002 in. in.	
21. Investment casting	60–125	—	±0.002 in. in.	Nonferrous
			±0.004 in./in.	Ferrous
22. Sintered metal	40–100	±0.005	±0.001	
23. Sintered ceramics	30–250	±0.030 in.	±0.020 in. in.	
24. Fusion welding	100–250	$\pm \frac{1}{8}$ in.	±0.010	
25. Spot welding	Same as parent metal	$\pm \frac{1}{16}$	±0.010	
26. Heat treating	Same as parent metal	±0.030	±0.010	Grind stock

Source: J. Datsko [51].

Table 4.15 Limiting Tolerances for Various Gear-manufacturing Processes

Process	Assumed size of Part	Degree[a] of Care	Tooth-to-tooth Spacing	Concentricity Full Indicator	Profile Inv. Var.	Helix	Composite Check Total Tooth Error	Tooth Thickness	Tooth Surface Finish
Sintering	0–1½ in. dia.								
Sintering	0–¾ in. face width	Commercial	0.0007	0.0025	0.0005	0.0007	0.003	0.001	45
Sintering	10–32 DP								
Sintering	1½–4 in. dia.								
Sintering	0–1 in. face width	Commercial	0.001	0.004	0.001	0.0007	0.004	0.002	45
Sintering	10–20 DP								
Stamping	0–1 in. dia.								
Stamping	0–1/16 in. face width	Commercial	0.001	0.003	0.001	0.0005	0.004	0.001	128
Stamping	20–128 DP								
Extruding	0–½ in. dia.								
Extruding	0–¼ in. face width	Commerical	0.001	0.005	0.002	0.001	0.007	0.001	64
Extruding	20–128 DP								
Hobbing	0–6 in. dia.	Commercial	0.0005	0.003	0.001	0.001	0.0035	0.002	63
Hobbing	0–1 in face width	Precision	0.0003	0.002	0.0005	0.0005	0.0025	0.001	32
Hobbing	20–12S DP	Very best	0.0002	0.001	0.0004	0.0003	0.0012	0.0005	16
Grinding	0–8 in. dia.	Commercial	0.0005	0.002	0.001	0.0005	0.0025	0.0015	40
Grinding	0–1 in. face width	Precision	0.0003	0.0012	0.0004	0.0003	0.0015	0.001	25
Grinding	8–64 DP	Very best	0.0002	0.0006	0.0003	0.0001	0.001	0.0003	13

[a] Commercial—quality abtainable in good job shape.
Precision—quality obtainable with new or first-class machine tools.
Very best—quality at 25 to 50% extra cost.

Source: N. P. Chironis, Gear Design and Applications, McGraw-Hill, New York, 1967.

Table 4.15a Surface Finish and Applicable Dimension Tolerance

Symbol	Quality class	Description	Maximum RMS value (in.)	Suitable range of total tolerance	Typical fabrication methods	Approximate relative cost to produce
1000√	Extremely rough	Extremely crude surface produced by rapid removal of stock to nominal dimension	1000	0.063–0.125	Rough sand casting, flame cutting	1
500√	Very rough	Very rough surface unsuitable for mating surfaces	500	0.015–0.063	Sand casting, contour sawing	2
250√	Rough	Heavy toolmarks	250	0.010–0.015	Very good sand casting, saw cutting, very rough machining	3
125√	Fine	Machined appearance with consistent toolmarks	125	0.005–0.010	Average machining—turning, milling, drilling; rough hobbing and shaping; die casting, stamping, extruding	4
63√	Fine	Semismooth without objectional tool marks	63	0.002–0.005	Quality machining—turning, milling, reaming; hobbing, shaping; sintering, stamping, extruding, rolling	6
32√	Smooth	Smooth, where tool marks are barely discernible	32	0.0005–0.002	Careful machining; quality hobbing, and shaping; shaving; grinding; sintering	10
14√	Ground	Highly smooth finish	16	0.0002–0.0005	Very best hobbing and shaping; shaving; grinding, burnishing	15
8√	Polish	Semimirrorlike finish without any discernible scratches or marks	8	0.0001–0.0002	Grinding, shaving, burnishing, lapping	20
4√	Superfinish	Mirrorlike surface without tool, grinding, or scratch marks of any kind	4	0.00001–0.0001	Grinding, lapping, and polishing	25

Source: D. Dudley, *Gear Handbook*, McGraw-Hill, New York, 1967.

148 Describing Component Geometry with Random Variables

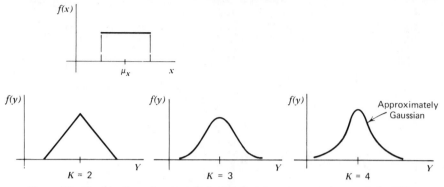

Figure 4.6 Combination of rectangularly distributed component characteristics [35].

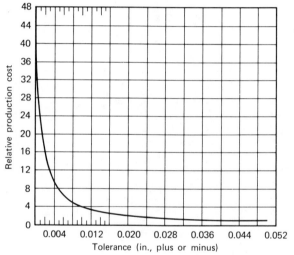

Figure 4.7 Relative cost of accuracy. [From F. W. Bolz, "Design of Considerations," *Transact. ASME* paper No. 49-A-53 (1949)].

Example F

The location constraint is illustrated in this example, where the model is

$$L = (a+b) \pm c$$

Using the dimensions given in Figure 4.8, determine the (1) maximum $(+3_s)$ dimension between the upper face of the piston and the crankshaft center line and (2) minimum distance.

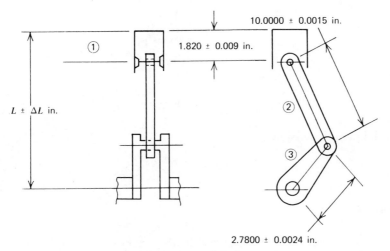

Figure 4.8 Piston–crankshaft geometry.

Solution. The rationale discussed in Section 3.3.2 leads to the following solution:

1. 1.820 ± 0.009 in.
   $(\mu_1, s_1) \approx (1.820, 0.003)$ in.
2. 10.000 ± 0.0015 in.
   $(\mu_2, s_2) \approx (10.000, 0.0005)$ in.
3. 2.780 ± 0.0024 in.
   $(\mu_3, s_3) \approx (2.780, 0.0008)$ in.

Using equation (2.19), and assuming conditions 1, 2, and 3 to be statistically independent:

$$\bar{L}_{max} = 1.820 + 10.000 + 2.780 \text{ in.} = 14.600 \text{ in.}$$

$$\bar{L}_{min} = 1.820 + 10.000 - 2.780 \text{ in.} = 9.040 \text{ in.}$$

Equation (2.21) yields

$$s_{L_{max}} = s_{L_{min}} = \sqrt{(0.003)^2 + (0.0005)^2 + (0.0008)^2} = 0.0032 \text{ in.}$$

The bilateral tolerance is

$$\Delta \approx (s) \cdot (3) = (0.0032)(3) = 0.0095 \text{ in.}$$

Thus

$$\left(\bar{L}_{max}, s_{L_{max}}\right) \approx (14.6, 0.0032) \text{ in.}$$

150 Describing Component Geometry with Random Variables

The maximum dimension is

$$L_{max} = 14.6 \pm 0.0095 \text{ in.}$$

Also,

$$(\bar{L}_{min}, s_{L_{min}}) \approx (9.04, 0.0032) \text{ in.}$$

The minimum dimension is

$$L_{min} = 9.04 \pm 0.0095 \text{ in.}$$

Example G

The clearance constraint is described in this example, which illustrates the likelihood of interference between a cartridge case and the restraining shoulder of a firing chamber, as shown in Figure 4.9. The model in this instance is

$$y = d_t - d_c$$

The estimated mean outside diameter of the cartridge case is $\bar{d}_c \approx (0.454 + 0.4/2) = 0.451$ in. on the basis of data in Chapter 2 Table 2.8. The standard deviation of the cartridge-case outside diameter is estimated as $s_{d_c} \approx (0.006/6) = 0.001$ in.; thus $(\bar{d}_c, s_{d_c}) \approx (0.451, 0.001)$ in. The mean inside diameter of the chamber at the shoulder is $\bar{d}_t \approx (0.455 + 0.457/2) \approx 0.456$ in. diameter. The

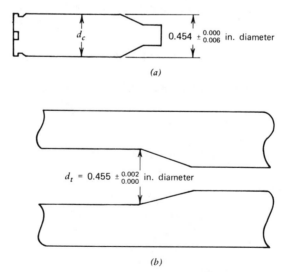

Figure 4.9 Chamber and cartridge case geometry. (a) Cartridge case. (b) Firing chamber.

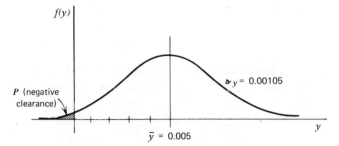

Figure 4.10 Clearance probability density model.

estimated standard deviation of the chamber diameter, s_t is $s_t \approx (0.002/6) = 0.00033$ in.; thus $(\bar{d}_t, s_t) \approx (0.456, 0.00033)$-in. diameter.

Assuming that the cartridge case and rifle firing chamber diameters are approximately normally distributed and independent, what is the probability of interference between a randomly selected rifle firing chamber and a randomly selected cartridge case, at the shoulder? Of interest is the difference random variable y, in equation (4.1). Thus from equations (2.22) and (2.24):

$$\bar{y} = \bar{d}_t - \bar{d}_c = 0.456 - 0.451 = 0.005 \text{ in.}$$

$$s_y = \sqrt{s_t^2 + s_c^2} = \sqrt{(0.00033)^2 + (0.001)^2}$$

$$= 0.00105 \text{ in.}$$

It is necessary to determine the probability of negative clearance, as shown in Figure 4.10.
Solving for the limits of integration associated with the standardized normal distribution in this problem:

$$P_{\text{interference}} = 1 - \int_0^\infty f(y)\,dy$$

$$= 1 - \int_a^b \frac{e^{-u^2/2}}{\sqrt{2\pi}}\,du$$

The lower limit a at $y = 0$ is

$$a = \frac{y - \bar{y}}{s_y} = \frac{0 - \bar{y}}{s_y} = \frac{-0.005}{0.00105} = -4.75$$

The upper limit at $y = \infty$ is

$$b = \frac{\infty - \bar{y}}{s_y} = \infty$$

152 Describing Component Geometry with Random Variables

Thus

$$P_{interference} = 1 - \int_{-4.75}^{\infty} \frac{e^{-u^2/2}}{\sqrt{2\pi}} du$$

From standard normal area tables (Appendix 1 and Tables or Normal Probability Functions [40]), the density (area) over the interval $(-4.75, \infty)$ is .99999967. Thus

$$P_{interference} = 1 - .99999967 = .00000033$$

Clearly, an interpretation of this probability indicates that the number of cases involving interference should be negligible.

4.4.1 Confidence Limits on Sum and Difference Random Values

The statistics of random variables X_1 and X_2 are usually estimated from samples of size n_1 and n_2. Given $(\bar{x}_1, s_{x_1}, n_1)$ and $(\bar{x}_2, s_{x_2}, n_2)$, limits can be estimated for X_1 and X_2 individually. With sums and differences, $y = (x_1 \pm x_2)$; however, although the mean $\bar{y}$ and the standard deviation s_y can be calculated, n_y cannot be estimated by a sampling procedure. The estimation of confidence limits on Y rests on the determination of an equivalent sample size n_y, as explained in Chapter 12.

4.5 NONLINEAR DIMENSIONAL COMBINATIONS

The model in the following example is

$$t = \sqrt{D_1^2 + D_2^2} \tag{a}$$

Example H

Estimate the nominal distance and the tolerance on the distance between the hole centers at P_1 and P_2 such that the risk α of being out of tolerance is no more than $54/10^4$ for parts drilled to the specifications in Figure 4.11.

Solution.

$$\alpha \approx \frac{54}{10,000} = 0.0054$$

The following conditions are derived from Figure 4.13:

$$D_1 = 14.375 \pm 0.0036 \text{ in.}$$
$$(\bar{D}_1, s_{D_1}) \approx (14.37, 0.0012) \text{ in.}$$
$$D_2 = 8.287 \pm 0.0078 \text{ in.}$$
$$(\bar{D}_2, s_{D_2}) \approx (8.287, 0.0026) \text{ in.}$$

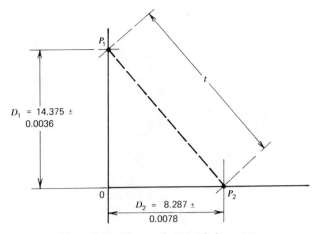

Figure 4.11 Distance between hole centers.

Utilization of equations (2.72) and (4.2) yields

$$\bar{t} \approx \sqrt{(14.375)^2 + (8.287)^2} = 16.593 \text{ in.}$$

From equation (2.73) the estimated value of s_t is

$$s_t \approx \sqrt{\frac{\bar{D}_1^2 s_{D_1}^2 + \bar{D}_2^2 s_{D_2}^2}{\bar{D}_1^2 + \bar{D}_2^2}}$$

$$s_t \approx \sqrt{\frac{(14.375)^2 \cdot (0.0012)^2 + (8.287)^2 \cdot (0.0026)^2}{(14.375)^2 + (8.287)^2}} = 0.00166 \text{ in.}$$

$$(\bar{t}, s_t) \approx (16.6, 0.00166) \text{ in.}$$

The estimated nominal value and tolerance on t is

$$t \approx 16.6 \pm 0.005 \text{ in.}$$

4.6 TOLERANCING BY COMPUTER (MONTE CARLO SIMULATION) [9]

Tight manufacturing tolerances are often specified although not always needed. Conventional tolerance stack-up calculations may show the possibility of an interference condition, saying nothing about the likelihood of interference. Although there is a possibility of interference, tight tolerances may not be needed if the probability of this condition is sufficiently low. Computer simulation of the assembly operation can estimate the probability

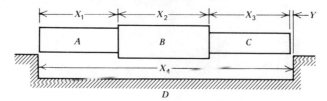

Figure 4.12 A situation with possible interference.

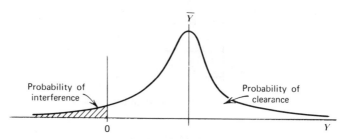

Figure 4.13 Frequency distribution of Y.

of an interference condition and should be tried before reducing manufacturing tolerances for fear of interference. If this technique is used, an entire design can be analyzed step by step and optimum tolerances can be arrived at, usually resulting in more economical products. Some applications are forgiving, even with liberal tolerances, whereas others demand tight tolerances for satisfactory performance. Monte Carlo simulation* is a design tool for determination of whether a given tolerance can be increased (see Figure 4.7 for the cost impact).

Suppose it is necessary to determine whether the assembly of items A, B, and C can fit in the opening in D in the following configuration, where X_1, X_2, and so on represent the four dimensions. We want to determine the clearance Y. A simple model,

$$Y = X_4 - (X_1 + X_2 + X_3) \tag{a}$$

describes the problem (see Figure 4.12).

The problem can be solved if the statistics (or samples) of X_1 through X_4 are known. Assume that dimensions of the mating parts are statistically independent. A computer program is needed to compute Y, using equation (a), several hundred or several thousand times, where random values of X_1, X_2, X_3, and X_4 are combined at each iteration to yield a random value of Y. The random values of Y are used in estimating the frequency distribution of Y. From this frequency distribution the probability of Y becoming negative can be estimated (see Figure 4.13).

*See Section 2.9.

Figure 4.14 Minimum, maximum, and desirable clearance.

Simulation by computer closely resembles an assembly operation, where an operator randomly picks items 1 through 3, assembles them, and puts the assembly into item 4. Choosing a random variate of X_1, following distribution 1, is similar to reaching into the bin of item 1, and similarly for X_2, X_3 and X_4. Computing

$$Y = (X_4) - [(X_1) + (X_2) + (X_3)]$$

is, in essence, "assembling them together" and "measuring the clearance."

Example I

Can the assembly of items 1, 2, and 3 fit into item 4 as shown in Figure 4.12? A minimum clearance of 0.020 in. is desired for ease in installation (see Figure 4.14). Dimensions X_1 through X_4 have normal distributions. What do conventional tolerance stack-up calculations indicate?

Given.

$$x_1 = 1.000 \pm .010$$
$$x_2 = 2.000 \pm .015$$
$$x_3 = 3.000 \pm .020$$
$$x_4 = 6.060 \pm .015 \qquad \text{(b)}$$

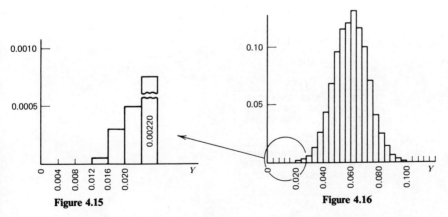

Figure 4.15   Figure 4.16

Histogram of clearance developed from a computer simulation.

Figures 4.15 and 4.16 show the frequency histogram of clearance Y developed from a Monte Carlo simulation of 20,000 iterations. It shows that the probability of insufficient (>0.20 in.) clearance is about .00035. Since a clearance of 0.020 in. is desired for installation ease only, this probability is acceptable.

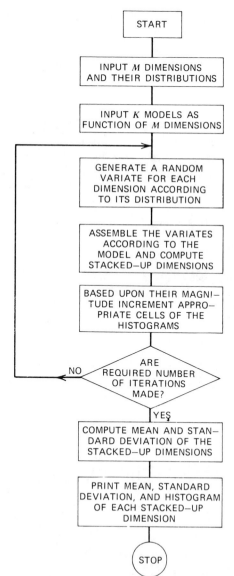

Figure 4.17 Flow diagram of a Monte Carlo program.

MONTE CARLO COMPUTER PROGRAM

It is simple to develop a Monte Carlo computer program to stack up tolerances and determine the statistics of the stacked-up dimensions. Later in Chapter 7 a general Monte Carlo (FORTRAN) computer program is given, which can accommodate a general expression with up to 10 random variables. At this point in time most engineering organizations include Monte Carlo simulation programs among their computer software. Figure 4.17 shows the program logic as it applies to Example I.

4.7. GEOMETRY VERSUS STRENGTH

With geometric dimensional mean values fixed, relaxing geometric tolerances and thereby increasing random variability often has the effect of increasing variability in applied load stress in a component. The result is either degradation in reliability or a requirement to increase component mean strength to maintain a specified reliability.

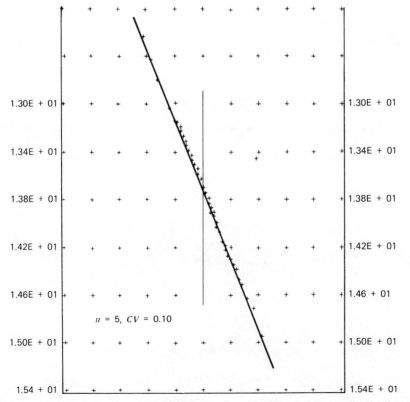

Figure 4.18 Analysis for sum of X_1 through X_5, where X values are lognormally distributed.

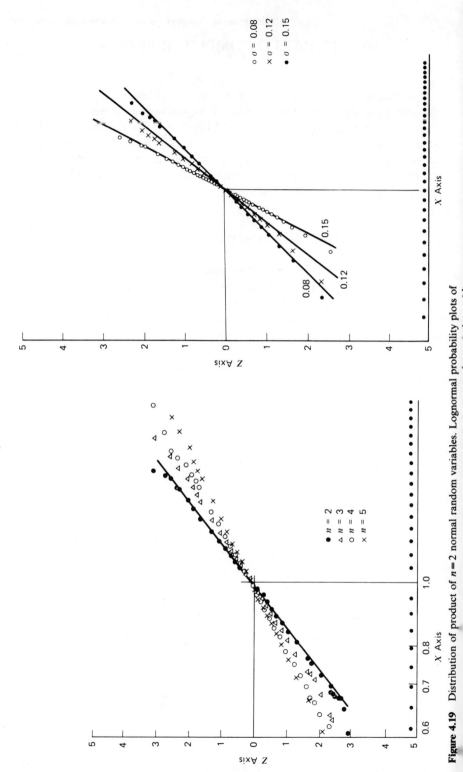

Figure 4.19 Distribution of product of $n=2$ normal random variables. Lognormal probability plots of

The way in which a dimensional variable contributes to distributional form may also contribute to the impact of applied stress. The following illustrates ways in which individual variables influence the distributional form of functions in which they appear:

1. From the central limit theorem for sums (Section 3.3.1) if $z = X_1 \pm X_2 \pm \cdots \pm X_n$:

$$Z \rightarrow \text{Normal, as } n \text{ increases (Figure 4.18)}$$

2. By the central limit theorem for products, if $Z = X_1 \cdot X_2 \cdot \cdots \cdot X_n$:

$$Z \rightarrow \text{Lognormal, as } n \text{ increases (Figure 4.19)}$$

The magnitude of geometric variability effects on component or system failure probability is a function of the power of a dimensional random variable in the stress function. Powers of random variables greater than 1 tend to magnify the effect of their variabilities, as Table 4.16 illustrates. In each case of $f(d)$ illustrated, $(\bar{d}, s_d) = (1.00, 0.05)$ and the coefficient of variation is $C_d = 0.05$. Column 5 gives $C_{f(d)}$, and column 6 indicates the magnification as the ratio $C_{f(d)}/C_d$.

Because of the randomness in component geometry, the distribution of performance to be expected of assembled functional component groups is random, influenced by the magnitude of component cross-sectional area variability, radius of gyration variability, and moment of inertia variability. Length variability effects stability, deflection magnitudes, axial strain, and angular twist magnitudes of components.

Table 4.16

1	2	3	4	5	6
Function of random value d	Power Function Example	Mean Estimator	Standard deviation Estimator	$V_{f(d)}$	Magnification, $C_{f(d)}/C_d$
$f(d) = d$	Circumference $C = \pi d$	$\bar{d} \approx 1.00$	$s_d \approx 0.05$	0.05	1
$f(d) = d^2$	Circular area $A = (\pi d^2/4)$	$\overline{f(d)} \approx 1.0025$	$s_d^2 \approx 0.10$	0.099	~2
$f(d) = d^3$	Section mod. sq. sect. $S = (d^3/6)$	$\overline{f(d)} \approx 1.0075$	$s_d^3 \approx 0.150$	0.148	~3
$f(d) = d^4$	Moment of inertia $I = (\pi/64) \cdot d^4$	$\overline{f(d)} \approx 1.0150$	$s_d^4 \approx 0.20$	0.19	~4

CHAPTER 5

The Stress Random Variable

The following topics are considered in this chapter:

Static stress

Definition of stress and variations with orientation
Invariance of normal stress sums
Static biaxial stresses
Principal stresses
Torsional, bending, and shearing stresses
Superposition

Stress and strain

Unit strain, elasticity, and stress–strain relations

Dynamic stress

Constant amplitude and random (zero mean)
Constant amplitude and random (nonzero mean)
Multiaxial dynamic stress
Estimating stress statistics with finite elements
Lewis stress model
Surface stresses

5.1 INTRODUCTION

Consider a machine component in equilibrium under the action of a system of external loads acting in various directions and in different planes. To examine the effect of such loading, imagine a plane passed through the body of the component. One of the two parts is removed and the other held in equilibrium by the system of internal (reactive) forces and couples, acting on the cut surface to replace the effect of the portion removed.

Noting the magnitude and direction of these variables over the surface of the cut, a small differential area dA is selected. The resultant of all internal

---

*For conversion to SI units, see Appendix 4.

forces acting on dA is designated by dF. The model for stress is

$$(\bar{s}, \text{\tiny$\succ$}_s) = \lim_{\Delta(\bar{A}, \text{\tiny$\succ$}_A) \to 0} \left[\frac{\Delta(\bar{F}, \text{\tiny$\succ$}_F)}{\Delta(\bar{A}, \text{\tiny$\succ$}_A)} \right] = \frac{d(\bar{F}, \text{\tiny$\succ$}_F)}{d(\bar{A}, \text{\tiny$\succ$}_A)} \text{ psi*} \quad \text{(a)}$$

If variability in F and A could be reduced to negligible magnitudes, equation (a) would converge to the familiar classical model for stress.

Stress, a vector, involves the concepts of direction, magnitude, area, and location. Strength of materials has provided a theory for modeling stress fields in elementary components subject to assumptions concerning the character of a component and its behavior under the action of loading and the elastic properties of the material [52].

Elementary theories have not attempted to predict the stress conditions in the neighborhood of loads, supporting members, or boundary irregularities [36].

The general theory of elasticity provided a more comprehensive understanding of the distribution of stresses in elastically deformed bodies. The possible theoretical solution of the stress distribution in loaded mechanical components was extended, as it was not subject to some of the limiting conditions and assumptions of elementary theories. Mathematical and mechanical difficulties, however, have been encountered in attempts to apply the theory [36, p. 9]. Finite-element analysis now provides a method for estimating stress fields in complicated mechanical components, and experimental stress analysis permits the direct determination of strains in elastically or plastically deformed models, from which stresses may be inferred. Finite-element analysis has recently been coupled with probabilistic concepts in design (see Madayag [26] and Section 5.18).

In many stress distribution problems the exact loading on a member is not known. Or loads may be known rather precisely, but stress determination may be difficult to approach theoretically because of the shape of a component. The strains (hence stresses) in such members can now be studied by the use of variable-resistance strain gauges and by photoelasticity.

RANDOM STATIC STRESS

5.2 ANALYSIS OF STRESS

A simple mechanical component subjected to the action of pure tensile, compressive, or shearing forces generates an internal reactive stress:

$$s_x = \frac{F}{A}$$

or

$$\tau = \frac{F}{A}, \quad (5.1)$$

162 The Stress Random Variable

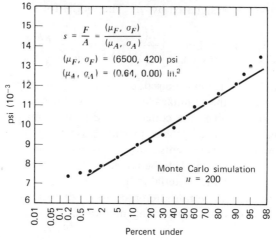

Figure 5.1 Fit of random variable $s = (F/A)$ to Gaussian probability density function model.

where s_x denotes tensile or compressive stress (psi), τ denotes shear stress (psi), F denotes force (lb), and A denotes cross-sectional area (in.^{2}).

If F and A are normally distributed, s_x and τ are approximately normally distributed (as indicated in Fig. 5.1), subject to scatter constraints (Section 2.6.12). If F and A are lognormally distributed, s_x and τ both have lognormal distributions.

When the exact form of the stress probability density function is not known, the asymptotic limiting distributional form may be used (i.e., is often lognormal).

Given the mean-value estimators $\bar{F}$ and $\bar{A}$ and the standard deviation estimators s_F and s_A from equation (2.28), the mean-value estimator of τ and s_x in terms of the statistics of F and A is

$$\bar{s}_x = \bar{\tau} \approx \frac{\bar{F}}{\bar{A}} \text{ psi*} \tag{5.1a}$$

and from equation (2.30), the standard deviation estimator of τ and s_x in terms of component statistics is

$$s_{s_x} = s_\tau \approx \sqrt{\frac{\bar{F}^2 s_A^2 + \bar{A}^2 s_F^2}{\bar{A}^4}} \text{ psi} \tag{5.1b}$$

Only if $s_F = s_A = 0$ will s_x and τ be deterministic.

Example A

Estimate the statistics of stress, assuming that F and A are statistically independent. The model used is equation (5.1).

*To convert from psi to pascal, multiply by $4.895 \cdot 10^3$.

Given.

$$(\bar{F}, \text{\textit{s}}_F) = (6500, 420) \text{ lb}; \quad C_F = \frac{420}{6500} = 0.065$$

$$(\bar{A}, \text{\textit{s}}_A) = (0.64, 0.06) \text{ in.}^2; \quad C_A = \frac{0.06}{0.64} = 0.095$$

The mean value estimator of stress s_x is, from equation (5.1a):

$$\bar{s}_x \approx \frac{6500}{0.64} = 10{,}160 \text{ psi.} = \frac{\bar{F}}{\bar{A}}$$

and the standard deviation estimator of $\text{\textit{s}}_x$ is, from equation (5.1b),

$$\text{\textit{s}}_{s_x} \approx \left(\frac{1}{0.64}\right)^2 \cdot \sqrt{(6500)^2(0.06)^2 + (0.64)^2(420)^2} = 1160 \text{ psi}$$

$$\approx \left[\frac{\bar{F}^2 \text{\textit{s}}_A{}^2 + \bar{A}^2 \text{\textit{s}}_F{}^2}{\bar{A}^4}\right]^{1/2}$$

$$(\bar{s}_x, \text{\textit{s}}_{s_x}) \approx (10{,}160, 1160) \text{ psi}$$

The coefficient of variation of s_x is

$$C_{s_x} = \frac{\text{\textit{s}}_{s_x}}{\bar{s}_x} \approx \frac{1160}{10{,}160} = 0.114$$

The scatter in stress s_x exceeds that in either load or cross-sectional area, as measured by the coefficient of variation.

The fit of the random variable s_x to the normal model, assuming F and A normally distributed, is shown in Figure 5.1. Figures 5.2 and 5.3 provide statistical analogues of the classical solid mechanics representation of a bar loaded in tension.

Assumptions applying to equation (5.1) are as follows:

1. Load is a random variable. When a load is moving, the inertia of a body affects the stress magnitude.

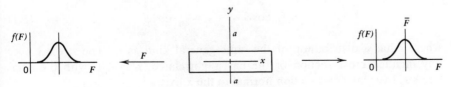

Figure 5.2 Loading normal to axis of a bar.

164   The Stress Random Variable

Figure 5.3 Free-body diagram of a bar.

2. The bar is free from inital stresses. Random residual stresses, if present, should be considered in addition to load-induced stresses [27].

3. The random stress intensities are distributed over the cross section of the bar, requiring microscopically homogeneous material, a section remote from abrupt changes in cross section and remote from the point of application of the load.

4. Cross-sectional area is a random variable.

5.3 VARIATION OF STRESS WITH ORIENTATION OF CROSS SECTION

In considering stresses in a bar under axial tensile loading F, only cross sections perpendicular to the axis of the bar are considered in Section 5.2. The term F denotes the random realization of a static load.

In Figure 5.4 the cross section pq, perpendicular to the plane of the figure, is inclined to the axis of the bar. If the bar is considered an aggregate of longitudinal fibers, each fiber experiences the same elongation, and the random value of force, representing the action of the right portion of the bar on its left portion, is distributed over the cross section pq. The left portion of the bar is in equilibrium under the action of an internal response to a random realization of external force. Thus the resultant distributed over pq equals any random F. If A is the cross-sectional area normal to the axis of the bar and ϕ is the angle, considered deterministic, between the x axis and the normal n to the cross section pq, the cross-sectional area of pq will be $A/\cos\phi$. The model for stress over this cross section is [53]

$$s = \frac{F}{A/\cos\phi} = \frac{F\cos\phi}{A} = s_x \cos\phi \tag{a}$$

where it is seen that the model for s is the product of a random variable (s_x) and a constant ($\cos\phi$), and the statistics of s will be estimated in terms of the statistics of s_x and $\cos\phi$, utilizing equations (2.8) and (2.10),

$$\bar{s} = \bar{s}_x \cdot \cos\phi; \qquad \mathfrak{d}_s = \mathfrak{d}_{s_x} \cdot \cos\phi \tag{5.2}$$

where s has a distribution of the same general kind as s_x. The stress s over any inclined cross section of the bar is correlated, $\rho^* = +1$, with the axial stress s_x over the cross section normal to the x axis.

*See Section 3.5.

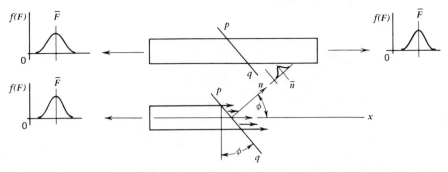

Figure 5.4 Variation of stress with orientation.

For $\phi = (\pi/2)$, the section pq is parallel to the x axis, and the statistics of s are

$$(\bar{s}, \mathfrak{d}_s) \cdot \cos\frac{\pi}{2} \approx (0, 0)$$

where s has the direction of the vector F and is usually resolved into two components, s_n and τ, as shown in Figure 5.5.

The component perpendicular to pq is the normal stress s_n, and the statistics of s_n in terms of the statistics of s_x and $\cos^2\phi$ are, from equations (2.8) and (2.10):

$$(\bar{s}_n, \mathfrak{d}_{s_n}) = (\bar{s}_x, \mathfrak{d}_{s_x}) \cdot \cos^2\phi \tag{5.3}$$

The statistics of the tangential component τ are

$$(\bar{\tau}, \mathfrak{d}_\tau) = \tfrac{1}{2}(\bar{s}_x, \mathfrak{d}_{s_x}) \cdot \sin 2\phi \tag{5.4}$$

where s_n and τ will be correlated with s_x and s, with correlation coefficients $\rho = +1$.

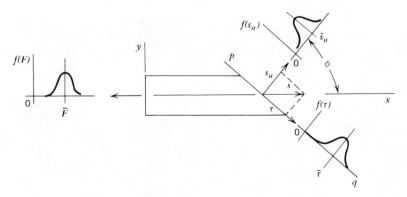

Figure 5.5 Distribution of normal and shearing stress.

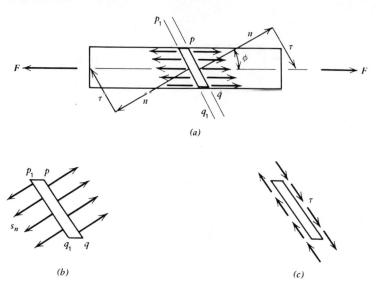

Figure 5.6 Element extension due to normal stress. (b) Normal stress components. (c) Tangential stress components.

If s_x is normally distributed, then s, s_n, and τ will be normally distributed. From Figure 5.6b, c, it is seen that the normal stresses s_n produce extension of the element in a direction normal to the cross section pq and the shearing stresses produce sliding of section pq with respect to $p_1 q_1$. Random variables are modeled by μ and σ and hence cannot be ordered. However, it is convenient to speak of maximum normal stress, by which is meant maximum average. The maximum average normal stress acts over cross sections normal to the axis of the bar, thus the maximum average effect occurs when $\phi = 0$

$$(\bar{s}_n)_{\max}, \, {}^{\circ}(s_n)_{\max} = (\bar{s}_x, \, {}^{\circ}s_x) \cdot \cos\phi = (\bar{s}_x, \, {}^{\circ}s_x)$$

Maximum average shearing stress acts over cross sections inclined at $\phi = 45°$ to the axis of the bar, since $\sin 2\phi = 1.00$ [see equation (5.4)]:

$$(\tau_{\max}, \, {}^{\circ}\tau_{\max}) = \tfrac{1}{2}(\bar{s}_x, \, {}^{\circ}s_x) \tag{5.5}$$

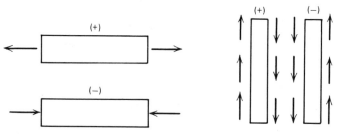

Figure 5.7 Stress sign conventions.

Equations (5.3) and (5.4), derived for tensile loads, can also be used for compressive loads. Figure 5.7 illustrates the rules of signs for mean normal and shearing stresses. Note that standard deviation is always treated as positive, since a negative standard deviation has no useful interpretation.

5.3.1 Confidence Limits on Stress

The statistics of F and A are usually estimated from samples of size n_F and n_A. Given $(\bar{F}, s_F, n_F)$ and $(\bar{A}, s_A, n_A)$, confidence limits can be estimated for F and A individually as discussed in Chapter 12. With the quotient of two random variables such as $s = (F/A)$, however, although $\bar{s}$ and s_s can be calculated, n_s cannot be determined by a counting procedure. The determination of confidence limits on s rests on the estimation of an equivalent sample size n_s, for which no method is known. For some cases, however, first approximations of confidence limits on s can be estimated, that is, when A is approximately deterministic.

5.4 SUM OF RANDOM NORMAL STRESSES

If the angle ϕ is greater than $\pi/2$, a cross section cut by a plane mm is obtained, as shown in Figure 5.8. Consider the case when mm is perpendicular to the cross section PQ. The mean stress components $\bar{s}_{n_1}$ and $\bar{\tau}_1$ at the plane mm are

$$\bar{s}_{n_1} = \frac{\bar{s}_x}{2} - \frac{\bar{s}_x}{2}\cos 2\phi = \bar{s}_x \sin^2\phi \tag{5.6}$$

$$\bar{\tau}_1 = \frac{\bar{s}_x}{2}\sin 2\phi \tag{5.7}$$

Using these results with equations (5.3) and (5.4), it is seen that

$$s_n + s_{n_1} = s_x \cos^2\phi + s_x \sin^2\phi = s_x \tag{5.8}$$

and

$$\tau = \frac{s_x}{2}\sin 2\phi = -\tau_1 \tag{5.9}$$

Equation (5.8) indicates that the sum of the normal stresses acting on the two perpendicular planes is invariant and equal to s_x. The shearing stresses acting on two perpendicular planes are numerically equal and of opposite sign.

The question remains as to whether the sum of $s_n + s_{n_1}$ is invariant, when in general s_n, s_{n_1}, and s_x are random variables. From equations (5.3) and (5.6) it is seen that s_n and s_{n_1} are linear transformations of s_x, hence correlated,

168 The Stress Random Variable

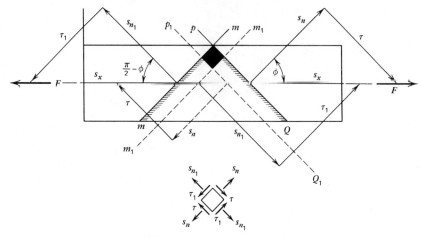

Figure 5.8 Sum of random normal stresses.

$\rho = +1$. Thus

$$(\bar{s}_n, s_{s_n}) + (\bar{s}_{n_1}, s_{s_{n_1}}) = (\bar{s}_x, s_{s_x})\cos^2\phi + (\bar{s}_x, s_{s_x})\sin^2\phi$$

$$= (\bar{s}_x, s_{s_x})$$

The sum of mean normal stresses $\bar{s}_n + \bar{s}_{n_1}$ is invariant and equal to $\bar{s}_x$. The invariance of the standard deviations can also be shown. From equation (2.37):

$$s_w = \sqrt{s_u^2 + s_v^2 + 2\rho s_u s_v} \tag{a}$$

where $w = u + v$. Let, $u = s_x \cos^2\phi$, $v = s_x \sin^2\phi$, and $\rho = +1$. Then, substituting into equation (a):

$$s_w = \sqrt{(s_x \cdot \cos^2\phi)^2 + (s_x \cdot \sin^2\phi)^2 + 2(1) s_x \cos^2\phi s_x \sin^2\phi}$$

$$s_w = s_x \cdot \sqrt{(\cos^2\phi + \sin^2\phi)^2} = s_x$$

Example B

To illustrate the invariance of sums of normal stresses, let $\phi = 30°$ (see Figure 5.8):

$$\sin 30° = 0.500; \qquad \cos 30° = 0.866$$

$$\sin^2 30° = 0.250; \qquad \cos^2 30° = 0.750$$

First, using equation (2.37) and utilizing the results in equation (a):

$$s_{(s_n+s_{n_1})} = \sqrt{(s_x \cdot \cos^2\phi)^2 + (s_x \cdot \sin^2\phi)^2 + 2s_x \cdot \cos^2\phi s_x \cdot \sin^2\phi}$$

$$= \sqrt{(0.75 s_x)^2 + (0.25 s_x)^2 + 2 s_x^2 \cdot (0.25)(0.75)}$$

$$= s_x \sqrt{0.5625 + 0.0625 + 0.3750} = s_x$$

Invariance is valid for the sum $s_n + s_{n_1}$ when $\rho = +1$.

Second, consider s_n and s_{n_1} statistically independent, $\rho = 0$. Determine whether $s_n + s_{n_1}$ is invariant. With $u = s_x \cos^2\phi$ and $v = s_x \sin^2\phi$ and substituting in equation (a):

$$s_w = \sqrt{s_u^2 + s_v^2} = \sqrt{(s_x \cdot \cos^2\phi)^2 + (s_x \cdot \sin^2\phi)^2}$$

$$= s_x \cdot \sqrt{(0.75)^2 + (0.25)^2}$$

$$= 0.7905 s_x$$

It is concluded that invariance does not hold for the case of statistically independent random normal stresses.

By taking the adjacent cross sections $m_1 m_1$ and $p_1 q_1$ parallel to mm and pq (Figure 5.8), an element is isolated and the directions of stresses acting on this element are identified. The shearing stresses acting on the sides of the element parallel to pq produce a clockwise couple $(+)$. The shearing stresses acting on the other two sides produce a counterclockwise couple $(-)$. Since the correlations ρ between the stress s_x and τ_1 is $+1$ and that between s_x and s_{n_1} is $+1$, the element may be in equilibrium for any particular value of the random variable s_x. It should also be noted that

$$s_x = \frac{s_n}{\cos^2\phi}; \quad s_x = \frac{2\tau}{\sin^2\phi}$$

$$s_n = \frac{2\tau \cos^2\phi}{\sin 2\phi}$$

and letting

$$k = \frac{2\cos^2\phi}{\sin 2\phi}$$

$$s_n = k\tau$$

where s_n and τ are correlated with a coefficient of correlation $\rho = +1$.

5.5 STATIC BIAXIAL TENSION OR COMPRESSION

The following analysis of a pressurized tube, biaxially stressed, is presented because cases of biaxial stresses imposed on mechanical components are common.

Consider the stresses in the cylindrical wall of a tube (ends sealed) resulting from a pressure difference of P lb/in^2. Let a small element from the wall, bounded by two adjacent axial sections and two adjacent circumferential sections, be removed, as shown in Figure 5.9.

The tube will be strained both circumferentially and axially, with accompanying circumferential and axial stresses. The tensile stress s_y (circumferential) is

$$(\bar{s}_y, \, {}^{\circ}s_y) \approx \frac{1}{2} \cdot \frac{(\bar{P}, \, {}^{\circ}s_p)(\bar{d}, \, {}^{\circ}s_d)}{(\bar{t}, \, {}^{\circ}s_t)} \text{ psi} \tag{5.10}$$

where d is the inside diameter (inches), t the wall thickness (inches), and P is the pressure difference (psi).

To calculate the axial tensile stress s_x, imagine the tube cut by a plane perpendicular to the x axis. Considering the equilibrium of one portion of the tube, the tensile force producing a longitudinal stress is the resultant of the pressure acting at the ends, that is,

$$F = P\left(\frac{\pi d^2}{4}\right)$$

The wall cross-sectional area is $A = \pi d t$, and the model for stress is

$$s_x = \frac{F}{A} = \frac{Pd}{4t}$$

Thus

$$(\bar{s}_x, \, {}^{\circ}s_x) \approx \frac{1}{4} \cdot \frac{(\bar{P}, \, {}^{\circ}s_p)(\bar{d}, \, {}^{\circ}s_d)}{(\bar{t}, \, {}^{\circ}s_t)} \text{ psi} \tag{5.11}$$

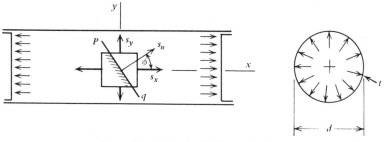

Figure 5.9 Pressurized thin-walled tube.

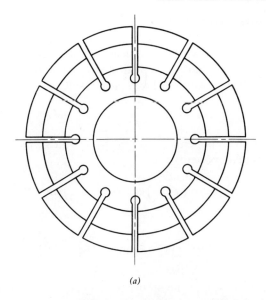

(a)

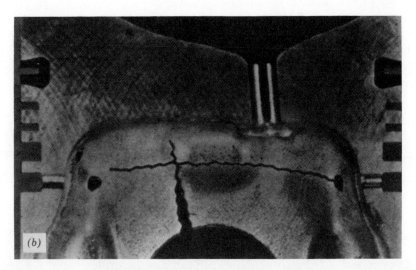

(b)

Figure 5.10 Drawing of equibiaxial specimen.

The element of tubular wall experiences both stresses s_x and s_y. Figure 5.10 illustrates an equibiaxially stressed component. Equations (5.10) and (5.11) indicate that s_x and s_y are correlated, and $\rho = +1$ and their distributions will be of the same type.

By utilizing equations (5.3) and (5.4), it is seen that tensile axial stress s_x acting on the plane pq produce normal and shearing stresses:

$$s'_n = s_x \cos^2 \phi; \quad \tau' = \tfrac{1}{2} s_x \sin \phi$$

To estimate the stress components acting on the plane pq due to the tensile stress s_y, note that the angle between s_y and n is $(\pi/2)-\phi$, measured clockwise from the y axis, where ϕ is measured counterclockwise from the x axis. From this it is seen that s_y is substituted in equations (5.3) and (5.4) for s_x and $-[(\pi/2)-\phi]$ is substituted for ϕ. The result is

$$s_n'' = s_y \sin^2\phi; \qquad \tau'' = -\tfrac{1}{2} s_y \sin^2\phi$$

Summing the stress components produced by s_x and s_y, the resultant normal and shearing stress components are

$$(\bar{s}_n, s_{s_n}) = (\bar{s}_x, s_{s_x})\cos^2\phi + (\bar{s}_y, s_{s_y})\sin^2\phi \qquad (5.12)$$

and

$$(\bar{\tau}, s_\tau) = \frac{1}{2}\left[(\bar{s}_x - \bar{s}_y); \sqrt{s_{s_x}^2 + s_{s_y}^2}\,\right]\sin 2\phi \qquad (5.13)$$

It is now useful to express the statistics of s_n and τ in terms of the known statistics of P, d, and t.

Noting that $s_x = \tfrac{1}{4}(Pd/t)$ and $s_y = \tfrac{1}{2}(Pd/t)$, it is necessary only to estimate the statistics of $w = (Pd/t)$.

The mean-value estimator $(\bar{w})$ is, from equation (2.59): $\bar{w} \approx (\overline{Pd}/\bar{t})$. To estimate s_w requires two steps. First, the mean and standard deviation of $u = Pd$ is calculated from equations (2.25) and (2.27):

$$s_u = \sqrt{(\bar{P}^2)s_d^2 + (\bar{d})^2 \cdot s_P^2 + s_P^2 \cdot s_d^2}$$

$$\bar{u} = \bar{P}\bar{d}$$

Next, the standard deviation of $w = (u/t)$ is calculated from equation (2.30)

$$s_w \approx \sqrt{\frac{(\bar{u})^2 s_t^2 + \bar{t}^2 s_u^2}{\bar{t}^4}}$$

The expression for s_n [equation (5.12)] can now be written

$$s_n = \frac{1}{4}\frac{Pd}{t}\cos^2\phi + \frac{1}{2}\frac{Pd}{t}\sin^2\phi$$

$$s_n = \frac{w}{2}\left[\left(\frac{\cos^2\phi}{2}\right) + \sin^2\phi\right]$$

$$(\bar{s}_n, s_{s_n}) = \frac{(w, s_w)}{2} \cdot \left[\frac{\cos^2\phi}{2} + \sin^2\phi\right]$$

where $\bar{s}_n$ is a maximum when $\phi = (\pi/2)$.

Equation (5.13) for shear (τ) can be rewritten

$$\tau = (s_x - s_y)\frac{\sin 2\phi}{2}$$

$$= \left(\frac{1}{4}\frac{Pd}{t} - \frac{1}{2}\frac{Pd}{t}\right)\frac{\sin 2\phi}{2}$$

$$\tau = -\frac{1}{4}\frac{Pd}{t}\frac{\sin 2\phi}{2} = -\frac{1}{8}w\sin 2\phi$$

and

$$(\bar{\tau}, \circ_\tau) = -\frac{1}{8}(\bar{w}, \circ_w)\sin 2\phi$$

where $\bar{\tau}$ is a maximum when $\phi = (\pi/4)$.

5.6 INVARIANCE OF SUMS OF COMBINED STRESSES

Equations (5.12) and (5.13) for mean values represent the normal and shearing stress components, s_x and s_y, that is, the stresses acting on the sides of the element in Figure 5.9. To represent the mean values of the stress components on any inclined plane, defined by some angle ϕ:

$$\frac{s_x + s_y}{2} - \frac{s_y - s_x}{2}\cdot\cos 2\phi = s_x\cos^2\phi + s_y\sin^2\phi$$

When the plane pq is rotated counterclockwise with respect to an axis perpendicular to the xy plane (Figure 5.9), the corresponding point is moving counterclockwise so that for each value of ϕ the corresponding values of the mean components, s_n and τ are obtained. It follows that the maximum mean normal stress component is equal to s_y.

Taking two perpendicular planes defined by the angles ϕ and $(\pi/2)+\phi$, which the normals n and n_1 make with the x axis, the corresponding mean τ_{max} stress components are given in Figure 5.11. From equations (5.12) and the biaxial effect of s_y:

$$s_n = s_x\cos^2\phi + s_y\sin^2\phi \qquad \text{(a)}$$

Furthermore,

$$s_{n_1} = s_x\sin^2\phi + s_y\cos^2\phi \qquad \text{(b)}$$

Adding equations (a) and (b) yields

$$s_n + s_{n_1} = s_x(\cos^2\phi + \sin^2\phi) + s_y(\sin^2\phi + \cos^2\phi) \qquad (5.14)$$

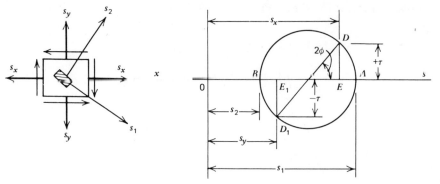

Figure 5.11 Geometric model of s_1 and s_2.   **Figure 5.12** Circle of stress.

and
$$s_n + s_{n_1} = s_x + s_y \tag{5.15}$$

Also,
$$\tau_1 = -\tau \tag{5.16}$$

Invariance holds for biaxial stresses s_x and s_y. Shearing stresses acting on two perpendicular planes are numerically equal but of opposite signs.

If s_x and s_y are generated in some manner from a common load source F, that is, by static pressure (or pressure pulses delivered by fluid in a hydraulic system or through air from a source of white noise), then s_x may equal s_y, and correlation is $\rho = +1$. If s_x and s_y are biaxial stresses generated by *statistically independent* loads F_x and F_y, then from equation (2.36):

$$\bar{s}_n + \bar{s}_{n_1} = \bar{s}_x + \bar{s}_y$$

and from equation (2.21):

$$\mathcal{s}(s_n + s_{n_1}) = \sqrt{\mathcal{s}_{s_x}^2 + \mathcal{s}_{s_y}^2}$$

5.7 PRINCIPAL STRESSES AS RANDOM PHENOMENA

It is important in engineering to carefully scrutinize all inequality relationships. The reason for the scrutiny is that two-parameter models, which describe the bulk of engineering variables, are not orderable. It cannot be said that one random variable so described is larger or smaller than another.

It can be seen that these remarks apply in particular to the concept of principal stresses. With biaxial loading these are usually designated as s_1 and

s_2, acting in directions perpendicular to planes on which shear stresses are nominally zero; s_1 has been designated the maximum principal stress and s_2, the minimum principal stress, suggesting that $s_1 > s_2$ (see Figure 5.12). The models for s_1 and s_2 as functions of s_x, s_y, and τ are

$$s_1, s_2 = \frac{s_x + s_y}{2} \pm \sqrt{\left(\frac{s_x - s_y}{2}\right)^2 + \tau^2} \tag{5.17}$$

From equations (5.17) and (2.59), the estimator of the mean of s_1 is

$$\bar{s}_1 \simeq f(\bar{s}_x, \bar{s}_y, \bar{\tau}_{xy}) = \frac{\bar{s}_x + \bar{s}_y}{2} + \sqrt{\left(\frac{\bar{s}_x - \bar{s}_y}{2}\right)^2 + \bar{\tau}^2} \tag{5.18a}$$

According to equations (5.17) and (2.62), the estimator of the standard deviation of s_1 is

$$s_{s_1} \simeq \sqrt{\left(\frac{\partial s_1}{\partial s_x}\right)^2 \cdot s_{s_x}^2 + \left(\frac{\partial s_1}{\partial s_y}\right)^2 \cdot s_{s_y}^2 + \left(\frac{\partial s_1}{\partial \tau}\right)^2 \cdot s_\tau^2} \tag{5.18b}$$

The mean-value estimators $\bar{s}_1$ and $\bar{s}_2$ and the inequality $\bar{s}_1 > \bar{s}_2$ are meaningful. However, with random variables s_1 and s_2, ordering is usually not valid in a strict sense.

Angle *DCA* in Figure 5.12 is the double angle between the stress s and the x axis, and since 2ϕ is measured from D to A in the clockwise direction, the direction of s_1 must be as shown in the figure. Note the element in the shaded area, Figure 5.11, with sides normal and parallel to s_1. Only normal stresses s_1 and s_2 act on the sides. The sign of ϕ must be considered negative because it is measured clockwise from the x axis

$$\tan 2\phi = \frac{2\tau}{s_x - s_y} \tag{5.19}$$

The mean of maximum shearing stress is given by the magnitude of the radius of the geometric model (Figure 5.12):

$$\tau_{max} = \frac{s_1 - s_2}{2} = \sqrt{\left(\frac{s_x - s_y}{2}\right)^2 + \tau^2} \tag{5.20}$$

Equations (5.17) through (5.20) solve the problem of estimation of the maximum average normal and maximum average shear stresses if the normal and the shearing stresses acting on any two perpendicular planes are given. Equations (5.18a) and (5.18b) provide estimations of the statistics of s_1.

5.8 TRIAXIAL STRESSES

Machine components in which triaxial stresses must be determined are infrequent in mechanical design. However, there should be an awareness of the existance of a third dimension. Six distinct quantities are sufficient to completely specify the state of stress. As in the case of biaxial stress, a particular orientation of the stress element may be found such that the shear stresses are nominally zero. This orientation defines the principal directions and stresses s_1, s_2, and s_3. The mean magnitude of these stresses may be found by solving the cubic equation

$$s^3 - (s_x + s_y + s_z)s^2 + (s_x s_y + s_x s_z + s_y s_z - \tau_{xy}^2 - \tau_{yz}^2 - \tau_{xz}^2)s$$
$$- (s_x s_y s_z + 2\tau_{xy}\tau_{xz}\tau_{yz} - s_x \tau_{yz}^2 - s_y \tau_{xz}^2 - s_z \tau_{xy}^2) = 0$$

5.9 TORSIONAL STRESSES

A moment that results in the twisting of a bar is called *torsion*. An external twisting moment, τ, is called the *external torque*.

A cylindrical solid bar subjected to a particular value of random torque, τ, experiences shearing stresses that vary linearly from zero at the center to a maximum at the surface. The surface stress is

$$\tau = \frac{T \cdot r}{J} \qquad (5.21)$$

where τ is the shearing stress (psi), T is the torque (in. lb), r is the radius (inches), and J is the polar moment of inertia (see Appendix 5). The statistics of τ estimated in terms of the statistics of T, r, and J are

$$\bar{\tau} \simeq \frac{\bar{T} \cdot \bar{r}}{\bar{J}} \text{ psi} \qquad (5.21a)$$

and

$$s_\tau \simeq \left[\left(\frac{\bar{r}}{\bar{J}} \right)^2 \cdot s_T^2 + \left(\frac{\bar{T}}{\bar{J}} \right)^2 \cdot s_r^2 + \left(\frac{\bar{T} \cdot \bar{r}}{\bar{J}^2} \right)^2 \cdot s_J^2 \right]^{\frac{1}{2}} \text{ psi} \qquad (5.21b)$$

The following assumptions apply to equation (5.21):

1. The bar experiences pure torque, that is, the application of a couple (Figure 5.13).
2. The shearing stress on a cross section perpendicular to the axis of symmetry of the bar is proportional to the distance from the centroidal axis;

Torsional Stresses   177

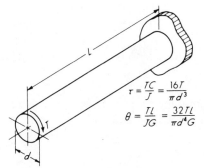

$$\tau = \frac{TC}{J} = \frac{16T}{\pi d^3}$$

$$\theta = \frac{TL}{JG} = \frac{32TL}{\pi d^4 G}$$

Figure 5.13   Torsion spring.

thus circular sections remain circular after twisting. Also, Hooke's law must apply.
3. There are no prior residual stresses in the bar.
4. Sections under consideration must be remote from diameter changes and the locale of couple application.

Example C

A shaft resists a torque of $(10{,}000, 1100)$ in. lb at the location B shown in Figure 5.14. If the ends of the shaft are fixed against rotation, estimate the statistics of the torque reactions T_1 and T_2.
Viewed probabilistically, there are three possible cases:

1. If dimensions of AB and BC are very closely controlled, they may be considered approximately deterministic.

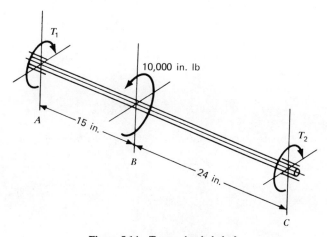

Figure 5.14   Torque loaded shaft.

2. If the dimension AC is closely controlled and the location of B is subject to random variation (tolerance on location is $\pm 0.375''$), AB and BC are negatively correlated $\rho \approx -1$.

3. If AB and BC are random variables, statistically independent, then AC is also a random variable.

Solution (for Case 2). The angular deformation in AB is equal to the angular deformation in BC. The value for ϕ_1 and ϕ_2 should be equated.

$$\phi_1 = \frac{15 T_1}{JG} = \phi_2 = \frac{24 T_2}{JG}$$

$$15 T_1 - 24 T_2 = 0 \qquad (a)$$

$$T_1 + T_2 = 10{,}000 \qquad (b)$$

From equation (a):

$$l_1(\overline{T}_1, s_{T_1}) = l_2(\overline{T}_2, s_{T_2}) \qquad (c)$$

$$(\overline{T}_1, s_{T_1}) = l_2(\overline{T}_2, s_{T_2})(\overline{T}_1, s_{T_1}) + (\overline{T}_2, s_{T_2}) = (10{,}000, 1100) \text{ in. lb}$$

$$\frac{l_2}{l_1}(\overline{T}_2, s_{T_2}) + (\overline{T}_2, s_{T_2}) = (10{,}000, 1100) \text{ in. lb} \qquad (d)$$

$$AB = l_1 = 15 \pm 0.375; \qquad s_{l_1} \approx \frac{\Delta l_1}{3} = 0.125 \text{ in.}$$

$$BC = l_2 = 24 \pm 0.375; \qquad s_{l_2} \approx \frac{\Delta l_2}{3} = 0.125 \text{ in.}$$

Thus

$$(\bar{l}_1, s_{l_1}) \approx (15, 0.125) \text{ in.}$$

$$(\bar{l}_2, s_{l_2}) = (24, 0.125) \text{ in.} \qquad (e)$$

From equation (c):

$$(15, 0.125) \text{ in.} \cdot (\overline{T}_1, s_{T_1}) = (24, 0.125) \text{ in.} \cdot (\overline{T}_2, s_{T_2}) \qquad (f)$$

Since l_1 and l_2 are correlated ($\rho = -1$), equation (2.45) is applied in estimating the mean of $(l_2 / l_1) = z$:

$$\bar{z} \approx \frac{\bar{l}_2}{\bar{l}_1}\left(1 + \frac{s_{l_1}}{\bar{l}_1}\right) \cdot \left[\frac{s_{l_1}}{\bar{l}_1} - \rho \cdot \frac{s_{l_2}}{\bar{l}_2}\right]$$

$$\bar{z} = \left(\frac{24}{15} + \frac{(0.125)24}{(15)^2}\right) \cdot \left[\frac{0.125}{15} + \frac{0.125}{24}\right] = 1.600$$

Torsional Stresses 179

The values l_1 and l_2 are correlated ($\rho = -1$), because B in Figure 5.14 is subject to random variations. Thus a random positioning of B at $l_1 + \Delta l$ is accompanied by a value $l_2 - \Delta l$. According to equation (2.47), the standard deviation of Z is

$$s_z \approx \frac{\bar{l}_2}{\bar{l}_1}\left[\frac{s_{l_1}^2}{\bar{l}_1^2} + \frac{s_{l_2}^2}{\bar{l}_2^2} - 2\rho \cdot \frac{s_{l_1} \cdot s_{l_2}}{\bar{l}_1 \cdot \bar{l}_2}\right]^{1/2}$$

$$s_z \approx \left(\frac{24}{15}\right) \cdot \left[\left(\frac{0.125}{15}\right)^2 + \left(\frac{0.125}{24}\right)^2 + 2\frac{(0.125)^2}{(15)(24)}\right]^{1/2} = 0.0216$$

and substituting for z in equation (f):

$$(\bar{T}_1, s_{T_1}) = (1.600; 0.0216)(\bar{T}_2, s_{T_2}) \qquad (g)$$

Writing the statistical equality with the use of equation (b), we obtain

$$(\bar{T}_1, s_{T_1}) + (\bar{T}_2, s_{T_2}) = (10{,}000, 1100) \text{ in. lb} \qquad (h)$$

Substituting equation (g) into equation (b) yields

$$(1.6000, 0.0216)(\bar{T}_2, s_{T_2}) + (\bar{T}_2, s_{T_2}) = (10{,}000, 1100) \text{ in. lb} \qquad (i)$$

If equations (2.45) and (2.47) are applied with equation (i), the statistics of T_1 and T_2 are

$$(\bar{T}_2, s_{T_2}) = (3850, 420) \text{ in. lb}$$

$$(\bar{T}_1, s_{T_1}) = (1.6000, 0.0216)(3850, 420)$$

$$= (6150, 680) \text{ in. lb}$$

Torsional shearing stress statistics must often be estimated from recorded measurements of torque (T) delivered to a rotating shaft. Horsepower statistics, in turn, are estimated from the statistics of T and n. The functional model is

$$HP = \frac{2 \cdot Tn}{(33{,}000)(12)} = \frac{FV}{33{,}000}$$

The model for estimating T as a function of horsepower (HP) measured by a recording dynamometer, is

$$T = \frac{63{,}025(HP)}{n} \qquad (5.22)$$

where HP is horsepower, T is torque (in. lb), n is shaft speed (rpm), F is force (pounds), and V is velocity (fpm).

Mean torque is [from equation (2.59)]

$$\bar{T} = 63{,}025 \cdot \left(\frac{\overline{HP}}{\bar{n}}\right) \tag{5.22a}$$

The standard deviation s_T is [from equation (2.62)]

$$s_T = \left[\left(\frac{\partial T}{\partial HP}\right)^2 \cdot s_{HP}^2 + \left(\frac{\partial T}{\partial n}\right)^2 \cdot s_n^2\right]^{1/2}$$

$$\frac{\partial T}{\partial HP} = \frac{63{,}025}{n}$$

$$\frac{\partial T}{\partial n} = -63{,}025\left(\frac{HP}{n^2}\right)$$

$$s_T = \left[\left(\frac{63{,}025}{\bar{n}}\right)^2 s_{HP}^2 + \left(\frac{63{,}025\,\overline{HP}}{\bar{n}^2}\right)^2 s_n^2\right]^{1/2} \tag{5.22b}$$

5.10 BENDING STRESSES

Most mechanical systems of structures contain components subject to random bending stresses. A model for bending stresses is shown in Figure 5.15. A beam loaded as shown in Figure 5.16 reacts with the outermost fibers compressed at the top, whereas those at the bottom are stretched. The fibers on plane A–A (Figure 5.17) remain unchanged. Stresses in the fibers at a distance c from the neutral plane are

$$s = \frac{Mc}{I}* \tag{5.23}$$

The statistics of s are in terms of the statistics of M, c, and I. According to equation (2.59), $\bar{s}$ is

$$\bar{s} \approx \frac{\overline{M}\,\overline{c}}{\overline{I}} = f(\overline{M}, \overline{c}, \overline{I}) \tag{5.23a}$$

*Alternatively:

$$s = \frac{Mc}{I} \pm \frac{P}{A}$$

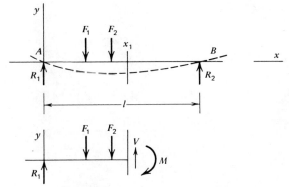

Figure 5.15 Bending-stress model.

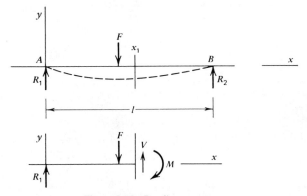

Figure 5.16 Bending stresses.

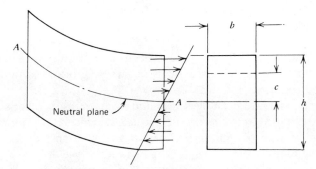

Figure 5.17 Rectangular bean under bending action.

The Stress Random Variable

From equation (2.62), the expression for $\mathfrak{s}_s$ is

$$\mathfrak{s}_s \approx \left[\left(\frac{\partial s}{\partial M}\right)^2 \cdot \mathfrak{s}_M^2 + \left(\frac{\partial s}{\partial c}\right)^2 \cdot \mathfrak{s}_c^2 + \left(\frac{\partial s}{\partial I}\right)^2 \cdot \mathfrak{s}_I^2 \right]^{1/2} \qquad (a)$$

The partial derivatives of M, c, and I with respect to s are

$$\frac{\partial s}{\partial M} = \frac{\bar{c}}{\bar{M}}$$

$$\frac{\partial s}{\partial c} = \frac{\bar{M}}{\bar{I}} \qquad (b)$$

$$\frac{\partial s}{\partial I} = \frac{\bar{M}\bar{c}}{\bar{I}^2}$$

Substitution of equation (b) into equation (a), $\mathfrak{s}_s$ yields

$$\mathfrak{s}_s \approx \left[\left(\frac{\bar{c}}{\bar{I}}\right)^2 \cdot \mathfrak{s}_M^2 + \left(\frac{\bar{M}}{\bar{I}}\right)^2 \cdot \mathfrak{s}_c^2 + \left(\frac{\bar{M}\bar{c}}{\bar{I}^2}\right) \cdot \mathfrak{s}_I^2 \right]^{1/2} \qquad (5.23\text{b})$$

For design purposes, where the expressions for $\bar{s}$ and $\mathfrak{s}_s$ contain unknowns, it is often easier to work with statistics of s estimated by means of the algebras given in Sections 2.5 and 2.6.

Statements applying to equation (5.23) are as follows:

1. Cross sections at which stresses are calculated are remote from load points and supports and also from changes in cross section.
2. The material behaves as though it follows Hooke's law, subject to random variations.
3. Cross sections through the beam before loading remain in the plane after load-induced bending. For cases where shearing stresses are zero, this is not strictly true. For constant shear stresses, warping of the sections is nearly constant and equation (5.23) can be applied.
4. The stress magnitude (subject to random variations) is otherwise proportional to the distance from the neutral surface, reaching a maximum value for a particular instance of random loading at the outermost fibers. Failure in bending is more probable than buckling or local failure.
5. Bending forces usually act normal to the centroidal axis of the beam and in the plane containing the centroidal axis.

For a rectangular section:

$$I = \frac{bh^3}{12} \qquad (5.24)$$

The statistics of I in terms of the statistics of b and h are, from equation (2.59),

$$\bar{I} \approx \frac{\bar{b}\bar{h}^3}{12} = f(\bar{b},\bar{h}) \qquad (5.24a)$$

According to equation (2.62), the expression for s_I is

$$s_I \approx \left[\left(\frac{\partial I}{\partial b}\right)^2 \cdot s_b^2 + \left(\frac{\partial I}{\partial h}\right)^2 \cdot s_h^2\right]^{1/2} \qquad (a)$$

The partial derivatives of b and h with respect to I are

$$\frac{\partial I}{\partial b} = \frac{h^3}{12}$$

$$\frac{\partial I}{\partial h} = \frac{bh^2}{4} \qquad (b)$$

If equation (b) is substituted into equation (a), s_I is

$$s_I \approx \left[\left(\frac{\bar{h}^3}{12}\right)^2 s_b^2 + \left(\frac{\bar{b}\bar{h}^2}{4}\right)^2 s_h^2\right]^{1/2} \qquad (5.24b)$$

5.11 SHEARING STRESSES

If the bending moment varies along the length of a beam, a shearing stress τ is induced. Its value (subject to random variations) is a function of the rate of variation of the moment

$$\tau = \frac{V}{Ib} \int_{y_0}^{c} y \, dA \qquad (5.25)$$

In equation (5.25) shearing stress is zero when $y_0 = c = (h/2)$. As y_0 decreases in value, shearing stress increases with the maximum occurring when $y_0 = 0$ (Figure 5.18).

Example D

Given.

$$\text{Shear force } (\bar{V}, s_V) = (3600, 200) \text{ lb}$$

$$\text{Shear area } (\bar{A}, s_A) = (0.097, 0.006) \text{ in.}^2 \qquad (a)$$

184 The Stress Random Variable

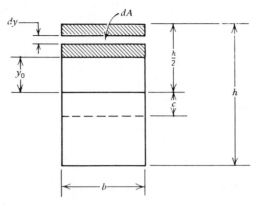

Figure 5.18 Moment as a function of distance from neutral axis.

Estimate the statistics of shearing stress τ by using the model,

$$\tau_{max} = \frac{3V}{2A} \qquad (b)$$

From equation (2.59):

$$\bar{\tau} \approx \frac{3}{2}\left(\frac{3600}{0.097}\right) = 55{,}670 \text{ psi}$$

According to equations (2.62) and the preceding model, s_τ is estimated. Taking partial derivatives of V and A with respect to τ:

$$\frac{\partial \tau}{\partial V} = \frac{3}{2\bar{A}}; \qquad \frac{\partial \tau}{\partial A} = \frac{3\bar{V}}{2}(A^{-2}) \qquad (c)$$

$$s_\tau \approx \left[\left(\frac{3}{2\bar{A}}\right)^2 \cdot s_V^2 + \left(\frac{3\bar{V}}{2\bar{A}^2}\right)^2 \cdot s_A^2\right]^{1/2} \qquad (d)$$

Substituting equation (a) into equation (d) yields

$$s_\tau \approx 2270 \text{ psi}$$

As might be expected, variability in V accounts for most of the variability in τ.

The following are models for predicting maximum shearing stress, τ_{max}:

1. $\tau_{max} = \dfrac{3V}{2A}$ (rectangular section beam)
2. $\tau_{max} = \dfrac{4V}{3A}$ (solid cylinder beam) $\qquad$ (5.26)
3. $\tau_{max} = \dfrac{2V}{A}$ (tubular beam)
4. $\tau_{max} = \dfrac{V}{A_w}$ (I beam and wF beams), where A_w is the area of the web cross section.

Since V/A is common to all expressions for τ_{max}, the statistics for $\left(\dfrac{V}{A}\right) = x$ are as follows. According to equation (2.59), $\bar{x}$ is

$$\bar{x} \approx \dfrac{\bar{V}}{\bar{A}}$$

From equation (2.62), the expression for s_x is

$$s_x \approx \left[\left(\dfrac{\partial x}{\partial v}\right)^2 s_v^2 + \left(\dfrac{\partial x}{\partial A}\right)^2 s_A^2\right]^{1/2} \qquad (a)$$

The partial derivatives of V and A with respect to x are

$$\dfrac{\partial x}{\partial v} = \dfrac{1}{\bar{A}}$$

$$\dfrac{\partial x}{\partial A} = \dfrac{\bar{V}}{\bar{A}^2} \qquad (b)$$

If equation (b) is substituted into equation (a), s_x is

$$s_x \approx \left[\left(\dfrac{1}{\bar{A}}\right)^2 s_V^2 + \left(\dfrac{\bar{V}}{\bar{A}^2}\right)^2 s_A^2\right]^{1/2}$$

Assumptions applying to equation (5.26) are as follows:

1. Shearing stress over a cross section has the same direction as the shearing force.
2. Shearing forces (subject to random variations) are uniformly distributed across the width of a beam (investigations have shown that this assumption is reasonable where b is not large relative to h).

5.12 SUPERPOSITION OF RANDOM VARIABLES

When probabilistic considerations are introduced in the determination of stresses due to two (or more) force components, two possible cases must be considered: (1) the additive effect of two (or more) independent forces acting simultaneously on a component and (2) the effect when the force components are correlated. The difference in effects is illustrated in Example E.

Example E

Consider a cantilever beam subjected to statistically independent forces X and Y (Figure 5.19). Both X and Y produce tensile stresses in the upper fibers of the beam.

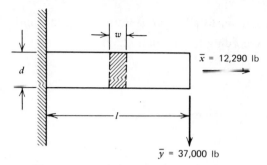

Figure 5.19 Independent force components.

Next, consider the same beam loaded as shown in Figure 5.20 by force vector Z.

Given the *statistically independent force components* X and Y (Figure 5.19), the statistics are determined to be

$$(\bar{x}, s_x) = (12{,}290, 1230) \text{ lb}$$

$$(\bar{y}, s_y) = (37{,}000, 3700) \text{ lb} \qquad (a)$$

With the beam loaded by force vector Z, the vector sum of independent $(\bar{x}, s_x)$ and $(\bar{y}, s_y)$, the resultant is [from equations (2.63) and (2.64)]:

$$(\bar{z}, s_z) = (39{,}000, 3530) \text{ lb}$$

Assume now that the beam in Figure 5.20 is loaded by a single vector, identical to $(\bar{z}, s_z)$ in the preceding equation. The decomposition of this vector random variable into *correlated X and Y* component vector random variables yields

$$(\bar{x}, s_x) = (12{,}290, 1110) \text{ lb}$$

$$(\bar{y}, s_y) = (37{,}000, 3350) \text{ lb} \qquad (b)$$

The mean-value estimators of X and Y in the independent and correlated cases are the same; however, the correlated random variables involve less scatter [i.e., compare equations (a) and (b)].

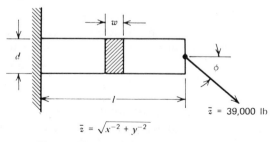

Figure 5.20 Correlated force components.

SUMMARY

To this point the discussion has centered around forces, force couples, moments, and geometry that induce stresses in components that are not dependent on the mechanical behavioral properties of the materials.

The quantitative estimation of stress* has involved:

1. Obtaining an appropriate functional model that identifies the relevant variables and formalizes their interaction.
2. Obtaining the statistics describing the behavior of individual variables, reflecting their multivalued and random character.
3. Determining the statistical independence or correlations among the variables.
4. Using appropriate formulae from the statistical algebras to estimate the statistics of stress in terms of the statistics of the individual variables.

STRESS AND STRAIN

Materials are usually elastic. They deform under loading, returning to essentially the original forms when unloaded, if the elastic limit has not been exceeded. In considering deformation under loading, the behavioral characteristics of the governing physical variables must be considered (see Table 5.1). The parameters usually encountered are (1) the elastic modulus E, often modeled by a normal distribution (see Figure 5.22), (2) the shear modulus of elasticity (G), and (3) the Poisson ratio, ν. The shear modulus G is a linear function of and is correlated with E, $\rho \simeq +1$, and the distributional form followed by its values is the same as for E.

5.13 UNIT STRAIN

A straight rod loaded in tension elongates. The magnitude of elongation is usually interpreted as the amount of elongation per unit length (i.e., unit strain):

$$\varepsilon = \frac{\delta}{l} \qquad (5.27)$$

where ε is in inches per inch and δ is the total elongation for a rod of length l inches. If a tensile load P is applied to a number of randomly selected

---

*The analogue of load stress in fracture mechanics is called the *stress-intensity factor*, designated K. It carries the units $KSI\sqrt{\text{in.}}$ (to convert from ksi to MPa/m, multiply by $\sqrt{1.099}$). Design utilizing probabilistic fracture mechanics methods to avoid brittle fracture is explained by examples in Chapter 10.

Table 5.1 Physical Parameters of Common Metals

Material	Modulus of Elasticity ($E \times 10^6$ psi [1])			Modulus of Rigidity (psi $\times 10^6$)	Density (lb/in²)
	Mean	σ	Sample		
Aluminum alloy 0.33[a]	10.3[c]			3.85[c]	0.10[c]
2014-T651 Plate	9.9	0.27	20		
2219-T87 Plate	9.7	0.27	15		
7039-T6 Plate	9.4		8		
7075-T73 Forging	10.3	See note [d]			
Beryllium	42[c]				0.066[c]
Beryllium copper 0.29[a]	18[c]			7[c]	0.30[c]
Brass, bronze 0.34[a]	16[c]			6[c]	0.30[c]
Magnesium alloy 0.35[a]	6.5[c]			2.4[c]	0.065[c]
Miscellaneous alloys					
Hastelloy X 0.30[a]	31.29	0.64	4		
K-Monel 0.32[a]	26.0	See note [d]			
Nickel alloy 0.30[a]	30[c]			11[c]	0.30[c]
Steel, carbon and alloy 0.30[a]	29.5[c]			11.4[c]	0.28[c]
AISI 4340 0.29[a]	29.0[e]	See note [d]			
Invar 36	21.0	See note [d]			
Steel, stainless 0.31[a]	28[c]			10.6[c]	0.28[c]
Type 440c 0.284[a]	29.2	See note [d]			
22-13-5 0.2850[a]	29.63[f]	0.54[f]	4[f]		
Titanium alloy 0.33[a]	16[c]			6.2[c]	0.16[c]
Ti-4Al-3Mo-1V[g]	16.7	0.4	113		
Ti-2.5Al-16V[g]	15.3	0.4	109		
Ti-5Al-2.5Sn[f]	18.05	0.35			
Zinc, die-cast alloy 0.33[a]	2[c]				0.24[c]

[a] Poisson ratio.
[b] Ambient temperature. Longitudinal direction for rolled sheet and plate where applicable.
[c] From published tables.
[d] Typical data variability ±5%.
[e] Estimated to equal properties of AISI 9310.
[f] Dynamic modulus of elasticity psi $\times 10^6$.
[g] Data are for compressive modulus of elasticity (E).
Source: values from Haugen et al. [54].

nominally identical rods, l will vary statistically, δ will vary statistically, and hence ε will be a random variable. The statistics of ε in equation (5.27) are estimated by utilizing equations (2.28) and (2.30), and

$$\bar{\varepsilon} = \frac{\bar{\delta}}{\bar{l}}$$

$$\sigma_\varepsilon \simeq \left[\frac{\bar{\delta}^2 \sigma_l^2 + \bar{l}^2 \sigma_\delta^2}{\bar{l}^4} \right]^{\frac{1}{2}}$$

The assumptions applying to equation (5.27) are

1. The cross section of a given bar (subject to random variations) remains essentially constant.
2. The material behaves as though it were homogenous in a microscopic sense, subject to random variations.
3. The pattern of strain is essentially uniform.

In the absence of bending or buckling, equation (5.27) will apply also for compression.

5.14 THE QUALITY OF ELASTICITY

5.14.1 Modulus of Elasticity

The modulus of elasticity E is illustrated in Figures 5.21 and 5.22. A material that regains its original shape and dimensions on the removal of an applied load is said to be "elastic." Hooke's law implies that the stress in an ideal material is proportional to the strain that produced it, "within certain limits." The limit is a random variable. Some elastic materials do not follow Hooke's law since they regain their original shapes without the limiting condition that strain (ε) be proportional to stress (s).

Since strain has been shown to be proportional to stress,

$$s = E \cdot \varepsilon \qquad (5.28)$$

within the elastic limit.

Estimators of the statistics of E are available and ε is measurable. Clearly, E and ε are statistically independent; thus estimates of stress statistics are feasible without prior knowledge of the loading. Tensile strength, S_μ, of Gray Iron is illustrated in Table 5.2.

By utilizing equation (2.25), $\bar{s}$ can be calculated in terms of $\bar{E}$ and $\bar{\varepsilon}$:

$$\bar{s} \approx \bar{E} \cdot \bar{\varepsilon} \qquad (5.29)$$

The standard deviation s_s can be estimated in terms of the statistics of E and ε by utilizing equation (2.27):

$$s_s \approx \sqrt{\bar{E}^2 s_\varepsilon^2 + \bar{\varepsilon}^2 s_E^2 + s_E^2 \cdot s_\varepsilon^2} \qquad (5.30)$$

Substituting $s = (F/A)$ and equation (5.27) into equation (5.28) yields

$$\delta = \frac{Fl}{AE} \qquad (5.31)$$

The statistics of δ in terms of the statistics of F, l, A, and E are, according to equations (2.59) and (5.31):

$$\bar{\delta} \approx \frac{\bar{F}\bar{l}}{\bar{A}\bar{E}} \qquad (5.32)$$

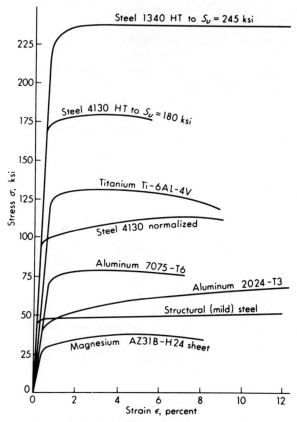

Figure 5.21 Stress–strain curves for metal alloys.

According to equation (2.62), the expression for s_δ is

$$s_\delta \approx \left[\left(\frac{\partial \delta}{\partial F}\right)^2 \cdot s_F^2 + \left(\frac{\partial \delta}{\partial l}\right)^2 \cdot s_l^2 + \left(\frac{\partial \delta}{\partial A}\right)^2 \cdot s_A^2 + \left(\frac{\partial \delta}{\partial E}\right)^2 \cdot s_E^2\right]^{1/2} \quad (a)$$

The partial derivatives of F, l, A, and E with respect to δ are

$$\frac{\partial \delta}{\partial F} = \frac{l}{AE}$$

$$\frac{\partial \delta}{\partial l} = \frac{Fl}{AE}$$

$$\frac{\partial \delta}{\partial A} = \frac{Fl}{A^2 E} \quad (b)$$

$$\frac{\partial \delta}{\partial E} = \frac{Fl}{AE^2}$$

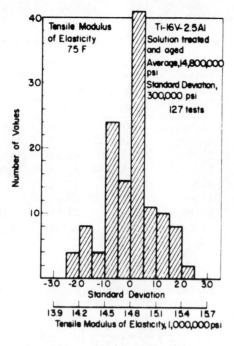
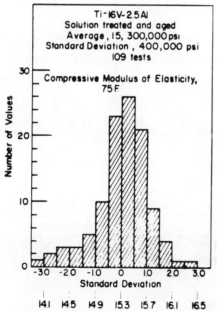

Figure 5.22 Modulus of elasticity.

Table 5.2. Properties of Gray Iron (Deterministic)

SAE No.	ASTM Class	Ultimate Tensile Strength (psi)	Compression Strength, (psi)	Torsional Shear Strength (psi)	Mean Fatigue Strength (psi)	BHN
G2000	20	22,000	83,000	26,000	10,000	156
—	25	26,000	97,000	32,000	11,500	174
G3000	30	31,000	109,000	40,000	14,000	201
G3500	35	36,500	124,000	48,500	16,000	212
G4000	40	42,500	140,000	57,000	18,500	235
G4500	—	45,000	—	—	—	217–269
—	50	52,500	164,000	73,000	21,500	262
—	60	62,500	187,000	88,500	24,500	202

Source: Juvinall [27, p. 559].

If equation (b) is substituted into equation (a), s_δ is

$$s_\delta \approx \left[\left(\frac{\bar{l}}{\overline{AE}}\right)^2 \cdot s_F^2 + \left(\frac{\overline{Fl}}{\overline{AE}}\right)^2 \cdot s_l^2 + \left(\frac{\overline{Fl}}{\overline{A^2E}}\right)^2 \cdot s_A^2 + \left(\frac{\overline{Fl}}{\overline{AE^2}}\right)^2 \cdot s_E^2\right]^{1/2} \quad (5.33)$$

5.14.2 Poisson Ratio

Measurements in experimental stress analysis show that when a material is loaded in tension, not only is an axial strain induced, but also a lateral strain of opposite sign (see Table 5.3).

$$\nu = \frac{\text{Lateral strain}}{\text{Axial strain}} \quad (5.34)$$

where ν represents the Poisson ratio.

Table 5.3. Poisson Ratio

Alloy	Poisson Ratio
AISI 4340[b]	$\bar{\nu} = 0.287$[a]
Type 440c stainless	$\bar{\nu} = 0.284$[a]
22-13-5 stainless	$\bar{\nu} = 0.285$
	$s_\nu = 0.0046$
Hastelloy X	$\bar{\nu} = 0.297$
	$s_\nu = 0.0031$
K-Monel 500	$\bar{\nu} = 0.320$
	Data variability ±10%

[a] Typical data variability ±5%.
[b] Close approximation for low-alloy steels.
Source: Haugen et al. [54].

Since axial strain is a random variable, lateral strain is a random variable, correlated with ε.

The shearing modulus of elasticity is related to E by the following equation:

$$E = 2G \cdot (1 + v) \tag{5.35}$$

5.15 STRESS-STRAIN RELATIONS

5.15.1 Uniaxial Stress

A uniform straight rod of constant cross section (subject to random variations) is loaded axially in tension by force F (see Figure 5.24). The coordinate axes will be numbered to correspond with the mean principal stress directions; thus

$$s_1 = \frac{F}{A}; \quad s_2 = 0; \quad s_3 = 0 \tag{5.36}$$

The statistics of s_1 in terms of the statistics of F and A are given by equations (5.1a) and (5.1b). The mean principal strain random variables are measured in directions normal to the nominal zero shear planes. From equation (5.34) the predicted principal strains in a uniaxial stress state are

$$\varepsilon_1 = \frac{s_1}{E}; \quad \varepsilon_2 = -v\varepsilon_1; \quad \varepsilon_3 = -v\varepsilon_1 \tag{5.37}$$

Thus ε_2 and ε_3 are correlated with random variable ε_1. The state of strain is triaxial.

5.15.2 Static Biaxial Stress

With biaxial stress, s_1 and s_2 take on nonzero values, whereas $s_3 = 0$ (see Figure 5.24). Variables s_1 and s_2 may be statistically independent or correlated, producing two possible biaxial equivalent stress (s_{eq}) effects:

$$s_{eq} = \sqrt{s_1^2 - s_1 s_2 + s_2^2}$$

The models of principal strain from solid mechanics are

$$\varepsilon_1 = \frac{s_1}{E} - \frac{v s_2}{E}; \quad \varepsilon_2 = \frac{s_2}{E} - \frac{v s_1}{E}; \quad \varepsilon_3 = -\frac{v s_1}{E} - \frac{v s_2}{E} \tag{5.38}$$

By means of equations (5.38), the principal strain statistics may be estimated in terms of the principal stress statistics. More often, the statistics of s_1 and s_2 may be estimated because strains are readily determined by experimental measurements:

$$s_1 = \frac{E(\epsilon_1 + \nu\epsilon_2)}{1-\nu^2} \qquad (5.39)$$

$$s_2 = \frac{E(\epsilon_2 + \nu\epsilon_1)}{1-\nu^2}$$

Consider equation (5.39). The statistics of s_1 in terms of the statistics of E, ν, ϵ_1, and ϵ_2 are as follows. From equation (2.59):

$$\bar{s}_1 \approx \frac{\bar{E}(\bar{\epsilon}_1 + \nu\bar{\epsilon}_2)}{1+\nu^2} \qquad (5.39a)$$

If ϵ_1 and ϵ_2 are correlated, $\rho = +1$, then $\epsilon_2 = c\epsilon_1$, where c is a linear relationship constant. Since the scatter in ν is small, it is assumed that the effect is negligible; thus

$$s_1 \approx \left[\frac{1+c\nu}{1+\nu^2}\right] \cdot E\epsilon_1$$

From equations (2.10) and (2.27):

$$\nu_{s_1} \approx \left(\frac{1+c\nu}{1+\nu^2}\right) \cdot \left[\bar{E}^2 \nu_{\epsilon_1}^2 + \bar{\epsilon}_1^2 \nu_E^2 + \nu_E^2 \nu_{\epsilon_1}^2\right]^{1/2} \qquad (5.39b)$$

DYNAMIC STRESS

5.16 RANDOM FATIGUE STRESSES

Basic materials fatigue behavior data are developed by testing specimens in simulated frequently observed dynamic load environments. Such data are needed to design components whose performance will be predictable subject to any one of a wide variety of random dynamic peak stress regimes.

Materials (depending on fatigue strength) must, in differing circumstances, resist one of the following conditions:

1. Constant-amplitude sinusoidal loads (nonrandom)
2. Narrow-band random loading amplitudes (a random phenomenon) (Figure 5.23)
3. Broad-band random loading
4. Combinations of static and superimposed random dynamic loading
5. Mixes of random dynamic loading with randomly occurring shock impulses

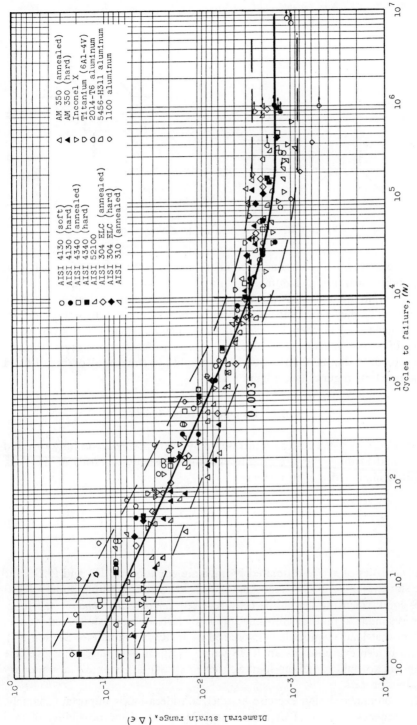

Figure 5.23 Effect of diametral strain range on fatigue life. (Swanson et al.)

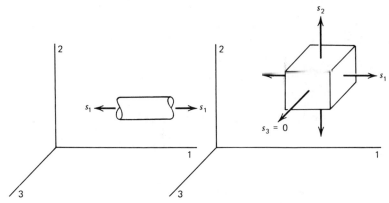

Figure 5.24 Uniaxial (left) and biaxial (right) stress.

The cycle lives of a component population of a given engineering material, responding to random loadings having the same root mean square (RMS) stress amplitude but with pulses occurring in random orders, may differ greatly since they represent a variety of combinations of interacting random phenomena. It has been established experimentally that for a given RMS level of stress, narrow-band random stress produces more severe fatigue strength degradation than do constant-amplitude sinusoidal or broad-band random stress regimes (see Figure 8.41).

5.16.1 Dynamic Stress Models

Zero Mean. First, with cases of essentially *constant-amplitude* fully reversed sinusoidal loading, the stress representation is usually in terms of the peak tensile component, s_a; for instance:

$$\left(\bar{s}_a, \text{\tiny >}s_a\right) = \frac{\left(\bar{F}_a, \text{\tiny >}F_a\right)}{\left(\bar{A}, \text{\tiny >}A\right)} \quad (5.40)$$

The statistics of s_a are estimated by utilizing equations (5.1a) and (5.1b) and the statistics of F_a and A. Note that if F_a is ideal constant amplitude, $\text{\tiny >}F \simeq \cdot 0$.

Second, when the loading is fully reversed *narrow-band random*, the function is similar to that for constant amplitude fully reversed sinusoidal loading except that load magnitude is expressed as an RMS_L value (see Section 3.7). Stress is also expressed as an RMS value. Swanson [41] states that over the long run to failure, fatigue loading (hence stress) is nominally a stationary random process.

In a complete design exercise another consideration is important, namely, the frequency domain. This requires consideration of the possibility of random resonant responses of the structure at one or more of its natural frequencies, which could lead to narrow-band stress regimes that would be much higher than for the static case. Hence maximum damping is essential if the natural frequencies also are to be minimized in the range of the excitation spectra. Further, the natural frequency of a component or system must not fall near a critical stress response frequency, which would lead to destructive resonance.

Nonzero Mean. Third, for cases of fluctuating loading, which lead to fluctuating stresses, the mean (s_m) and alternating (s_a) stress components (*constant amplitude*) are represented by s_R, a random vector, as indicated in Figure 5.25a:

$$s_R = \sqrt{s_a^2 + s_m^2} \qquad (5.41)$$

The statistics of s_R are expressed in terms of the statistics of s_a and s_m, by means of equations (2.63) and (2.64):

$$\bar{s}_R = \sqrt{\bar{s}_a^2 + \bar{s}_m^2} \qquad (5.41a)$$

$$\vartheta_{s_R} = \left[\frac{\bar{s}_a^2 \vartheta_{s_a}^2 + \bar{s}_m^2 \vartheta_{s_m}^2}{\bar{s}_a^2 + \bar{s}_m^2} \right]^{1/2} \qquad (5.41b)$$

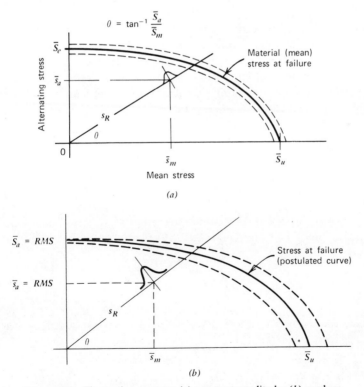

Figure 5.25 Fluctuating stresses: (*a*) constant amplitude; (*b*) random.

198 The Stress Random Variable

Fourth, instances of fluctuating loading, in which the alternating component is *narrow-band random*, are numerous. For example, consider the parabola depicted in Figure 5.25b. It is postulated that for random excitation, the curve follows the same general shape as the Gerber parabola for the constant-amplitude case (Figure 5.25a).

If S_a is the RMS (in psi) of stress at failure for n cycles, s_a is the alternating random stress component, and the stress vector in Figure 5.25b is

$$s_R = \sqrt{s_m^2 + s_a^2} \tag{5.42}$$

the statistics of which are, utilizing equations (2.63) and (2.64):

$$\bar{s}_R = \sqrt{\bar{s}_m^2 + \bar{s}_a^2} \tag{5.42a}$$

$$s_{s_R} \approx \left[\frac{\bar{s}_m^2 s_{s_m}^2 + \bar{s}_a^2 s_{s_a}^2}{\bar{s}_m^2 + \bar{s}_a^2} \right]^{1/2} \tag{5.42b}$$

5.17 MULTIAXIAL DYNAMIC STRESS

Although there are some exceptions, most fatigue failures are surface initiated. Since the stress normal to an unloaded surface is zero, it is seldom necessary to consider fatigue failure of a mechanical component in the triaxial stress state. The following considerations apply only to dynamic biaxial stress situations, including combined dynamic tension and torsion.

Employing the distortion energy theory to estimate equivalent stress for biaxially stressed components, for zero means ($\bar{s}_{1m} = \bar{s}_{2m} = 0$), the limiting values of s_{1a} and s_{2a} are given by

$$s_{eq} = \sqrt{s_{1a}^2 - s_{1a} s_{2a} + s_{2a}^2} \quad * \tag{5.43}$$

where s_a refers to a constant-amplitude alternating stress component. The mean value of s_{eq} is, from equation (2.59) (with the model), equation (5.43):

$$\bar{s}_{eq} = \sqrt{\bar{s}_{1a}^2 - \bar{s}_{1a} \bar{s}_{2a} + \bar{s}_{2a}^2} \tag{5.43a}$$

*The design criterion, in cases of biaxially stressed components (to avoid fatigue failure) is

$$P(S_f > s_{eq}) \geq R$$

where S_f is unaxial stress at failure and s_{eq} is equivalent unaxial stress.

The standard deviations of s_{eq} for the two identifiable cases are:

1. Stress components s_{1a} and s_{2a} are statically independent, $\rho=0$. Applying equation (2.62) with equation (5.46), the standard deviation $\sigma_{s_{eq}}$ is

$$\sigma_{s_{eq}} = \left[\left(\frac{\partial s_{eq}}{\partial s_{1a}}\right)^2 \cdot \sigma_{s_{1a}}^2 + \left(\frac{\partial s_{eq}}{\partial s_{2a}}\right)^2 \cdot \sigma_{s_{2a}}^2\right]^{\frac{1}{2}} \qquad (a)$$

The partial derivatives of s_{1a} and s_{2a} with respect to s_{eq} are

$$\frac{\partial s_{eq}}{\partial s_{1a}} = \frac{1}{2} \cdot \frac{2s_{1a} - s_{2a}}{\sqrt{s_{1a}^2 - s_{1a}s_{2a} + s_{2a}^2}}$$

$$\frac{\partial s_{eq}}{\partial s_{eq}} = \frac{1}{2} \cdot \frac{2s_{2a} - s_{1a}}{\sqrt{s_{1a}^2 - s_{1a}s_{2a} + s_{2a}^2}} \qquad (b)$$

Substitution of equation (b) into equation (a) yields

$$s_{eq} = \frac{1}{\sqrt{\bar{s}_{1a}^2 - \bar{s}_{1a}\bar{s}_{2a} + \bar{s}_{2a}^2}} \left[\left(4\cdot\bar{s}_{2a}^2 - 4\bar{s}_{1a}\bar{s}_{2a} + \bar{s}_{2a}^2\right) \cdot \sigma_{s_{1a}}^2 s_{1a} + \left(4\bar{s}_{2a}^2 - 4\bar{s}_{1a}\bar{s}_{2a}\right.\right.$$

$$\left.\left. + s_{1a}^2\right) \cdot \sigma_{s_{2a}}^2\right]^{\frac{1}{2}} \qquad (5.43b)$$

2. Stress components s_{1a} and s_{2a} are correlated, $\rho = +1$, and in phase. Let the statistics of s_{1a} and s_{2a} be $(\bar{s}_{1a}, \sigma_{s_{1a}})$ and $(\bar{s}_{2a}, \sigma_{s_{2a}})$; then

$$s_{2a} = cs_{1a}$$

Then

$$s_{eq} = \sqrt{s_{1a}^2 - s_{1a} \cdot cs_{1a} + (c \cdot s_{1a})^2} = \sqrt{s_{1a}^2 \cdot (1 - c + c^2)}$$

$$s_{eq} = s_{1a}\sqrt{c^2 - c + 1} \qquad (5.44)$$

Equivalent stress statistics [from equations (2.8) and (2.10)] are

$$\bar{s}_{eq} \simeq [c^2 - c + 1]^{1/2} \cdot \bar{s}_{1a} \qquad (5.44a)$$

$$\sigma_{s_{eq}} \simeq [c^2 - c + 1]^{1/2} \cdot \sigma_{s_{1a}} \qquad (5.44b)$$

Extensive computer simulation studies (Monte Carlo) indicate that when s_{1a} and s_{2a} are normally distributed, the hypothesis that s_{eq} is normally distributed cannot be rejected at the $\delta = .05$ level of significance. This has

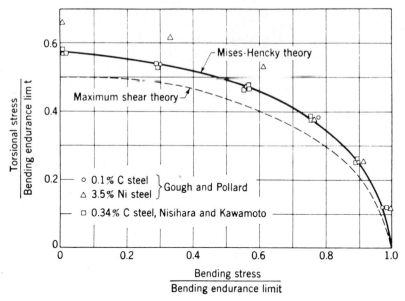

Figure 5.26 Fatigue strength of materials subject to combined stresses.

been shown for variables s_{1a} and s_{2a} having coefficients of variation to $c_s = 0.15$. Furthermore, the χ^2 test indicates that this normality postulation cannot be rejected at the $\delta = .05$ level of significance.

Dolan* has compiled data for fatigue failures due to combined torsion and bending; these are shown in Figure 5.26, where the distortion-energy theory closely predicts stress at failure. The data for Figure 5.26 were from tests in which reversed torsion and reversed bending were synchronous and in phase. Thus directions of the mean principal stresses did not vary during testing. Figure 5.26 suggests that a specimen or component subjected to combined stresses, completely reversed and in phase, will fail whenever the distortion energy reaches the value required for failure of a rotating beam specimen.

Example F

A section of a thick-walled hydraulic cylinder subjected to hydraulic pressure cycling P_i and having random peak pulses is shown in Figure 5.27. The model for this is

$$\left(\overline{P}_i, s_{P_i}\right) = (7000, 620) \text{ psi}$$

The cross-sectional dimensions are closely controlled and may be considered approximately deterministic, with values $(\bar{r}_i, s_{r_i}) = (6.00, \sim 0)$ in. and $(\bar{r}_0, s_{r_0}) \approx (6.375, \sim 0)$ in. Estimate s_{eq}, as follows. The models of correlated and

*T. J. Dolan, *Stress Range*, ASME Handbook Sec. 6–2, p. 97 (New York, McGraw-Hill, 1953).

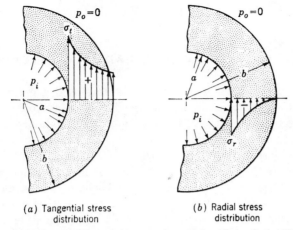

(a) Tangential stress distribution

(b) Radial stress distribution

Figure 5.27 Distribution of stresses in a thick-walled cylinder [1].

in-phase tangential and radial stresses due to applications of P_i are

$$s_t = P_i\left(\frac{r_0^2 + r_i^2}{r_0^2 - r_i^2}\right) = (16.51)\cdot P_i \text{ psi}; \qquad s_r = -P_i \text{ psi} \qquad (a)$$

The mean value and standard deviation of s_t and s_r, utilizing the models in equation (a), are

$$(\bar{s}_t, \mathbin{\char`\^}s_t) = (115{,}500, 10{,}240) \text{ psi}.$$

$$(\bar{s}_r, \mathbin{\char`\^}s_r) = (-7000, 620) \text{ psi}.$$

Letting

$$\frac{s_r}{s_t} = c = \frac{P_i}{16.51 P_i} = 0.0606; \qquad c^2 = 0.0037 \qquad (b)$$

yields

$$[c^2 - c + 1]^{1/2} = [0.943]^{1/2} = 0.971$$

1. Equivalent stress, utilizing equations (5.44a) and (5.44b) for correlated tangential and radial stress with $(\bar{P}_i, \mathbin{\char`\^}{P_i})$, is

$$(\bar{s}_{eq}, \mathbin{\char`\^}s_{eq}) \approx 0.971 \cdot (115{,}500, 10{,}240) \text{ psi} = (112{,}200, 9950) \text{ psi}$$

2. For the case of s_t and s_r statistically independent:

$$\bar{s}_{eq} \approx 112{,}160 \text{ psi}; \qquad \mathbin{\char`\^}s_{eq} = \sqrt{(10{,}240)^2 + (620)^2} = 10{,}300 \text{ psi}$$

In this case the radial stress makes a minimal contribution to equivalent stress.

FINITE ELEMENTS

5.18 ESTIMATING STRESS STATISTICS WITH FINITE ELEMENTS

If the designer of a complex component employed a presently available finite-element computer code in the usual way, a complete deterministic analysis could be obtained by the following sequence, outlined in Figure 5.28.

The building and testing of hardware is often too expensive and time consuming to serve as a feasible method for achieving a high degree of reliability. Finite-element methods provide a much needed approach to assessing the integrity of a complex design that is being considered. It is an analysis methodology for estimating stress intensity and deflections in components and structures and may also provide information useful in design synthesis.

To illustrate the probabilistic method in this text, only simple analytical formulas for idealized shapes such as beams and shafts have been used thus far to relate probabilistic quantities to stresses and other failure-inducing mechanisms. Since computer techniques for analyzing various aspects of complex systems have advanced greatly in the past few decades, it is now feasible to estimate the requirements for reliable designs before hardware has actually been constructed. In practice, where stresses are high and the consequence of failure severe, designers and analysts use large digital computer programs to compute stresses and deflections in complex structural shapes. Although these programs are extremely powerful, they have not been in a convenient form useful in the support of probabilistic design or analysis.

Existing finite-element stress analysis programs can, however, be adapted to predict mean stresses throughout a structure and the variances in the stresses when the stress producing influences (loads, preloads, temperatures, pressures, material properties, etc.) are defined statistically. Given the stresses throughout a part and the strength of the material of which it is to be made, described statistically, it is possible to find the probability of no failure by

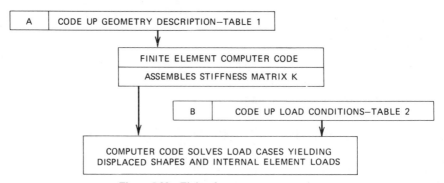

Figure 5.28 Finite-element computer code.

Figure 5.29 Pressure vessel–nozzle connection.

methods described in Chapters 9 and 10. In this section we estimate the stress statistics throughout the flange connecting a pressure vessel to a nozzle (see Figure 5.29).

Figure 5.30 shows a computer plot of the grid in a finite-element program used to study stresses in the flange joint. In this case the three-dimensional problem has been simplified by axial symmetry. The loading is very complex; weight of the nuclear core results in a loading; there are thermal stresses resulting from radiation and coolant flow and the different materials used in the pressure vessel, nozzle, and bolt; and there is internal pressure, nozzle thrust, and a bolt preload.

Different material properties will exist for the structure (aluminum for the pressure vessel and steel for the nozzle). Nonlinear effects may occur if mating surfaces displace.

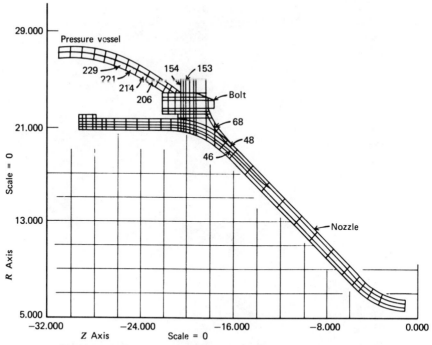

Figure 5.30 Plot illustrating pressure vessel–flange joint connection.

The task of laying out a reasonable grid and assuring that the loads and boundary conditions are compatible is a requirement. Once this is accomplished, however, the problem still remains to adapt the computer program (which finds deterministic stresses and deflections) to produce a probabilistic description of the stresses.

The entire finite-element program may be regarded as a numerical machine for computing stresses (or functions of the stresses) at each element in the structure given a number of input loads, geometric quantities, and so on.

The stresses at each element are combined (Section 5.19.2) to yield an equivalent stress. In the text that follows, the stress variable s is equivalent dynamic stress. Strength S to be used in analysis would be the uniaxial tensile fatigue strength. In this example the statistics of 10 variables are used as basic input quantities to the finite-element program: pressure P, core weight C, bolt preload BT, temperature T, nozzle thickness t_N, pressure vessel thickness t_{pv}, modulus of elasticity of nozzle E_N, modulus of elasticity of pressure vessel E_{pv}, coefficient of thermal expansion of nozzle α_N, coefficient of thermal expansion of pressure vessel α_{pv}.

The analogue of evaluating equation (2.59) for the mean stress involves running the finite-element program once with all parameters at their mean values. The result is that the stresses estimated in all elements approximate the mean stresses. After it is generally found that only a fraction of the elements covering a structure are highly stressed, thus justifying study of only those elements that have a reasonably high degree of stress. In Figure 5.30, 11 element numbers have been identified as the more highly stressed elements in the pressure vessel, the flange, and the nozzle.

For each element for which the variance is required, a relationship analogous to equation (2.62) exists. In the present example there were 10 parameters to vary. Since the stress at each element may require study, the partial derivatives must be evaluated numerically (as slopes) since there is no analytic expression for stress (see Figure 5.31). Referring to Figure 5.31, consider the partial derivatives of the stress function in element 65, as eight of the 10 parameters are varied. Tabulations of the computations involved in finding the partial derivatives for pressure and temperature are given in Tables 5.4 and 5.5 for several elements. Also, the contributions to the stress variance, assuming that all the parameters are independent, are tabulated.

Table 5.6 summarizes the results of the calculations and shows the standard deviations of the stresses in the elements. Note that only 11 runs are necessary to generate the data to compute all the partial derivatives for all the elements necessary to predict the variance in all stresses due to the 10 probabilistic parameters. One run establishes all the mean stresses; and 10 runs, in which a single parameter is varied each time, establishes all the partial derivatives.

Table 5.6 contains useful information even before it is used to compute reliability. The entries in Table 5.6 point out which parameter is most

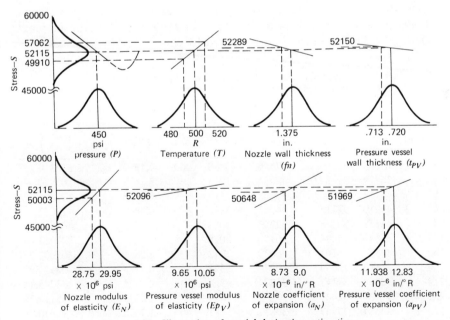

Figure 5.31 Illustration of partial derivative estimation.

Table 5.4 Stress Variance Due to Pressure

$$\left(\frac{\bar{s}-s_p}{\Delta p}\right)^2 s_p^2 = (\partial s/\partial p)^2 s_p^2$$

Element	$\bar{s}$	s_p	$\bar{s}-s_p$	Δ_p	$\dfrac{\bar{s}-s_p}{\Delta_p}$	s_p	$\left(\dfrac{\partial s}{\partial p}\right)^2 s_p^2$
46	46569.	29328.	17241.	450	38.3133	2.59	9846.7332
48	37734.	19579.	18155.		40.344		10918.4145
53	44036.	26278.	18658.		41.4622		11531.8169
65	52115.	57292.	−5177.		−11.5044		287.812
73	50168.	57421.	−7250.		−16.1179		1742.6492
163	30444.	46116.	−7092.		−17.5178		2060.2093
189	44161.	51726.	−7003.		−16.0067		1404.7763
206	32655.	2069.	25596.		56.800		21702.6202
214	36143.	6860.8	28202.2		62.8490		26496.8115
221	32463.	6128.9	26004.1		58.5202		22972.0094
229	32767.	5102.	27665.	450	61.4778	2.59	25050.0194

significant for each element in producing variations in stress (see Section 2.7.2).

In practice, the stress information in Table 5.6 is used with information from subsequent chapters to estimate reliabilities and modify elements to achieve near-optimum designs.

The literature of mechanical engineering provides numerous functional models with which to predict stress. The stress models in the following sections are examples from the literature. In some instances the empirical models may fall short of being rigorously complete. However, as more

Table 5.5 Stress Variance Due to Temperature

$$\left(\frac{s'-s''}{2\Delta T}\right)^2 s_T^2 = \left(\frac{\partial s}{\partial T}\right)^2 s_T^2$$

Element	s'	$\bar{s}$	s''	$s'-s''$	$2\Delta T$	$\dfrac{s'-s''}{2\Delta T}$	s_T	$\left(\dfrac{\partial s}{\partial T}\right)^2 s_T^2$
46	44668.	46569.	47546.	−2878.	400	7.195	66.7°	230310.25
48	36479.	37.034	37994.	−1515.		3.7875		63820.02
53	43152.	44906.	45665.	−2513.		6.2825		175596.83
65	49910.	52115.	54062.	−4152.		10.38		479042.90
73	48139.	50168.	52222.	−4083.		10.2075		460543.45
153	36842.	38444.	50170.	−3028.		8.32		307962.84
159	42527.	44161.	46030.	3012.		8.28		305008.78
206	30601	12655.	30274.	327.		0.8175		2973.22
214	30208.	39141.	35871.	367.		0.9175		3745.10
221	11701.	32461.	30272.	429.		1.0725		5112.36
229	34604.	32747.	30871.	481.	400	1.7075	66.7°	12910.99

Table 5.6 Determination of the Stress Function Variance[a]

Sect.	El.	$\left(\dfrac{\partial s}{\partial p}\right)^2 s_p^2$	$\left(\dfrac{\partial s}{\partial c}\right)^2 s_c^2$	$\left(\dfrac{\partial s}{\partial BT}\right)^2 s_{BT}^2$	$\left(\dfrac{\partial s}{\partial T}\right)^2 s_T^2$	$\left(\dfrac{\partial s}{\partial t_N}\right)^2 s_{t_N}^2$	$\left(\dfrac{\partial s}{\partial t_{po}}\right)^2 s_{t_{po}}^2$	$\left(\dfrac{\partial s}{\partial E_N}\right)^2 s_{E_N}^2$	$\left(\dfrac{\partial s}{\partial E_{po}}\right)^2 s_{E_{po}}^2$	$\left(\dfrac{\partial s}{\partial a_N}\right)^2 s_{a_N}^2$	$\left(\dfrac{\partial s}{\partial a_{po}}\right)^2 s_{a_{po}}^2$	s_s
Muzzle	46	9847	1602	6288	200110	5273	3285	172780	2433	100461	987	751
	48	10918	3025	9610	60820	21873	6289	72568	3069	87487	216	528
	50	11502	3025	9990	175597	54067	6184	152643	3244	108244	416	745
	65	888	169	7	479141	13407	544	616162	4707	60560	2382	1005
	70	1741	1	379	461611	980	36	674182	420	29264	2790	1083
Flange	153	2063	1	502	102963	2810	64	376815	1846	29722	10532	251
	159	1895	9	29	306000	70518	196	484082	1870	19897	12392	941
	206	21700	108900	24100	2973	4438	36	5548	37875	90672	30449	572
	214	26497	23716	38921	3215	4527	—	615	51008	112090	22629	533
	221	22972	34909	23994	5117	1545	821	4416	48487	143501	2900	537
	229	25050	4226	11257	12971	710	1546	538	67775	341713	12022	691

[a] $s_s = \left[\sum_{j=1}^{n}\left(\dfrac{\partial s}{\partial x_j}\right)^2 \cdot s_{x_j}^2\right]^{1/2}$. Underlined variances constitute 10% of the total element variance.

complete models are evolved, they can be utilized in probabilistic applications. The modified Lewis model, presented next, is extensively used to predict dynamic stresses in gear teeth.

5.19 LEWIS MODEL FOR DYNAMIC STRESSES

The kinematics of gear teeth and gear trains (and the force analysis of gears and gear trains) have been developed in the literature. Also, mathematical models, establishing the functional relationships among the relevant engineering variables, to estimate gear tooth stresses are found in mechanical design texts [1, 27].

The previously mentioned theory is not repeated here. Rather, the probabilistic considerations that influence random gear stresses, bearing on specified life requirements, are treated. An alternative probabilistic approach for estimating gear-tooth stress field statistics is also available, that is, the procedure coupling finite-element methods and probabilistic concepts, outlined in Section 5.18. Comparative studies indicate that predictions based on finite-element analysis are usually conservative. Common gear-tooth terminology is indicated in Figure 5.32.

The two most likely modes of spur gear failure have been found to be (1) fatigue failures due to repeated bending stresses and (2) surface failures due to repeated contact stress cycling (also the principal mode of failure for ball and roller bearings; see Section 5.20.2).

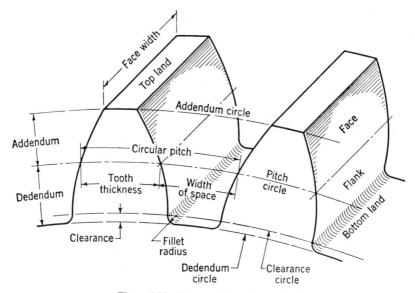

Figure 5.32 Nomenclature of gear teeth.

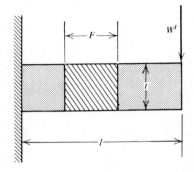

Figure 5.33 Gear-tooth stress models.

5.19.1 Bending Stresses

Figure 5.33 indicates a cantilever beam having depth t, width F, and length l, all of which are dimensional random values. With P_t, a random transmitted load applied as shown, a rough approximation of tensile stress s is given by

$$s = \frac{Mc}{I} = \frac{6P_t l}{Ft^2} \text{ (psi)} \tag{a}$$

In Figure 5.34 maximum tensile stress is indicated at point a, where

$$\frac{\frac{t}{2}}{x} = \frac{l}{\frac{t}{2}}; \quad x = \frac{t^2}{4l} \tag{b}$$

Substitution of equation (b) into equation (a) and multiplication by p_c/p_c [circular pitch, $p_c = (\pi d/N)$] yields

$$s = \frac{P_t}{F \cdot p_d \cdot y} \tag{c}$$

where $y = 2x/3p_c$, diametral pitch is $P = \pi/P_c$, and $Y = \pi y$. The classical Lewis model for stress is

$$s = \frac{P_t \cdot P}{FY} \text{ psi} \tag{d}$$

where s refers to static tensile stress.

5.19.2 Modified Lewis Model

To closely approximate the dynamic tensile fillet stress in a spur gear tooth, it is necessary to consider the (1) stress concentration (in the fillet area), (2) contact ratio effect on division of tooth loading, and (3) dynamic effects.

210 The Stress Random Variable

Incorporation of these above factors in equation (d) yields

$$s = \frac{P_t \cdot P}{K_v \cdot F \cdot J} \tag{5.45}$$

where K_v is an empirical dynamic factor, $(78/78+\sqrt{V}\,)$; J is a geometry factor, including K_t or K_f:

1. The factor J is defined by the American Gear Manufacturers' Association (AGMA) as

$$J = \frac{Y}{K_f}$$

2. For tip loading

$$Y = \pi y = \pi \frac{2x}{3p_c}$$

3. For a 20° involute spur gear

$$K_t = 0.18 + \left(\frac{t}{r_f}\right)^{0.15} \cdot \left(\frac{t}{l}\right)^{0.45}$$

Usually $K_f \simeq K_t$ because for gear materials the notch sensitivity factor (q) is large. The fillet radius (r_f) is a random variable. The equation for K_t (condition 3, above) may be solved by simulation. To develop the statistics of particular cases, see Chapter 7.

Example G

In the following example the statistics of dynamic stresses are estimated by utilizing the modified Lewis model. The following design specifications apply: (1) high commercial quality, (2) ground finish, (3) 20° involute, full depth, (4) pitch diameter, $d = 12.00$ in., (5) number of teeth, $N = 24$, and (6) AISI 1045 steel, heat treated to BHN 420.

Estimate $\bar{s}$ and s_s to satisfy these specifications and for the loading (P_t) given as follows. [utilize equation (5.45)]:

$$\text{Diametral pitch} = p = \frac{N}{d} = \frac{24}{12} = 2 \text{ teeth/in.} = \frac{\pi}{p_c}$$

$$\text{Circular pitch} = p_c = \frac{\pi d}{N} = \frac{3.1416 \cdot (12.00)}{24} = 1.5708$$

$$\text{Transmitted load} = \left(\bar{P}_t, s_{P_t}\right) = (10{,}000, 500) \text{ lb.}$$

Lewis Model for Dynamic Stresses

The face width or tooth face thickness (F) (Figure 5.32) will be the unknown to be determined at the beginning of Section 9.3.10. All other design variables are assigned mean values and standard deviations (based on machine and process capabilities). By utilizing AGMA recommendations to establish gear geometry (Table 5.7) and assuming that the geometric random values are approximately normally distributed, the following steps are performed.

1. The statistics of l are estimated from equation (2.59):

$$\bar{l} = \frac{2.000}{P} = \frac{2.000}{2.000} = 1.000$$

The variability in l is estimated from Table 4.14:

$$\pm \Delta_l = \pm 0.002 \text{ in.}; \quad s_l \simeq 0.002/3 = 0.00067 \text{ in.}; \quad (\bar{l}, s_l) = (1.000, 0.00067) \text{ in.}$$

(e)

2. $$\bar{t} = \bar{p}_c/2 = (1.57)/2 = 0.785 \text{ in.}$$

Table 5.7 Standard AGMA and USASI tooth systems for spur gears

Quantity	Coarse pitch[a] (<20P) Full Depth		Fine pitch (>20P) Full Depth
Pressure angle ϕ	20 deg	25 deg	20 deg
Addendum a	$\dfrac{1.000}{P}$	$\dfrac{1.000}{P}$	$\dfrac{1.000}{P}$
Dedendum b	$\dfrac{1.250}{P}$	$\dfrac{1.250}{P}$	$\dfrac{1.200}{P} + 0.002$ in.
Working depth h_k	$\dfrac{2.000}{P}$	$\dfrac{2.000}{P}$	$\dfrac{2.000}{P}$
Whole depth h_t (Milliinch)	$\dfrac{2.25}{P}$	$\dfrac{2.25}{P}$	$\dfrac{2.200}{P} + 0.002$ in.
Circular tooth thickness t	$\dfrac{\pi}{2P}$	$\dfrac{\pi}{2P}$	$\dfrac{1.5708}{P}$
Fillet radius of basic rack r_f	$\dfrac{0.300}{P}$	$\dfrac{0.300}{P}$	Not standardized
Basic clearance c (Milliinch)	$\dfrac{0.250}{P}$	$\dfrac{0.250}{P}$	$\dfrac{0.200}{P} + 0.002$ in.
Clearance c (shaved or ground teeth)	$\dfrac{0.350}{P}$	$\dfrac{0.350}{P}$	$\dfrac{0.3500}{P} + 0.002$ in.
Minimum number of pinion teeth	18	12	18
Minimum number of teeth per pair	36	24	
Minimum width of top land t_0	$\dfrac{0.25}{P}$	$\dfrac{0.25}{P}$	Not standardized

[a]But not including 20P.
Source: Shigley [1].

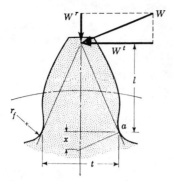

Figure 5.34 Gear-tooth stress models.

The variability in t is $s_t = 0.0067$ in., based on machining capabilities, [i.e., see equation (e)]; thus

$$(\bar{t}, s_t) \simeq (0.785, 0.00067) \text{ in.} \tag{f}$$

Estimate $(\bar{x}, s_x)$ (Figure 5.34), utilizing equations (b), (e), and (f) with normal function algebra:

$$x = \frac{t^2}{4l}$$

$$\bar{x} \simeq \frac{(\bar{t})^2}{4(\bar{l})} = \frac{(0.785)^2}{4(1.000)} = 0.1541 \text{ in.}$$

$$s_x \simeq \frac{(0.616, 0.000825)}{(4.000, 0.00268)} = 0.00053 \text{ in.}$$

$$(\bar{x}, s_x) \simeq (0.154, 0.00053) \text{ in.} \tag{g}$$

Estimate the statistics of y

$$y = \frac{2}{3} \cdot \frac{x}{P_c} = \frac{x}{3}$$

$$(\bar{y}, s_y) \simeq (0.0514, 0.000177) \text{ in.} \tag{h}$$

$$(\bar{Y}, s_Y) \simeq 3.1416(0.0514, 0.000177) = (0.162, 0.000556). \tag{i}$$

$$\bar{K}_t^* \simeq 0.18 + \left(\frac{\bar{t}}{\bar{r}_f}\right)^{0.15} \cdot \left(\frac{\bar{t}}{\bar{l}}\right)^{0.45}$$

$$\simeq 0.18 + \left[\left(\frac{0.785}{0.150}\right)^{0.15}\right] \cdot \left[\left(\frac{0.785}{1.000}\right)^{0.45}\right] = 1.329.$$

*See Shigley [1, p. 799].

From studies of typical K_t variability:

$$s_{K_t} \simeq 0.06 \cdot \overline{K}_t = 0.0666$$

$$(\overline{K}_t, s_{K_t}) \simeq (1.33, 0.0666) \simeq (\overline{K}_f, s_{K_f}) \qquad (j)$$

The geometry factor J is

$$(\overline{J}, s_J) \simeq \frac{(0.162, 0.000556)}{(1.33, 0.0666)} = (0.148, 0.009).$$

If it is given that the spur gear rotates at $(900, 45)$ rpm, the dynamic factor K_v may be estimated from the empirical formula

$$K_v \simeq \frac{78}{78 + \sqrt{v}}$$

$$(\overline{v}, s_v) = (2830, 142) \text{ fpm} = (900, 45)\frac{2\pi r}{12}$$

where $r = 6$. Utilizing normal function algebra:

$$(\overline{\sqrt{v}}, s_{\sqrt{v}}) \simeq (53.2, 0.0212)$$

$$(\overline{K}_v, s_{K_v}) \simeq (0.595, \sim 0)$$

Utilizing equation (d'):

$$(\bar{s}, s_s) \simeq \frac{(20{,}000, 1000)}{(0.5954, \sim 0)(\overline{F}, 0.01\overline{F})(0.148, 0.009)} = \frac{(272{,}300, 14{,}100)}{(\overline{F}, 0.01\overline{F})}. \qquad (k)$$

If equation (k) is utilized together with equations (2.28) and (2.30), gear tooth bending stress, as a function of F,* is

$$(\bar{s}, s_s) \simeq \left(\frac{272{,}300}{F}, \frac{14{,}320}{F}\right) \text{ psi}. \qquad (l)$$

In Section 9.3.10 design synthesis of the spur gear, described in the prior example, is completed. The face width, $(\overline{F}, s_F)$, is so calculated that the probability of surviving an indefinite number of load cycles will satisfy the requirement $R \geq 0.999$.

*Gear face width.

5.20 SURFACE STRESS

5.20.1 Spur Gear Stresses

Surface failure of spur gear teeth may result from repeated cycles of contact stresses (Figure 5.35). Thus gears must be designed such that the surface-endurance limit of the material will exceed the contact stress intensity with a suitably high degree of probability. In many cases where a surface failure will eventually occur, the first evidence appears near the pitch line since maximum dynamic loads occur near this location.

A model for estimating the statistics of the contact stress s_c at the interface of the interacting gears, is given in Table 5.8e. If the radius of curvature of the gear tooth profile at the pitch point K_1 is taken as a mean value, then

$$R_1 = R_2 = \frac{d}{2} \sin \phi$$

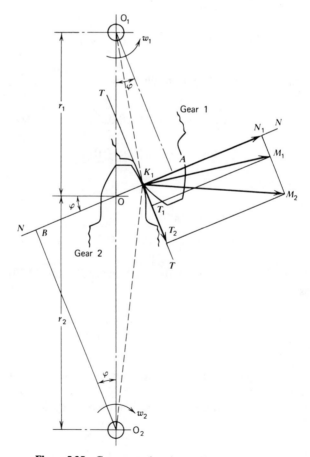

Figure 5.35 Geometry of an interacting gear pair [49].

Table 5.8. Contact Stress Equations, s_C

Spheres	Cylinders
(a) $s_c = 0.616_3 \sqrt{\dfrac{P}{R^2}\left(\dfrac{E_1 E_2}{E_1 + E_2}\right)^2}$ $a = 0.880_3 \sqrt{PR\left(\dfrac{1}{E_1}+\dfrac{1}{E_2}\right)}$ $\delta = 0.775_3 \sqrt{\dfrac{P^2}{R}\left(\dfrac{1}{E_1}+\dfrac{1}{E_2}\right)^2}$	(d) $s_c = 0.591 \sqrt{\dfrac{P_1\, E_1 E_2}{R(E_1 + E_2)}}$ $a = 1.076 \sqrt{\dfrac{P_1 R(E_1 + E_2)}{E_1 E_2}}$ $\delta = \dfrac{0.579 P_1}{E}\left(\dfrac{1}{3}+\ln\dfrac{2R}{a}\right)$
(b) $s_c = 0.616_3 \sqrt{P\left(\dfrac{1}{R_1}+\dfrac{1}{R_2}\right)^2\left(\dfrac{E_1 E_2}{E_1+E_2}\right)^2}$ $a = 0.880_3 \sqrt{\dfrac{PR_1 R_2}{(R_2-R_1)}\left(\dfrac{1}{E_1}+\dfrac{1}{E_2}\right)}$ $\delta = 0.775_3 \sqrt{P^2\left(\dfrac{1}{E_1}+\dfrac{1}{E_2}\right)^2\left(\dfrac{1}{R_1}+\dfrac{1}{R_2}\right)^2}$	(e) $s_c = 0.591 \sqrt{\dfrac{P_1 E_1 E_2}{(E_1 + E_2)}\left(\dfrac{1}{R_1}+\dfrac{1}{R_2}\right)}$ $a = 1.076 \sqrt{\dfrac{P_1 R_1 R_2}{(R_1 - R_2)}\left(\dfrac{1}{E_1}+\dfrac{1}{E_2}\right)}$
(c) $s_c = 0.616_3 \sqrt{P\left(\dfrac{1}{R_1}+\dfrac{1}{R_2}\right)^2\left(\dfrac{E_1\, E_2}{E_1-E_2}\right)^2}$ $a = 0.880_3 \sqrt{\dfrac{PR_1 R_2}{R_2-R_1}\left(\dfrac{1}{E}+\dfrac{1}{E_2}\right)}$ $\delta = 0.775_3 \sqrt{P^2\left(\dfrac{1}{E_1}+\dfrac{1}{E_2}\right)\left(\dfrac{1}{R_1}-\dfrac{1}{R_2}\right)}$ Spherical cavity	(f) $s_c = 0.591 \sqrt{\dfrac{P_1 R_1 R_2}{(R_2 - R_1)}\left(\dfrac{1}{R_1}-\dfrac{1}{R_2}\right)}$ $a = 1.076 \sqrt{\dfrac{P_1 R_1 R_2}{R_2 - R_2}\left(\dfrac{1}{E_1}+\dfrac{1}{E_2}\right)}$

where d is pitch diameter. In Example H the contact stress (s_c) statistics of the gear described in Example G are estimated as functions of F.

Example H

In this example the pitch diameters d of gears 1 and 2 are assumed equal, and $\phi = 20°$ is the pressure angle. In Example G, $\bar{d}$ was taken as 12.00 in., and variability in d due to machining tolerances (Table 4.14) was $\pm \Delta = \pm 0.010$ in.

The radii of curvature statistics of the interacting gears at the point of contact are equal, although R_1 and R_2 are statistically independent (Table 5.8):

$$\bar{R}_1 = \bar{R}_2 = \frac{12.000}{2}\sin 20° = 2.052 \text{ in.}$$

An estimation of s_{R_1} and s_{R_2} based on the range of values described by the tolerances is

$$s_{R_1} = s_{R_2} \simeq \frac{0.010}{3} = 0.0033 \text{ in.}$$

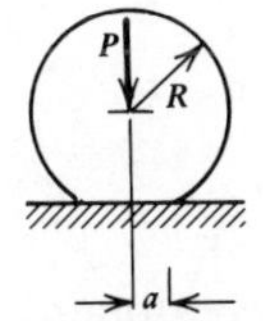

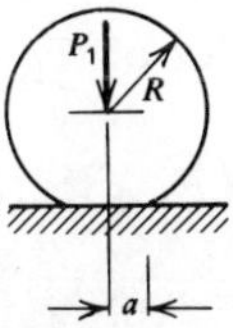

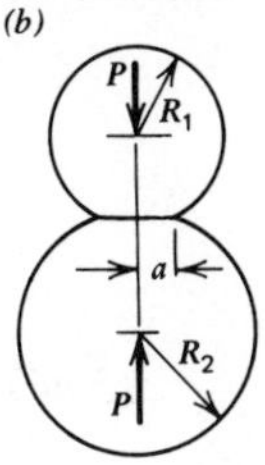

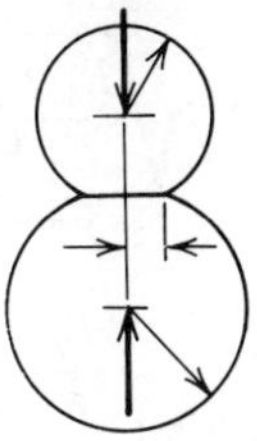

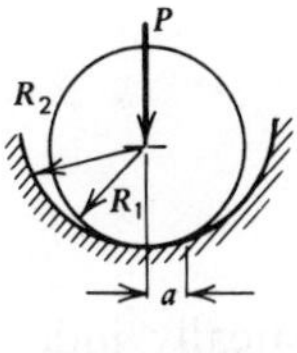

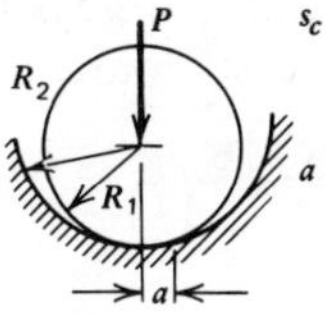

In the following examples the modulus of elasticity for the gear pair (AISI 1045 steel) is taken as (Table 5.1; see also Ref. 56)

$$E_1 = E_2 = 29 \cdot 10^6 \pm 10^6 \text{ psi}$$

Estimators of the statistics of E_1 and E_2 are

$$(\bar{E}, s_E) \simeq (29 \cdot 10^6, 3.33 \cdot 10^5) \text{ psi}$$

Utilizing the expression for contact stress, s_c, from Table 5.8e, we obtain

$$s_c = 0.591 \cdot \sqrt{\frac{P_1 \cdot E_1 \cdot E_2}{E_1 + E_2}\left(\frac{1}{R_1} + \frac{1}{R_2}\right)}$$

$$s_c = 0.591 \cdot \sqrt{\frac{P_t}{F} \cdot \frac{\frac{1}{R_1} + \frac{1}{R_2}}{\frac{1}{E_1} + \frac{1}{E_2}}} \quad ; \quad \left(P_1 = \frac{w_t}{F}\right) \quad (5.46)$$

where P_t is total transmitted gear load.

The mean value and standard deviation of P_t from Example G are

$$(\bar{P}_t, s_{P_t}) = (10{,}000, 500) \text{ lb}$$

Assuming that the variables are normally distributed and statistically independent, let

$$w_1 = \frac{1}{R_1} + \frac{1}{R_2}$$

The mean and standard deviation of $1/R_1$ and $1/R_2$ [using equations (2.31) and (2.32)] are

$$\left(\overline{\frac{1}{R_1}}, s_{1/R_1}\right) = (0.487, 0.0000025)$$

and using equations (2.19) and (2.21a):

$$(\bar{w}_1, s_{w_1}) = (0.975, 0.0011)$$

Let

$$w_2 = \frac{1}{E_1} + \frac{1}{E_2}$$

$$\left(\overline{\frac{1}{E_1}}\right) \simeq \frac{1}{29 \cdot 10^6} = 3.448 \cdot 10^{-8}$$

$$s_{1/E} \simeq 3.960 \cdot 10^{-10}; \qquad (\bar{w}_2, s_{w_2}) = (6.90 \cdot 10^{-8}, 5.550 \cdot 10^{-10})$$

From equations (2.28) and (2.30):

$$\frac{(\bar{w}_1, s_{w_1})}{(\bar{w}_2, s_{w_2})} = (14.1 \cdot 10^6, 1.160 \cdot 10^5)$$

and

$$(\bar{s}_c, s_{s_c}) = 0.591 \cdot \sqrt{\frac{(14.1 \cdot 10^{10}, 7.16 \cdot 10^9)}{(\bar{F}, s_F)}}$$

From equation (2.47):

$$\bar{s}_c \simeq \frac{2.22 \cdot 10^5}{\bar{F}^{1/2}} \text{ psi}$$

The value $s_F \simeq 0.01 \cdot \bar{F}$ is a conservative estimate of face width variability. From equation (2.50):

$$s_{s_c} \simeq \frac{1.14 \cdot 10^4}{\bar{F}^{1/2}}$$

The contact stress in the gear pair is

$$(\bar{s}_c, s_{s_c}) \simeq \left[\frac{2.22 \cdot 10^5}{\bar{F}^{1/2}}, \frac{1.14 \cdot 10^4}{\bar{F}^{1/2}} \right] \text{ psi}$$

5.20.2 Ball-bearing Stresses

The problem of selecting bearings to satisfy specified load and life requirements (from vendors' catalogs) is briefly considered in Section 10.7.2. Bearing design to satisfy specified load, life, and reliability is discussed with examples in Section 9.3.10. In this section contact stress statistics are estimated by utilizing Hertzian stress equations (see Table 5.8c), for a particular Conrad-type single-row radial ball bearing, namely, #207 in Table 5.9. The stress statistics are expressed as functions of loading P.

Example I

For a ball-bearing and race relationship at maximum developed pressure in a sphere and spherical socket, contact stress s_c is

$$s_c = (0.616) \cdot \sqrt[3]{PE^2 \left[\frac{d_2 - d_1}{d_1 d_2} \right]^2} \qquad (5.46)$$

218 The Stress Random Variable

Equation (5.46) is obtained from Table 5.8c. Assumptions in this example are

$$d_1 = 2R_1; \quad d_2 = 2R_2$$

$P = $ Dynamic applied load (pounds)

$E = $ Modulus of elasticity (psi)

$d_1 = $ Diameter of ball bearing (inches)

$d_2 = $ Diameter of bearing race (inches)

From Table 5.1 and Ref. 56 the mean value of E is estimated as $29 \cdot 10^6$ psi with a range of approximately $2 \cdot 10^6$ psi; thus

$$(\overline{E}, s_E) \simeq (29 \cdot 10^6, 3.33 \cdot 10^5) \text{ psi}$$

Bearing #207 in Table 5.9 has a mean diameter of $\overline{d}_1 = 0.4375$ in. From manufacturer's tolerance tables, a tolerance range of 0.0005 in. is reasonable, and

$$s_{d_1} = s_{d_2} \simeq 8.33 \cdot 10^{-5} \text{ in.}$$

With deep-groove radial ball bearings, maximum load capacity occurs when the diameter of the race equals that of the balls. However, with $d_2 = d_1$, friction and temperature are likely to become excessive at higher rpm values. With high loading and low rpm, a close fit may be desirable.

First, we estimate the race diameter such that $d_2 > d_1$ with a probability $R = .9999$. This is easily calculated by the methods described in Chapter 9. For $R = 0.9999$, $Z = -3.62$ (Appendix 1), from Table 5.9:

$$\overline{d}_1 = 0.4375 \text{ in.}$$

and from equation (9.10):

$$R = .9999 = \frac{1}{\sqrt{2\pi}} \int_Z^\infty e^{-t^2/2} dt = \frac{1}{\sqrt{2\pi}} \cdot \int_{-3.62}^\infty e^{-t^2/2} dt$$

Then, utilizing equation (9.11) and assuming that d_1 and d_2 follow Gaussian distributions approximately:

$$-3.62 = -\frac{\overline{d}_2 - 0.4375}{\sqrt{2(8.33 \cdot 10^{-5})^2}}$$

and

$$\bar{d}_2 = 0.4380 \text{ (minimum)}$$

Reducing friction requires increasing clearance and sacrificing some load-carrying capacity. In practice, manufacturers usually maintain a relationship such as, (Figure 5.36)

$$d_2 = 1.03 d_1 \quad \text{to} \quad d_2 = 1.08 d_1$$

The mean and the standard deviation of s_c are determined by the partial derivative method by utilizing equations (2.59) and (2.62) with equation (5.46):

$$s_c \simeq (0.616) \cdot \sqrt[3]{\bar{P}(29 \cdot 10^6)^2 \cdot \left[\frac{0.473 - 0.438}{(0.473) \cdot (0.438)} \right]^2}$$

$$\bar{s}_c \simeq (17{,}790) \cdot \sqrt[3]{\bar{P}}$$

and

$$\nu_{s_c} \simeq (612) \cdot \sqrt[3]{\bar{P}}$$

Thus

$$(\bar{s}_c, \nu_{s_c}) \approx \left[(17{,}790) \cdot \sqrt[3]{\bar{P}} \; ; \; (612) \cdot \sqrt[3]{\bar{P}} \right] \text{ psi} \qquad (a)$$

The contact stress statistics in equation (a) are utilized with other data to estimate the loading capacity of #207 type bearings for surviving 10^8 load cycles with a probability of 0.9999 (see Section 9.3.10).

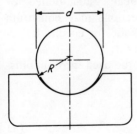

Figure 5.36 Ball-bearing geometry.

CHAPTER 6

Random-Deflection Variables

The following topics are considered in this chapter:

Deflection responses

Torsional stress, deflection, and k
Beam deflection and radius of curvature
Superposition
Indeterminate problems

Strain energy

Strain energy in tension
General strain energy expression
Castigliano's theorem

Instability

Euler model for pin-ended columns, including critical load, critical unit load and distribution of P_{cr}
Vibration

6.1 INTRODUCTION

Deflection of mechanical elements under loading is considered in this chapter. Often the component designed must satisfy a rigidity requirement such that the probability of exceeding a specified amount of deflection will be suitably small. All components of a loaded mechanical system deflect or deform, and in some members relatively small deflections may constitute failure. Alternatively, springs may be designed to provide for elastic deflections of relatively large magnitudes, as shown in Figure 6.1.
The *fundamental torsional formula* [equation (5.21)] can be used in probabilistic helical spring design (Figure 6.2):

$$\tau = \frac{Tr}{J} = \frac{8FD}{\pi d^3} \qquad (a)$$

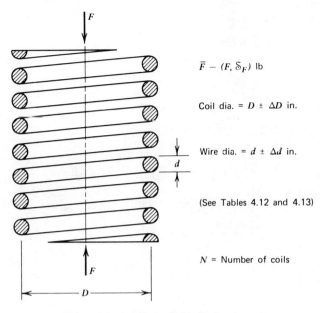

Figure 6.1 Axially loaded helical spring.

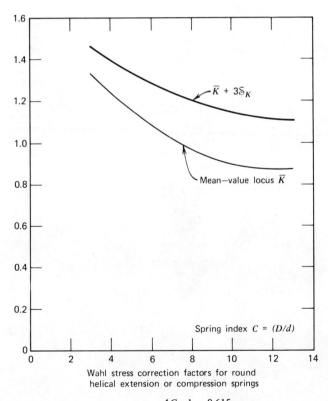

Figure 6.2 Wahl stress correction factors. $K = \dfrac{4C-1}{4C-4} + \dfrac{0.615}{C}$ where $C = D/d$, D = coil diameter, and d = wire diameter. *Note*: Values of D and d are from the tolerance tables in Chapter 4.

222 Random-Deflection Variables

The statistics of τ in terms of the statistics of F, D, and d are as follows. According to equation (2.59), $\bar{\tau}$ is

$$\bar{\tau} \approx \frac{8\,\overline{FD}}{\pi \bar{d}^3} \tag{a'}$$

From equation (2.62) the expression for s_τ is

$$s_\tau \approx \left[\left(\frac{\partial \tau}{\partial F}\right)^2 \cdot s_F^2 + \left(\frac{\partial \tau}{\partial D}\right)^2 \cdot s_D^2 + \left(\frac{\partial \tau}{\partial d}\right)^2 \cdot s_d^2\right]^{1/2} \tag{b}$$

The partial derivatives of F, D, and d with respect to τ are

$$\frac{\partial \tau}{\partial F} = \frac{8D}{\pi d^3}; \qquad \frac{\partial \tau}{\partial D} = \frac{8F}{\pi d^3}; \qquad \frac{\partial \tau}{\partial d} = \frac{-24FD}{\pi d^4} \tag{c}$$

Substituting (c) into (b), s_τ is

$$s_\tau \approx \left[\left(\frac{8\bar{D}}{\pi \bar{d}^3}\right)^2 s_F^2 + \left(\frac{8\bar{F}}{\pi \bar{d}^3}\right)^2 s_D^2 + \left(\frac{24\,\overline{FD}}{\pi \bar{d}^4}\right)^2 s_d^2\right]^{1/2}$$

$$J = \pi d^4/32; \qquad r = d/2; \qquad T = F \cdot D/2. \tag{a''}$$

Example A

Given are the following statistics describing the variables F, D, and d

$$(\bar{F}, s_F) = (80, 11.5) \text{ lb}$$
$$(\bar{D}, s_D) = (2.50, 0.125) \text{ in.}$$
$$(\bar{d}, s_d) = (0.1875, 0.005) \text{ in.}$$

Utilize equations (a') and (a'') to compute $\bar{\tau}$ and s_τ of a helical spring

$$\bar{\tau} \approx \frac{8\,\overline{FD}}{\pi \bar{d}^3} = \frac{8(80)(2.50)}{\pi(0.1875)^3} = 77{,}300 \text{ psi}$$

$$s_\tau \approx \left[\left(\frac{8\bar{D}}{\pi \bar{d}^3}\right)^2 \cdot s_F^2 + \left(\frac{8\bar{F}}{\pi \bar{d}^3}\right)^2 s_D^2 + \left(\frac{24\,\overline{FD}}{\pi \bar{d}^4}\right)^2 s_d^2\right]^{1/2}$$

$$\approx \left[\left(\frac{8(2.5)}{3.14(0.1875)^3}\right)^2 \cdot (11.5)^2 + \left(\frac{8(80)}{3.14(0.1875)^3}\right)^2 \cdot (0.125)^2\right.$$

$$\left. + \left(\frac{24(80)(2.50)}{\pi(0.1875)^4}\right)^2 (0.005)^2\right]^{1/2}$$

$$s_\tau \approx 13{,}000 \text{ psi}$$

The statistics of torsional stress are

$$(\bar{\tau}, s_\tau) \approx (77{,}300;\ 13{,}000)\ \text{psi}$$

The model for deflection, y, of a helical spring is

$$y = \frac{8FD^3N}{d^4 \cdot G}\ \text{in.} \qquad (d)$$

where

$$E = 2G(1 + \mu)$$

The statistics of y in terms of the statistics of F, D, d, and G are estimated by using equations (2.59) and (2.62) with equation (d):

$$\bar{y} = \frac{8\bar{f}\bar{D}^3 N}{\bar{d}^4 \bar{G}}\ \text{in.} \qquad (d')$$

$$s_y \approx \left[\left(\frac{8\bar{D}^3 N}{\bar{d}^4 \bar{G}}\right)^2 s_F^2 + \left(\frac{24\bar{F}\bar{D}^2 N}{\bar{d}^4 \bar{G}}\right)^2 \cdot s_D^2 \right.$$

$$\left. + \left(\frac{32\bar{F}\bar{D}^3 N}{\bar{d}^5 \bar{G}}\right)^2 \cdot s_d^2 + \left(\frac{8\bar{F}\bar{D}^3 N}{\bar{d}^4 \bar{G}^2}\right)^2 \cdot s_G^2 \right]^{1/2}\ \text{in.} \qquad (d'')$$

The *spring rate* k is

$$k = \frac{d^4 G}{8 D^3 N}\ \text{lb/in.} \qquad (e)$$

The statistics of k in terms of the statistics of d, G, and D are

$$\bar{k} = \frac{\bar{d}^4 \bar{G}}{8 \bar{D}^3 N} \qquad (e')$$

$$s_k \approx \left[\left(\frac{\bar{d}^3 \bar{G}}{2\bar{D}^3 N}\right)^2 \cdot s_d^2 + \left(\frac{\bar{d}^4}{8\bar{D}^3 N}\right)^2 \cdot s_G^2 + \left(\frac{-3\bar{d}^4 \bar{G}}{8\bar{D}^4 N}\right)^2 \cdot s_D^2 \right]^{1/2} \qquad (e'')$$

Example B

To study the distribution of k, we plot simulated random spring constant values on normal and lognormal papers. Assume that d, G, and D are normal random variables. The model for k is equation (e). The following data are

224   Random-Deflection Variables

given:

$$(\bar{d}, s_d) = (0.063, 0.00017) \text{ in.}, \quad (\bar{D}, s_D) = (2.00, 0.014) \text{ in.},$$
$$(\bar{E}, s_E) = (30 \cdot 10^6, 4.5 \cdot 10^5) \text{ lb/in.}^2, \quad (\bar{G}, s_G) = (1.15 \cdot 10^7, 1.73 \cdot 10^5) \text{ lb/in.}^2;$$
$$N = 10$$

From equations (e') and (e") the calculated values of $\bar{k}$ and s_k are

$$(\bar{k}, s_k) \approx (0.284, 0.0079) \text{ lb/in.}$$

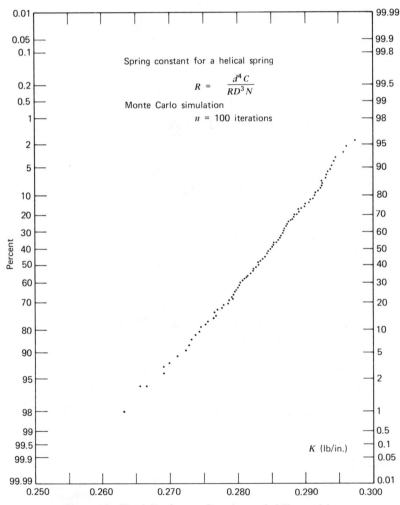

Figure 6.3 Fit of K values to Gaussian probability model.

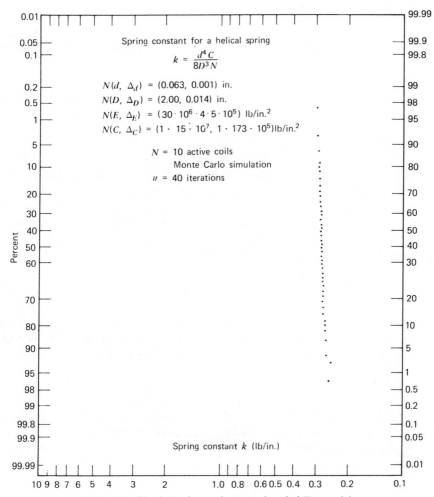

Figure 6.4 Fit of K values to lognormal probability model.

A Monte Carlo simulation of k,* using the data in equations (e), yields

$$(\bar{k}, s_k) \approx (0.285, 0.0078) \text{ lb/in.}$$

Figures 6.3 and 6.4 show plots of $n=100$ and $n=40$ values of k (on normal and lognormal probability paper) from the Monte Carlo simulation, indicating that either the normal or the lognormal distributions may provide reasonable models for K.

When rigidity is a design criterion, materials having high mean tensile strength may be only negligibly, if at all, more suitable than lower-strength

---

*See program Monti, Chapter 7, Section 7.4.6.

226 Random-Deflection Variables

materials having comparable statistical elastic properties, measured by the modulus of elasticity (E) and shear modulus of elasticity (G). Large variances can be as damaging in terms of performance of population as can lower mean values.

DEFLECTION

6.2 DEFLECTION OF BEAMS

Since a multitude of mechanical components are modeled as beams, the deflection of beams is considered next. Assume that the Curve AmB in Figure 6.5 is the neutral axis of a simply supported beam after bending and that the curvature of the deflection curve at a given point depends on the magnitude of the bending moment at the point. The relation between curvature and the moment is the same as a case of pure bending, the model for which is

$$\frac{1}{r_c} = \frac{M}{EI} \qquad (6.1)$$

The presence of I in equation (6.1) suggests that the radius of curvature r_c will be sensitive to variability in cross-sectional dimensions since higher powers of a dimensional random variable are involved in moment of inertia I.

Example C

Estimate the increase in radius of curvature standard deviation with increasing variability in cross-sectional radius in a beam. To achieve a reliability of $R = .99999$, a certain tubular cross-sectional beam requires an outside radius

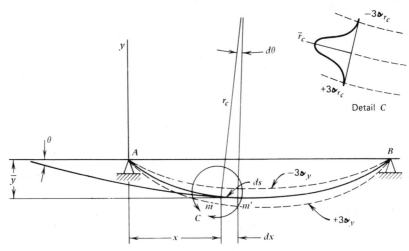

Figure 6.5 Beam deflection.

r_0 of

$$(\bar{r}_0, s_{r_0}) = (2.792, 0.0419) \text{ in.}$$

and a wall thickness of

$$\bar{t} = 0.0558 \text{ in.}$$

A model for moment of inertia I (Appendix 5), is

$$I \approx \pi r_0^3 \cdot t \tag{a}$$

The statistics of r_0^3, obtained from the statistics of r_0, are required. Since the scatter in r_0 is small, from equation (2.59):

$$E(r_0^3) \approx f[E(x_1), E(x_2), \ldots, E(x_n)] = f(\bar{r}_0) \approx \bar{r}_0^3$$

and from equation (2.62):

$$s_{r_0^3} \approx \left[\sum_{i=1}^{n} \left(\frac{\partial r_0^3}{\partial x_i}\right)^2 \cdot s_{x_i}^2\right]^{1/2}$$

$$\frac{\partial r_0^3}{\partial x_1} = \frac{dr_0^3}{dr_0} = 3\bar{r}_0^2$$

and

$$s_{r_0^3} \approx \left[(3\bar{r}_0^2)^2 \cdot s_{r_0}^2\right]^{1/2} = 3\bar{r}_0^2 s_{r_0}$$

Thus

$$(\bar{r}_0)^3, s_{r_0^3} \approx (\bar{r}_0^3, 3\bar{r}_0^2 \cdot s_{r_0}) \text{ in.}^3$$

Following the same procedure as for r_0, the variance on moment of inertia [equation (a)] is

$$s_I^2 \approx (3\pi \bar{r}_0^2 \bar{t})^2 \cdot s_{r_0}^2 + (\pi \bar{r}_0^3)^2 \cdot s_t^2$$

$$s_I^2 = 16 \cdot 80 (s_{r_0}^2) + 0.363$$

Given.

$$(\bar{M}, s_M) = (174{,}000;\ 5760) \text{ in. lb}$$

$$(\bar{E}, s_E) \approx (30 \cdot 10^6;\ 4.50 \cdot 10^5) \text{ lb/in.}^2$$

$$\bar{I} = 2.86 \text{ in.}^4 \tag{b}$$

228 Random-Deflection Variables

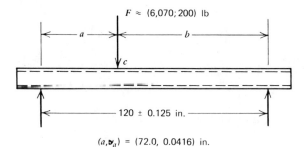

$(a, \mathbf{v}_a) = (72.0, 0.0416)$ in.

Figure 6.6 Tubular beam model.

If equation (6.1) is utilized, the radius of curvature r_c is

$$(\bar{r}_c, s_{r_c}) = \frac{(\bar{E}, s_E)(\bar{I}, s_I)}{(\bar{M}, s_M)} \tag{c}$$

and from equation (2.59), $\bar{r}_c$ at point c (Figure 6.6) is

$$\bar{r}_c \approx \frac{30 \cdot 10^6 \cdot 2.86}{174{,}000} = 494 \text{ in.} \tag{c'}$$

The variance on the radius of curvature as a function of s_I^2, from equation (c), is obtained by taking partial derivatives of M, E, and I with respect to r_c and substituting values into equation (2.62):

$$s_{r_c} \approx \left[\left(\frac{\partial r_c}{\partial M} \right)^2 \cdot s_M^2 + \left(\frac{\partial r_c}{\partial E} \right)^2 \cdot s_E^2 + \left(\frac{\partial r_c}{\partial I} \right)^2 \cdot s_I^2 \right]^{1/2}$$

$$\frac{\partial r_c}{\partial M} = \frac{\bar{E}\bar{I}}{\bar{M}^2}$$

$$\frac{\partial r_c}{\partial E} = \frac{\bar{I}}{\bar{M}}$$

$$\frac{\partial r_c}{\partial I} = \frac{\bar{E}}{\bar{M}}$$

$$s_{r_c}^2 = \left(\frac{\bar{E}\bar{I}}{\bar{M}^2} \right)^2 s_M^2 + \left(\frac{\bar{I}^2}{\bar{M}^2} \right) s_E^2 + \left(\frac{\bar{E}^2}{\bar{M}^2} \right) s_I^2 \tag{c''}$$

Substitution from equation (b) into equation (c'') yields

$$s_{r_c}^2 = 267.2 + 54.9 + 29{,}730 \cdot (s_I^2)$$

Table 6.1 Standard Deviation of Curvature Radius versus Standard Deviation in Cross-sectional Outside Radius

s_{r_0} (in.)	s_I^2 (in.^{4})	s_{r_c} (in.)
0.042	0.065	47.6
0.125	0.298	95.8
0.374	2.39	267.2

The standard deviation of radius of curvature s_{r_c} versus standard deviation in cross-sectional outside radius s_{r_0} is given in Table 6.1.

6.3 DEFLECTION OF A SIMPLY SUPPORTED BEAM WITH CONCENTRATED LOAD

To formalize the relationships among the random variables F, a, b, and l that influence deflection y, the following models from solid mechanics are useful. Expressions for bending moment corresponding to portions AC and CB in Figure 6.7 are

$$M_{AC} = -\frac{Fb}{l}x; \quad M_{BC} = -\frac{Fb}{l}x + F(x-a) \tag{6.2}$$

Also

$$EI\left(\frac{d^2y}{dx^2}\right) = -M \tag{6.3}$$

Equating equations (6.2) and (6.3) yields

$$EI\left(\frac{d^2y}{dx^2}\right) = -\frac{Fb}{l}x \quad \text{for} \quad x \leqslant a$$

$$EI\left(\frac{d^2y}{dx^2}\right) = -\frac{Fb}{l}x + F(x-a) \quad \text{for} \quad x \geqslant a \tag{a}$$

Thus

$$EI\frac{dy}{dx} = \frac{Fb}{6l}(l^2 - b^2 - 3x^2) \quad \text{for} \quad x \leqslant a$$

$$EI\frac{dy}{dx} = \frac{Fb}{6l}(l^2 - b^2 - 3x^2) + \frac{F(x-a)^2}{2} \quad \text{for} \quad x \geqslant a \tag{b}$$

and

$$EI_y = \frac{Fb_x}{6l}(l^2 - b^2 - x^2) \quad \text{for} \quad x \leq a$$

$$EI_y = \frac{Fb_x}{6l}(l^2 - b^2 - x^2) + \frac{F(x-a)^3}{6} \quad \text{for} \quad x \geq a \tag{c}$$

At point A:

$$\theta_1 = \left(\frac{dy}{dx}\right)_{x=0} = \frac{Fb(l^2 - b^2)}{6lEI} \tag{d}$$

and at point B:

$$\theta_2 = \left(\frac{dy}{dx}\right)_{x=l} = \frac{F \cdot ab(l+a)}{6lEI} \tag{e}$$

By utilizing the functional models, equations (b), (c), (d), and (e), the statistics of slope (dy/dx) and deflection y can be estimated.

For example, consider equation (c). The statistics of y may be estimated in terms of the statistics of E, I, F, b, and l at a selected point x. From equation (2.59) with equation (c):

$$\bar{y} \approx \frac{\overline{Fb_x}}{6\overline{E}\overline{I}\,\overline{l}} \cdot (\bar{l}^2 - \bar{b}^2 x^2) \tag{c'}$$

According to equation (2.62), the expression for s_y is

$$s_y \approx \left[\left(\frac{\partial y}{\partial F}\right)^2 s_F^2 + \left(\frac{\partial y}{\partial b}\right)^2 s_b^2 + \left(\frac{\partial y}{\partial E}\right)^2 s_E^2 + \left(\frac{\partial y}{\partial I}\right)^2 s_I^2 + \left(\frac{\partial y}{\partial l}\right)^2 s_l^2 + \left(\frac{\partial y}{\partial x}\right)^2 s_x^2\right]^{1/2} \tag{c''}$$

Maximum deflection occurs at a statistically described point where $(dy/dx) = 0$. If $a > b$ in Figure 6.7, the maximum deflection occurs between A and C; thus equating equation (b) with 0 yields

$$l^2 - b^2 - 3x^2 = 0; \quad x = \frac{\sqrt{l^2 - b^2}}{\sqrt{3}} \tag{f}$$

Since l and b are dimensional and location random variables, from equations (2.59), (2.62), and (f), the location statistics of y_{max} are

$$\bar{x} \approx \frac{\sqrt{\bar{l}^2 - (\bar{b})^2}}{\sqrt{3}} \tag{6.4}$$

and

$$\sigma_x \approx \frac{1}{\sqrt{3}} \left[\frac{\bar{l}^2 s_l^2 + (-\bar{b})^2 s_b^2}{\bar{l}^2 - \bar{b}^2} \right]^{1/2} \tag{6.5}$$

Thus the model for y_{max} is

$$y_{max} = \frac{Fb(l^2 - b^2)^{3/2}}{9\sqrt{3} \cdot EI} \tag{6.6}$$

and for loading at midspan,

$$y_{(b=l/2)} = \frac{Fl^3}{48 \cdot EI} \tag{6.7}$$

6.4 DEFLECTION WITH LOAD SUPERPOSITION

With a simply supported beam, subjected to both uniform and concentrated loads, deflection at any point is described by the sum statistics of the random deflections attributable to each load acting separately. Positions of the acting loads are immaterial; however, deflections must be summed for the same point on the beam. Loads independence or correlation must be considered (Figure 6.7).

Example D

A 1.250-in.-diameter (mean) steel shaft on which two gears are mounted is shown in Figure 6.8. Since locations are closely controlled, geometric variability can be ignored in this case. Estimate the mean shaft deflection $\bar{y}$ and the two-sided tolerance Δ_y at point A due to forces F_1 and F_2 (see Figure 6.9).

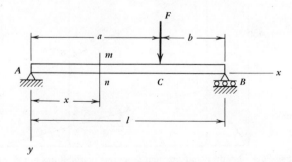

Figure 6.7 Simply supported beam model.

232 Random-Deflection Variables

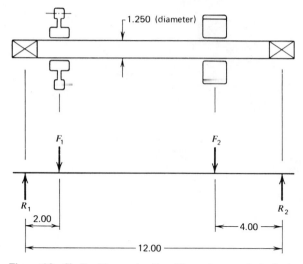

Figure 6.8 Shaft with gear loading (dimensions are in inches).

Forces acting on the shaft are

$$\left(\overline{F}_1, s_{F_1}\right) = (120, 15) \text{ lb}$$

$$\left(\overline{F}_2, s_{F_2}\right) = (90, 12) \text{ lb}$$

The shaft is simply supported by self-aligning bearings, with gears assumed to act in a single plane. Forces F_1 and F_2 are independent, and $E \approx (29.5 \cdot 10^6, 5.0 \cdot 10^5)$ psi describes the modulus of elasticity (Table 5.1).

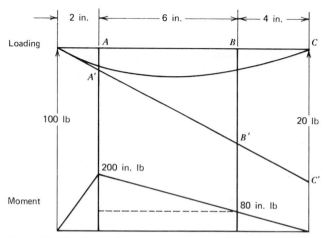

Figure 6.9 Loading and moment diagrams (variabilities not shown).

The statistics of moment of inertia for a shaft of circular cross section (Appendix 5) are

$$\bar{I} \approx \frac{\pi \bar{d}^4}{64}$$

$$s_I \approx \frac{\pi \bar{d}^3}{16} \cdot s_d$$

In this example $s_d \approx 0.015\bar{d}$; thus

$$\bar{I} \approx \frac{3.1416(1.250)^4}{64} = 0.120 \text{ in.}^4$$

$$s_I \approx 0.00719 \text{ in.}^4$$

$$\overline{EI} \approx (29.5 \cdot 10^6) \cdot (0.120) = 3.42 \cdot 10^6$$

Utilizing the concept of superposition, calculate the deflection statistics at F_1 due to the action of forces F_1 and F_2. Consider all dimensions as deterministic.

1. Estimate the deflection y_1 due to $(\bar{F}_1, s_{F_1})$ at point A (Figure 6.9). From equation (c):

$$EI \cdot y_1 = \frac{xF_1 b_1}{6l}(l^2 - b_1^2 - x^2)$$

$$(\bar{y}_1, s_{y_1}) = \frac{2(120, 15)10}{6(3.42 \cdot 10^6, 9.43 \cdot 10^4)12}\left[(12.00)^2 - (10.00)^2 - (2.00)^2\right]$$

$$= \frac{11.111 \cdot (120, 15)}{(3.42 \cdot 10^6; 2.16 \cdot 10^5)} \text{ in.}$$

$$(\bar{y}_1, s_{y_1}) = (3.89 \cdot 10^{-4}; 4.99 \cdot 10^{-5}) \text{ in.}$$

2. Estimate the deflection y_2 due to (F_2, s_{F_2}), at point A (Figure 6.9):

$$EI \cdot y_2 = \frac{xF_2 b_2}{6l} \cdot (l^2 - b_2^2 - x^2)$$

$$(\bar{y}_2, s_{y_2}) = \frac{4(90, 12)2}{6 \cdot 12(EI)}\left[(12.00)^2 - (4.00)^2 - (2.00)^2\right]$$

$$= \frac{13.78(90, 12)}{(3.42 \cdot 10^6, 9.43 \cdot 10^4)}$$

$$(\bar{y}_2, s_{y_2}) = (3.63 \cdot 10^{-4}, 3.58 \cdot 10^{-5}) \text{ in.}$$

3. By the principal of superposition, *total deflection y* is

$$(\bar{y}, s_y) = (\bar{y}_1, s_{y_1}) + (\bar{y}_2, s_{y_2}) = (7.52 \cdot 10^{-4}, 6.14 \cdot 10^{-5}) \text{ in.}$$

The tolerance is

$$\pm 3 s_y \approx \pm \Delta_y = \pm 1.843 \cdot 10^{-4} \text{ in.}$$

6.5 INDETERMINATE PROBLEMS

Situations arise in design where there is insufficient information to directly determine all unknowns. A greater number of unknowns may exist than there are equations of equilibrium. For an indeterminate beam [53], the maximum moment cannot be determined from the equations of static equilibrium, making it necessary to first determine the deflections.

Consider a propped cantilever beam supporting a uniformly distributed load, w lb/in. (Figure 6.10). Beam weight imposes a negligible effect. At the free end the reaction is R_1 and at the fixed end, reaction R_2 and moment M exist. The vertical forces and the moments about any axis must sum to zero. In this case there are two equilibrium relationships and three unknowns. To calculate M, the conditions of zero deflection at point A and B and zero slope at point B must be utilized.

The model [equation (6.1)] is

$$\frac{1}{r_c} = \frac{M}{EI} = \frac{d^2y}{dx^2} \quad \text{(a)}$$

Integrating (a) and substituting for M, the expression for slope as a function of distance x from the left end is

$$EI \frac{dy}{dx} = \frac{R_1 x^2}{2} - \frac{wx^3}{6} + c_1 \quad \text{(b)}$$

With slope 0 at point B, the condition at $x = l$ is $(dy/dx) = 0$, and

$$c_1 = \frac{wl^3}{6} - \frac{R_1 l^2}{2} \quad \text{(c)}$$

Substituting equation (c) into equation (b) and integrating,

$$EI \cdot y = \frac{R_1 x^3}{6} - \frac{wx^4}{24} + \frac{wl^3 x}{6} - \frac{R_1 l^2 x}{2} + c_2 \quad \text{(d)}$$

Indeterminate Problems 235

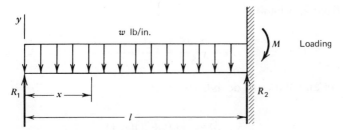

Figure 6.10 Uniformly loaded cantilever beam.

From Figure 6.11, $y=0$ at point B, and when $x=l$, $y=0$. For these conditions, equation (d) yields

$$\frac{R_1 l^3}{6} + \frac{wl^4}{24} + \frac{wl^4}{6} - \frac{R_1 l^3}{2} = 0 \tag{e}$$

The mean value and standard deviation of $wl = F_T$ are, from equations (2.21) and (2.23):

$$\bar{F}_T = \bar{w} \cdot \bar{l}; \qquad s_{F_T} = \sqrt{\bar{w}^2 s_l^2 + \bar{l}^2 s_w^2 + s_w^2 \cdot s_l^2}$$

From equation (e):

$$R_1 = \frac{3wl}{8} = \frac{3}{8} \cdot F_T$$

$$R_2 = F_T - R_1 = F_T - \frac{3}{8} F_T = \frac{5}{8} F_t$$

The transformations from F_T to R_1 and R_2 are linear, thus resulting in distributions of the same general type. Further, if $(s_l/\bar{l}) \ll 1, l$ may be treated as an approximate constant, and the distributions of F_T, R_1, and R_2 would approximate that of w.

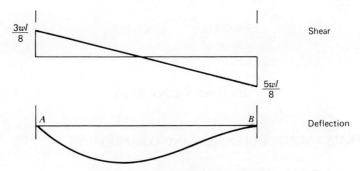

Figure 6.11 Shear and deflection diagrams.

The fixed end bending moment M is

$$M = \frac{3wl^2}{8} - \frac{wl^2}{2} = -\frac{wl^2}{8} \tag{f}$$

The mean and standard deviations of M are

$$\overline{M} = \frac{\overline{w}}{8}(\overline{l}^2 + s_l^2)(-1)$$

and

$$s_M + \frac{1}{8}\sqrt{4\overline{w}^2\overline{l}^2 \cdot s_l^2 + 2\overline{w}^2 \cdot s_l^4 + \overline{l}^2 \cdot s_w^2 + s_l^2 \cdot s_w^2 + 4\overline{l}^2 \cdot s_l^2 \cdot s_w^2 + 2s_l^4 \cdot s_w^2}$$

The model for maximum deflection may now be completed. Substitution of $R_1 = \frac{3}{8}wl = \frac{3}{8}F_T$ into equation (d) with $c_2 = 0$ yields

$$EI \cdot y = \frac{R_1 x^3}{6} - \frac{wx^4}{24} + \frac{wl^3 x}{6} - \frac{R_1 l^2 x}{2}$$

$$y = \frac{w}{48EI}(3lx^3 - 2x^4 - l^3 x) \tag{g}$$

Taking the derivative of equation (g) yields

$$48EI\frac{dy}{dx} = \frac{3}{16}wlx^2 - \frac{wx^3}{6} - \frac{wl^3}{48}$$

where $dy/dx = 0$ yields the point of maximum deflection

$$\frac{3wlx^2}{16} - \frac{wx^3}{6} - \frac{wl^3}{48} = 0$$

Utilizing the preceding model, the statistics of x are

$$\bar{x} = 0.421\bar{l}; \qquad s_x = 0.421 s_l$$

STRAIN ENERGY

6.6 STRAIN ENERGY IN TENSION AND COMPRESSION

Since a material body in motion possesses kinetic energy, changes in motion mean changes in the kinetic-energy content. Both stationary and moving components of a mechanical system are often subjected to dynamic

Strain Energy in Tension and Compression

loads. Dynamic forces from rotating gears, flywheels, pulleys, cams, and so on are transferred to frames or supports through bearings, linkages, and springs. Absolute rigidity never exists, and dynamic loads require energy absorption.

According to the principle of energy conservation, external work done on a body may be stored internally as strain energy. This concept is useful in determining stresses and deflections due not only to energy loads, but also to static forces.

During a simple tension test under monotonic loading, work is done on the bar, which is partially or completely converted to potential energy of strain. If strain remains within the elastic limit, the work is completely converted into potential energy that can be recovered during gradual unleading.

If F is a load of random magnitude and the corresponding random elongation is δ, the strain energy diagram will be as shown in Figure 6.12.

The term F_1 indicates a mean intermediate load and δ_1, the mean elongation resulting from it. An increase dF_1 causes an increase $d\delta_1$ in elongation. The work done by F_1 during this elongations is $F_1 d\delta_1$. If allowance is made for the increase of F during elongation, the work done is represented by the area $abcd$. Total work in increasing the load from O to F is the summation of all the elemental areas enclosed in OAB. It represents the mean total energy U stored in the bar during strain:

$$\text{Work} = \text{total energy} = U = \frac{F\delta}{2} \qquad \text{(a)}$$

It is assumed that F and δ are linearly related, hence correlated, $\rho = +1$. If the distribution of F is known, the distribution of δ is defined. Utilization of equation (a) with equation (2.42) yields

$$\overline{U} = \tfrac{1}{2}\bigl(\overline{F}\overline{\delta} + \rho \cdot s_F s_\delta\bigr) \qquad \text{(a')}$$

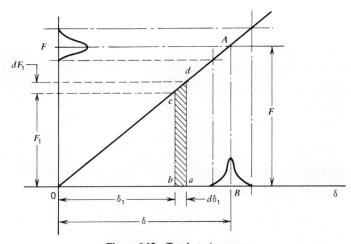

Figure 6.12 Total strain energy.

Application of equation (2.43) with equation (a) yields

$$s_U = \tfrac{1}{2}\left[\bar{F}^2 s_\delta^2 + \bar{\delta}^2 s_F^2 + s_F^2 \cdot s_\delta^2 + 2\rho \bar{F}\bar{\delta} s_F s_\delta + \rho^2 s_F^2 s_\delta^2\right]^{1/2} \quad (a'')$$

Utilization of equation (5.31) yields

$$\delta = \frac{Fl}{AE} \quad (b)$$

The statistics of δ are:

$$\bar{\delta} \simeq \frac{\bar{F}\bar{l}}{\overline{AE}}. \quad (b')$$

$$s_\delta \simeq \left[\left(\frac{\bar{l}}{\overline{AE}}\right)^2 s_F^2 + \left(\frac{\bar{F}}{\overline{AE}}\right)^2 s_l^2 + \left[\frac{\bar{F}\bar{l}}{\overline{EA}^2}\right]^2 s_A^2 + \left[\frac{\bar{F}\bar{l}}{\overline{AE}^2}\right]^2 s_E^2\right]^{1/2} \quad (b'')$$

If equations (a) and (b) are combined, the model for strain energy in a bar is

$$U = \frac{F^2 l}{2AE} \quad (c)$$

$$U = \frac{AE\delta^2}{2l} \quad (d)$$

Since $\varepsilon = (\delta/l)$ [equation (5.27)] and $s = \varepsilon E$ [equation (5.28)]:

$$U = \frac{s^2 A l}{2E} \quad (e)$$

The statistics of U [equation (e)], expressed in terms of the statistics of s, A, l, and E are as follows. According to equation (2.59), the mean value estimator $\bar{U}$ is

$$\bar{U} \approx \frac{\bar{s}^2 \bar{A}\bar{l}}{2\bar{E}} \quad (e')$$

From equation (2.62) the standard deviation estimator s_U is

$$s_U \approx \left[\left(\frac{\partial U}{\partial s}\right)^2 s_s^2 + \left(\frac{\partial U}{\partial A}\right)^2 s_A^2 + \left(\frac{\partial U}{\partial l}\right)^2 s_l^2 + \left(\frac{\partial U}{\partial E}\right)^2 s_E^2\right]^{1/2}$$

The required partial derivatives are

$$\frac{\partial U}{\partial s} = \frac{sAl}{E}$$

$$\frac{\partial U}{\partial A} = \frac{s^2 l}{2E}$$

$$\frac{\partial U}{\partial l} = \frac{s^2 A}{2E}$$

$$\frac{\partial U}{\partial E} = \frac{s^2 Al}{2E^2}$$

$$s_U \approx \left[\left(\frac{\bar{s}\bar{A}\bar{l}}{\bar{E}}\right)^2 \cdot s_s^2 + \left(\frac{\bar{s}^2 \bar{l}}{2\bar{E}}\right)^2 \cdot s_A^2 + \left(\frac{\bar{s}^2 \bar{A}}{2\bar{E}}\right)^2 \cdot s_l^2 + \left(\frac{\bar{s}^2 \bar{A}\bar{l}}{2\bar{E}^2}\right)^2 \cdot s_E^2 \right]^{1/2} \quad (e'')$$

Example E

Strain energy in a statically loaded wire is described in this example. Utilizing equation (c), estimate the statistics of U by partial derivative methods and by Monte Carlo simulation. Plot the simulation results on normal and lognormal probability paper. Given:

$$(\bar{F}, s_F) = (250, 25) \text{ lb}$$

$$(\bar{l}, s_l) = (10.00, 0.125) \text{ in.}$$

$$(\bar{E}, s_E) = (30, 0.45) 10^6 \text{ psi.}$$

$$(\bar{d}, s_d) = (0.063, 0.00016) \text{ in.} \left(A = \frac{\pi}{4} d^2\right)$$

From equations (2.59) and (2.62) with (c):

$$(\bar{U}, s_U) \approx (3.342, 0.672) \text{ in. lb}$$

Based on $n = 200$ Monte Carlo (see Section 4.6) simulated values of U, the mean and the standard deviation estimators are

$$(\bar{U}, s_U) \simeq (3.381, 0.687) \text{ in. lb}$$

Figures 6.13 and 6.14 show plots of the simulated values of U on normal and lognormal probability paper. Indications are that the parent population of U may be either normally or lognormally distributed. However, χ^2 and

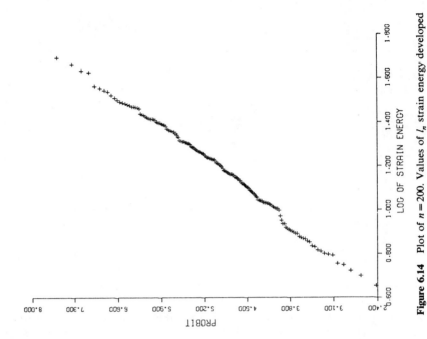

Figure 6.14 Plot of $n = 200$. Values of l_n strain energy developed by Monte Carlo simulation (lognormal plot).

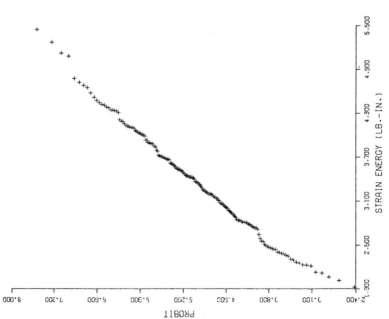

Figure 6.13 Plot of $n = 200$. Values of strain energy developed by Monte Carlo simulation (normal plot).

Strain Energy in Tension and Compression 241

Kolmogornov–Smirnov tests favor the normal model (see Bayes procedures in Section 3.2.2).

When compared with the equations for static stress, $s=(F/A)$ and $\tau=(F/A)$, the expression $U=(s^2 lA/2E)$ provides interesting implications. The average static stress in a body, subject to a force, depends on the area alone. Capacity for absorbing energy, however, depends on the length and cross-sectional area of the component and the value of the modulus of elasticity E [see equation (c)].

To absorb large amounts of energy, a mechanical component should be long, have small cross-sectional area, and have a low modulus of elasticity. The largest part of the energy will be stored in components at highly stressed local points of discontinuity, sometimes conducive to premature failure. Components subjected to energy loads should be designed for uniform stress distribution throughout, thus providing maximum energy absorption at minimum stress values.

Example F

The two tension bolts shown in Figure 6.15 are intended to resist sinusoidal fully reversed loading. Both are 1.000-in.-diameter 8NC having minor (root) diameters of 0.850 in. The effective stressed length of each bolt is 16.0 in. The material is AISI C1020 steel (1 in. round, cold drawn). The mean modulus of elasticity is $\bar{E}=30 \cdot 10^6$ psi. (Table 5.1). Determine the mean energy that each bolt can absorb.

In Figure 6.15 the mean root area $\bar{A}$ of the threads of the bolts is 0.565 in.² The stress in the threaded area of both bolts is 40,000 psi. The shank cross-sectional area $\bar{A}_1$ of bolt a is 0.785 in.² The mean stress $\bar{s}_2$ in the shank area of bolt b is 40,000 psi; thus mean stress $\bar{s}_1$ in the shank of bolt a is

$$\bar{s}_1 = \frac{\bar{A}_2}{\bar{A}_1} \cdot \bar{s}_2 = \frac{0.565}{0.785}(40,000) = 28,800 \text{ psi}.$$

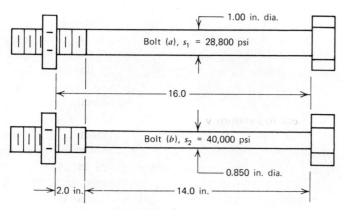

Figure 6.15 Tension bolts.

The energy absorbable by bolt a is

$$\bar{U}_{(a)} = \frac{\bar{s}_1^2 \bar{A}_1 \bar{l}_1}{2\bar{E}} + \frac{\bar{s}_2^2 \bar{A}_2 \bar{l}_2}{2\bar{E}}$$

$$= \frac{(4 \cdot 10^4)^2 (0.565)(2)}{2(30 \cdot 10^6)} + \frac{(2.88 \cdot 10^4)^2 (0.785)(14)}{2(30 \cdot 10^6)}$$

$$= 182 \text{ lb in.}$$

The energy absorbable by bolt b is

$$\bar{U}_b \approx \frac{\bar{s}_2^2 \bar{A}_2 \bar{l}}{2\bar{E}} = \frac{(4 \cdot 10^4)^2 (0.565)(16)}{2(30 \cdot 10^6)} = 241.0 \text{ lb in.}$$

$$\frac{\bar{U}_b}{\bar{U}_a} = 1.32$$

which implies that the reduction of bolt b shank diameter effectively increased the mean energy load capacity by 32% without increasing the average level of stress. Variability in U_a and U_b is estimated by using equations (2.62) and (e).

6.7 GENERAL EXPRESSION FOR STRAIN ENERGY

In cases of axial tension and compression, twist, and bending, the model for energy of strain can be a function of the second degree in external forces or a function of the second degree in displacements. This holds for the general deformation of an elastic body provided that (1) the material follows Hooke's law, subject to random variations, and (2) conditions are such that there is a high probability that displacements due to strain do not effect the action of the external forces. With these provisions, the displacements of an elastic system may be considered approximately linear functions of the external loads (see Figure 6.16).

Consider a mechanical element subjected to the action of external forces having average values $\bar{F}_1, \bar{F}_2, \ldots, \bar{F}_n$ and supported in such a manner that movement as a rigid body is negligible. Displacements are due to elastic deformations alone. Let $\bar{\delta}_1, \bar{\delta}_2, \ldots, \bar{\delta}_n$ denote the average displacements of the points of application of the forces, each measured in the direction of the corresponding force. If the external forces increase gradually, remaining always in equilibrium with resulting internal elastic forces, the work done during deformation will equal the strain energy stored in the deformed body.

General Expression for Strain Energy 243

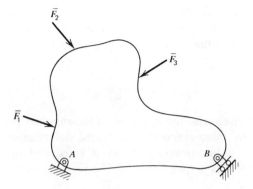

Figure 6.16 Deformation of an elastic body.

The amount of energy does not depend on the order in which the forces are applied, only on their final magnitudes.

Assume that all external forces $F_1, F_2, \ldots, F_n$ increase simultaneously in the same ratio. Then the relation between each force and its corresponding displacement can be represented by a diagram analogous to that shown in Figure 6.16. The work done by $F_1, F_2, \ldots, F_n$ equals the strain energy stored in the element

$$U = \frac{F_1 \delta_1}{2} + \frac{F_2 \delta_2}{2} + \cdots +$$

Total energy of strain is equal to half the sum of the products of each external force and its corresponding displacement. In these determinations it is assumed that the average displacements $\bar{\delta}_1, \bar{\delta}_2, \ldots, \bar{\delta}_n$ are linear functions of the average forces $\bar{F}_1, \bar{F}_2, \ldots, \bar{F}_n$.

The mean value of U_i is, from equations (2.42) and (2.8):

$$\bar{U}_i = \tfrac{1}{2}\left(\bar{F}_i \bar{\delta}_i + \rho s_{F_i} \cdot s_{\delta_i}\right)$$

Hence

$$2\bar{U}_i > \bar{F}_i \bar{\delta}_i$$

The standard deviation of U_i is given by equations (2.43) and (2.10):

$$s_{U_i} \approx \tfrac{1}{2}\left[\bar{F}_i^2 s_{\delta_i}^2 + \bar{\delta}_i^2 s_{F_i}^2 + s_{F_i}^2 s_{\delta_i}^2 + 2\rho \bar{F}_i \bar{\delta}_i s_{F_i} s_{\delta_i} + \rho^2 s_{F_i}^2 s_{\delta_i}^2\right]^{1/2}$$

$U = (U_1 + U_2 + \cdots + U_n)$, and in cases where $F_1, F_2, \ldots, F_n$ are mutually statistically independent, $U_1, U_2, \ldots, U_n$ are statistically independent:

$$\bar{U} = \left(\bar{U}_1 + \bar{U}_2 + \cdots + \bar{U}_n\right)$$

$$s_U = \left[s_{U_1}^2 + s_{U_2}^2 + \cdots + s_{U_n}^2\right]^{1/2}$$

244 Random-Deflection Variables

When two or more of the F_i values are correlated, the statistics of each such group is evaluated separately, considering the correlation and producing a resultant $[U]_i$. Since the resultants $[U]_i$, are independent, they are summed to yield U_{tot}:

$$U_{tot} = [U]_1 + [U]_2 + \cdots +$$

In the preceding discussion the reactions at the supports were not considered. The work done by these reactions during the deformation is zero since the displacement of an immovable support is zero, and the displacement of the movable support is perpendicular to the reaction, with any friction at the supports neglected.

Example G

Consider a mechanical component modeled as a beam, bearing supported at the ends, and loaded at the midpoint by a force F and a couple M applied at A, as shown in Figure 6.17.

Obtain an expression for the strain energy in the beam resulting from the applications of F and M. Models for deflections δ_1 and δ_2 at $x=(l/2)$, due to F and M, are

$$Y_{x=l/2} = \delta_1 = \frac{F \cdot l^3}{48 EI} ; \qquad Y_{x=l/2} = \delta_2 = \frac{M \cdot l^2}{16 EI}$$

By superposition, the total deflection δ is

$$\delta = \delta_1 + \delta_2 = \frac{F \cdot l^3}{48 EI} + \frac{M \cdot l^2}{16 EI}$$

The model for slope at point A due to the load F is

$$\theta_1 = \left(\frac{dy}{dx}\right)_{x=0} = \frac{Fb(l^2 - b^2)}{6lEI} = \frac{Fl^2}{16 EI}$$

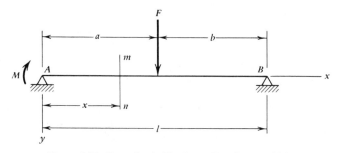

Figure 6.17 Beam loaded by force F and a couple M.

The model for θ_2 at point A due to the couple M is

$$\theta_2 = \left(\frac{dy}{dx}\right)_{x=0} = \frac{Ml}{3EI}$$

The total slope at point A is

$$\theta = \theta_1 + \theta_2 = \frac{Fl^2}{16EI} + \frac{Ml}{3EI}.$$

The strain energy equals the work done by the force F and by the couple M:

$$U = \frac{F\delta}{2} + \frac{M\theta}{2} = \frac{1}{EI}\left[\frac{F^2l^3}{96} + \frac{M^2l}{6} + \frac{MFl^2}{16}\right]$$

The value s_u is calculated by utilizing equation (2.62).

6.8 CASTIGLIANO'S THEOREM

For the various expressions of strain energy, a method may be developed for calculating the displacement of points of an elastic body during deformation. In the case of simple tension.

$$U = \frac{F^2l}{2AE} \tag{a}$$

The derivative of U with respect to F is

$$\frac{dU}{dP} = \frac{Fl}{AE} = \delta$$

where dU/dP yields the displacement due to the load at its point of application, in the direction of action. For a cantilever beam with the external force P applied at the free end:

$$U = \frac{F^2l^3}{6EI}$$

The deflection is

$$\delta = \frac{dU}{dF} = \frac{Fl^3}{3EI}$$

With a torqued circular shaft:

$$U = \frac{M^2l}{2GI}; \qquad \frac{dU}{dM} = \frac{Ml}{GL} = \theta$$

246 Random-Deflection Variables

The following example illustrates the application of Castigliano's theorem in estimating deflection in a two-member frame.

Example H

Determine the preliminary specifications. A system consisting of two steel bars of equal length and nominally identical cross sections supports a load F, as shown in Figure 6.18. Calculate the vertical displacement of H due to load F.

Given.

$$l = 20 \pm \tfrac{1}{8} \text{ in.} = CH = BH$$

$$\text{Load} = (4000;\ 400) \text{ lb.} \tag{a}$$

Mean cross section $\bar{A}$ of members $= 0.500$ in.^{2}

$$\frac{s_A}{\bar{A}} \simeq 0.050 \tag{b}$$

The material is nickel steel, for which the modulus of elasticity is (Table 5.1.)

$$(\bar{E}, s_E) \approx (30 \cdot 10^6;\ 4.5 \cdot 10^5) \text{ psi} \tag{c}$$

Assume that buckling of the compression member is not likely.

Estimate the statistics of deflection δ. The tensile force in the member HB and the compressive force in the member HC are each functions of the

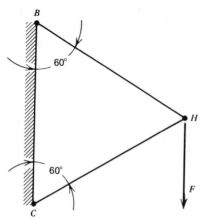

Figure 6.18 Two-member frame.

geometry and orientation of the members, in addition to F; thus

$$U = \frac{F^2 l}{2AE} \quad \text{(d)}$$

By superposition, the model for strain energy of the system is as expressed by equation (d):

$$U = U_1 + U_2 = 2\frac{F^2 l}{2AE} = \frac{F^2 l}{AE} \quad \text{(e)}$$

The vertical displacement of A is given by the derivative of U with respect to F:

$$\delta = \frac{dU}{dF} = \frac{2Fl}{AE} \quad \text{(f)}$$

Assume that F, l, A, and E follow undefined distributions and are statistically independent. Then deflection δ is calculated by applying the algebra of expectation to equations (f) and substituting the statistics of F, l, A, and E:

$$(\bar{\delta}, s_\delta) \approx 2 \frac{(4000, 400)(20, 0.042)}{(0.50, 0.025)(30 \cdot 10^6, 4.5 \cdot 10^5)}$$

From equations (2.25) and (2.27), the numerator and denominator product statistics are

$$(\bar{\delta}, s_\delta) \approx \frac{(160{,}000,\ 16{,}000)}{(15 \cdot 10^6, 7.5 \cdot 10^5)} \text{ in.}$$

and from equations (2.28) and (2.30), the vertical displacement is (see Figure 6.19)

$$(\bar{\delta}, s_\delta) \approx (1.07 \cdot 10^{-2}, 1.07 \cdot 10^{-3}) \text{ in.} \quad \text{(g)}$$

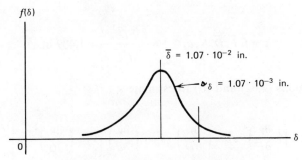

Figure 6.19 Deflection density function.

248 Random-Deflection Variables

The next example illustrates the estimation of deflection in a loaded helical spring, using strain-energy methods.

Example I

The strain energy stored in a helical compression spring may be estimated with the use of equation (a):

$$U = \frac{4F^2 D^3 N}{d^4 G} \tag{a}$$

Employing equation (a) and Castigliano's theorem, the model for estimating y is

$$y = \frac{dU}{dF} = \frac{8FD^3 N}{d^4 G} \tag{b}$$

where F, D, N, d, and G are as defined in Section 6.1.

$$(\bar{F}, s_F) = (63, 6) \text{ lb}$$

$$(\bar{D}, s_D) = (2.5, 0.009) \text{ in.}$$

$$(\bar{d}, s_d) = (0.25, 0.002) \text{ in.}$$

$$(\bar{N}, s_N) = (8, 0)$$

$$(\bar{G}, s_G) = (11.45 \cdot 10^6, 0.15 \cdot 10^6) \text{ psi} \tag{c}$$

Estimate the statistics of y, and the tolerance $\pm \Delta y$, and perform a study to identify the variables contributing most to the variability in y.

Utilize equations (2.59) and (b) to estimate $\bar{y}$:

$$\bar{y} = f(\bar{F}, \bar{D}, N, \bar{d}, \bar{G}) = \frac{8 \overline{FD}^3 \overline{N}}{\bar{d}^4 \bar{G}} = 1.409 \text{ in.} \tag{b'}$$

Utilize equations (2.62) and (b) to estimate s_y:

$$s_y \approx \left[\left(\frac{\partial y}{\partial F}\right)^2 \cdot s_F^2 + \left(\frac{\partial y}{\partial D}\right)^2 \cdot s_D^2 + \left(\frac{\partial y}{\partial d}\right)^2 \cdot s_d^2 + \left(\frac{\partial y}{\partial G}\right)^2 \cdot s_G^2 \right]^{1/2} \tag{b''}$$

The partial derivatives required in equation (b″) are

$$\left(\frac{\partial y}{\partial F}\right)^2 = \left[\frac{8ND^3}{d^4G}\right]^2 = 0.0005$$

$$\left(\frac{\partial y}{\partial D}\right)^2 = \left[\frac{24FN \cdot D^2}{d^4G}\right]^2 = 2.857$$

$$\left(\frac{\partial y}{\partial d}\right)^2 = \left(-\frac{32FD^3N}{d^5G}\right)^2 = 507.9$$

$$\left(\frac{\partial y}{\partial G}\right)^2 = \left[\frac{8FD^3N}{d^4G^2}\right]^2 \approx 0 \tag{d}$$

Substitution of equations (c) and (d) into equation (b″) yields

$$s_y \approx \left[\begin{array}{cccc} F & D & d & G \\ 0.018 + & 0.00023 + & 0.00203 + & 0.00005 \end{array}\right]^{1/2}$$

$$s_y \approx 0.143 \text{ in.} \tag{e}$$

Thus

$$(\bar{y}, s_y) = (1.409, 0.143) \text{ in.}$$

The coefficient of variation is $C_y = 0.101$. Examination of equation (a) indicates that scatter in F is the dominant contributor to scatter in y.

INSTABILITY

6.9 RANDOM COLUMN BEHAVIOR

The column is the simplest of the structural elements subject to instability. The underlying question to the engineer, is the significance of column buckling in a both mathematical and physical sense. Euler published the theory of perfect elastic columns in 1759. The following treatment assumes no perfection. Rather, a considerable scatter in performance is predicted in terms of supportable loads.

A column is loaded in compression and fails due to disturbing of an unstable equilibrium. An element that fails by bulging or crushing is not considered a column because at stresses less than yield, Hooke's law is operative, and unloading of the axial forces results in a return to the original configuration. The relative slenderness of a compression member is the ratio of length to radius of gyration, l/r_g, called the *slenderness ratio*. Radius of gyration r_g is (see Appendix 5)

$$r_g = \sqrt{\frac{I}{A}} \quad ; \quad r_g^2 = \frac{I}{A} \text{ in.}^2 \tag{6.8}$$

Moment of inertia (I) and cross-sectional area (A) statistics are needed to describe radius of gyration that must be related to the centroidal axis about which the column tends to bend, that is, usually the least radius of gyration. There is no positive way of distinguishing between a column and a compression member because there is no particular mean value of l/r_g above which a component becomes a column, as is indicated in Figure 6.26. This dichotomy would be described by a distribution of values. There are four identifiable modes of column failure: (1) elastic buckling (over the length of the column), (2) inelastic buckling (over the length of the column), (3) elastic or inelastic buckling (over a small portion of the column length), and (4) twisting or torsional failure, involving rotation about the longitudinal or centroidal axis

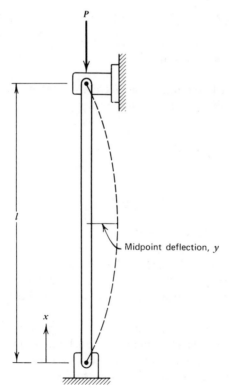

Figure 6.20 Pin-ended column.

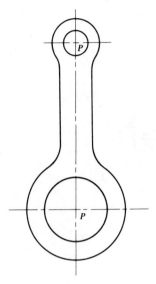

Figure 6.21 Axially loaded connecting rod.

[1, p. 77]. Modes 3 and 4, not considered here, are associated with thin-walled compression members.

A pin-ended column having length l and with ends free to rotate is shown in Figure 6.20. It is assumed that no frictional forces act at the ends. In developing theory, the longitudinal axis is assumed straight before the force acts. As the load is increased from zero to the full magnitude of P, the column follows Hooke's law and compresses. At the same time it remains in stable equilibrium. However, since statistical deviation from ideal axial straightness usually exists (see Tables 4.5 through 4.9), bending moments exist. As the force P is increased, however, the column axis retains essentially its original shape until P exceeds P_{cr}. An axially loaded connecting rod is shown in Figure 6.21.

CRITICAL LOAD

When loading exceeds P_{cr}, a condition of unstable equilibrium exists, and a small lateral force or the existence of eccentricity or crookedness will initiate buckling. Impending buckling is not easily detected. Equation (6.9) is called *Euler's formula for pin-ended columns*:

$$P_{cr} = \frac{\pi^2 EI}{l^2} \tag{6.9}$$

The models for estimating the statistics of P_{cr} are, utilizing equation (2.59):

$$\bar{P}_{cr} \approx f(\bar{E}, \bar{I}, \bar{l}) = \frac{\pi^2 \bar{E}\bar{I}}{\bar{l}^2} \qquad (6.9a)$$

and utilizing equation (2.62):

$$s_{P_{cr}} \approx \left[\left(\frac{\partial P_{cr}}{\partial E}\right)^2 s_E^2 + \left(\frac{\partial P_{cr}}{\partial I}\right)^2 s_I^2 + \left(\frac{\partial P_{cr}}{\partial l}\right)^2 s_l^2 \right]^{1/2} \qquad (a)$$

The required partial derivatives are

$$\frac{\partial P_{cr}}{\partial E} = \frac{\pi^2 \bar{I}}{\bar{l}^2}$$

$$\frac{\partial P_{cr}}{\partial I} = \frac{\pi^2 \bar{E}}{\bar{l}^2}$$

$$\frac{\partial P_{cr}}{\partial l} = -\frac{2\pi^2 \bar{E}\bar{I}}{\bar{l}^3} \qquad (b)$$

Substitution of equation (b) into equation (a) yields

$$s_{P_{cr}} \approx \left[\left(\frac{\pi^2 \bar{I}}{\bar{l}^2}\right)^2 \cdot s_E^2 + \left(\frac{\pi^2 \bar{E}}{\bar{l}^2}\right)^2 \cdot s_I^2 + \left(\frac{\pi^2 \bar{E}\bar{I}}{\bar{l}^3}\right)^2 \cdot s_l^2 \right]^{1/2} \qquad (6.9b)$$

Some indication of the validity of equations (6.9a) and (6.9b) as predictors of elastic column behavior can be gained by examining Figure 6.26. For l/r_g ratios of 100 and greater, the statistics of equation (6.9) appear to be somewhat conservative.

To avoid elastic buckling, with an acceptable probability R, the criterion is

$$P(P_{cr} > P) \geq R$$

where P is the applied load and the model for P_{cr} is equation (6.9).

Example J

This example illustrates the relative contributions of variables r, l, t, and E to variability in P_{cr} as indicated by $s_{P_{cr}}$. The column is assumed to be a round hollow tube, for which moment of inertia I is (Appendix 5):

$$I \approx 2\pi r^3 t$$

In this example the following statistics describe the variables:

$$(\bar{r}, s_r) = (0.750, 0.0063) \text{ in.}$$

$$(\bar{l}, s_l) = (22.00, 0.0625) \text{ in.}$$

$$(\bar{t}, s_t) = (0.040, 0.004) \text{ in.}$$

$$(\bar{E}, s_E) = (10.3 \cdot 10^6, 0.515 \cdot 10^6) \text{ psi.} \qquad (a)$$

When equation (6.9) is utilized as the model for predicting P_{cr}, the mean value estimator is

$$\bar{P}_{cr} = \frac{\pi^2 \bar{E} \bar{I}}{\bar{l}^2} = 2\pi^3 \cdot \left(\frac{\bar{E} \bar{r}^3 \bar{t}}{\bar{l}^2} \right) = 3600 \text{ lb} \qquad (b)$$

Utilization of equations (2.62) and (b) yields

$$s_{P_{cr}} \approx 2\pi^3 \left[\left(\frac{\partial P_{cr}}{\partial E} \right)^2 \cdot s_E^2 + \left(\frac{\partial P_{cr}}{\partial r} \right)^2 \cdot s_r^2 + \left(\frac{\partial P_{cr}}{\partial t} \right)^2 \cdot s_t^2 + \left(\frac{\partial P_{cr}}{\partial l} \right)^2 \cdot s_l^2 \right]^{1/2} \qquad (c)$$

The required partial derivatives are

$$\frac{\partial P_{cr}}{\partial E} = \frac{r^3 t}{l^2}$$

$$\frac{\partial P_{cr}}{\partial r} = \frac{3 E r^2 t}{l^2}$$

$$\frac{\partial P_{cr}}{\partial t} = \frac{E r^3}{l^2}$$

$$\frac{\partial P_{cr}}{\partial l} = -\frac{2 E r^3 t}{l^3} \qquad (d)$$

Substitution of equation (d) into equation (c) yields

$$s_{P_{cr}} \approx 2\pi^3 \left[\left(\frac{\bar{r}^3 \bar{t}}{\bar{l}^2} \right)^2 s_E^2 + \left(\frac{3 \bar{E} \bar{r}^2 \bar{t}}{\bar{l}^2} \right)^2 s_r^2 + \left(\frac{\bar{E} \bar{r}^3}{\bar{l}^2} \right)^2 s_t^2 + \left(\frac{2 \bar{E} \bar{r}^3 \bar{t}}{\bar{l}^3} \right)^2 s_l^2 \right]^{1/2} \qquad (e)$$

Substitution of values from equation (a) into equation (e) gives

$$s_{P_{cr}} \approx [186 \cdot 10^2 + 40.1 \cdot 10^2 + 12,400 \cdot 10^2 + 3330 \cdot 10^2]^{1/2} = 1260 \text{ lb} \qquad (f)$$

254   Random-Deflection Variables

Examination of equation (f) indicates that wall thickness t of the column is the largest contribution to variability in P_{cr}.

Since $\overline{P}_{cr} - 2 \cdot s_{P_{cr}} \approx 3600 - 2(1260) = 1080$ lb (see Appendix 1) it is seen that on the average the probability will be 0.977 that a column as specified in this example will support 1080 lb or more.

It is seen from equation (6.9), that critical load is not a function of material strength. It depends on the modulus of elasticity of the material, with variability in E an important factor.

CRITICAL UNIT LOAD

Substitution of equation (6.8) into equation (6.9) yields a model for predicting critical unit load:

$$s_{cr} = \frac{P_{cr}}{A} = \frac{\pi^2 E}{(l/r_g)^2} \tag{6.10}$$

The expressions for the statistics of s_{cr} in terms of the statistics of E, I, l, and A are from equation (2.59):

$$\bar{s}_{cr} \approx \pi^2 \cdot \frac{\overline{E} \cdot \overline{I}}{\bar{l}^2 \overline{A}} \tag{6.10a}$$

From equation (2.62):

$$s_{s_{cr}} \approx \pi^2 \left[\left(\frac{\partial s_{cr}}{\partial E} \right)^2 \cdot s_E^2 + \left(\frac{\partial s_{cr}}{\partial I} \right)^2 \cdot s_I^2 + \left(\frac{\partial s_{cr}}{\partial A} \right)^2 \cdot s_A^2 + \left(\frac{\partial s_{cr}}{\partial l} \right)^2 \cdot s_l^2 \right]^{1/2}$$

The required partial derivatives are

$$\frac{\partial s_{cr}}{\partial E} = \frac{\pi^2 I^2}{(l^2 \cdot A)}$$

$$\frac{\partial s_{cr}}{\partial I} = \frac{2\pi^2 EI}{(l^2 A)}$$

$$\frac{\partial s_{cr}}{\partial A} = -\frac{\pi^2 EI^2}{A^2 l^2}$$

$$\frac{\partial s_{cr}}{\partial l} = -\frac{2\pi^2 EI^2}{A^2 l^3}$$

$$s_{s_{cr}} = \left[\left[\frac{\pi^2 \bar{I}^2}{(\bar{l}\overline{A})^2} \right]^2 \cdot s_E^2 + \left[\frac{2\pi^2 \overline{E}\,\bar{I}}{(\bar{l}\overline{A})^2} \right]^2 \cdot s_I^2 + \left(\frac{2\pi^2 \overline{E}\,\bar{I}^2}{\overline{A}^3 \bar{l}^2} \right)^2 \cdot s_A^2 + \left(\frac{2\pi^2 \overline{E}\,\bar{I}^2}{\overline{A}^2 \bar{l}^3} \right)^2 \cdot s_l^2 \right]^{1/2}$$

$$\tag{6.10b}$$

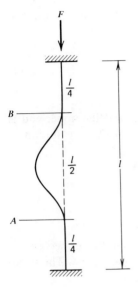

Figure 6.22 Column with two fixed ends.

EULER MODELS—MISCELLANEOUS END CONDITIONS

Expressions for P_{cr} for columns having various end conditions have been obtained from equation (6.9) (see Figures 6.22 and 6.23).

Equations (6.11) through (6.14) provide useful models for which the statistics can be calculated, as for equation (6.9).

In Figure 6.22, the column ends are fixed, resulting in inflection points at A and B, $l/4$ (an average distance) from each end. The column from A to B

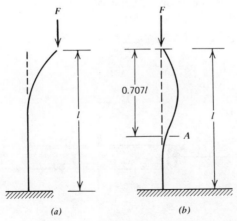

Figure 6.23 Additional column models. (a) One end free, one end fixed. (b) One end rounded, one end fixed.

256 Random-Deflection Variables

assumes a curve similar to that of Figure 6.20. Substitution of $l/2$ for l in equation (6.9) yields, for the fixed-end case,

$$P_{cr} = \frac{\pi^2 EI}{(l/2)^2} = \frac{4\pi^2 \cdot EI}{l^2} \qquad (6.11)$$

In Figure 6.23a, ι the column is fixed at the lower end and the upper end is free, amounting to one half of the case shown in Figure 6.22. Substitution of $2l$ for l in equation (6.9) yields

$$P_{cr} = \frac{\pi^2 \cdot EI}{(2l)^2} = \frac{\pi^2 \cdot EI}{4l^2} \qquad (6.12)$$

In column b of Figure 6.23 the lower end is fixed and the upper end is rounded, a condition frequently approached in mechanical design. The mean inflection point occurs at A, that is, at $0.707l$ from the upper end:

$$P_{cr} = \frac{\pi^2 \cdot EI}{(0.707l)^2} = \frac{2\pi^2 EI}{l^2} \qquad (6.13)$$

The Euler equation may be written as

$$P_{cr} = N \frac{\pi^2 \cdot EI}{l^2} \qquad (6.14)$$

in which N is associated with end conditions, such as $\tfrac{1}{4}$, 1, 2, or 4. End fixity of actual columns does not approach the ideal, and the lack of ideal end fixity brings into question the proper value of l. One result is to increase the variability due to l.

$$P_{cr} = \frac{\pi^2 \cdot ET}{(2l)^2} = \frac{\pi^2 \cdot EI}{(4l)^2} \qquad (6.12)$$

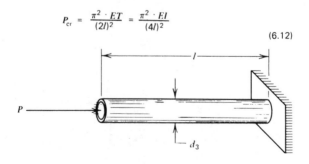

Example K

Estimate the statistics of P_{cr} and a distributional model for P_{cr} of a randomly selected circular aluminum alloy compression rod.

Given.

$$\text{From Table 5.1, } (\overline{E}, s_E) = (10.3, 0.515) \cdot 10^6 \text{ psi} \tag{a}$$

$$\text{Length} = 22 \pm 0.1875 \text{ in.} \approx (22.0, 0.0625) \text{ in.} \tag{b}$$

$$\text{Diameter (Table 4.5)} = 0.500 \pm 0.0125 \text{ in.} \approx (0.500, 0.0041) \text{ in.} \tag{c}$$

Equation (6.9):
$$P_{cr} = \frac{\pi^2 EI}{l^2} \tag{d}$$

From Appendix 5:
$$I = \frac{\pi d^4}{64} \tag{e}$$

Substitution of equation (e) into equation (d) yields

$$P_{cr} = \frac{\pi^2 EI}{l^2} = \left(\frac{\pi^3}{64}\right)\left(\frac{Ed^4}{l^2}\right) \tag{f}$$

1. Estimate the statistics of P_{cr}:

$$\overline{P}_{cr} \approx \left(\frac{\pi^3}{64}\right) \cdot \frac{10 \cdot 3 \cdot 10^6 \cdot (0.5)^4}{(22)^2} = 592 \text{ lb}$$

Utilize equations (2.62) and (f):

$$s_{P_{cr}} \approx \left[\left(\frac{\partial P_{cr}}{\partial E}\right)^2 \cdot s_E^2 + \left(\frac{\partial P_{cr}}{\partial d}\right)^2 \cdot s_d^2 + \left(\frac{\partial P_{cr}}{\partial l}\right)^2 \cdot s_l^2\right]^{1/2}$$

$$\frac{\partial P_{cr}}{\partial E} = \frac{\pi^3}{64} \frac{d^4}{l^2}; \quad \frac{\partial P_{cr}}{\partial d} = 4\left(\frac{\pi^3}{16}\right)\frac{Ed^3}{l^2}; \quad \frac{\partial P_{cr}}{\partial l} = \frac{\pi^3}{32}\frac{Ed^4}{l^3}$$

$$s_{P_{cr}} \approx \frac{\pi^3}{64}\left[\left(\frac{\overline{d}^4}{\overline{l}^2}\right)^2 \cdot s_E^2 + \left(\frac{4\overline{E}\overline{d}^3}{\overline{l}^2}\right)^2 \cdot s_d^2 + \left(\frac{2\overline{E}\overline{d}^4}{\overline{l}^3}\right)^2 \cdot s_l^2\right]^{1/2} = 33 \text{ lb}$$

Thus $(\overline{P}_{cr}, s_{P_{cr}}) = (592, 33)$ lb

2. Plot the results of a simulation study of P_{cr} on normal and lognormal probability paper, to assist in selecting a reasonable distributional model.

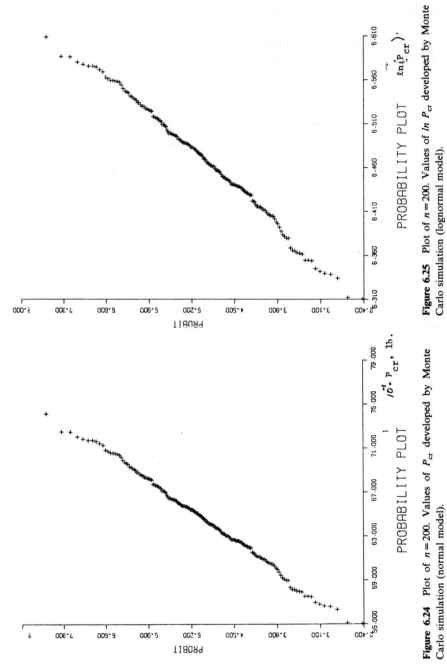

Figure 6.24 Plot of $n = 200$. Values of P_{cr} developed by Monte Carlo simulation (normal model).

Figure 6.25 Plot of $n = 200$. Values of $\ln P_{cr}$ developed by Monte Carlo simulation (lognormal model).

Utilizing equation (6.9) as the model and assuming that l, E, and d given in the preceding equations are normally distributed, a population of $n=200$ values of P_{cr} was generated by a Monte Carlo simulation, utilizing the preceding data. Figures 6.24 and 6.25 show the plots of the 200 simulated values on normal and lognormal probability paper. Both χ^2 and KS test results favor the normal distribution as the most likely model of the data (see Bayes decision procedures in Section 3.2.2).

The Euler formula represents a mathematically derived model that is simple and direct. However, in some respects this model is not complete. Consider equation (6.10)

$$S_{cr} = \frac{P_{cr}}{A} = \frac{\pi^2 \cdot E}{(l/r_g)^2} \tag{6.15}$$

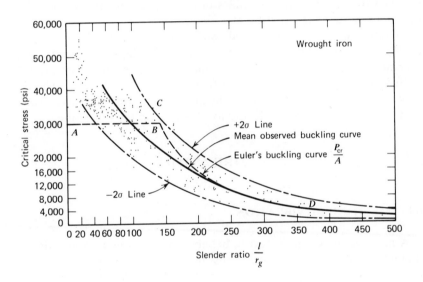

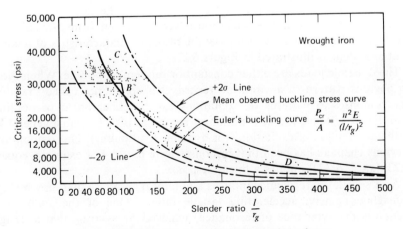

Figure 6.26 Behavior of hinged and flat-ended columns.

and the plot of the relation between P_{cr}/A and l/r_g, shown in Figure 6.26. As with other equations in mechanics, the Euler equation implies adherence to Hooke's law, which imposes no restrictions. As indicated in Figure. 6.26 the mean proportionality scatter band appears to be centered about the line DB. For mean unit loads greater than A in magnitude, there is increasing likelihood that stress is no longer proportional to strain (with increases above AB); thus Hooke's law may not apply. The segment BC of the mean value curve may, therefore, be increasingly meaningless for $l/r_g < B$, that is, an increasing probability that crushing would occur before a load represented by a point on BC could be applied.

Many experiments indicate a considerable scatter of values around point B (Figure 6.26). Often, failure in the neighborhood of B indicates plastic yielding, thus suggesting that elastic buckling may be more appropriately modeled by a distribution of values of l/r_g, often somewhat larger than the theoretical values suggested by B.

As discussed previously, a small variation in d can result in a large effect on P_{cr}. Random deviations from initial straightness, concentric loading, nominal cross-sectional dimensions, and effective length effect must be expected. The logical approach is to first establish the statistical variabilities. When a column does experience buckling, the stresses on a cross section are axial compressive on one side and axial tensile on the other side.

VIBRATION

6.10 MECHANICAL VIBRATION

A mechanical system is said to be vibrating when its components undergo motions that change in time; thus the response of mechanical parts to dynamic loading is considered vibration. The responses of interest in mechanical design are stresses and deflections, and the excitations (dynamic loadings) are forces and accelerations. Records taken from transducers or accelerometers or strain gauges show the response characteristics of a component or system as illustrated in Figure 6.27.

The dynamic loads are either constant or random amplitude. When there is no obvious pattern in a vibration record, it is sometimes called "random." The responses to patterns of excitations can be either linear or nonlinear. With linear systems, the distribution of responses is of the same general kind as the distribution of excitations (or dynamic load pulses). The assumption in previous chapters has been that responses were linear. For example, equation (5.23) $[s = (Mc/I)]$ implies a linear response.

As is seen later, resonance induces a nonlinear response that may be highly damaging, namely, accelerating fatigue failure. One of the goals in this section is the avoidance of resonance, achieved by assuring that a designed

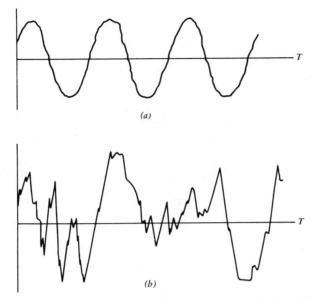

Figure 6.27 Vibration records: acceleration or strain as a function of time: (a) predominantly sinusoidal sample; (b) predominantly random sample.

system responds linearly, with a suitable probability. In other words, if the performance of a system is linear, stresses and strains will remain within their predicted envelopes, such that $P(S > s) \geq R$.

Nonlinear superimposed effects on the use environment distribution of working stresses may increase the likelihood of early fatigue failure. In vibration analysis the distribution of natural frequency or frequencies, deflection, and induced forces are of interest [45 and 57].

The literature of both harmonic and random vibration is extensive and is not discussed in detail here; rather, models that can be reduced to spring-mass systems are considered in this section. It is assumed that frictional forces are small; thus damping of the motion can be neglected and the system treated as having a single degree of freedom (Figure 6.28).

The differential equation of a spring-mass system yields the model

$$x = X \cos(\omega_n t - \phi) \tag{6.17}$$

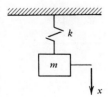

Figure 6.28 Spring-mass system.

262 Random-Deflection Variables

where natural frequency ω_n is

$$\omega_n = k^{1/2} m^{-1/2} \tag{a}$$

where k is the spring-mass constant and m is the mass. From equation (2.59) the statistic of mean natural frequency is

$$\bar{\omega}_n \approx \left[\frac{\bar{k}}{\bar{m}} \right]^{1/2} \tag{a'}$$

and from equations (2.62) and (a):

$$s_{\omega_n} \approx \left[\left(\frac{\partial \omega_n}{\partial k} \right)^2 \cdot s_k^2 + \left(\frac{\partial \omega_n}{\partial m} \right)^2 \cdot s_m^2 \right]^{1/2} \tag{b}$$

The required partial derivatives are

$$\frac{\partial \omega_n}{\partial k} = 1/2 \, k^{-1/2} m^{-1/2}$$

$$\frac{\partial \omega_n}{\partial m} = -1/2 \, k^{1/2} m^{-3/2} \tag{c}$$

Substitution of equation (c) into equation (b) yields

$$s_{\omega_n} \approx \left[\frac{s_k^2}{4 \, \bar{k} \bar{m}} + \frac{\bar{k} s_m^2}{4 \bar{m}^3} \right]^{1/2} \tag{a''}$$

Refer to Figure 6.29, where X is the amplitude of vibration, ϕ is the phase angle (i.e., the starting condition), and ω_n is called the *natural circular frequency*; which takes on a range of values in a population of spring-mass systems, since both k and m are random variables. Frequency in cycles per second is

$$f = \frac{\omega_n}{2\pi} \text{ (hertz)}$$

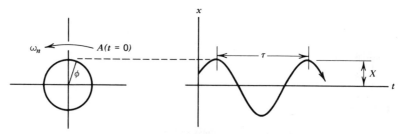

Figure 6.29 Constant-amplitude sinusoidal vibration.

The motion is shown schematically in Figure 6.29, with vector X rotating counterclockwise. Projection of X on the x axis represents the motion of the system. Frequently, X is a random variable.

Motion continuing with the same amplitude is called *free vibration*. Velocity and acceleration models are obtained by taking successive derivatives of equation (6.17).

In equation (6.18) the spring constant (k) is defined as $k=(F/x)$ [57, p. 21], representing the stiffness of a system. Considering a collection of nominally identical systems, k is a random variable with units of lb/in. It is also seen that the maximum spring force is a random variable

$$F_{s,\max} = kx \qquad (6.18)$$

6.10.1 Vibration and Damping

If vibration ceases in a short period of time, it is said to be *damped*. The usual representation of a damped system is as shown in Figure. 6.30.

A first analysis is usually made by assuming that friction can be lumped and its effect represented by the viscous action of a dashpot. The friction effect is represented by the coefficient of viscous damping (c) defined as follows:

$$c = \frac{F}{\dot{x}} \qquad (6.19)$$

According to equation (2.59), with equation (6.19), the statistic $\bar{c}$ is

$$\bar{c} \simeq \frac{\bar{F}}{\bar{\dot{x}}} \qquad (6.19a)$$

and from equation (2.49):

$$s_c \simeq \left[\frac{s_{F}^2 \cdot \bar{\dot{x}}^2 + s_{\dot{x}}^2 \cdot \bar{F}^2}{\bar{\dot{x}}^4} \right]^{1/2} \qquad (6.19b)$$

where F is the frictional force and $\dot{x}$ the velocity.

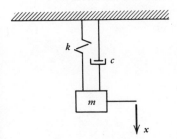

Figure 6.30 Damped vibration.

The critical damping constant, c_{cr} is the amount of damping that would result in zero vibration. The ratio of actual to critical damping (ζ) is

$$\zeta = \frac{c}{c_{cr}} \qquad (6.20)$$

Equations (2.28) and (2.30) with equation (6.20) yield

$$\bar{\zeta} \simeq \frac{\bar{c}}{\bar{c}_{cr}} \qquad (6.20a)$$

and

$$s_\zeta \simeq \left[\frac{s_c^2 \cdot \bar{c}_{cr}^2 + s_{c_{cr}}^2 \cdot \bar{c}^2}{\bar{c}_{cr}^4} \right]^{1/2} \qquad (6.20b)$$

Damping is usually small and does not often exceed 0.15 unless friction is introduced. A simple way to estimate the amount of damping in an element or a system is to disturb it physically (with a hammer blow) and observe the decay of the motion on an oscilloscope. The test is repeated a number of times and the resulting data reduced to statistics. The model for damped free vibration is

$$x = X e^{-\zeta \omega_n t} \cos(\omega_n \cdot t - \phi) \qquad (6.21)$$

where

$$\omega_{nd} = \omega \sqrt{1 - \zeta^2} \qquad (6.22)$$

Free vibration is represented by a decaying exponential function.

6.10.2 Steady-state Vibration

When a vibrating force $F = F_0 \sin \omega t$ is applied to the lumped mass (see Figure 6.33) after transients have decayed to zero, the system is said to have steady-state vibration.

Equations (6.23) through (6.26) provide useful models for which the statistics can be calculated by utilizing the algebra of normal functions.

$$m\ddot{x} + c\dot{x} + kx = F_0 \sin \omega t \qquad (6.23)$$

where ω is the circular frequency.

The steady-state model is

$$x = X \sin(\omega t - \phi)$$

where

$$X = \frac{F_0/k}{\sqrt{\left(1-\frac{\omega^2}{\omega_n^2}\right)^2 + \left(2\zeta\frac{\omega}{\omega_n}\right)^2}}, \qquad (6.24)$$

$$\phi = \tan^{-1}\frac{2\cdot\zeta\left(\frac{\omega}{\omega_n}\right)}{1-\left(\frac{\omega^2}{\omega_n^2}\right)} \qquad (6.25)$$

where ϕ is the phase-angle difference between force and motion and X is the amplitude of motion of the system. Or, rearranged as

$$\frac{X}{F_0/k} = \frac{1}{\sqrt{\left(1-\frac{\omega^2}{\omega_n^2}\right)^2 + \left(2\zeta\frac{\omega}{\omega_n}\right)^2}} \qquad (6.26)$$

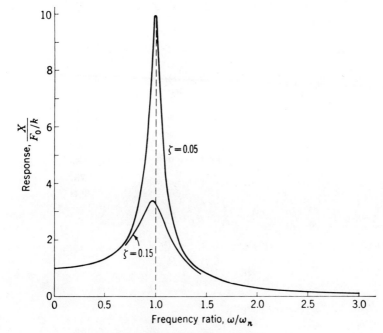

Figure 6.31 Frequency ratio ω/ω_n.

From Figure 6.31 it is seen that damping reduces response, especially near resonance, that is, when ω/ω_n is near unity. Devices operated at high-frequency ratios $\omega/\omega_n \geqslant 3$ tend to have minimal vibration problems. Probabilistically, the criterion is often specified as

$$P\left(\frac{\omega}{\omega_n} \geqslant 3\right) \geqslant R$$

where ω_n is natural frequency and R is a specified probability.

6.10.3 Natural Frequencies

When the distribution of operating speed and of natural frequency of a device are known, the probability of troublesome vibration may be estimated. The natural frequency of components and systems must be calculated. To avoid resonance and maintain response linearity, reduction or increase of geometric magnitudes may be required to adjust frequencies. Thus ω_n may be calculated as

$$\omega_n = \frac{g}{W/k} \tag{6.27}$$

where W is the weight of the mass, or

$$\omega_n = \frac{g}{x} \tag{6.28}$$

where x is the static deflection of the spring due to W. Since x is usually expressed in inches, g should be in in./sec^2:

$$g = 32.16 \text{ ft/sec}^2 = 386 \text{ in./sec}^2$$

In the analysis of a device, certain components may be treated as a system of springs in a parallel or a series relationship.

1. *Springs in parallel* must have the same deflection. If x is the deflection and F is the total spring force, then for three springs in parallel:

$$F = k_1 x + k_2 x + k_3 x \tag{6.29}$$

The model for equivalent spring constant is

$$k_e = \frac{F}{x} = k_1 + k_2 + k_3 \tag{6.30}$$

2. If three *springs in series* are loaded with a weight W, than

$$\frac{W}{k_e} = \frac{W}{k_1} + \frac{W}{k_2} + \frac{W}{k_3} \tag{6.31}$$

$$\frac{1}{k_e} = \frac{1}{k_1} + \frac{1}{k_2} + \frac{1}{k_3} \tag{6.32}$$

For examples of springs in parallel and in series, see Example L.

Example L

1. Determine the preliminary specifications. Consider a rotating mass on a shaft, supported by bearings mounted on two steel angles. The shaft is 1.250 ± 0.0025 in. in diameter and rotates at 2200 ± 180 rpm. The weight of the rotating mass is 64 ± 0.75 lb, located at the center of the shaft. The eccentricity is 0.006 ± 0.0009 in. The support angles are $4 \times 4 \times \frac{1}{4}$ in. structural steel. Estimate the statistics of natural frequencies and the likelihood of resonance of a system assembled from parts selected at random as shown in Figure 6.32.

Amplitudes will be estimated by treating the system shown in Fig. 6.32 as a single-degree-of-freedom system. It is assumed that all variables are mutually independent and normally distributed.

The forcing frequency in this example will be

$$(\bar{\omega}, s_\omega) = \frac{2\pi}{60}(2200, 60) = (230, 6.27) \text{ rad/sec}$$

2. The magnitude of the force generated due to the eccentricity e is

$$F = me\omega^2 \qquad (a)$$

Substitution of the statistics of m, e, and ω into (a) yields

$$(\bar{F}, s_F) = \frac{(64, 0.25)(0.006, 0.0003)(230, 6.27)^2}{386} \text{ lb}$$

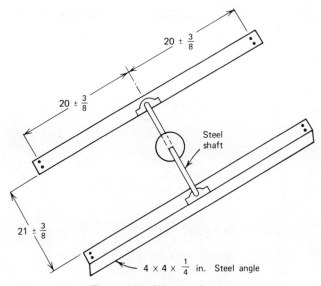

Figure 6.32 Vibrating System.

268 Random-Deflection Variables

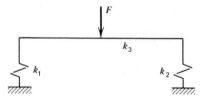

Figure 6.33 Spring system.

Equations (2.25) and (2.27) as well as (2.8) and (2.10) applied successively yield

$$(\bar{F}, s_F) = (52.6, 2.85) \text{ lb}$$

The two $4 \times 4 \times \tfrac{1}{4}$ angles are treated as springs in parallel having constants k_1 and k_2, with identical statistics. The equivalent spring is considered as in series with the rotating shaft (Figure 6.33).
Deflection of a simply supported beam with a concentrated load at the center is (equation 6.7)

$$y = \frac{Fl^3}{48EI}$$

Thus the model of the constant for each spring (angles and shaft) is

$$k = \frac{F^*}{y} = \frac{48EI}{l^3} \tag{b}$$

To utilize equation (b), it is necessary to estimate the statistics of I for the $4 \times 4 \times \tfrac{1}{4}$ angles and I for the shaft.

3. The moment of inertia (from AISC tables [58]) for $4 \times 4 \times \tfrac{1}{4}$ angles is

$$\bar{I} = 2.94 \text{ in.}^4$$

The value s_I is estimated as $s_I \simeq 0.05 \cdot \bar{I}$ in.^{4}; thus

$$(\bar{I}, s_I) \simeq (2.94, 0.147) \text{ in.}^4 \tag{c}$$

4. The moment of inertia of the shaft is (Appendix 5) $I = (\pi d^4/64)$ from given data and since $s_d \simeq (0.0025/3) = 0.0008$ in.:

$$\bar{I} - \frac{\pi}{64}(\bar{d})^4 = 0.120 \text{ in.}^4$$

and

$$s_I \simeq \frac{\pi}{64}(4\bar{d}^3 \cdot s_d) = 0.00031 \text{ in.}^4$$

$$(\bar{I}, s_I) = (0.120, 0.00031) \text{ in.}^4 \tag{d}$$

*Weights of the shaft and the angles are not considered in this example.

5. From Table 5.1, for carbon steel, $\bar{E} = 29.5 \cdot 10^6$ psi and estimated $s_E \simeq 0.0166(\bar{E}) = 5.0 \cdot 10^5$ psi.

$$(\bar{E}, s_E) \simeq (29.5 \cdot 10^6, 5.0 \cdot 10^5) \text{ psi} \qquad (e)$$

6. Determination of the spring constants (steel angles) are

$$k_1 = k_2 = 48 \frac{EI}{l^3} \qquad (f)$$

Stiffness (EI) of the angles is

$$(\bar{E}, s_E) \cdot (\bar{I}, s_I) = (29.5 \cdot 10^6, 5.0 \cdot 10^5)(2.94, 0.147) = (86.7 \cdot 10^6, 4.50 \cdot 10^6) \qquad (g)$$

According to equations (2.59) and (2.62), the statistics of l^3 are

$$(\bar{l^3}, s_{l^3}) = (40.0, 0.25)^3 = (64{,}000, 1200) \text{ in.}^3 \qquad (h)$$

Estimate spring constant k_1 and k_2 statistics by substituting equations (g) and (h) into equation (f):

$$(k_1, s_{k_1}) = (k_2, s_{k_2}) = 48 \frac{(83.8 \cdot 10^6, 4.50 \cdot 10^6)}{(64{,}000, 1200)} \text{ lb/in.}$$

From equation (2.28):

$$\bar{k}_1 = \bar{k}_2 = 62{,}900 \text{ lb/in.}$$

From equation (2.30):

$$s_{k_1}^2 = s_{k_2}^2 \simeq \frac{(83.8 \cdot 10^6)^2 (1.2 \cdot 10^3)^2 + (6.4 \cdot 10^4)^2 (4.501 \cdot 10^6)^2}{(6.4 \cdot 10^4)^4} (48)^2$$

$$\simeq 5550 \cdot 10^3 \cdot (48)^2$$

$$s_{k_1} = s_{k_2} = 3580 \text{ lb/in.}$$

The statistics of k_1 and k_2 are

$$(\bar{k}_1, s_{k_1}) = (\bar{k}_2, s_{k_2}) \simeq (62{,}900, 3580) \text{ lb/in.} \qquad (i)$$

7. The model of the spring constant k_3 is

$$k_3 = 48 \frac{EI}{l^3} \qquad (a')$$

$$(\bar{E}, s_E)(\bar{I}, s_I) = (29.5 \cdot 10^6, 5.0 \cdot 10^5)(0.120, 0.000307)$$

$$= (3.42 \cdot 10^6, 6.88 \cdot 10^4) \qquad (b')$$

$$(\bar{l}, s_l)^3 = [(21.0, 0.125) \text{ in.}]^3 \qquad (c')$$

270 Random-Deflection Variables

From equations (2.59) and (2.62), the statistics of l^3 are

$$\bar{l}^3 = 9261 + 0.983 = 9260 \text{ in.}^3$$

$$s_{l^3} \approx 3\bar{l}^2 \cdot s_l = 3(141)(0.125) = 165 \text{ in.}^3$$

$$\left(\bar{l}^3, s_{l^3}\right) \approx (9260, 165) \text{ in.}^3$$

$$\left(\bar{k}_3, s_{k_3}\right) = 48 \frac{(3.42 \cdot 10^6, 6.88 \cdot 10^4)}{(9260, 165)}$$

From equations (2.28) and (2.8):

$$\bar{k}_3 = 48 \frac{3.42 \cdot 10^6}{9260} = 17{,}700 \text{ lb/in.}$$

From equations (2.30) and (2.10):

$$s_{k_3}^2 = (48)^2 \frac{(3.42 \cdot 10^6)^2 \cdot (165)^2 + (6.88 \cdot 10^4)^2 \cdot (9262)^2}{(9261)^4} = 101.2(48)^2$$

$$s_{k_3} = 510 \text{ lb/in.}$$

$$\left(\bar{k}_3, s_{k_3}\right) = (17{,}700, 510) \text{ lb/in.} \tag{d'}$$

8. The equivalent spring constant [equation (6.32)] for springs 1 and 2 acting in parallel and then in series with spring 3 is

$$\frac{1}{k_e} = \frac{1}{k_1 + k_2} + \frac{1}{k_3}$$

Then

$$\frac{1}{k_e} = \frac{k_3 + k_1 + k_2}{(k_1 + k_2)k_3}$$

and

$$k_e = \frac{(k_1 + k_2)k_3}{k_1 + k_2 + k_3} \tag{e'}$$

The k values of springs 1 and 2 are statistically independent. Determination of the standard deviation of k_e requires utilization of equation (2.62) with equation (e'):

$$s_{k_e} \approx \left[\left(\frac{\partial k_e}{\partial k_1}\right)^2 s_{k_1}^2 + \left(\frac{\partial k_e}{\partial k_2}\right)^2 s_{k_2}^2 + \left(\frac{\partial k_e}{\partial k_3}\right)^2 s_{k_3}^2 \right]^{1/2} \tag{f'}$$

Taking the partial derivatives of equation (e′) with respect to k_e and evaluating at the means, see equations (i) and (d′):

$$\frac{\partial k_e}{\partial k_1} = \frac{(k_1+k_2+k_3)k_3 - (k_1+k_2)k_3 \cdot 1}{(k_1+k_2+k_3)^2} = \frac{k_3^2}{(k_1+k_2+k_3)^2} = 0.0152$$

$$\frac{\partial k_e}{\partial k_2} = \frac{(k_1+k_2+k_3)k_3 - (k_1+k_2)k_3 \cdot 1}{(k_1+k_2+k_3)^2} = \frac{k_3^2}{(k_1+k_2+k_3)^2} = 0.0152$$

$$\frac{\partial k_e}{\partial k_3} = \frac{k_1^2 + k_1 k_2 + k_1 k_3 + k_1 k_2 + k_2^2 + k_2 k_3 - k_1 k_3 - k_2 k_3}{(k_1+k_2+k_3)^2}$$

$$\frac{\partial k_e}{\partial k_3} = \frac{(k_1+k_2)^2}{(k_1+k_2+k_3)^2} = 0.316 \tag{g′}$$

Note that k_3 is much the largest contributor to variability in k_e. If equations (i), (d′), and (g′) are substituted into equation (f′), the estimator of s_{k_e} is

$$s_{k_e} \approx [2832 + 2832 + 25{,}973]^{1/2} = 178 \text{ lb/in.} \tag{h′}$$

Estimation of $\bar{k}_e$ by equation (2.59) with equation (e′) yields

$$\bar{k}_e = \frac{(\bar{k}_1 + \bar{k}_2)\bar{k}_3}{\bar{k}_1 + \bar{k}_2 + \bar{k}_3} = \frac{(125{,}800)(17{,}700)}{143{,}500} = 15{,}500 \text{ lb/in.}$$

Thus

$$(\bar{k}_e, s_{k_e}) = (15{,}500 \; 178) \text{ lb/in.} \tag{i′}$$

9. The natural frequency of the undamped system is [from equation (6.34)]

$$\omega_n = \sqrt{\frac{k_e}{m}} = (386)^{1/2} k_e^{1/2} m^{-1/2}$$

$$(\bar{\omega}_n, s_{\omega_n}) = \sqrt{\frac{(\bar{k}_e, s_{k_e})}{(64.0, 0.25)} \cdot 386} = \sqrt{\frac{(15{,}500, 178)}{(64.0, 0.25)} \cdot 386} \tag{j′}$$

From equations (2.59) and (j′):

$$\bar{\omega}_n \approx \sqrt{\frac{15{,}500 \cdot 386}{64.0}} = 306 \text{ rad/sec}$$

272 Random-Deflection Variables

From equations (2.62) and (j'):

$$s_{\omega_n} \approx \left[\left(\frac{\partial \omega_n}{\partial k_e}\right)^2 \cdot s_{k_e}^2 + \left(\frac{\partial \omega_n}{\partial m}\right)^2 \cdot s_m^2\right]^{1/2} \tag{k'}$$

Taking partial derivatives of j (j') with respect to ω_n, we obtain

$$\frac{\partial \omega_n}{\partial k_e} = (386)^{1/2}\left(\frac{1}{2}\right)k_e^{-1/2}m^{1/2} = 0.0986$$

$$\frac{\partial \omega_n}{\partial m} = (386)^{1/2}\left(\frac{1}{2}\right)k_e^{1/2}m^{-3/2} = 2.39 \tag{l'}$$

Substitution of equation (l') into equation (k') yields

$$s_{\omega_n} \approx \sqrt{386} \cdot [(0.0986)^2 \cdot (178)^2 + (2.34)^2 \cdot (0.25)^2]^{1/2} = 1.85 \text{ rad/sec}$$

The natural frequency of the system expressed in rpm is*

$$n_{cr} = \frac{60\omega}{2\pi}$$

$$(\bar{n}_{cr}, s_{n_{cr}}) = \frac{60}{2\pi}(306, 1.85) = (2920, 11.7) \text{ rpm} \tag{m}$$

The rate of rotations (n) of the mass (forcing frequency) is given as

$$(2200; 60) \text{ rpm}. \tag{n}$$

It is easily shown, by methods developed in Chapter 9, that resonance of this system is very unlikely. Utilization of equation (9.11) with equations (m) and (n) yields

$$\frac{\bar{n} - \bar{n}_{cr}}{\sqrt{s_n^2 + s_{n_{cr}}^2}} \approx z = \frac{2200 - 2920}{\sqrt{(60)^2 + (11.7)^2}} = \frac{-720}{63.6} = -11.3$$

From equation (9.10) and Appendix 1, the probability of resonance is

$$P(n = n_{cr}) = 1 - \frac{1}{\sqrt{2\pi}} \int_z^\infty e^{-z^2/2} dz = 1 - \int_{-11.3}^\infty \frac{e^{-z^2/2}}{\sqrt{2\pi}} dz \approx 0$$

The amplitude $\bar{y}$ of vibration is estimated with equations (a) and (i'):

$$(\bar{y}, s_y) = \frac{(\bar{F}, s_F)}{(\bar{K}_e, s_{K_e})} = \frac{(52.6, 2.85) \text{ lb}}{(15,500, 178) \text{ lb/in.}} \tag{o}$$

From equations (2.28) and (2.30):

$$\bar{y} = \frac{52.6}{15,500} = 0.0034 \text{ in.}$$

$$s_y \ll \bar{y}$$

---

*Natural frequency in rpm (n_{cr}) is called *critical speed*.

CHAPTER 7

Geometric Stress Concentration

In this chapter we consider the estimation of stress concentration statistics a priori because these are needed in probabilistic stress functions utilized in typical design synthesis. The elementary stress formulas coupled multiplicatively with design factors (of which "stress concentration" is one) are utilized to estimate the statistics of significant (actual) stresses in real components.

7.1 INTRODUCTION

The elementary stress formulas discussed in Chapter 5 were based on components that have a constant section or a section with gradual change of contour. Such conditions, however, are rarely attained in actual machine parts or structural members. Various geometric discontinuities result in a modification of simple stress distributions. Furthermore, the surface of a mechanical or a structural component is inherently weak, highly stressed, and subjected to the most abuse in processing, fabrication, and use. It follows that surface discontinuities and defects are of prime importance in determining fatigue behavior. This localization of high stresses is called *stress concentration*.

Finite-element analysis is one effective means of estimating localized stresses in existing designs of parts. As discussed in Section 5.20, both mean values and standard deviations may be assessed. Applied stress magnitudes in the localized areas of geometric discontinuities are functions of local geometry, largely independent of the nature of applied forces, and material behavior of a mechanical component. Due to the nature of manufacturing processes, geometric dimensions and hence stress concentrations vary randomly over a population of components.

Stress concentration, when coupled with dynamic loading, assumes major importance because virtually all fatigue failures of actual mechanical components occur at localized areas of stress concentration. In automotive applications K_t is typically found to be on the order of 3 to 4 [33]. Previously,

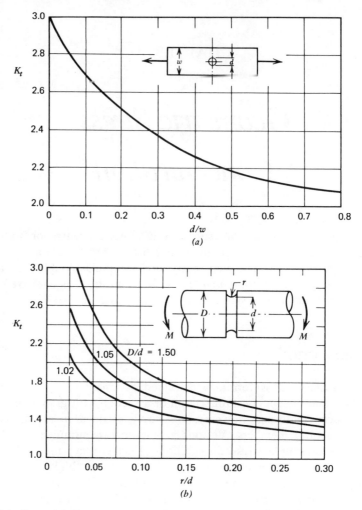

Figure 7.1 Deterministic geometric stress concentration curves [59]. (a) Bar in tension or simple compression with a transverse hole: $\sigma_0 = F/A$, where $A = (w-d)t$ and where t is the thickness [1]. (b) grooved round bar in bending: $\sigma_0 = Mc/I$, where $c = d/2$ and $I = \pi d^4/64$.

experimentally determined stress concentration factors were modeled deterministically (see Figure 7.1).

The intensity of stress may be inferred from strains determined by means of photoelastic methods. An experimental stress concentration factor can be determined from photographs such as Figure 7.2, by line counts proportional to s_{max} in a notched beam and s in an identical unnotched beam under the same loading [37]. The ratio of the maximum stress s_{max} (within the elastic limit) to the nominal stress s^* has a very definite value and is called the *stress*

---

*Generally, s is the stress calculated by using the elementary stress equations and the net cross section.

Introduction 275

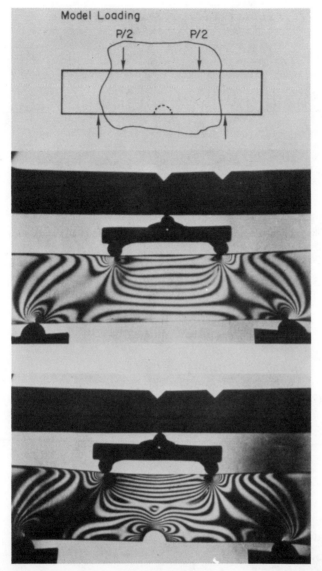

Figure 7.2 Pictorial representation of stress distribution in a simple beam in bending; isochromatics, 0° rotation, 18 second exposure.

concentration factor, designated K_t:

$$\left(\overline{K}_t, \,_{^sK_t}\right) = \frac{\left(\bar{s}_{\max}, \,_{^s s_{\max}}\right)}{\left(\bar{s}, \,_{^s s}\right)} \tag{7.1}$$

Certain geometric shapes have been analyzed by methods based on the theory of elasticity, to estimate stress concentration factors. For example, Figure 7.3 indicates an infinite plate uniformly stressed in tension s. An

Geometric Stress Concentration

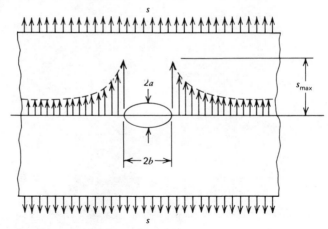

Figure 7.3 Stress distribution near an elliptical hole in an infinite plate loaded in tension [1, p. 221].

elliptical hole in the plate results in increased stress intensity at the hole's edges [60]:

$$s_{max} = s \cdot \left(1 + \frac{2b}{a}\right) \quad \text{(a)}$$

where s_{max}, s, a, and b are random variables.

If the hole is circular, a and b will be correlated $\rho = +1$. Then from equations (2.45) and (a):

$$\frac{\bar{b}}{\bar{a}} = \frac{\bar{b}}{\bar{a}} \left[1 + \frac{s_a}{\bar{a}} \left(\frac{s_a}{\bar{a}} - \rho \cdot \frac{s_b}{\bar{b}} \right) \left(1 + 3 \cdot \frac{s_a^2}{\bar{a}^2} + \cdots + \right) \right] \quad \text{(a')}$$

$$\frac{\bar{b}}{\bar{a}} \approx \bar{b}/\bar{a} = 1.00$$

From equation (2.47), the standard deviation $s_{b/a}$ is

$$s_{b/a} \approx \frac{\bar{b}}{\bar{a}} \left[\frac{s_a^2}{\bar{a}^2} + \frac{s_b^2}{\bar{b}^2} - 2\rho \frac{s_a s_b}{\bar{a} \cdot \bar{b}} \right]^{1/2} \approx 0 \quad \text{(a'')}$$

Thus

$$(\bar{b}/\bar{a}, s_{b/a}) \simeq (1, 0)$$

Furthermore, s_{max} will be

$$s_{max} \simeq s \cdot (3, 0); \quad (\bar{s}_{max}, s_{s_{max}}) \simeq 3(\bar{s}, s_s) \quad \text{(b)}$$

Example A

A mechanical member experiences peak stresses, $(\bar{s}, s_s) = (34{,}000; 2{,}040)$ psi. At the edge of a circular hole the maximum stress is [from equation (b)]

$$(\bar{s}_{max}, s_{s_{max}}) \simeq 3(34{,}000, 2040) = (102{,}000, 6120) \text{ psi}$$

Consequently, not only is localized mean stress much larger, but the upper $3s$ value is

$$\bar{s}_{max} + 3 s_{s_{max}} \approx 120,\ 360 \text{ psi}$$

The value s_{max} can be estimated at the edge of a transverse crack by setting $(a/b) \ll 1$. In such a case a and b will very likely be statistically independent, and the random variable K_t will tend to have a large mean value and standard deviation. Together with considerations from fracture mechanics, this helps to explain the frequent observation of small fatigue cracks in mechanical components propagating rapidly to complete failure.

For the case of a shallow elliptical notch in a plate of infinite width (Figure 7.4), Neuber's [61] equation for K_t is

$$K_t = \frac{s_{max}}{s} = 1 + 2\sqrt{\frac{h}{r_0}} \tag{a}$$

Since r_0 and h are very likely independent dimensional random variables, $[(\bar{r}_0, s_{r_0})$ and $(\bar{h}, s_h)]$, the mean value estimator of K_t from equations (2.59) and (a) is

$$\bar{K}_t \approx 1 + 2\sqrt{\frac{\bar{h}}{\bar{r}_0}} \tag{a'}$$

The standard deviation s_{K_t} from equations (2.62) and (a) is

$$s_{K_t} \approx \left[\frac{s_h^2}{\bar{h}\bar{r}_0} + \frac{\bar{h} \cdot s_{r_0}^2}{\bar{r}_0^3} \right]^{1/2} \tag{a''}$$

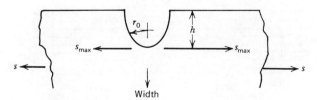

Figure 7.4 Elliptical edge notch in an infinite plate.

Figure 7.5 Force-flow analogy [1].

If $(\bar{h}/\bar{r}) \ll 1$, the mean value of $\bar{K}_t \to 1$, and the standard deviation converges to

$$s_{K_t} \to s \sqrt{h/r_0}$$

Although $\bar{K}_t \to 1$ (i.e., $\bar{s}_{max} \to \bar{s}$), s_{K_t} will be nonzero if s and s_{max} are independent and $s_{s_{max}}$ and s_s are nonzero. Thus in the limit, $(K_t, s_{K_t}) = (1, s_{s'})$, where $s' = s_{max}/s$.

The method of photoelasticity provides a means for experimentally checking estimates of local strain intensities. Many of the results published by Peterson [59] were obtained by photoelastic methods [62, 63].

Variable-resistance strain gauges are commonly utilized to experimentally verify point (or small area average) surface strain, usually after a whole field brittle lacquer study (perhaps supported by a finite-element analysis) has been made to establish high stress areas and directions of the mean principal stresses. Many writers point to the value of flow analogy in visualizing stress concentration (Figures 7.5 and 7.6) and suggesting possible ways by which stress concentration may be reduced or eliminated.

On shafts any gears, sheaves, bearings, and similar components are often seated against shoulders, and it is in the shoulder fillets that failures usually occur. Also, when gears, pulleys, or the like are interference fit over shafts, discontinuities are created near the boundaries. Figure 7.7 shows a component with force flow lines. The fillet radius r should be as large as possible to minimize K_t. However, r may be limited by the shape of a mating part. Thus undercutting the shoulder will provide a larger fillet.

Introducing additional notches as shown in Figure 7.7b,c is also often effective in reducing stress concentration. When grooves and notches cannot

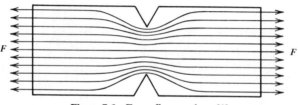

Figure 7.6 Force-flow analogy [1].

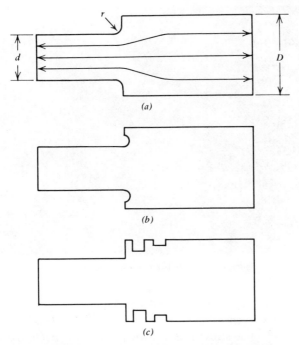

Figure 7.7 Means of reducing stress concentration.

be avoided, they should be located in low-stress areas. Dimensional tolerances associated with fillet radii should be as small as practical since the effect of K_t is related to dimensional variability.

7.2 STRESS CONCENTRATION AND STATIC LOADS

The full value of geometric stress concentration factors may not reflect actual performance of a mechanical component, as such factors are determined solely on the basis of macrogeometry. Microstructure and mechanical properties of the materials do not enter into the determination of K_t. The effect is, however, found to depend on the degree of ductility of the material.

The stress–strain curve of an ideal brittle material is similar to that shown in Figure 8.11a, that is, $\varepsilon_p \ll \varepsilon_e$ (ε_p = plastic strain, ε_e = elastic strain). The full value of K_t is used with this class of materials. For designs wherein materials in a possible brittle condition are specified for load-carrying components, see Section 10.11.

As a practical expedient, the maximum stress at a point of discontinuity is usually estimated by calculating nominal stress and then multiplying this value by K_t [27, p. 237], see Section 10.5.1. There is some experimental photoelastic justification for this practice, see Figure 7.8. With static loads,

Figure 7.8 Simple beam stress concentration.

the influence of stress raisers depend on the nature of the material. For materials that have even a small amount of ductility, local yielding largely nullifies the effect of a stress raiser. Stress based on net cross section is a sufficient estimation, and stress concentration factors need not be employed.

7.3 DETERMINING THEORETICAL K_t

The equations applying to Figure 7.9 (configuration 6, Appendix 9) are utilized for determining the theoretical stress concentration factor of a single geometric configuration and loading situation. Comparable equations for other geometric configurations are given in Appendix 9:

$$\alpha_1 = \frac{2[(a/r)+1]\sqrt{a/r}}{[(a/r)+1]\tan^{-1}\sqrt{a/r} + \sqrt{a/r}}$$

$$\alpha_2 = \frac{4a/r\sqrt{a/r}}{\left[[(a/r)-1]\tan^{-1}\sqrt{a/r} + \sqrt{a/r}\right]}$$

$$\alpha_3 = \frac{2[(a/r)+1] - \alpha_1\sqrt{(a/r)+1}}{\dfrac{4}{\alpha_2}[(a/r)+1] - 3\alpha_1}$$

$$K_{tk} = \alpha_1 + 0.8\alpha_3 \qquad K_{fk} = 1 + 2\sqrt{\dfrac{t}{r}}$$

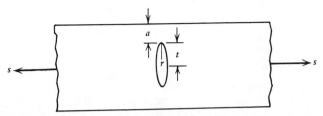

Figure 7.9 Flat plate with hole loaded in tension (see Appendix 9).

K_t is given by

$$K_t = 1 + \frac{(K_{tk}-1)(K_{fk}-1)}{\sqrt{(K_{tk}-1)^2+(K_{fk}-1)^2}}$$

The FORTRAN equivalent is as follows [where VAR (1) = a, VAR (2) = t, and VAR (3) = r; and VALUE is the resulting K_t factor]:

B = VAR(1)/VAR(3)
C = SQRT(B)
A1 = (2.*(B+1.)*C)/((B+1.)*ATAN(C)+C)
A2 = (4.*B*C)/(3.*((B−1.)*ATAN(C)+C))
A3 = (2.*(B+1.)−A1*SQRT(B+1.))/((4./A2)*(B+1.)−(3.*A1))
AKTX = A1+.8*A3
AKFK = 1.+2.*SQRT(VAR(2)/VAR(3))
VALUE = 1.+((AKTK−1.)*(AKFK−1.))/SQRT((AKTK−1.)**
2.+(AKFK−1.)**2.))

7.4 PROBABILISTIC APPROACH

The basic relationships needed to determine a theoretical stress concentration factor are shown in Section 7.3. In obtaining statistics describing stress concentration factors, Monte Carlo simulation methods* [9] may be utilized for a profile such as Figure 7.9.

7.4.1 Estimating K_t Statistics by Monte Carlo Simulation

Program Monti is a Monti Carlo computer simulation program (FORTRAN) designed to estimate the statistics (mean and standard deviation) of an arbitrary algebraic function (up to 10 random variables) from the statistics of the variables appearing in the function. In this section, program Monti is used to estimate the statistics of the stress concentration factor K_t from its defining algebraic equation. Figure 7.10 shows an example of the program as used for the geometry of a flat plate with a hole loaded in tension. As with all

---

*There were discussed in Section 2.9.

282 Geometric Stress Concentration

```
            PROGRAM MONTI(INPUT,OUTPUT)
            DIMENSION IR(2),VALUE(1000),AVG(10),SD(10),VAR(10)
C
C           READ IN THE NUMBER OF SAMPLE POINTS DESIRED AND THE NO. OF VARIABLES
            READ 10, NDP,NVR
     10     FORMAT(2I10)
            SEED=4.0
            CALL RANSET(SEED)
C           READ IN MEAN AND STD. DEV. OF THE RANDOM VARIABLES
            DO 12 I=1,NVR
     12     READ 11,AVG(I),SD(I)
       11   FORMAT(2F10.0)

            DO 200 JJ=1,6
            PRINT 201, AVG(1), SD(1), AVG(2), SD(2), SD(3)
     201    FORMAT(1H1,//////////,10X,"THE MEAN VALUE OF A IS  ". F4.2, 2X,
            2"INCHES",5X, "THE STANDARD DEVIATION OF A IS ". F7.5,
            3/, 10X, " THE MEAN VALUE OF T IS   " F4.2, 2X,  "INCHES",
            45X, "THE STANDARD DEVIATION OF T IS  ", F7.5,/,
            552X, "THE STANDARD DEVIATION OF R IS  "F7.5)
            PRINT 6
      6     FORMAT(/, 6X, "RADIUS R", 11X, "R OVER SD", 10X, " BD OVER SD",
            210X, "MEAN OF KT", 7X, "STD. DEV. OF KT")

            DO 100 II=1,10
C
C           CALCULATE RANDOM VALUES OF A, T, AND R
            DO 1 I=1,NDP
            DO 5 J=1,NVR
            CALL RANNOR(RN)
            VAR(J)=RN*SD(J)+AVG(J)
            IF (VAR(J) .LT. 0.0) GOTO 7
            GOTO 5
      7     VAR(J)=1.E-50
      5     N=N+1

C           EQNS TO BE USED FOR PARTICULAR CASE BEING STUDIED
            B=VAR(1)/VAR(3)
            C=SQRT(B)
            A1=(2.*(B+1.)*C)/((B+1.)*(ATAN(C))+C)
            A2=(4.*B*C)/(3.*((B-1.)*(ATAN(C))+C))
            A3=(2.*(B+1.)-A1*(SQRT(B+1)))/((4./A2)*(B+1.)-(3.*A1))
            AKTK=A1+1.B*A3)
            AKFK=1.+(2.*SQRT(VAR(2)/VAR(3)))
      1     VALUE(I)=1.+(((AKTK-1.)*(AKFK-1.))/(SQRT((AKTK-1.)**2.+(AKFK-1.)
            2**2.)))
```

Figure 7.10 Program Monti illustrating a hole in a flat plate loaded in tension.

Monti Carlo routines, a random number generator is required. This is the purpose of subroutine RANNOR. A flowchart is shown in Figure 7.11.

In Figure 7.12, stress concentration factors have been calculated using typical choices of statistics for the variables (assumed to be normally distributed). In design applications the engineer will specify mean values and variances that apply to the problem at hand. Some K_t statistics may be estimated by interpolation of the graphs in Figure 7.12. It is often necessary to compute K_t factors for particular design situations.

The operation of program Monti can best be described by an example. For a flat plate with a hole loaded in tension (configuration 6, Appendix 9) the random variables involved are a, t, and r. If this geometry were employed in the design of a component, the physical situation would control the selection

```
            BD=2.*AVG(1)+2.*AVG(2)
            ASD=2.*AVG(I  )
            ROSD=VAR(3)/ASD
            BDOSD=BD/ASD
C
            COMPUTE THE MEAN
            X=0.0
            DO 2 J=1.NDP
      2     X=X+VALUE(J)
            XBAR=K/FLOAT(NDP)
C
C           COMPUTE THE STANDARD DEVIATION
            SSD=0.0
            DO 3 I=1/NDP
      3     SSD=SSD+((VALUE(I)-(BAR)**2)
            SIGMA2=SSD/FLOAT(NDP)
            SIGMA=SQRT(SIGMA2)
C
C           PRINT THE RADIUS. VALUES OF R/LD AND BD/LD. MEAN AND ST. DEV. OF
C           KT
            PRINT 4. AVG(3). ROSD. BDOSD. XBAR. SIGMA
      4     FORMAT(/. 8X. F4.2. 15X. F6.3. 17X. F6.3. 10X. F5.3)
C
C           INCREMENT VALUE OF RADIUS R
            IF(AVG(3) .GE. .05) GOTO 20
            AVG(3)=AVG(3)+.01
            GOTO 100
     20     AVG(3)=AVG(3)+.01
    100     CONTINUE
C
C           INCREMENT OUTSIDE DIMENSION
    200     AVG(1)=AVG(1)+0.5

            STOP
            END

            SUBROUTINE RANNOR(RN)
            C=3.1415926536
            A=RANF(0.0)
            B=RANF(SEED)
            E=SQRT(-2.0*ALOG(A))
            RN=(E*COS(2*C*B))

            RETURN
            END
```

Figure 7.10 *Continued*

of values of the variables within specific ranges. After specific values have been selected for the mean values of these random variables, the manufacturing process to be employed can be considered in estimating variances.

After the geometric configuration and loading conditions for the design situation have been determined, the equations that model the stress concentration factor are selected from Appendix 9 and their FORTRAN code equivalent inserted into program Monti under the comment card "equations to be used for particular case being studied." The program now independently selects a set of random values of the input variables to calculate a random value of K_t. This random procedure is repeated by the computer the number of times specified by the variable NDP (number of data points). The program was originally written for 1000 iterations. With a sample of 1000

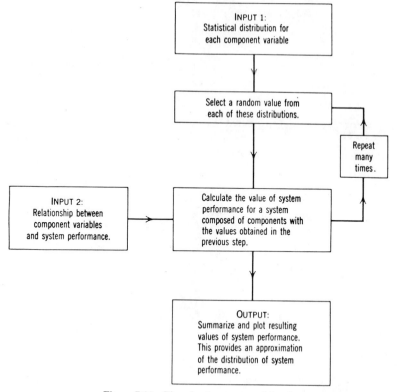

Figure 7.11 Flowchart—program Monti.

values of K_t the sample mean ($\bar{K}_t$) and standard deviation (σ_{K_t}) are computed.* The program requires one data card for each variable that appears in the defining algebraic equation. The data cards in the present example are:

Card 1: Number of data points (NDP) desired and number of variables (NVR) to be used in 2I10 format.

Card 2: Initial value of the mean of the first variable and the standard deviation of the first variable, in this case a in (2F10.0 format).

Card 3: Initial value of mean of variable t and standard deviation of variable t in 2F10.0 format.

Card 4: Initial value of mean of variable r and standard deviation of variable r in 2F10.0 format.

*In instances where a complete K_t envelope is to be developed, the values of the variable AVG(3) are then incremented and the procedure is repeated over a range of $\bar{r}$ values.

>AVG(1) Denotes a mean value of a
>AVG(2) Denotes a mean value of t
>AVG(3) Denotes a mean value of r

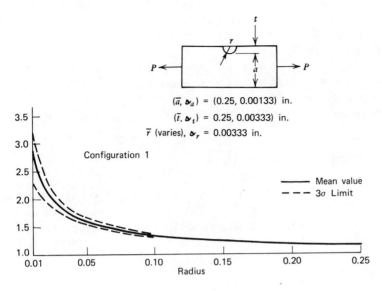

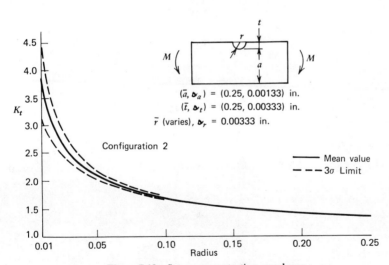

Figure 7.12 Stress concentration envelopes.

To calculate a "sample population" of values of K_t, random values of the variables a, t, and r must be supplied to the defining equation of K_t. This is done under the comment, "calculate random values of a, t, and r." These random values of a, t and r are given the new variable names VAR(1), VAR(2), and VAR(3).

Although this program was developed for the purpose of evaluating the statistical variation of the stress concentration factor, its utility is not limited to this one application. For any function of several variables the program will estimate its mean value as well as its corresponding standard deviation.

7.5 STATISTICS OF K_t ENVELOPES

The following K_t plots* have been obtained by means of the computer program presented in Section 7.4.1. Edge distance a and t were arbitrarily established as 0.25, and notch radii varied from 0.05 to 0.25 in the examples.* The standard deviations of the variables a, t, and r used in the examples were based on typical machining capabilities, as follows:

$$\sigma_a = 0.00133 \text{ in.}$$

$$\sigma_t = 0.00333 \text{ in.}$$

$$\sigma_r = 0.00333 \text{ in.}$$

In the machining tolerances on each of the variables a, t, and r in the example are

$$\pm \Delta_a = \pm 0.004 \text{ in.}$$

$$\pm \Delta_t = \pm 0.010 \text{ in.}$$

$$\pm \Delta_r = \pm 0.010 \text{ in.}$$

In actual design situations a, r, and t values are assigned to suit the situation.

---

*See Figure 7.12 and Appendix 9.

CHAPTER 8

Describing Materials Behavior with Random Variables

The following topics are considered in this chapter:

Static strength

Tensile ultimate, yield, and shear strength statistics
Proportional limit and yield point
Torsional strength
Tensile ductility and fracture toughness
Hardness versus mechanical properties

Dynamic strength

Materials behavior subject to constant-amplitude sinusoidal loading with zero and nonzero mean and S–N and Gerber statistical envelopes
Narrow-band random loading with zero and nonzero mean
Strength, a random process

8.1 INTRODUCTION

In this chapter the static and dynamic behavior of polycrystalline metallic engineering materials are found to be multivalued because of random differences in chemistry, processing, producers, and so on.

The deformation and fracture of materials as a consequence of applied surface tractions are the important phenomena associated with the mechanical behavior of materials [64]. By far the greatest amount of metals and their alloys is used for structural purposes in which resistance to deformation is of primary importance. Since quantitative engineering descriptions of materials behavior have a macroempirical basis, the considerations of this chapter are confined to the macrolevel of most design engineering activity.

By describing materials behavior statistically, the influences of variable metallurgical makeups, varying results of heat treating, and mechanical processing variability are taken into account in the model.

Even in establishing deterministic design allowables, sample mean strength $(\bar{S})$ is the starting point from which is subtracted the factor $K_\alpha \cdot \text{\textbar}_S$, where K_α is the one-sided tolerance limit factor corresponding to a proportion at least $1 - \alpha$ of a normal distribution and a confidence coefficient of $\gamma = 0.95$ (Appendix 3).

Example A

Estimate the lower 95% one-sided confidence limit L for S_u from information in Figure 8.1:

$$L = \bar{S}_u - k_\alpha \cdot \frac{\sigma_S}{\sqrt{n}}$$

1. If the standard deviation is assumed known and $\sigma_{S_u} = 2.63$ ksi, then from Appendix 1, $k_\alpha = 1.645$ is the 5% point of an (assumed) normal distribution:

$$L = 63.7 - 1.645 \cdot \frac{2.63}{\sqrt{1000}} = 63.5 \text{ ksi}$$

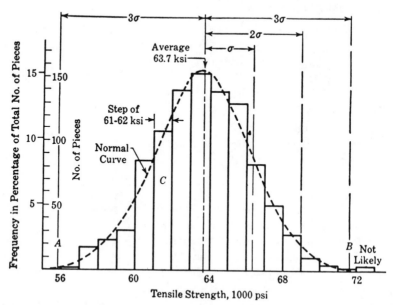

Figure 8.1 AISI 1020 Steel histogram (1-in. round bars from 25 heats and two vendors) [65]. $\bar{S}_u, {}_{\text{\textbar}S_u} \simeq (63.7, 2.63)$ ksi and $n \approx 1000$. To convert from strength in ksi to MPa, multiply by $9.807 \cdot 10^3$.

2. If standard deviation is assumed unknown and $s_{S_u} = 2.63$ ksi, then from Appendix 3, $t_{\alpha;n-1} = 1.645$ is the 5% point of the t distribution:

$$L = 63.7 - K(t_{\alpha;n-1}) \cdot \frac{s_{S_u}}{\sqrt{n}} = 63.5 \text{ ksi}$$

The equivalence of L in 1 and 2 is attributable to the large sample [66].

Materials selection is usually the design engineer's responsibility and must be completed before component dimensions can be specified. The proportioning of components is delineated so that the likelihood of internal stresses and strains being compatible with materials behavior characteristics is equal to or exceeds an a priori specified probability.

STATIC STRENGTH

8.2 DEFINITIONS

By "strength" is meant the maximum unit stress that can exist within a material without impairing the usefulness or ability of that material to perform its function in a mechanical system. For a specific individual component, strength is unique but usually unknown. What can be known are the parameter estimators that define the distribution of strength, determined from tests of specimens (see Bayes decision procedures, Section 3.2.2).

For any specific material, there are several stress-magnitude random variables, each of which may be considered as the strength of the material, that is, shear or yield strength.

Static strength is the strength displayed by a material subject to static loading. Based on the findings from numerous studies, the distribution of static tensile (ultimate) strength is usually normal or approximately so (Figure 8.2). The examination of numerous histograms, [67] and [54], indicates that the distributions of static yield strength are usually approximately normal [68, 69] (Figure 8.3).

In Sections 8.8 and 10.2 it is inferred that static yield in shear is a linear function of static tensile yield and hence very likely has the same general type of distribution.

8.3 TENSION TESTING

Test specimens are machined to conform to stated dimensions and tolerances, in accord with established ASTM standards. The original cross-sectional area of each specimen is determined and recorded, and then variable

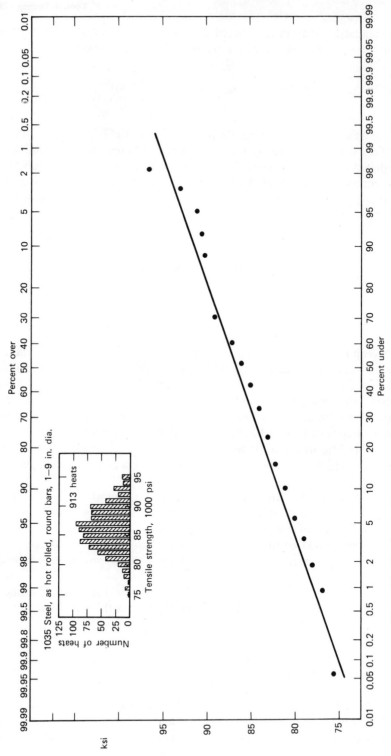

Figure 8.2 Fit of typical ultimate strength test values to Gaussian probability density function model [67].

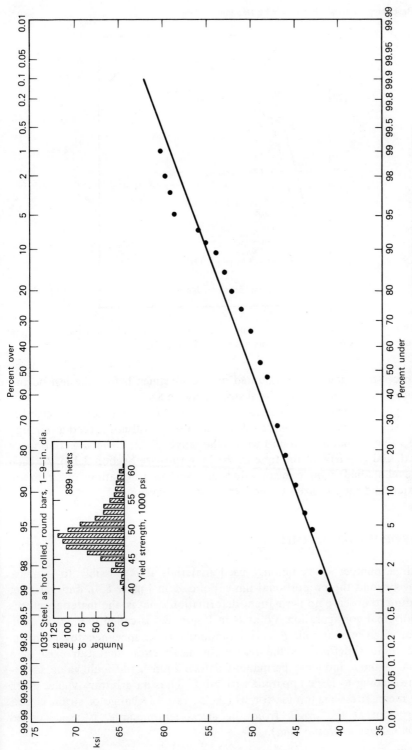

Figure 8.3 Fit of typical yield strength test values to Gaussian probability density function model [67].

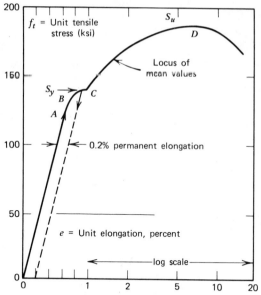

Figure 8.4 Stress vs strain. Sketch of result of typical tension test of a material, to illustrate definitions of yield strength S_y and ultimate strength S_u [1, p. 17].

resistance strain gauges are mounted on the specimen before the test begins, with observations of load and total strain (Figure 8.4).

Enough specimens (n) of material x are tested to assure that the probability is at least $1 - \alpha$ that the absolute difference $|\bar{x} - \mu_x|$ will not exceed a required ε. The value μ_x should be found within the range $(\bar{x} - \varepsilon; \bar{x} + \varepsilon)$, the confidence interval, $100 \cdot (1 - \alpha)\%$ of the time in the long run (see Section 3.3.1). At each load level studied, sample mean values and standard deviations of strain are estimated and then utilized to plot stress–strain envelopes.

8.4 PROPORTIONAL LIMIT

With a number of commonly used materials it is possible to increase loading beyond the proportional limit (point A in Figure 8.4), and yet after unloading to observe no permanent deformation; that is, the material returns to its original configuration. Point B in Figure 8.4 is called the *elastic limit*, and for a given material, B varies randomly from specimen to specimen. Elastic limit is defined as the maximum mean stress that can exist in a material without inducing permanent deformation, a difficult value to determine leading to large apparent variability. Thus an arbitrary value, taken at the mean stress magnitude for which the rate of change of strain is 50% greater than at the origin from the stress–strain curve, is used (called *Johnson's apparent elastic limit*).

8.5 YIELD POINT

With low-carbon steel there is a stress value above which strain increases rapidly, without a corresponding increase in stress, called the *yield point* (point C in Figure 8.4). The material has passed from dominantly elastic to a condition where the plastic state dominates. For ductile materials, at the yield point a sudden drop in load is observed on the tensile test load indicator, so intense with some materials that the slope of the curve reverses. Although yield point is clearly observable, the apparent scatter in values is much larger than that associated with tensile ultimate strength.

8.6 TENSILE YIELD STRENGTH, S_y

The stress–strain diagrams of many materials display no sudden change of slope. Because the stress associated with yield point is often a criterion in design, there is need for a value that can be used for materials having no clearly identifiable yield point. Thus an arbitrary value called *yield strength* (S_y) (Figure 8.5) corresponding to a fixed amount of permanent deformation (usually 0.10% or 0.20% of the original gauge length) is used. For typical statistics of yield strength, see Table 8.1.

8.7 ULTIMATE TENSILE STRENGTH

The ultimate or tensile strength (S_u) of a material is the distribution of maximum stress (based on original cross sections) observed on the stress–strain curve, (point D in Figure 8.4).

Materials such as mild steel tend to tail off after the maximum stress is reached. Certain materials exhibit essentially the same properties in compression as in tension.

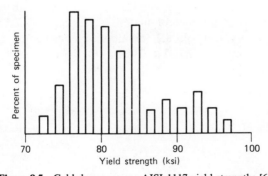

Figure 8.5 Cold-drawn versus AISI 1117 yield strengths [65].

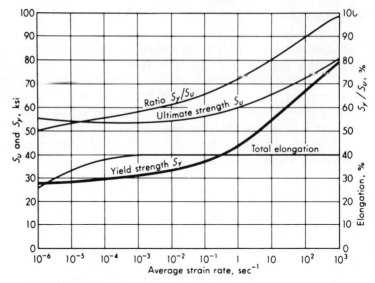

Figure 8.6 Effect of strain rate on static yield and ultimate strengths.

Achieving true axial loading is often a problem; also, ultimate compressive stress of ductile materials cannot be determined by tests because with plastic deformation the increase in area reduces the true stress. Also, with increased loading a ductile material bulges and no definite point of failure can be identified. Note that yield and ultimate strengths may be functions of strain rate as indicated in Figure 8.6.

STATIC STRENGTH DATA

Table 8.1 provides typical examples of the tensile ultimate and yield strength statistics of a number of materials frequently used in engineering. The A and B values in Table 8.2 for aluminum, titanium, and magnesium alloys, based on samples $n \geqslant 100$ [66] may also be utilized to estimate mean values and standard deviations:

$$A = \bar{S} - k_1 s_S = \bar{S} - 2.686^* \cdot s_S$$

*See Appendix 3.

Table 8.1 Static Strength of Aluminum Alloys[a]

Material	Reference	Remarks	Condition	Ultimate Tensile, S_u, KSI			Tensile Yield, S_y, KSI		
				Mean	s	Sample	Mean	s	Sample
2014 (AMS 4135)	54		Forging	70.0	1.89	36	63.0	2.23	36
2014-T651	127	Plate, L–T	0.50–1.50 in.	85.0	2.76	20	75.0	1.75	20
	127	Plate, T–L	0.50–1.50 in.	83.0	1.34	20	72.0	2.07	19
2014-T651	127	Plate L–T	Ambient temp.	69.0	1.82	20	63.0	1.35	20
	127	Plate T–L	Ambient temp.	69.0	1.42	20	61.0	1.65	20
2024-0	68	Clad sheet L–T	0.023–0.067	28.1	1.73	69	13.0	0.89	69
		Sheet	Bare	67.0	0.93	248			
2024-T3	54		Sheet	50.6	2.94	61			
2024-T4		Clad sheet	0.014–0.022 T–L	60.1	2.53	62	36.1	1.90	61
			0.023–0.028 L–T	66.4	2.42	153	41.4	2.14	153
			L–T	60.9	2.27	73	36.6	1.91	73
	68		0.029–0.035 L–T	64.9	2.07	257	40.1	2.02	257
			T–L	61.6	2.17	116	37.3	1.91	116
			0.068–0.076 L–T	65.8	1.59	64	41.5	2.46	64
2024-T6	68	Clad sheet	0.029–0.035 L–T	63.6	2.51	299	50.1	2.85	299
			0.058–0.067	65.9	2.56	194	51.8	1.82	194
2024-T81	68	Clad sheet	0.029–0.057 L–T	65.9	2.34	18	59.7	3.25	18
			0.068–0.140 L–T	69.9	1.70	15	64.4	2.41	15
2024-T86	68	Clad sheet	0.036–0.045 L–T	71.0	1.48	57	66.9	2.20	57
			0.086–0.096 L–T	75.2	1.53	69	70.5	2.84	69
2219-T87	127	Plate	0.50–1.5 L–T	67.0	1.64	17	54.0	1.53	17
			T–L	67.0	1.85	17	54.0	2.01	17
6061-T4	68	Sheet	0.032–0.125	36.6	1.69	1461	20.1	2.30	1461
6061-T6	68	Sheet	0.032–0.125	45.6	1.91	1648	41.7	2.89	1648
7039-T6	127	Plate—320°F	0.50–1.50 L–T	78.9	0.56	8	65.9	0.94	8
			T–L	80.6	0.65	7	65.3	0.41	7

[a]For additional data, see Appendix 10A.

where k_1 is $k_{0.99, 0.95, n}$ and α is the 1% significance level (see Chapter 12):

$$1 - \alpha = 1.00 - 0.01 = 0.99$$

$$\underline{\gamma = 0.95 = \text{confidence};} \qquad n \geqslant 100$$
$$B = \bar{S} - k_2 s_S = \bar{S} - 1.528 * s_S$$

where k_2 is $k_{0.90, 0.95, n}$ and α is the 10% significance level:

$$1 - \alpha = 1.00 - 0.10 = 0.90$$

$$\underline{\gamma = 0.95 = \text{confidence};} \qquad n \geqslant 100$$

Thus

$$\bar{S} = 2.320 B - 1.320 A$$

$$s_S = \frac{B - A}{1.158}$$

The test data on which A and B values were based, are on file at the Battelle Memorial Institute in Columbus, Ohio. Tables 8.3a and 8.3b list static strength statistics of ASTM A325 structural bolts.

8.8 SHEAR STRENGTH

Since bending forces are usually present, pure shear is seldom found; however, direct shear tests can be performed as indicated in Figure 8.7. In applying force it is not feasible to read the induced shearing strains; thus the elastic limit or yield for shear cannot be directly determined. The force at failure divided by specimen cross section is defined as shear ultimate strength. Statistics of ultimate shear strength for a structural titanium alloy (at various temperatures) are given in Figure 8.8.

Rough approximations of direct shear strength may be obtained as follows [1]:

Material	$\frac{S_{su}}{S_u}$
Aluminum alloys	0.60
Steel	0.75
Copper	0.90
Malleable iron	0.90
Cast Iron	1.30

Table 8.2
Design Mechanical and Physical Properties of Ti-6Al-4V (Sheet, Strip, and Plate)

Specification	MIL-T-9046							
Form	Sheet, strip, and plate							
Condition	Annealed				Solution-treated and aged			
Thickness, in.	≤0.1875		0.1875–4.000	<0.1875	0.1875 to 0.750	0.751 to 1.000	1.001 to 2.000	2.001 to 4.000
Basis	A	B	S	S	S	S	S	S
Mechanical properties:								
F_{tu}, ksi	134	139	130	160	160	150	145	130
F_{ty}, ksi	126	131	120	145	145	140	135	120
F_{cy}, ksi:								
L	132	138	126	154				
LT	132	138	126	162				
F_{su}, ksi	79	81	76	100				
F_{bru}, ksi:								
(e/D = 1.5)	197	204	191	238				
(e/D = 2.0)	252	261	245	286				
F_{bry}, ksi:								
(e/D = 1.5)	171	178	163	210				
(e/D = 2.0)	208	216	198	232				
e, percent:								
In 2 in.	8[a]	...	10	5[b]	8	6	8	6
E, 10^3 ksi				16.0				
E_c, 10^3 ksi				16.4				
G, 10^3 ksi				6.2				
μ				0.31				
Physical properties:								
ω, lb/in.^{3}				0.160				
C, Btu/(lb)(F)				See Figure 5.4.3.0				
K, Btu/[(hr)(ft^2)(F)/ft]				See Figure 5.4.3.0				
α, 10^{-6} in./in./F				See Figure 5.4.3.0				

[a] 8-0.025 to 0.062 in.
10-0.063 in. and above.

[b] 5-0.050 in. and above.
4-0.033 to 0.049 in.
3-0.032 in. and below.

Table 8.2 Continued

Design Mechanical and Physical Properties of Clad 7178 Aluminum Alloy (Sheet and Plate)

Specification	QQ-A-250/15 and QQ-A-250/22																							
Form	Sheet and Plate																							
Temper	T6								T651							T76			T7651		
Thickness, in.	0.015–0.044		0.045–0.062		0.063–0.187		0.188–0.249		0.250–0.499		0.500–1.000		1.001–1.500		1.501–2.000	0.045–0.062	0.063–0.187	0.188–0.249	0.250–0.499	0.500–1.000[a]
Basis	A	B	A	B	A	B	A	B	A	B	A	B	A	B	S	S	S	S	S	S
Mechanical properties:																				
F_{tu}, ksi:																				
L	76	76	78	78	80	80	82	82	81	83	81	83	81	83	77	69	70	72	71	70
LT	76	76	78	78	80	80	82	82	82	84	82	84	82	84	78	70	71	73	72	71
F_{ty}, ksi:																				
L	67	66	69	68	71	70	72	71	73	71	72	71	71	71	67	60	61	62	60	59
LT	66	68	69	70	71	72	73	73	75	73	74	73	73	73	68	59	60	61	50	59
F_{cy}, ksi:																				
L	67	66	69	68	71	70	72	71	73	71	73	71	71	71	68	60	61	62	50	59
LT	71	73	73	75	75	77	76	78	76	78	74	76	73	75	69	63	64	65	63	62
F_{su}, ksi	45	47	47	48	48	49	49	50	47	48	45	46	42	43	37	41	42	43	42	42
F_{bru}, ksi[b]:																				
($e/D = 1.5$)	118	121	121	124	124	127	127	130	125	128	121	124	115	118	104	108	110	113	112	106

	152	156	160	164	168	154	158	148	152	140	143	126	140	142	146	144	138
$(e/D = 2.0)$																	
F_{bry}, ksi[b]:																	
$(e/D = 1.5)$	99	102	105	108	109	107	110	104	106	98	101	89	91	93	95	93	86
$(e/D = 2.0)$	115	119	122	126	128	127	130	122	126	116	120	106	106	108	110	108	100
e, percent:																	
L	...	...	...	...	...	...	...	...	...	...	...	...	...	...	...	...	...
LT	7	8	8	8	8	8	8	6	4	...	...	3	8	8	8	8	6
E, 10^3 ksi																	
Primary	10.3		10.3	10.3				10.3					10.3	10.3	10.3	10.3	
Secondary	9.5		9.8	10.0				10.0					9.5	9.8	10.0	10.0	
E_c, 10^3 ksi																	
Primary	10.5		10.5	10.5				10.6					10.5	10.5	10.5	10.6	
Secondary	9.7		10.0	10.2				10.3					9.7	10.0	10.2	10.3	
G, 10^3 ksi	...		...	...				...					...	...	...	...	
μ	0.33		0.33	0.33				0.33					0.33	0.33	0.33	0.33	

Physical properties:
- ω, lb/in.^{3} 0.102
- C, Btu/(lb)(F) .... 0.23 (at 212 F)
- K, Btu/[(hr)(ft^2)(F)/ft] ...... 72 (at 77 F)
- α, 10^{-6} in./in./F .... 13.0 (at 68 to 212 F)

Physical properties:
- ω, lb/in.^{3} 0.102
- C, Btu/(lb)(F) .... 0.23 (at 212 F)
- K, Btu/[(hr)(ft^2)(F)/ft] ...... 86 (at 77 F)
- α, 10^{-6} in./in./F .... 13.0 (68 to 212 F)

[a]These values have been adjusted to represent the average properties across the whole section, including the 1-½ percent per side nominal cladding thickness.

[b]See Table 3.1.2.1.1. Bearing values are "dry pin" values per Section 1.4.7.1.

Table 8.3a Static Strength Tables for Structural Steels,
ASTM A325 Structural Bolts: Direct Tension Calibrations

Number of Specimens Tested	Ultimate Load		Spec. Vlt. (%)	Rupture Load		Elongation at Proof Load		Elongation at Ult. Load		Elongation at Rupt. Load	
	Mean (kips)	Standard Deviation (kips)		Mean (kips)	Standard Deviation (kips)	Mean (in.)	Standard Deviation (in.)	Mean (in.)	Standard Deviation (in.)	Mean (in.)	Standard Deviation (in.)
Regular Head Bolts with Rolled Threads											
5	54.3	1.22	102.1	45.2	1.89	0.010	0.0001	0.134	0.0370	0.31	0.047
5	53.3	0.30	100.0	45.2	1.05	0.023	0.0050	0.230	0.0201	0.31	0.044
5	56.0	0.55	107.0	45.0	1.15	0.010	0.0001	0.128	—	0.27	0.050
0	—	—	—	—	—	—	—	—	—	—	—
5	60.4	2.45	113.5	49.0	4.19	0.010	0.0003	0.135	—	0.24	0.040
5	73.7	0.96	105.7	63.2	2.40	0.011	0.0002	0.134	0.0050	0.36	0.030
5	73.1	0.04	104.9	62.5	1.77	0.011	0.0008	0.220	—	0.39	0.033
5	91.2	1.91	113.9	76.6	3.15	0.010	0.0000	0.200	—	0.30	—
Regular Head Bolts with Cut Threads											
3	53.9	0.14	101.3	42.0	2.64	0.012	0.0005	0.171	0.0134	0.37	0.036
3	52.2	0.09	98.1	35.0	—	0.010	0.0008	0.257	0.0072	0.34	—
3	53.3	0.52	101.1	43.3	1.45	0.021	0.0025	0.323	0.0182	0.53	0.164
3	54.8	1.04	103.0	45.3	0.53	0.014	0.0005	0.164	0.0316	0.33	0.050
3	55.2	0.72	103.8	46.7	1.00	0.015	0.0008	0.175	0.0297	0.39	0.026
3	53.6	0.51	101.1	49.9	1.21	0.016	0.0006	0.181	0.0051	0.33	0.026
3	55.2	0.95	103.8	46.7	1.16	0.017	0.0004	0.171	0.0070	0.34	0.040
Heavy Head Bolts with Rolled Threads											
7	58.3	1.29	109.6	51.6	2.73	0.018	0.0002	0.132	0.018	—	—
4	65.2	1.04	122.6	60.6	1.63	0.011	0.0001	0.109	0.0086	0.19	—
5	67.5	1.01	126.9	62.1	1.43	0.010	0.0001	0.112	0.0234	0.18	—
5	59.4	1.12	111.7	51.6	1.70	0.010	0.0002	0.112	0.0141	0.31	—
—	—	—	—	—	—	—	—	—	—	—	—
5	55.6	1.00	104.5	49.0	0.81	0.010	0.0003	0.077	0.0021	0.19	—
5	54.1	1.60	101.7	48.0	1.41	0.011	0.0001	0.009	0.0059	0.19	—

Table 8.3b Static Strength Tables for Structural Steels, ASTM A325 Structural Bolts: Torqued Tension Calibrations

Number of Specimens Tested	Ultimate Load Mean (kips)	Ultimate Load Standard Deviation (kips)	Minimum Spec. Vlt. (%)	Rupture Load Mean (kips)	Rupture Load Standard Deviation (kips)	Elongation at Proof Load Mean (in.)	Elongation at Proof Load Standard Deviation (in.)	Elongation at Ult. Load Mean (in.)	Elongation at Ult. Load Standard Deviation (in.)	Elongation at Rupt. Load Mean (in.)	Elongation at Rupt. Load Standard Deviation (in.)	Load at $\frac{1}{2}$ turn (kips)	Elongation at $\frac{1}{2}$ turn (in.)
						Regular Head Bolts with Rolled Threads							
3	45.7	0.65	87.2	38.5	2.00	0.009	0.0002	0.072	0.0067	0.17	0.014	41.3	0.0234
3	45.0	0.51	86.3	34.3	4.01	0.031	0.0016	0.087	0.0102	0.21	0.021	39.3	0.0330
4	51.1	0.765	95.9	40.5	0.58	0.010	0.0012	0.069	0.0112	0.20	0.036	46.2	0.0212
3	50.9	2.69	95.7	47.1	0.62	0.009	—	0.052	0.0129	0.10	0.027	46.3	0.0198
8	55.4	1.84	101.1	47.9	1.74	0.009	0.0010	0.076	0.0118	0.16	0.017	49.9	0.0236
3	56.0	2.00	30.3	47.9	2.69	0.017	0.0043	0.060	0.0079	0.12	0.011	50.3	0.0232
0	—	—	—	—	—	—	—	—	—	—	—	—	—
2	75.7	2.48	94.5	66.0	—	0.012	0.000	0.077	0.020	0.15	—	64.9	0.0246
						Regular Head Bolts with Cut Threads							
3	49.8	0.28	93.6	41.4	1.10	0.014	0.0009	0.141	0.0116	0.22	0.023	41.7	0.0251
3	47.2	0.77	88.7	40.3	2.52	0.026	0.0018	0.204	0.0119	0.30	0.051	36.0	0.0255
3	46.3	0.58	87.0	43.1	1.10	0.033	0.0059	0.281	0.0453	0.38	0.052	34.8	0.0244
3	52.0	0	97.7	43.8	1.94	0.015	0.0011	0.122	0.0116	0.20	0.011	40.8	0.0234
3	51.7	0.58	97.2	46.3	1.16	0.015	0.0028	0.169	0.0374	0.24	0.022	41.2	0.0250
3	56.2	0.20	94.1	44.2	1.76	0.017	0.0004	0.163	0.0161	0.25	0.019	39.8	0.0253
3	51.5	1.32	95.7	45.2	2.25	0.018	0.0006	0.159	0.0210	0.23	0.017	40.5	0.0242
						Heavy Head Bolts with Rolled Threads							
6	48.4	1.69	91.0	40.3	2.61	0.019	0.0005	0.066	0.0144	0.11	0.029	41.9	0.0233
3	55.3	0.72	103.9	40.0	5.38	0.012	0.0001	0.055	0.0047	0.10	0.085	42.3	0.0139
3	55.8	2.86	104.9	49.4	7.74	0.011	0.0012	0.059	0.024	0.10	—	45.3	0.0144
5	52.2	1.90	98.1	43.8	3.28	0.010	0.0009	0.055	0.0222	0.09	0.038	47.7	0.0190
3	49.4	1.35	91.0	36.3	1.53	0.010	—	0.083	0.0112	0.18	0.029	39.8	0.0182
5	49.3	1.69	92.7	37.7	3.10	0.012	0.0005	0.052	0.0063	0.09	0.014	46.2	0.0221
4	46.7	0.33	87.8	41.4	1.68	0.013	0.0005	0.055	0.0106	0.10	0.008	41.8	0.0165

Source: Rumpf and Fisher [70].

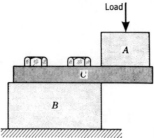

Figure 8.7 Shear test.

8.9 TORSIONAL STRENGTH

Torsional strength of materials is determined by applying twisting couples to cylindrical rods. Torque magnitude is plotted against angle of twist similarly to stress–strain diagrams as indicated in Figure 8.9. The elastic limit is usually obtained by finding the torque corresponding to point A. If this value is designated T_e, the elastic limit (S_{se}) is

$$S_{se} = \frac{T_e \cdot r}{J} \tag{8.1}$$

where J is the polar moment of inertia (see Appendix 5). The statistics of S_{se} in terms of the statistics of T_e, r, and J are as follows. From equation (2.59),

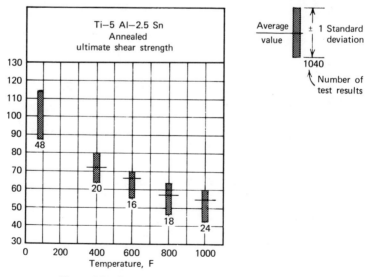

Figure 8.8 Shear strength statistics, Ti-5Al–2.5 Sn [7].

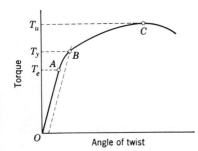

Figure 8.9 Twist angle versus torque.

S_{se} is

$$\bar{S}_{se} \approx \frac{\bar{T}_e \cdot \bar{r}}{\bar{J}} \tag{8.1a}$$

From equation (2.62), the expression for $s_{S_{se}}$ is

$$s_{S_{se}} \approx \left[\left(\frac{\partial S_{se}}{\partial T_e}\right)^2 s_{T_e}^2 + \left(\frac{\partial S_{se}}{\partial r}\right)^2 s_r^2 + \left(\frac{\partial S_{se}}{\partial J}\right)^2 s_J^2 \right]^{1/2} \tag{a}$$

The partial derivatives of T_e, r, and J with respect to S_{se} are

$$\frac{\partial S_{se}}{\partial T_e} = \frac{r}{J}$$

$$\frac{\partial S_{se}}{\partial r} = \frac{T_e}{J}$$

$$\frac{\partial S_{se}}{\partial J} = \frac{T_e \cdot r}{J^2} \tag{b}$$

If equation (b) is substituted into equation (a), $s_{S_{se}}$ is

$$s_{S_{se}} \approx \left[\left(\frac{\bar{r}}{\bar{J}}\right)^2 s_{T_e}^2 + \left(\frac{\bar{T}_e}{\bar{J}}\right)^2 \cdot s_r^2 + \left(\frac{\bar{T}_e \cdot \bar{r}}{\bar{J}^2}\right)^2 \cdot s_J^2 \right]^{1/2} \tag{8.1b}$$

The approximations $\bar{S}_{se} \simeq 0.288 \bar{S}_u$ is usually somewhat conservative for carbon and low-alloy steels.

Torsional yield strength is usually specified as that corresponding to a specified permanent twist angle deformation. Point B in Figure 8.9 was obtained by drawing a dashed line parallel to point A, offset from point O by the amount of the permanent deformation angle. Torsional yield strength is then calculated by utilizing the torque value T_y:

$$S_{sy} = \frac{T_y \cdot r}{J} \qquad (8.2)$$

The statistics of S_{sy} are analogue estimates of those of S_{se} earlier. *Ultimate torsional strength* is that corresponding to T_u, that is, to point C in Figure 8.9:

$$S_{su} = \frac{T_u \cdot r}{J} \qquad (8.3)$$

Equation (8.3) has been derived by assuming that the material follows Hooke's law, subject to random variations. With torque values in the plastic range, equation (8.3) is not valid. The value calculated is called the *modulus of rupture*. The statistics of S_{su} are analogue estimates of those of S_{se} earlier.

8.10 TENSILE DUCTILITY

Two metallic materials may have very similar hardness and strength properties, and one material may display much greater ability to sustain overloading because of a property called *ductility* [47, p. 150]. Ductility may be defined as [16]:

$$D = \ln \frac{A_0}{A_f} = -\ln(1 - RA).$$

where A_0 and A_f are initial and final fracture cross-sectional areas and in the tensile test:

$$\frac{A_0 - A_f}{A_0} = RA \qquad (a)$$

The statistics of area reduction (RA) in terms of the statistics of A_0 and A_f are as follows. From equations (2.59) and (a), $\overline{RA}$ is

$$\overline{RA} = \frac{\overline{A_0} - \overline{A_f}}{\overline{A_0}}^* \qquad (a')$$

*Also see Figure 1.16.

Tensile Ductility

According to equation (2.62), the expression for $\mathfrak{s}_{RA}$ is

$$\mathfrak{s}_{RA} \approx \left[\left(\frac{\partial RA}{\partial A_0} \right)^2 \cdot \mathfrak{s}_{A_0}^2 + \left(\frac{\partial RA}{\partial A_f} \right)^2 \cdot \mathfrak{s}_{A_f}^2 \right]^{1/2} \quad (b)$$

The partial derivatives of A_0 and A_f with respect to RA are

$$\frac{\partial RA}{\partial A_0} = \frac{A_f}{A_0^2}$$

$$\frac{\partial RA}{\partial A_f} = \frac{1}{A_0} \quad (c)$$

If equation (c) is substituted into equation (b), $\mathfrak{s}_{RA}$ is

$$\mathfrak{s}_{RA} \approx \left[\left(\frac{\overline{A}_f}{\overline{A}_0^2} \right)^2 \cdot \mathfrak{s}_{A_0}^2 + \left(\frac{1}{\overline{A}_0} \right)^2 \cdot \mathfrak{s}_{A_f}^2 \right]^{1/2} \quad (a'')$$

Brittle material behavior is indicated in Figure 8.10, where very little plastic deformation precedes rupture. Ductile behavior is indicated in 8.11. Ductility has been indicated by the percentage of elongation that has occurred at fracture, with an arbitrary division between brittleness and ductility considered as 5% elongation. Materials displaying more than 5% elongation in a tensile test [3] or if $S_y > (E/150)$, are considered ductile.

The capacity of ductile materials to tolerate overloads provides conservatism in design and is desirable in permitting cold working of a material. Metal forming operations employing power brakes, stretch presses, drop hammers, and hydraulic presses also require materials in a ductile state.

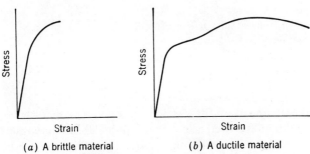

(a) A brittle material

Figure 8.10 Brittle material.

(b) A ductile material

Figure 8.11 Ductile material.

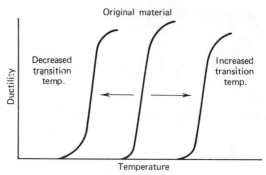

Figure 8.12 Effect of transitional temperature of several material and test parameters [47].

Many materials that are considered brittle, such as tungsten and molybdenum, are more properly described as having *transition temperatures* (T_0) that are higher than normal ambient [3, p. 332].

The general feature of a *transition temperature* is diagrammatically shown in Figure 8.12, with ductility, as measured by reduction in area (random variable) plotted on the ordinate against temperature. Transition temperatures of materials provide important information in certain applications of fracture mechanics analysis.

Factors that reduce transition temperature are:

1. Grain refinement, cold work
2. Purification
3. Surface smoothness
4. Alloying
5. Heat treatment to high ductility

Factors that increase transition temperature are:

1. High strain rate
2. Recrystallization
3. Impurities
4. Large size
5. Stress concentrations
6. Physical and chemical surface effects
7. Hydrogen embrittlement
8. Residual stress
9. Precipitation
10. Multiaxial stress

Below a certain temperature the ductility of a given material tends to be very low; however, ductility increases very rapidly with increasing T_0 in the neighborhood of the transition temperature. Thus the behavior of a material should not be classed as either brittle or ductile, unless a temperature is

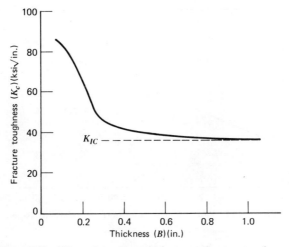

Figure 8.13 Effect of specimen thickness on fracture toughness.

specified. For example, at temperatures no lower than ambient, iron behaves in a highly ductile manner. The transition temperature of Molybdenum is near room temperature, and rather small temperature increases or decreases result in either ductile or brittle material. The transition temperature of tungsten is near 400°F. All transition temperatures are in general random variables.

8.11 FRACTURE TOUGHNESS

Few statistical descriptions of the behavior of brittle materials have appeared in the literature prior to publication of the *Damage Tolerance Design Handbook* [71] (which also contains data on some aluminum alloys). Much recent progress in the analysis of high-strength, hard materials used in mechanical devices and structures has resulted from developments in fracture mechanics theory and methodology.

In fracture mechanics the materials behavior analogue of static strength is a random variable, called *fracture toughness* (K_c), carrying units of ksi$\sqrt{\text{in}}$. * Fracture toughness is the minimum level of stress in a material capable of producing unstable crack propagation. It is important to note that K_c is a function of thickness (B) of the stressed material as indicated in Figure 8.13. The lower limiting value of fracture toughness, for a given material is represented by K_{1c}.

The higher and transitional values of fracture toughness (plane stress) are associated with small and intermediate values of B, usually designated K_c.

---

*To convert from ksi$\sqrt{\text{in}}$ to MPa$\sqrt{\text{m}}$, multiply by 1.099.

308 Describing Materials Behavior with Random Variables

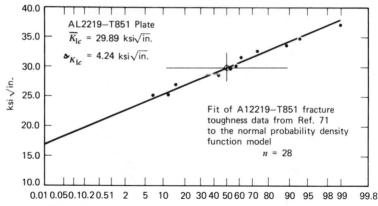

Figure 8.14 Plot of K_{1C} data on normal probability paper.

Lower bounding values (plane strain) of fracture toughness, usually designated K_{Ic}, are associated with thick sections (large B) of a material. The parameter associated with dynamic loads environments, called *threshold fracture toughness*, is usually designated $(\Delta K)_{th}$.

The distributional models describing plane strain and plane stress fracture toughness of common alloys are not discussed in the literature. However, as with other static properties of materials, the normal may be postulated as a likely (i.e., conservative) model pending more studies (see Figure 8.14).

FRACTURE TOUGHNESS DATA

Table 8.4 provides typical examples of plane strain K_{Ic} (minimum value associated with thick parts) and plane stress K_c (high value, accompanied by a shear lip, and associated with thin parts), fracture toughness, described by statistics. For the support of actual engineering design, more extensive sources [e.g., 71] should be consulted for specific behavioral statistics.

8.12 RESILIENCE

The resilience of a material is defined as its capacity for absorbing energy within the elastic range. Its measure is the modulus of resilience. Total strain energy is given by [Section 6.7, equation (e)]

$$U = \frac{s^2 lA}{2E} \tag{a}$$

Table 8.4 Typical Plane Strain Fracture Toughness (K_{Ic}), ksi$\sqrt{\text{in.}}$

Alloy	Thickness (B) × Width (W)	$\bar{K}_{Ic}$	Standard Deviation $s_{K_{Ic}}$	Sample n	Condition
\multicolumn{6}{l}{Examples—Aluminum Alloys (Thick Sections)}					
2014-T651	0.50×1.00 L-T	22.47	1.83	18	Plate
2014-T652	1.50×3.00 L-T	28.83	3.63	12	Forgings
2014-T652	1.50×3.00 T-L	21.88	3.21	13	Forgings
2024-T851	0.75×1.50 L-T	23.29	2.16	42	Plate
2024-T851	0.75×1.50 T-L	20.06	1.30	37	Plate
2024-T852	0.75×1.50 L-T	26.95	4.89	26	Forgings
2024-T852	0.75×1.50 T-L	18.75	3.12	17	Forgings
2124-T851	1.50×3.00 L-T	28.49	3.96	43	Plate
2124-T851	1.50×3.00 T-L	24.38	2.59	43	Plate
2214-T651	1.50×3.00 L-T	35.34	2.90	10	Plate
2214-T651	1.50×3.00 T-L	31.78	0.89	10	Plate
2219-T851	1.25×2.50 L-T	35.14	2.19	14	Plate
2219-T851	1.25×2.50 T-L	29.89	4.24	28	Plate
7049-T73	1.00×2.00 L-T	30.86	3.19	43	Forgings
7049-T73	1.00×2.00 T-L	22.12	2.75	27	Forgings
7075-T651	2.00×4.00 L-T	26.45	1.70	29	Plate
7075-T651	2.00×4.00 T-L	24.36	1.44	34	Plate
7075-T6	2.00×4.00 L-T	25.93	2.83	3	Bars and forgings
7075-T6	2.00×4.00 T-L	21.69	2.45	11	Bars and forgings
7075-T7351	1.00×2.00 L-T	33.27	0.67	7	Extrusions
7075-T7351	1.00×2.00 T-L	20.58	0.76	19	Extrusions
7079-T6	1.00×2.00 L-T	32.94	2.72	9	Plate
7178-T651	1.00×2.00 L-T	25.28	1.88	5	Plate
7178-T651	1.00×2.00 T-L	21.46	1.80	10	Plate
\multicolumn{6}{l}{Examples—Alloy Steels (Thick Sections)}					
AISI 4340	1.00×2.00 T-L	45.60	Undefined	3	Plate
D6AC	0.75×1.50 L-T	92.07	7.55	36	Plate
D6AC Intermediate tough	0.75×1.50 L-T	65.45	12.45	99	Plate
HP 9-4-.20 Austenitized	2.00×6.00 L-T	136.14	7.24	37	Forgings
HP 9-4-.20 Austenitized	2.00×6.00 T-L	125.78	10.39	18	Forgings
300M	1.00×2.00	49.20	Undefined	4	Plate
D6AC Intermediate tough	L-T	95.39	6.15	37	Forgings
D6AC	0.75×1.50 L-T	67.33	11.47	55	Forgings
18 Ni (200 Maraging Steel	4.00×8.00 L-T	82.00	Undefined	3	Plate
\multicolumn{6}{l}{Examples—Stainless Steel (Thick Sections)}					
PH 13-8 MO	1.00×2.0 L-T	55.00	Undefined	2	Plate or slab
15-5 PH Stainless	1.00×2.00 L-R	86.90	Undefined	1	Bar
17-4 PH Stainless	1.00×2.00 L-R	84.60	Undefined	1	Bar

Table 8.4 Continued

Alloy	Thickness (B) × Width (W)	K_{Ic}	Standard Deviation $s_{K_{Ic}}$	Sample n	Condition
Examples—Stainless Steel (Thick Sections)					
Custom 45 (VIM + Var) Code 1	.500 × 1.50 L-T	46.20	Undefined	3	Forgings
AFC (Course grain Code 1)	0.5 × 1.50 L-T	25.00	Undefined	1	Plate
AFC 260	0.5 × 1.50 L-T	47.00	Undefined	1	Plate
Examples—Titanium Alloys (Thick Sections)					
Ti-6AL-4V	2.00 × 5.00 L-T	84.07	5.29	37	Forgings
Ti-6AL-4V Heat C-3905	2.00 × 5.00 T-L	84.10	6.70	46	Forgings
Ti-6AL-4V-2SN (ELI) (Code 1)	0.25 × 5.00 L-T	29.80	Undefined	3	Plate
Ti-8MO-8V-2Fe-3AL	1.00 × 2.00 L-T	54.00	Undefined	3	Plate
Inconel 718	1.88 × 5.00 T-L	129.00	Undefined	1	Forgings
Examples—Aluminum Alloys—K_c (Thin Sections)					
2014-T6 −320°F	0.051 × 6.00	39.40	4.55	9	Sheet
2014-T6 78°F	0.063 × 18.00	65.18	1.39	10	Sheet
2024-T3 70°F	0.032 × 20.00	85.61	6.37	6	Sheet
2024-T3 70°F	0.080 × 20.00	83.44	2.07	6	Sheet
2024-T3 70°F	0.040 × 7.5	17.01	7.18	14	Clad sheet
2024-T4 70°F	0.040 × 7.5	16.31	6.44	12	Sheet
2024-T81 70°F	0.065 × 9.00	59.57	4.26	4	Sheet 0.063
2219-T87 70°F	0.10 × 6.00	78.80	11.13	8	Sheet 0.100
7075-T6 70°F	0.040 × 6.00	37.42	12.41	6	Sheet 0.040
7075-T6 70°F	0.063 × 24.00	67.59	7.60	13	Sheet 0.063
7075-T651 70°F	0.10 × 8.00	55.37	3.63	6	Sheet 0.100
7075-T7351 70°F	0.50 × 8.00	60.20	8.34	26	Plate 0.50
7475-T61	0.063 × 16.00	84.13	3.86	6	Sheet 0.063
7475-T61	0.063 × 16.00	76.45	6.80	8	Clad sheet 0.063
Examples—Alloy Steels—K_c					
300M 70°F	0.12 × 5.00	141.07	8.68	3	Sheet 0.125
300M 70°F	0.375 × 5.00	71.35	6.97	6	Plate 0.375
18 Ni, Maraging Steel	0.026 × 4.00	124.09	7.85	5	Sheet 0.025
Examples—Stainless Steels—K_c					
PH-14-8MO	0.025 × 24.00	225.67	10.34	4	Sheet 0.025
PH-14-8MO	0.050 × 24.00	294.16	Undefined	1	Sheet 0.050
PH-14-8MO	0.093 × 24.00	369.42	Undefined	1	Sheet 0.090
Examples—Titanium Alloys—K_c (Thin Sections)					
Ti-6AL-4V	0.025 × 4.00	107.17	3.04	5	Sheet
Ti-6AL-4V	0.025 × 18.00	136.49	10.45	5	Sheet
Ti-6AL-4V	0.040 × 8.00	121.40	7.05	7	Sheet
Ti-8AL-1MO-1V	0.020 × 12.00	98.37	9.98	4	Sheet
Ti-8AL-1MO-1V	0.050 × 20.00	158.15	11.79	9	Sheet

Table 8.4 Continued

Examples—Titanium Alloys—K_c (Thin Sections)

Alloy	Thickness $(B) \times$ Width (W)	Average $\bar{K}_{Ic}$	Standard Deviation $s_{K_{Ic}}$	Sample n	Condition
Ti-5AL-2.5SN	0.020 × 3.00	95.40	2.79	13	Sheet
Ti-5AL-2.5SN	0.020 × 6.00	94.54	1.81	14	Sheet
Ti-5AL-2.5SN	0.020 × 12.00	93.77	4.05	16	Sheet
Ti-5AL-2.5SN	0.100 × 2.00	77.64	1.05	3	Sheet
Example—Nickel Base Alloys					
Inconel 718 75°F	0.025 × 4.00	161.42	8.57	17	Sheet
Inconel 718 −320°F	0.025 × 4.00	180.40	8.82	15	Sheet
Inconel 718 −423°F	0.025 × 4.00	173.67	10.53	18	Sheet

Source: Damage Tolerance Design Handbook [71].

The modulus of resilience is the strain energy that can be absorbed by unit volume of a material, with stress intensity, s_p, at the proportional limit. Let

$$s_p = s \quad \text{and} \quad lA = 1 \text{ in}^3. \tag{b}$$

Substitution of equation (b) into equation (a) yields

$$U_p = \frac{s_p^2}{2E} \tag{8.4}$$

The interpretation of the modulus is shown in Figure 8.15, point A, indicates stress magnitude at the proportional limit. Line AB is drawn vertically. The shaded zone represents the quantity of energy that can be contained in unit volume of a material at the proportional limit. Since the proportional limit is

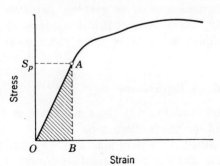

Figure 8.15 Modulus of resilience.

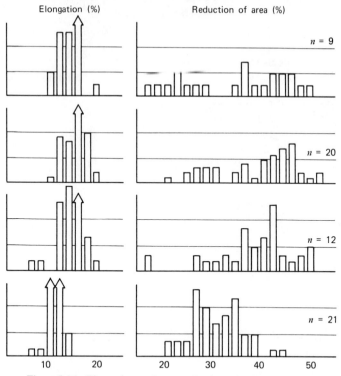

Figure 8.16 Elongation and area reduction of steel castings [1].

a random variable the amount of stored energy is a random variable. Large energy-storing capacity, that is, a large modulus AOB, implies high proportional limit and low modulus of elasticity.

8.13 HARDNESS

8.13.1 Introduction

When the material of a mechanical component must be wear resistant or erosion resistant, or where plastic deformation is to be avoided, hardness becomes a consideration. As is seen later, the measure of Brinell hardness of a material is often useful as an empirical estimator of other mechanical properties [1, p. 194; 72, p. 179].

8.13.2 Hardness Measuring Systems

Several methods of hardness testing are available, with those in greatest use being the Brinell, Rockwell, Vickers, and the Shore scleroscope.

Commonly, hardness testing systems utilize a standard load that is applied to a ball or pyramid in contact with the material being tested. Hardness is then expressed as a function of the dimension of the resulting indentation. Hardness numbers are not functionally convertible from one system to another. However, comparison between the numbers is often possible.

8.13.3 Hardness versus Mechanical Properties

Considering Vickers hardness (HV), the criterion for yield point has been deduced as [72, p. 7; 73]

$$HV \approx 3 \cdot (\text{yield stress of metal cold worked 8\%})$$

Empirical relationships have been found that enable approximate determination of tensile strength from Brinell hardness (see Figure 8.17).

Calculations based on test results for carbon and low-alloy wrought steels involving $n=111$ pairs of values (S_u, BHN), yielded the following statistics [65] [equation (8.5)]:

$$\left(\bar{S}_u, {}_{s_{S_u}}\right) \approx BHN \cdot (495, 20.5) \text{ psi} \qquad (8.5)$$

where $BHN = 200$ to 450.

Example B

Given that Brinell hardness of a carbon steel is 150 determine the approximate statistics of S_u. Then, from equations (2.8) and (2.10):

$$\bar{S}_u \approx 150(495) = 74{,}250 \text{ psi}$$

$${}_{s_{S_u}} \approx 150(20.5) = 3{,}080 \text{ psi}$$

Shigley [1, p. 209] states that estimated values are accurate within 3%. Studies at the University of Arizona indicate that $\bar{S}_u$ estimated from average Brinell hardness is slightly conservative.

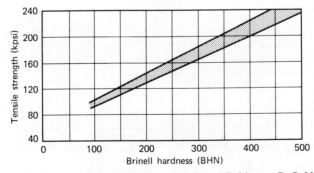

Figure 8.17 Relation between tensile strength and hardness (C. Lipson, G. C. Noll, and L. S. Clock).

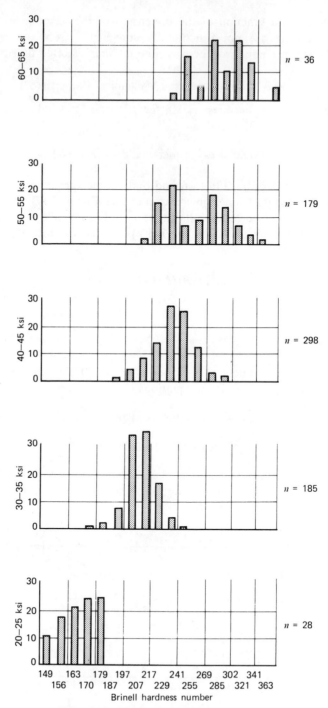

Figure 8.18 Distribution of hardness data [67].

Table 8.5 Rockwell C Hardness Measurements of AISI 4340 Steel, Plotted to Indicate Hardness as a Stationary and Ergodic Discrete Random Process.[a]

	1	2	3	4	5	6	7	8	9	10	11	12	13	14	Mean	Standard Deviation
1	17.0	20.2	21.0	20.6	20.8	21.2	20.8	20.8	22.0	19.4	21.2	21.6	20.6	22.0	20.6571	1.2538
2	21.0	19.8	21.4	20.2	21.2	21.6	20.8	20.4	20.2	21.4	22.2	21.4	21.8	22.0	21.1000	0.7307
3	21.6	21.8	21.6	21.6	22.0	22.2	22.0	22.0	20.8	20.6	21.4	20.6	21.0	20.8	21.4286	0.5649
4	21.2	21.0	21.2	21.4	21.2	21.2	21.4	21.0	20.0	21.0	21.2	17.6	20.6	20.4	20.7429	0.9874
5	20.2	20.2	20.8	20.4	20.6	20.6	20.4	20.6	17.8	20.6	19.8	19.8	20.2	18.6	20.0428	0.8492
6	20.0	19.8	20.0	19.0	19.0	20.8	19.4	20.4	19.0	20.6	19.0	19.4	20.2	20.8	19.8143	0.6860
7	21.0	21.0	19.6	20.8	19.8	21.0	20.4	20.8	21.0	20.2	20.2	21.0	20.4	20.8	20.6286	0.4763
8	21.0	21.2	20.8	21.4	21.6	21.2	21.2	21.2	20.8	20.2	20.8	20.6	21.2	21.2	21.0000	0.4076
9	20.8	21.0	21.6	20.8	21.0	20.6	20.6	20.2	21.4	19.6	20.6	20.6	20.6	20.4	20.7000	0.4883
10	22.2	21.0	21.2	20.6	21.6	19.4	20.8	20.0	19.6	19.2	19.0	20.2	19.4	18.4	20.1714	1.0781
11	20.2	20.0	21.2	20.8	20.6	18.2	19.6	20.0	20.2	21.0	20.2	20.0	19.6	20.0	20.1143	0.7305
12	20.4	19.8	19.6	19.6	19.6	21.8	21.8	21.0	21.6	20.8	21.8	21.8	21.2	21.6	20.8857	0.9139
13	19.0	20.6	21.4	20.0	22.2	21.0	20.8	21.2	21.2	21.0	21.4	20.8	21.2	20.4	20.8714	0.7467
14	21.4	19.4	21.6	20.4	20.0	20.0	20.6	20.8	21.4	20.8	20.6	20.8	20.4	20.6	20.6286	0.5967
Mean	20.5000	20.4857	20.9143	20.5429	20.8000	20.7714	20.7571	20.7429	20.5000	20.5143	20.6714	20.4143	20.6000	20.5714		
Standard deviation	1.2764	0.6960	0.7004	0.7079	0.9381	1.0224	0.7197	0.5403	1.1361	0.6689	0.9627	1.0597	0.6516	1.0637		

[a] For the 196 data points, the mean value is 20.63 and the standard deviation is 0.8737. Note that establishing ergodicity is desirable since thereafter a single sample is sufficient to establish process characteristics.

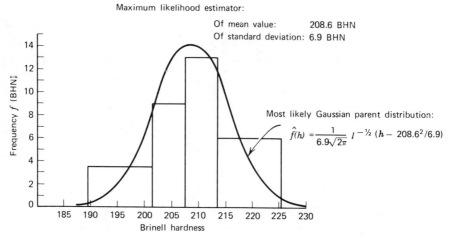

Figure 8.19 Distribution of Brinell hardness, histogram, and frequency function, AISI 4340 steel [7] (also see table 1.2).

8.13.4 Statistical Aspects of Hardness

Hardness measurements statistically reduced yield plots similar to those shown in Figures 8.18 and 8.19. An ensemble of samples plotted as a matrix suggests a random process. Table 8.5 consists of 14 samples (rows), each consisting of $n = 14$ hardness values for which the mean value and standard deviation estimators have been determined. The close similarity among the mean values of the rows and those of the columns, also among the values of the standard deviations, suggests that Brinell hardness may be viewed as a stationary and an ergodic random process.

DYNAMIC STRENGTH

8.14 FATIGUE

Often, applied loads and stresses are dynamic and vary or fluctuate over time. Consider the stresses in the surface fibers of a rotating shaft, under the action of bending forces (Figure 8.20). A particular axially aligned surface fiber experiences tension and compression during every complete revolution of the shaft. A turbine shaft rotating at 800 Hz produces tensile and compressive fiber stresses 800 times per second. Very likely the shaft will also be loaded axially due to blade reaction, thus resulting in superposition of an axial stress upon the bending stresses [74].

Repeated loads, random patterns of loading due to natural phenomena, and so on, were discussed in Chapter 3. It has long been observed that

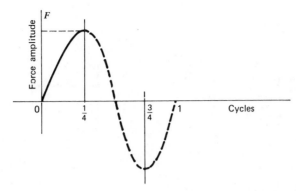

Figure 8.20 Loading in fully reversed dynamic test.

mechanical components fail under the action of fully reversed, repeated, fluctuating, or random loads, whose peak stress levels were less than static ultimate and very often less than static yield strength of material. The features common in such failures was that loading (and inferred stressing) was repeated many times (Figure 8.21a). For a comprehensive discussion of fatigue, see the literature [44, 46, 75, 76].

A fatigue failure is observed to start with one or more small cracks (Figure 8.21b), usually initiated at the surface of a component, where stresses are often the highest, imperfections may be present, and the grains are least supported. Cracks leading to subsequent failures initiate at points of geometric discontinuity in the material of a component. Changes in cross section, fillets, keyways, and holes are common sites of crack initiation, as are surface scratches, inclusions, gas pockets, and machining irregularities. A crack is initially too small for detection by visual or standard inspection techniques. With the development of a crack, local geometric stress concentration effects assist in its growth. At the macro level as the area is reduced, the stress intensity grows in magnitude, until sudden fracture over the remaining area occurs. Associated with fracture, two distinct phenomena can be identified: (1) initiation and progressive growth of the crack and (2) sudden failure due to fracture [77].

Large deflections are often observed in mechanical components prior to static fracture because stress intensity has exceeded the yield strength of the material. On detection of yield, the affected components may be either strengthened or replaced before failure occurs. There is often visual warning of impending static failure. With fatigue, however, there is seldom any evidence of yield or of impending fracture. Fatigue is a complicated and fascinating phenomenon for which no comprehensive theory exists.

The behavior of materials subject to constant amplitude sinusoidal (zero mean) loading modeled as $X(t) = A(\sin t)$ is considered in Section 8.14.1. The maximum value of $X(t)$ is of deterministic magnitude, A. Data describing materials behavior (statistically) subject to such loading can be found in

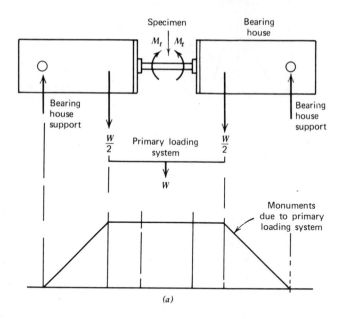

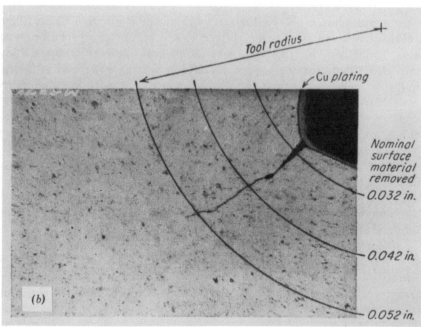

Figure 8.21 Linkage for R. R. Moore rotating-beam fatigue machine [4].

Appendix 10.* An example of design for constant-amplitude dynamic loading is given at the end of Section 9.3.9. Deterministic $X(t)$ models a special case of a narrow band random loading, that is, a single amplitude and a single frequency. In Section 8.14.2 we examine the behavior of materials subjected to a static mean load with a constant amplitude sinusoidal dynamic component associated (fluctuating). In Section 8.14.3 we discuss the behavior of materials subject to narrow-band random loading (zero mean). Data describing materials behavior subject to narrow-band random loading, can be found in Figures 8.41 and 8.42; also see Swanson [29].

The behavior of materials subject to dynamic narrow-band random loading (nonzero mean) is discussed in Section 8.14.4.

8.14.1 Endurance Limit and Fatigue Strength[†]

Two of the most widely used and simplest of fatigue testing machines have been the R. R. Moore rotating beam machine [44, p. 89] and the Sontag dynamic axial tester. The Moore machine subjects a specimen to pure transverse bending (constant-amplitude loading) with negligible shear or axial loading (Figure 8.21).

Previously, stress at failure (fatigue strength) of a material was estimated from lines determined by plots of cycle lives for one test to failure at each of a number of stress levels. For the rotating-beam test (Figure 8.21a), a constant-amplitude bending load was applied to each specimen, that is, a test at each of a number of controlled extreme fiber stress amplitudes. Data were plotted as cycle life versus tensile stress at failure, on a log–log scale, through which a best-fit line was drawn. The result was a conventional S–N diagram (Figure 8.22). An ordinate value S_f, on the S–N diagram just below the stress at failure line is called *fatigue strength*, corresponding to a number of cycles (n) of interest. In design the objective is assurance that applied load stress amplitude is less than the stress at failure for the material specified (with an acceptable probability).

One of the key problems with the constant amplitude test is that the prevailing failure mechanism is quite different from the one that prevails as a result of random-amplitude real-world loading.[‡] The location of cracking is often different. There also appears to be much greater secondary cracking with variable amplitude loading [32].

If the S–N curve became horizontal, as for steels, it was assumed that failure would not occur at stress below this limit. The (strength) stress at failure represented by the horizontal asymptote is called the *endurance limit* (S_e).

---

*For supplemental information, see Lipson et al. [78, 79].
[†]Fatigue (S_f) strength associated with n cycles of life.
[‡]In analyses employing Miner's linear cumulative damage rule, S–N diagrams developed from constant-amplitude test results have been utilized in attempts to predict cycle lives of components to be subsequently subjected to dynamic random-load regimes.

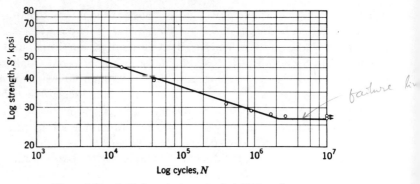

Figure 8.22 $S-N$ diagram for annealed 1040 steel.

With the accumulation of large amounts of test data, it was observed that points rarely conformed to best fit $S-N$ lines and that point density decayed as an inverse of departure from the $S-N$ line. Recognition of inherent differences in cycle lives among specimens subjected to fixed amplitude stress led to the development of more annotated $S-N$ curves to reflect scatter.

So-called composite $S-N$ curves [74] (see Figure 8.23) include the loci of mean values together with rough approximations of upper and lower probability bounds, as illustrated in Figure 8.26. The past decade has witnessed the development of probabilistic design theory and methods for mechanical components and systems. These methods require complete statistical modeling of design variables [9].

Well-defined statistical $S-N$ envelopes are needed to provide the data support for probabilistic design for unlimited life and for finite life. As suggested in Figure 8.27, the statistical $S-N$ envelope for a material may be developed by plotting distributions of cycles to failure at a number of fixed stress (s) levels. The envelope may be developed in another way, by plotting distributions of stress at failure (Figure 8.28) at a number of fixed cycle lives;

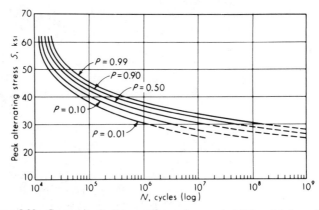

Figure 8.23 Composite $S-N$ curves for various probabilities of failure, P.

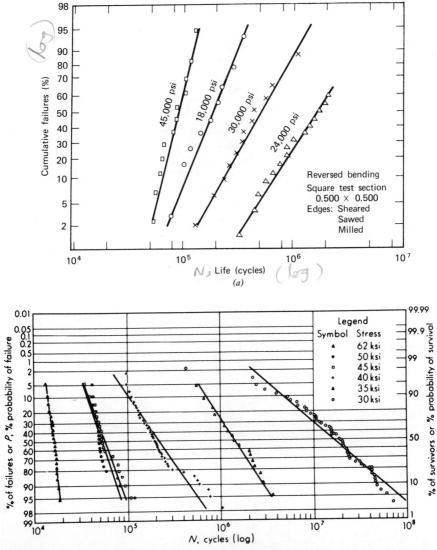

Figure 8.24 (a) Lognormal plot of SAE grade 950X steel. (b) Logarithmic normal probability diagrams for individual fatigue lifetimes at different stresses, 7075-T6 aluminum alloys [3].

for data, see Appendix 10. Monotonic decreasing stress at failure value with increasing cycle life for such distributional strength (S) envelopes is assumed.

Studies indicate that the distribution of cycles to failure at a fixed level of maximum bending stress may be approximately lognormal* [26, p. 152; 80] (Figure 8.24).

---

*See Hahn and Shapiro [13, p. 118] for further discussion of the lognormal as a model of cycles to failure. Also see Lipson and Sheth [81].

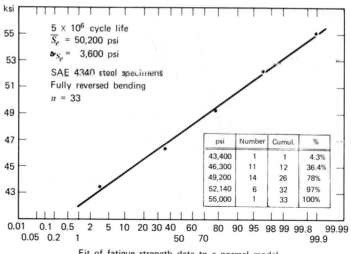

Fit of fatigue strength data to a normal model

Figure 8.25 Distribution of strength at $5 \cdot 10^6$ cycle life.

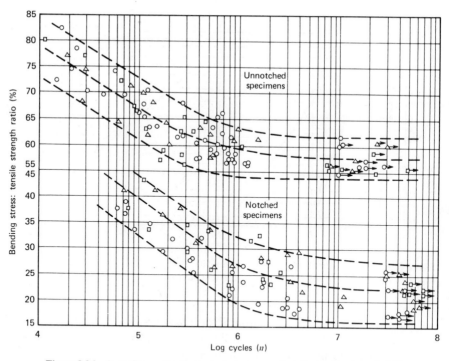

Figure 8.26 S–N Curves for three steels in terms of static tensile strength [44].

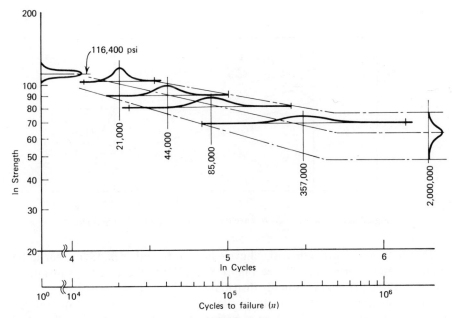

Figure 8.27 Statistical $S–N$ surface for SAE 4340 steel wire—cold drawn and annealed, at fixed stress levels [4].

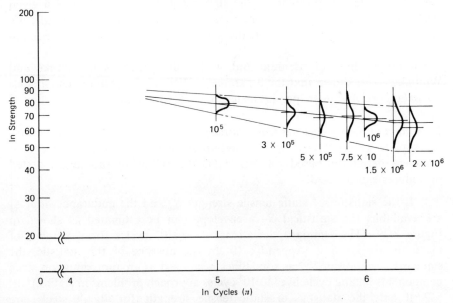

Figure 8.28 Statistical $S–N$ surface for SAE 4340 steel, cold drawn and annealed, at fixed cycle lives [4].

323

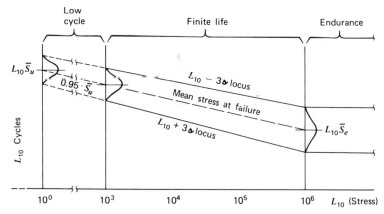

Figure 8.29 Statistical $S-N$ Envelope (constant-amplitude sinusoidal).

Distributions of strength at fixed cycle life, on the other hand, appear to be approximately normal [1, 27] (see Figure 8.25). Weibull presents a substantial amount of data that are normally distributed.*

The preceding remarks indicate an evolution in endurance limit and fatigue strength predictive modeling. It may be postulated that a particular alloy steel in a given condition possesses a unique statistical $S-N$ envelope, not influenced by the methods by which test data for its development are generated. Subject to this postulation, the statistics of stress at failure of a given steel alloy may be estimated by interpolation for any desired finite cycle life or for unlimited endurance from the statistical $S-N$ envelope (see Section 9.3.8).

For fatigue lives of titanium and steel with $S_u \leqslant 240$ ksi, Mischke and Whittaker have shown that the 3-parameter Weibull distribution also models the data closely, with shape parameter, $\beta = 3.0$ [13, 68, 82].

Estimation of S-N Envelopes from Partial Information. When it is necessary to estimate an $S-N$ envelope for a ferrous material in the absence of a complete definition based on test data, the following thoroughly tested procedures can be used:

1. If the statistics of static tensile strength S_u and the endurance limit S_e are available, the statistical $S-N$ envelope can be estimated as shown in Figure 8.29. (The standard deviations, $\flat$, at 10^3 cycles for a number of materials are given in Appendix 10. In the absence of this statistic, the standard deviation of S_u has sometimes been used as an approximation since components having cycle lives to 10^3 cycles approach problems in statics.)

2. If only the statistics of static tensile strength (for ductile steels) are available, the statistical $S-N$ envelope can be estimated as follows. The statistics of endurance limit are estimated by equation (8.6).

---

*W. Weibull, "Scatter of Fatigue and Fatigue Strength in Aircraft Structural Materials and Parts," in *Papers from Fatigue of Aircraft Structures*, Academic Press, New York, pp. 126–145.

With a statistical description of S_u and estimated values of $\bar{S}_e$ and $\circ_{S_e}$, proceed to sketch an estimated S–N envelope as shown in Figure 8.29.

CONSTANT-AMPLITUDE FATIGUE STRENGTH DATA

Appendix 10 provides typical fatigue data for materials subject to constant-amplitude dynamic loading. In addition, Lipson et al. [78, 79] provide a wealth of supplementary fatigue information (see Tables 8.6 and 8.7). The preceding reduced data are expressed in terms of Weibull parameters x_0, b, and θ, or means and standard deviations. Since the mean value and standard deviation estimators are often used in engineering practice the following transformations can be utilized:

$$\bar{S}_f = x_0 + (\theta - x_0)\Gamma(1+1/b)$$

$$c_{S_f} = \frac{\circ_{S_f}}{\bar{S}_f} = \frac{\bar{S}_f - x_0}{\bar{S}_f} b^{-0.926}, \quad \text{for} \quad 0.70 < b < 10.0$$

Figure 8.30 provides values of $\Gamma[1+1/b]$ as a function of b (due to Wirsching). Additional sources of fatigue data are the SAE Cumulative Fatigue Damage Test Program, SAE paper No. 750038 by Tucker and Bussa, and other works [39, 54, 66].

Tables 8.8 and 8.9 [3] list summary statistics of endurance limits for a number of irons, contrasting results in fully reversed bending against repeated bending tests. In instances where test based data are not available, for carbon and alloy steels, the following empirical formula may be utilized to obtain estimates of the endurance limit statistics:

$$(\bar{S}_e, \circ_{S_e}) = (0.498, 0.051)(\bar{S}_u, \circ_{S_u}) \tag{8.6b}$$

The preceding statistics were based on a sample of $n = 159$ values (see Figure 8.31). Equation (8.6b) yields estimations that appear to be consistent with statements in Shigley [1], if 40 to 60% of $\bar{S}_u$ represents a range of values equivalent to $\sim 4 \circ_{S_u}$. For aluminum alloy fatigue strength statistics, see Appendix 10.B.

When test-based data are not available (for aluminum alloys), rough estimates of fatigue strength statistics for required cycle lives may be obtained from diagrams such as Figure 8.32. Clearly, designs based on rough estimates of materials behavior statistics must subsequently be verified by testing.

Low-Cycle Fatigue. Low-cycle fatigue where plastic strain of metallic engineering materials dominates (for cycle lives $n < 10^3$ and often for $n < 10^4$)

Table 8.6 Fatigue Strength Distributions and Their Parameters $S_u = 61$–81 ksi; $S_y = 45$–61 ksi

2024 Aluminum (Bars)

Fatigue Strength Distributions and Their Parameters

Life in Cycles

Axial (Completely Reversed)
Effect of Stress Concentration

	1×10^3	1×10^4	5×10^4	1×10^5	5×10^5	1×10^6	5×10^6	1×10^7	1×10^8
σ	3.818*	3.020	2.564	2.389	2.028	1.890	1.604	1.495*	1.183
μ	62.30	49.28	41.84	38.99	33.10	30.84	26.18	24.40	19.30
b	1.729*	1.303*	1.434	1.372	1.717	1.680	1.609*	1.609*	1.609*
θ	49.61	38.46	32.24	29.87	25.07	23.23	19.46	18.03	14.00
X_c	43.98	34.68	28.95	26.88	22.24	20.64	17.35	16.07	12.48
b	1.850*	1.850	1.851	1.851	1.851	1.851	1.851	1.851*	1.851*
θ	42.70	31.77	25.84	23.64	19.23	17.59	14.31	13.09	9.743
X_c	34.71	25.83	21.01	19.22	15.63	14.30	11.63	10.64	7.921
b	0.9193*	0.7988	0.9826	1.182	0.7652	1.083	0.8076*	1.148*	0.9826*
θ	48.66	30.99	22.67	19.85	14.43	12.64	9.205	8.070	5.135
X	43.36	27.75	20.14	17.43	12.93	11.17	8.240	7.101	4.562
β	0.1486*	0.2151	0.2787	0.3115	0.4035	0.4510	0.5841	0.6530	0.9253*
M	76.31	52.70	40.69	36.40	28.11	25.14	19.41	17.37	11.99
β	0.2787	0.4035	0.5227	0.5842	0.7567	0.8458	1.095*	1.224*	1.772*
	41.32	28.54	22.04	19.71	15.22	13.62	10.51	9.408	6.498

Table 8.7 Rotary Bending Fatigue Strength of AISI 3140 Steel $S_u = 108, 109$ ksi; $S_y = 87, 75$ ksi[a]

H.T.	Spec.	Life (Cycles)	X_0 ksi				b				θ ksi			
			10^3	10^4	10^5	10^6	10^3	10^4	10^5	10^6	10^3	10^4	10^5	10^6
			Effect of Stress Concentration											
K_3	No-N		89	74	66	57	3.7	5.2	5	5.5	100.4	87.7	76.7	67.2
K_3	V-N			30	19	20		5.6	6.1	5.1		62.2	47.3	36
K_4	No-N		85	75	66.5	59.0	2.2	2.2	2.4	2.2	87.9	77.87	69.08	61.07
K_4	V-N			53	36.5	25		5	3.2	3.5		70.2	44.25	29.53
			Effect of Heat Treatment											
K_3	No-N		89	74	66	57	3.7	5.2	5	5.5	100.4	87.7	76.7	67.2
K_4	No-N		85	75	66.5	59.0	2.2	2.2	2.4	2.2	87.9	77.87	69.08	61.07
K_3	V-N			30	19	20		5.6	6.1	5.1		62.2	47.3	36
K_4	V-N			53	36.5	25		5	3.2	3.5		70.2	44.25	29.53

[a]Composition—see p. 198, item 1; heat treatment—A: K_3—OQ from 1520°F, tempered at 1300°F, $S_u = 108$ ksi, $S_y = 87$ ksi; B: K_4—air blast quenched from 1520°F, tempered at 1050°F, $S_u = 109$ ksi, $S_y = 75$ ksi; specimen condition—see p. 203; meaning of symbols—see p. 46. (See Appendix 10.B.)

Source: Lipson et al. [79].

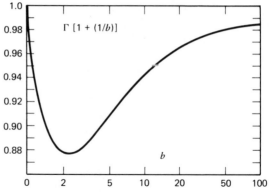

Figure 8.30 The term $\Gamma[1+(1/b)]$ as a function of b.

has received much attention in research recently [47]. However, some of the postulated theory is still in need of experimental verification. A considerable amount of data describing low-cycle fatigue behavior is available to support design and analysis; however, almost none is at present described by statistics [83].

When design for low-cycle fatigue is required, one approach for estimating the stress at failure (in the absence of test-based data) is by an interpolation of stress at failure (S_f) statistics from 0.25 to 1000 cycles. The means and standard deviations are known for many materials at $n=0.25$ and $n=1000$ cycles. Also known are the slopes of the mean and $\pm 2s$ loci at 0.25 and 1000 cycles.

The equations of the postulated mean-value locus and $\pm 2s$ loci are written [see equation (a)], thus producing the dashed-line curves in Figure 8.33 (where $L_{10}S = \log$ to base 10 of S). It is to be noted that an error of 50% in locating

Table 8.8 Fully Reversed Bending

Material Surface	Est. Mean Fatigue Limit, psi	Est. Std. Deviation, psi	Est. Std. Error in Estimate of Mean, psi	Estimated 95% Conf. Limits for TMFL[a]	
				Lower Limit, psi	Upper Limit, psi
As-cast Nodular					
As-cast	26,000	1,700	800	24,400	27,600
Machined	29,500	2,900	1,435	26,700	32,000
Notched	20,500	600	300	19,800	21,200
HT Nodular					
As-cast	27,500	1,100	700	26,200	28,800
Machined	30,000	5,200	2,500	25,200	34,800
Notched	22,500	1,100	700	21,200	23,800
Pearlitic Malleable					
As-cast	23,000	2,900	1,400	20,200	25,800
Machined	25,000	1,100	700	23,700	26,300
Notched	20,500	900	500	19,400	21,600
Ferritic Malleable					
As-cast	22,000	2,400	1,100	18,900	24,100
Machined	20,000	1,100	900	18,200	21,800
Notched	15,500	1,000	500	14,400	16,600

[a] TMFL = True mean fatigue limit.

Table 8.9 Repeated Bending

Summary of Statistical Analysis of Fatigue Results, Assuming Normal Distribution (49) $R = 0$

Material Surface	Est. Mean Fatigue Limit, psi	Est. Std. Deviation, psi	Est. Std. Error in Estimate of Mean, psi	Estimated 95% Conf. Limits for TMFL	
				Lower Limit, psi	Upper Limit, psi
As-cast Nodular					
As-cast	36,000	3,400	1,400	33,200	38,800
Machined	47,500	4,100	2,300	43,000	52,000
Notched	38,500				
HT Nodular					
As-cast	40,000	7,400	3,400	33,300	46,700
Machined	45,500	4,300	1,800	42,000	49,000
Notched	41,000				
Pearlitic Malleable					
As-cast	38,000	5,600	2,300	33,200	42,800
Machined	35,500	5,200	2,400	30,700	40,300
Notched	38,000	1,600	700	36,600	39,400
Ferritic Malleable					
As-cast	34,000	2,200	950	32,100	35,900
Machined	28,500	2,140	1,100	26,300	30,700
Notched	25,000				

Surface Correction Factor[a], Surface and Fatigue Ratio (49)

Material	Ult. Tensile Strength, psi	Mach. Surface Mean Fatigue Limit $R = -1$, psi	Fatigue Ratio	As-cast Surface Mean Fatigue Limit, psi	C_{surf}[a]
As-cast Nodular	89,000	29,500	0.33	26,000	0.88
HT Nodular	104,000	30,000	0.29	27,500	0.92
Pearlitic Malleable	83,000	25,000	0.30	23,000	
Ferritic Malleable	51,000	20,000	0.39	22,000	0.92

[a] Surface correction factor is the reduction of the mean fatigue limit due to surface condition. In this case, it is the ratio of the mean fatigue limit of a cast surface to one with a machined surface.

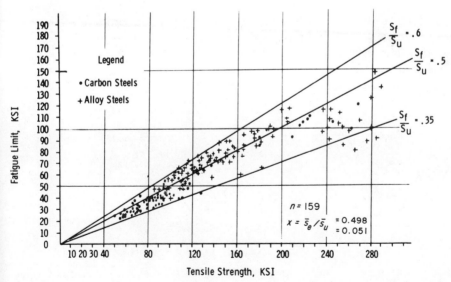

Figure 8.31 Relation between rotating bending fatigue limit and tensile strength of wrought steels [3, p. 42].

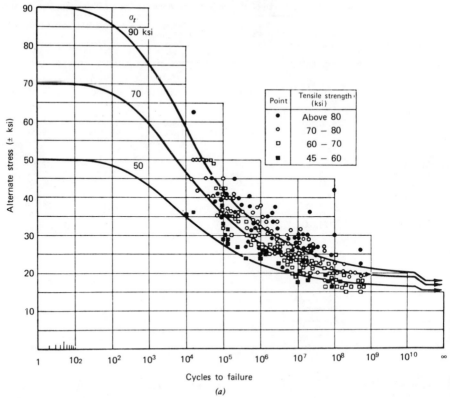

Figure 8.32 (a) Effect of cycles to failure on reversed axial fatigue strength of aluminum alloys.

the postulated mean value locus within its limits would produce an error of approximately 5% in the estimated mean value of stress at failure. A simple straight-line estimator between points A and B would usually result in conservative mean value estimates.

The parabolic approximation of the low-cycle mean value locus is as follows.

$$L_{10}n = \frac{2t\sqrt{L_{10}\bar{S}_u - L_{10}\bar{S}_f}}{\sqrt{L_{10}\bar{S}_u - L_{10}\bar{S}_{10^3}}} + \frac{3.0 - 2t}{L_{10}\bar{S}_u - L_{10}\bar{S}_{10^3}*} \cdot (L_{10}\bar{S}_u - L_{10}\bar{S}_f) \qquad (a)$$

where

$$t = 3 + \frac{L_{10}\bar{S}_{10^3} - L_{10}\bar{S}_u}{\tan\theta}$$

$$\tan\theta = -\frac{L_{10}\bar{S}_{10^3} - L_{10}\bar{S}_e}{3.0}$$

For any required life, $0 < n < 1000$, equation (a) may be solved for fatigue strength $\bar{S}_f$ and σ_{S_f}.

*In equation (a), S_{10^3} is the fatigue strength S_f associated with $n = 1000$ cycles.

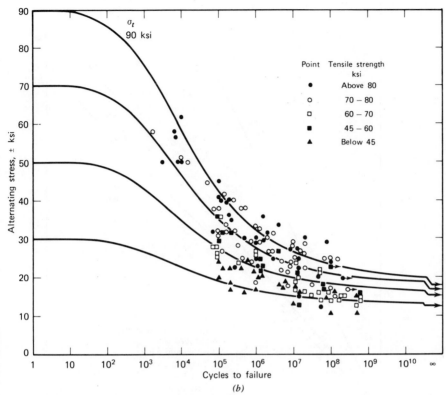

Figure 8.32 *continued* (b) Effect of cycles to failure on fatigue strength of aluminum alloys in bending [75].

TORSIONAL ENDURANCE LIMIT

A collection of data was assembled by Dolan to shed light on the relationship between torsional endurance limit and the endurance limit determined by rotating beam tests. Figure 8.34 shows a plot of the available data and in addition plots of the distortion energy and maximum shear stress theory predictions (see Chapter 10). From Figure 8.34 it is apparent that the

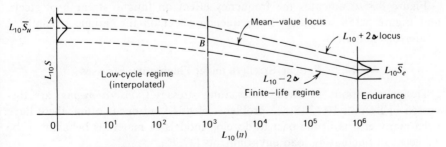

Figure 8.33 S–N Envelope for carbon and low-alloy steels.

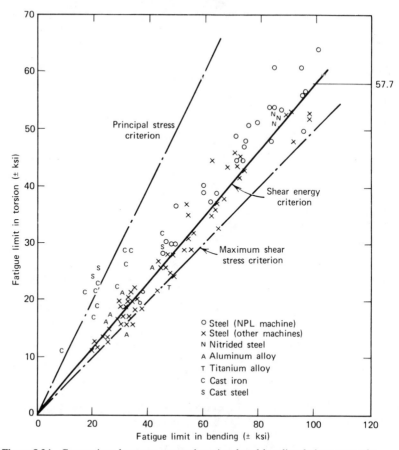

Figure 8.34 Comparison between reversed torsional and bending fatigue strengths.

distortion energy theory closely predicts fatigue strength in torsion, as expressed by equation (8.7):

$$S_{se} \simeq 0.577 S_e \qquad (8.7)$$

Since S_e is a random variable, S_{se} is also a random variable.

Figure 8.35a indicates the frequency effect on fatigue strength of steels, and Figure 8.35b indicates the frequency effect on the fatigue strength of nonferrous materials.

8.14.2 Fatigue Strength under Fluctuating Stresses

Design situations involving fluctuating stresses (*nonzero means*) are discussed in Section 10.13. Since problems of this kind occur often, there have been many efforts in the past to develop models of materials behavior to be expected in fluctuating load environments [27, 75].

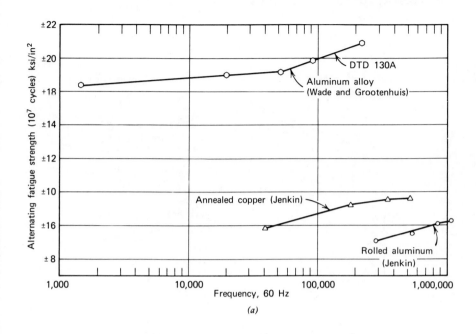

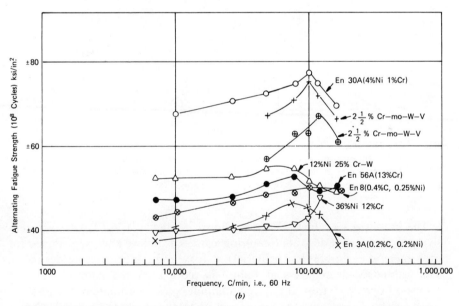

Figure 8.35 Effect of frequency on fatigue strength of steels and nonferrous materials [46].

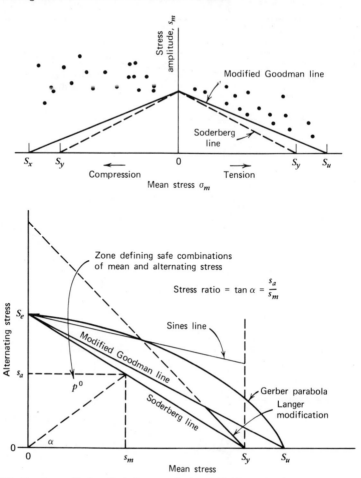

Figure 8.36 Goodman, Gerber, Soderberg, and other definitions of safe design. (*a*) Deterministic fatigue diagram showing typical combinations of mean stress and stress amplitude resulting in failure. Diagram shows modified Goodman line [1, p. 270]. (*b*) The Goodman, Gerber, Soderberg, Sines, and Langer definitions of the safe design region [84].

The modified Goodman diagram shown in Figure 8.36*a* has been extensively used, where a diagram (featuring a stress at failure line) is generated for each specific cycle life (and material), that is, endurance or some desired *n*. The estimated mean stress at failure is plotted on the abscissa and the estimated dynamic stress at failure amplitude, on the ordinate of the applicable diagram. Classically, the realization of intended life is assumed when a combination of static and dynamic stresses represented by a point plots below the straight (stress at failure) line joining S_e and S_u (Figure 8.36*b*).

In probabilistic design, statistical models are required. To determine the actual shape of the stress at failure (for endurance), mean value loci and $\pm 3s$ curves (Figure 8.37) for low-alloy steels requires extensive testing to provide

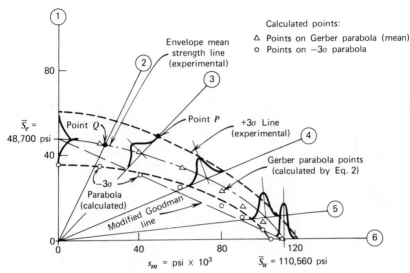

Figure 8.37 Resulting endurance stress envelope from test program (AISI 4340 steel) [84]. For an alternative method of plotting these curves, see Figure 10.29.

data from which to estimate the needed statistics [84]. Distributions of stress at failure for a spectrum of ratios s_a/s_m are required to define each envelope (see Table 8.10). For the AISI 4130 and 4340 steels tested, the plots of the mean values of stress at failure follow the parabolic model very closely, considering sample sizes. The $\pm 3_s$ values also plotted closely along comparable $\pm 3_s$ parabolic curves.

Since the fit to parabolic curves for the mean value loci and $\pm 3\sigma$ loci indicate conformity (and considering plots such as Figure 8.36a), the Gerber

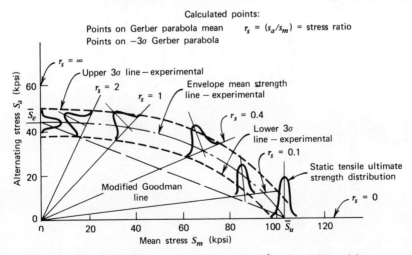

Figure 8.38 Distributional Goodman fatigue diagram for 10^6 cycles of life and for stress ratios $r_s = S_a|S_m$ of 0, 0.1, 0.4, 1, 2, and ∞ for SAE 4130 steel [84].

Table 8.10 Test Plan Parameters[a]

Test No.	Stress Ratio Alternating: Mean	Specimen Diameter (in.)	Assumed Mean Values of Stress to Start Staircase Method Analysis	Assumed Mean Values of Load[b] to Start Staircase Method Analysis	Increment Used for Staircase Method, Stress (lb)	Increment Used for Staircase Method, Stress (lb)
1	∞	$d = 0.0620 \pm 0.0005$	$S_a = 59,000$	$L_a^b = 178$	$\Delta S_a = 4,000$	$\Delta L_a = 12$
2	2	$d = 0.0740 \pm 0.0005$	Cycles to	None	None	None
3	1	$d = 0.0695 \pm 0.0005$	$S_m = 48,000$	$L_m = 182$	$\Delta S_m = 3,0000$	$\Delta L_m = 12$
4	1/2.5	$d = 0.0645 \pm 0.0005$	$S_m = 80,000$	$L_m = 247$	$\Delta S_m = 3,000$	$\Delta L_m = 9$
5	1/10	$d = 0.0645 \pm 0.0005$	$S_m = 100,000$	$L_m = 327$	$\Delta S_m = 2,500$	$\Delta L_m = 8$
6	0	Diameter varied	Stress calculated for each specimen	None (static test)	None	None

[a]See footnote, p. 339.
[b]Alternating load is L_a and mean load is L_m.
Source: Haugen and Hritz [84].

parabola is suggested as a fatigue model for sinusoidal fluctuating stressed low-alloy steels subject to further confirmation. The statistics utilized in design are the mean and standard deviation estimators of the radius vector S_R modeled by equation (8.8):

$$S_R = \sqrt{S_m^2 + S_a^2} \quad * \tag{8.8}$$

The magnitude of S_R is the random variable length (in psi units) of a line of slope $r_s = (\bar{s}_a/\bar{s}_m)$ from the origin to its intersection with the stress at failure envelope, as shown in Figure 8.38. The statistics of S_R are estimated by using equations (2.63) and (2.64) with equation (8.8). [see equation (5.45) for stress representation].†

Summary of Section 8.14.2. The following estimation methods may be used in design with carbon and low-alloy steels, subject to subsequent verification.

1. It appears that the deterministic parabola (proposed by Gerber in 1874) very closely approximates the mean-value locus of the stress at failure envelope, determined experimentally for AISI 4340 and 4130 steels subjected to fluctuating loading.

2. Theoretical parabolas approximate the $+3s$ and $-3s$ loci of the allowable combined stress envelopes, as experimentally determined with fluctuating loading.

3. It may be postulated that the mean strength line of the allowable combined stress envelope can be developed to a good approximation from the mean ultimate static strength $\bar{S}_u$ and the mean endurance limit $\bar{S}_e$, together with equation (8.9a). As a further postulation, the $\pm 3s$ lines may be developed to a close approximation if $\bar{S}_u \pm 3s_{S_u}$ and $\bar{S}_e \pm s_{S_e}$ are known, by using these values in writing the parabolic equations [83, 84] [i.e., equation (8.9b)].

4. If the endurance limit mean line follows a parabolic path and $\bar{S}_e \approx 0.498 \cdot \bar{S}_u$ [equation (8.6)], approximate strength envelopes can be developed from a knowledge of the distribution of S_u only and from the fact that the $\pm 3s_{S_e}$ endurance interval can be described by using equation (8.6), for steels to approximately 200 ksi ultimate strength (see Section 10.10).

5. It may be further postulated that materials behavior envelopes for finite life can be developed by plotting finite life strength (S_f) on the ordinate and ultimate tensile strength (S_u) on the abscissa, and then joining these points with a parabolic curve [i.e., equation (8.9)]. For utilization of the Gerber envelope in design, see Chapter 10. Constant life diagrams for numerous materials are found in Mil-Handbook 5B [66]. The equation of Gerber

---

*To generate data, the "staircase" (or "up and down") test method is used, with increase or decrease of test stress designed to maintain the ratio s_a/s_m constant from test to test [26].

At each stress ratio a sample of test results is obtained from which sample statistics are calculated. The statistics are then used to check parabolic models for stress at failure envelopes.
†The format for fluctuating stress must be consistent with equation (8.8) (see Equation 5.14).

parabola mean-value locus is

$$\frac{\bar{S}_a}{\bar{S}_e} + \left(\frac{\bar{S}_m}{\bar{S}_u}\right)^2 = 1 \qquad (8.9a)$$

Equations of Gerber parabola $\pm 3_s$ loci are

$$\frac{\bar{S}_a}{\bar{S}_e \pm 3_{>S_e}} + \left(\frac{\bar{S}_m}{\bar{S}_u \pm 3_{>S_u}}\right)^2 = 1 \qquad (8.9b)$$

In Section 8.14.1 the behavior of engineering materials subject to constant-amplitude sinusoidal loading was considered. Loading on mechanical systems in the real world is, however, generally random as well as dynamic. Thus it is necessary to have strength behavioral descriptions available that predict materials performance in frequently observed real-world operational situations. Fully reversed narrow-band random loading is more likely than constant amplitude sinusoidal fully reversed. However, in the past most of the materials tests employed constant-amplitude loading. Until computer-controlled servohydraulic testing systems became available, service (random) load tests could not be performed at all or not in an economical manner. Today, service and other types of random testing are used with increasing frequency.

8.14.3 Material Behavior Subject to Narrow-Band Random Loading, (Zero Mean)

Investigations [e.g., 29, 77, 86] suggest that the statistical $S-N$ envelopes describing materials behavior subjected to constant-amplitude (Figure 8.27) versus narrow-band random loading (Figures 8.39 and 8.40) differ considerably. Lipson and Sheth [87] (see also Chapter 9) concur with these findings.

The testing procedures for narrow-band random loading of material specimens are considerably more involved (requiring more sophisticated test equipment) than in constant-amplitude sinusoidal loading tests. A programmed random sequence of load amplitudes is applied to each specimen in a test, with each load sequence differing from all the others in a given random process (see Section 8.15) at a specified RMS level.

To describe fatigue strength in practice, the strength statistics of a material subject to random loading processes at a number of RMS levels are required (see Tables 8.11 and 8.12). From these data, $S-N$ surfaces similar to those for constant-amplitude sinusoidal loading are plotted (see Figures 8.41 and 8.42). It is to be noted that steels, subject to narrow-band random loading, do not display a knee in the $S-N$ envelope [77, 88, 89]. At higher cycle lives the scatter in values tends to be somewhat diminished. Probabilistic design of mechanical components, subject to narrow-band random loading is discussed in Section 10.14 with examples.

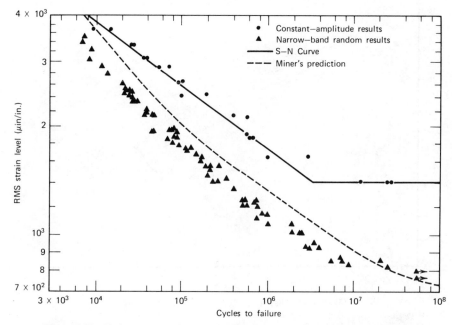

Figure 8.39 Fatigue test results for 2024-T3 aluminum alloy (vibrations exciter).

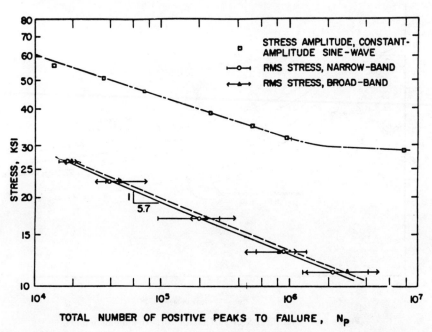

Figure 8.40 Total number of positive peaks to failure N for 2024-T3 aluminum alloy. R. Ikegami [77].

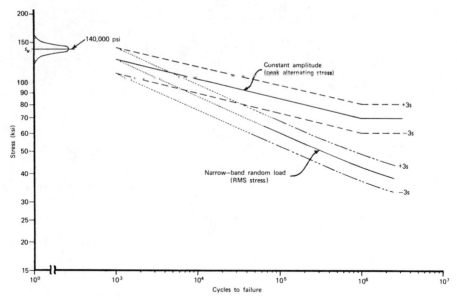

Figure 8.41 Statistical $S-N$ surface for 4340 steel alloy.

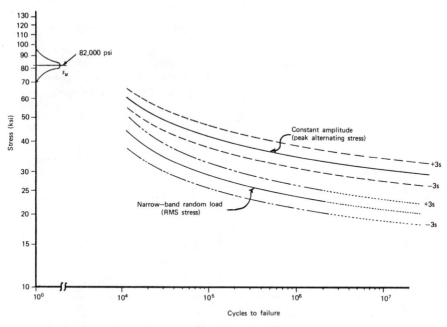

Figure 8.42 Statistical $S-N$ surface for 7075-T6 aluminum alloy.

Table 8.11 Random Load Fatigue Data For 7075-T6 Aluminum Alloy

Mean Stress (ksi)	Rms Stress (ksi)		Logmean Endurance (cycles)	Standard Deviation (log)	Coefficient of Variation	No. of Tests
0	B_H	40.36	17,980	0.090	0.021	20
	A_H	38.73	25,179	0.119	0.027	20
	C_H	37.14	30,809	0.077	0.017	20
	B''	33.66	26,934	0.098	0.022	20
	B	32.84	36,995	0.101	0.022	20
	B'	32.75	86,096	0.123	0.025	20
	A''	31.30	30,443	0.101	0.022	20
	A	29.61	49,307	0.091	0.019	20
	C''	29.56	67,864	0.082	0.017	20
	A'	28.53	181,654	0.117	0.022	20
	C	26.73	130,892	0.103	0.020	20
	B_L	24.08	1,688,366	0.156	0.025	20
	C''	21.04	214,037	0.079	0.015	20
	C_L	20.91	3,254,944	0.135	0.021	20

[a]Specifications: Room-temperature random-load fatigue data using randomized program constant-amplitude method (i.e., simulation of a random loading process); type of loading: rotating beam; specimen: 3 in. long, 0.3125 in. diameter, machined to 0.1875 in. diameter, unnotched (see Figure 8.42); ultimate tensile strength = $\bar{S}_u \approx 82$ ksi; tensile yield strength = $\bar{S}_y \approx 76$ ksi (see Table 8.1).
Source: Swanson [29, p.36].

Table 8.12 Random Load Fatigue Data for SAE 4340 Steel Alloy[a]

Mean Stress (ksi)	Rms Stress (ksi)	Logmean Endurance (cycles)	Standard Deviation (log)	Coefficient of Variation	No. of Tests
0	56.74	158,044	0.068	0.013	20
	55.60	206,066	0.155	0.029	20
	54.80	277,323	0.159	0.029	20
	46.96	243,845	0.118	0.022	20
	45.78	429,510	0.143	0.025	20
	44.91	1,800,000	0.188	0.030	19
	37.15	1,310,000	0.161	0.026	20
	35.71	2,750,000	0.244	0.038	20

[a]Specifications: Room-temperature random-load fatigue data using randomized program constant-amplitude method (i.e., simulation of a random loading process; type of loading: rotating beam; specimen: 3 in. long by 0.3125 in. diameter, machined to 0.185 in. diameter, unnotched (see Figure 8.41); ultimate tensile strength = $\bar{S}_u \approx 140$ ksi; tensile yield strength = $\bar{S}_y \approx 129.3$ ksi (see Table 8.1).
Source: Swanson [29, p. 37].

8.14.4 Materials Behavior Subject to Narrow-band Random Loading (Nonzero Mean)

Perhaps the most general (and least studied) dynamic load regime (imposed on materials) is that of narrow- or broad-band random associated with nonzero mean loads, for which it is possible to postulate dynamic stress at failure models. At the two limiting conditions, the following may be known:

1. Static ultimate strength, $(\bar{s}_u, \mathit{s}_{s_u})$ psi, with a zero dynamic component.
2. Dynamic strength, $(0, RMS)$ psi, with a zero static component. The vector representation, as in Section 8.14.2, may be taken as

$$S_R = \sqrt{S_m^2 + S_a^2} = \sqrt{S_u^2 + (RMS)^2},$$

where S_a is the random dynamic stress at failure, $S_a = RMS$, for the cycle life of interest.

The curve joining S_u on the abscissa and RMS on the ordinate may be postulated to approximate a parabola or, more conservatively, a straight line. Thus equations (8.9) may be utilized.

NARROW-BAND RANDOM FATIGUE DATA

Table 8.14 illustrates data for engineering materials subject to narrow-band random loading. The reduced data are expressed in terms of Logmean endurance cycles ($\tilde{N}$) and standard deviations (log s_N). Since the mean value $\bar{N}$ and standard deviation s_N estimators are most often used in engineering practice, the following transformations may be useful.

Given.

$$\text{Logmean endurance cycles} = \tilde{N}$$

$$\text{Standard deviation (logs)} = \mathit{s}_x$$

Then, let $\bar{x} = \log_{10} \tilde{N}$:

$$\bar{N} = 10^{\bar{x} + \frac{1}{2}(\mathit{s}_x^2/0.434)}$$

and

$$C_N = \frac{\mathit{s}_N}{\bar{N}} = \left[10^{(\mathit{s}_x^2/0.434)} - 1 \right]^{1/2}$$

For instance, for 18% nickel maraging steel (Table 8.13):

$$\tilde{N} = 87{,}026 \text{ cycles}$$

Table 8.13 Narrow-band Random Fatigue Data for 18% Nickel Maraging Steel

Type of Loading	Material UTS/0.2%YS/El	Specimen and K_T	Mean Stress (ksi)	RMS Stress (ksi)	Logmean Endurance Cycles	Standard Deviation (logs)	Coefficient of Variation	No. of Tests
Axial	18% Nickel maraging steel vacuum melt vacuum cast 250.0/ 240.0/10.0	7.5×0.7 in. dia. waisted to 0.25 in. dia. unnotched	50	45	64,992	0.085	0.018	7
				40	114,653	0.086	0.017	10
				35	762,653	0.256	0.044	7
				40	83,286[c]	0.191	0.039	10
				40	87,026[f]	0.195	0.039	10
			0	52	91,212	0.111	0.022	7
				45	211,411	0.067	0.013	7
				40	395,190	0.144	0.026	7

Table 8.14 Room-temperature Rayleigh Endurance Data F

Type of Loading	Material UTS/0.2%YS/El	Specimen and K_T	Filtered White Noise Frequencies	Excitation	Constan amplitude (?)
Reverse bending	24 ST Aluminum bar alloy extruded	413/16 × 1/2 in. dia. circular notched	$f_1 = 83$ cps	Direct electro-magnetic $R = .84$	Yes
Reverse bending	2024-T3 Sheet	Type 1 sheet bending (SF-10-U machine) 0.040 in. thick	$f_1 = 40$ cps	Electromagnetic direct to specimen with other end connected to antinode of a transverse steel beam	Yes

(Contains pilot information from NACA TN 4050, 1957)

Type of Loading	Material UTS/0.2%YS/El	Specimen and K_T	Filtered White Noise Frequencies	Excitation	Constan amplitude (?)
Cantilever reverse bending	7075-T6 Aluminum alloy	Notched cantilever double-edged notch 3 × 1 × 1/4 in.	Tape of subsonic cold air jet noise 6.4-sec. sample repeated $f_1 = 119$ cps	Direct electro-magnetic	Yes; not s machine
Cantilever bending	2024-T4 extruded aluminum bar alloy	Round bar with circumferential notch $K_T = 1.77$	Zero crossings = 25.2 cps with $R = .91$	Direct electro-magnetic	Yes; but progran constan amplitu
Cantilever bending	SAE 4130 normal steel sheet	3 × 1 × 3/16 in. double-edged notch 2 V notched $K_T = 7.25$ (axial)	$f_1 = 123$ cps	Direct electro-magnetic	Yes; not same machin
Cantilever bending	2024 Aluminum alloy bare and alclad 2024-T3	33/8 × 1 × 0.063 in. 2 holes (transverse) 1/8 in. dia. on each side of midbeam clamp 7 in. long to form center-clamped beam $K_T = 2$	$f_1 = 185$ cps	Direct electro-magnetic at midpoint of beam	No (from lit.)

Random-load Fatigue Tests Using Analogous Random Processes

Palmgren/: Miner Prediction (?)	Mean Stress (ksi)	RMS Stress (ksi)	Logmean Endurance (cycles)	Standard Deviation (log)	Coefficient of Variation	No. of Tests
Yes; test less than prediction	0	8.36	2.09×10^6	0.328	0.052	7
		9.12	1.144×10^6	0.1528	0.025	8
Yes; test greater than prediction	0	(14.76)	275,000 (Unnotched)			24
		(12.60)	530,000 (Unnotched)			24
Yes; test equaled prediction at low stress, test greater than prediction at high stress		(23.40)	20,000 (Notched)			24
		(19.80)	28,500 (Notched)			24
		(11.88)	110,000 (Notched)			24
		(10.44)	1,050,000 (Notched)			24
Yes; test greater than prediction	0	18.5	43,000	Estimates A total of 125 tests		
		15	75,000			
		10	278,000			
		5.5	1.71×10^6			
		4.33	6×10^6			
Yes; test equaled prediction	0	23	10,400	7-15		
		20	18,750			
		18	30,850			
		16	53,500			
		15	98,580			
		13	190,500			
		11.4	630,000			
Yes; test less than prediction	0	4	1.5×10^7	Estimates A total of 145 tests		
		5	8×10^6			
		10	10^6			
		15	270,000			
		20	102,000			
		25	57,000			
		28.5	40,000			
Yes; test less than prediction	0	13.2	145,000			3
		(61.9 Db)	bare[c]			
		11.25	290,000			3
		(60.5 Db)				
		9.8	437,000			3
		(59.3 Db)				
		6.5	813,000			3
		(56.0 Db)	clad[d]			
		11.5	85,100			3
		(61.0 Db)				

Table

Type of Loading	Material UTS/0.2%YS/E1	Specimen and K_T	Filtered White Noise Frequencies	Excitation	Constant amplitude D (?)
Bending	2024-T3 bare sheet 72.1/48.8/20	Tapered 0.040-in. thick sheet $K_T = 1$	$f_1 = 6$ cps (A)	E/H $R = .955$ (A)	Yes

(Note: L-73 is similar to 2014-T6 alclad)

Type of Loading	Material UTS/0.2%YS/E1	Specimen and K_T	Filtered White Noise Frequencies	Excitation	Constant amplitude D (?)
Axial	L-73 Aluminum alloy sheet 65.4/56.9/12	$K_T = 1$	$f_1 = 88$ cps	Shaker plus resonant bending beam; shaker connected to antinode specimen at one	Yes; clipping ratio = 2.6
Axial	2024-T4 Bar extruded 86.1/68.55/12.0	Unnotched waisted 7×0.68 in. dia. waisted to 0.28 in.	$f_1 = 45$ cps	Electromagnetic + rigid lever $R = .96$; clipping ratio = 3.6	Yes
Cantilever bending	Magnesium alloy AZ31B-H24 sheet 41.4/30.3/12	Triangular flexure specimen 2.4 in. long and 1.5 in. wide (maximum) 0.063 in. thick	Filtered to 50–500-cps band	Direct electromagnetic f_1: 100 cps	No
				200 cps	
Cantilever (Complex)	DTD 710, DTD 746	Riveted skin-web test piece 0.048 in. thick with 1/8 in. dia. 100 deg head with countersunk holes	$f_1 = 265$ cps (approx.)	Direct electromagnetic	Yes
					Yes
Axial	Beryllium nickel alloy (1.95% Be) 1/2-hard extruded rod 272.5/232.8/3	Unnotched waisted specimen (as 3-64-2)	$f_1 = 33.3$ cps	Electromagnetic + rigid lever $R = .96$	No

Continued

Palmgren/: Miner Prediction (?)	Mean Stress (ksi)	RMS Stress (ksi)	Logmean Endurance (cycles)	Standard Deviation (log)	Coefficient of Variation	No. of Tests
Yes; test less than prediction	0	A 22.0	46,300	0.0437	0.0094	20
		A 19.3	81,400	0.0446	0.0091	20
		A 15.0	262,000	0.0469	0.0087	20
		A 13.5	420,000	0.0587	0.0104	20
Yes; test greater than prediction	0	23.0	17,750			3
		20.2	63,000			4
		13.86	115,000			4
		11.88	141,000			4
		9.11	595,000			4
		6.34	3,590,000			1
Yes; test less than prediction	16.0	6.35	2.44×10^6	0.151	0.024	7
		6.70	952,792	0.130	0.022	6
		8.13	650,213	0.196	0.034	6
		9.5	459,247	0.055	0.010	6
		11.3	278,037	0.129	0.024	6
		12.0	119,699	0.129	0.025	6
		12.7	60,037	0.134	0.028	6
Yes (not P/M) K_T:	0					
1.00		10.3–5.3	4.2×10^6 to 493,000			54
2.04		18.0–8.5	1.3×10^6 to 160,000			44
1.84		17.2–10.1	94,000 to 626,000			52
1.90		19.0–12.7	48,000 to 200,000			52
2.04		11.4–8.5	1.4×10^6 to 5×10^6			50
1.84		11.0–8.3	665,000 to 2.3×10^6			54
1.90		11.3–8.7	590,000 to 1.6×10^6			51
No	0	2.5	649,100,000	DTD710		4
		4.0	25,120,000			4
		6.0	6,220,000			4
		8.0	3,215,000			2
		10.0	1,911,000			4
No	0	2.5	312,400,000	DTD746		4
		4.0	16,720,000			4
		6.0	5,814,000			4
		8.0	1,700,000			2
		10.0	1,001,000			4
. . .	48	33	199,412	0.040	0.0076	5

Table

Type of Loading	Material UTS/0.2%YS/E1	Specimen and K_T	Filtered White Noise Frequencies	Excitation	Constant-amplitude D (?)
Axial	18% Nickel maraging steel vacuum melt vacuum cast 250.0/ 240.0/10.0	7 1/2 × 0.7-in. dia. waisted to 0.25-in. dia. unnotched	$f_1 = 33.3$ cps	Electromagnetic shaker + rigid lever	Yes
Axial	18% Nickel maraging steel vacuum melt vacuum cast 250.0/ 240.0/10.0	7 1/2 × 0.7-in. dia. waisted to 0.25-in. dia.; unnotched with transverse weld at midlength Weld specimen in air	$f_1 = 33.3$ cps	Electromagnetic shaker + rigid lever	Yes
Flexure cantilever beam	6061-T6 Aluminum alloy plate (transverse)	2.69-in. free cantilever length × 1 1/2 × 0.062 in. thick $K_T = 1.05$	20-cps bandwidth $Q = 300-1100$ $f_1 = 300$ cps	Electromagnetic direct	Yes; same machine

Source: Swanson [29].

Continued

Palmgren/: Miner Prediction (?)	Mean Stress (ksi)	RMS Stress (ksi)	Logmean Endurance (cycles)	Standard Deviation (log)	Coefficient of Variation	No. of Tests
No	50	45	64,992	0.085	0.018	7
		40	114,653	0.086	0.017	10
		35	762,653	0.256	0.044	7
		40	83,286	0.191	0.039	10
		40	87,026	0.195	0.039	10
	0	52	91,212	0.111	0.022	7
		45	211,411	0.067	0.013	7
		40	395,190	0.144	0.026	7
No	48	33	149,893	0.349	0.067	7
		33	95,271	0.318	0.064	7
		33	85,484	0.112	0.023	7
	48	23	236,511	0.169	0.031	4
		23	467,739	0.610	0.108	4
		20.8	1.02×10^6	0.340	0.057	3
		20.8	802,542	1.032	0.175	4
Yes; test greater than prediction	0	19.88	72,300			8
		17.04	245,000			8
		14.20	642,000			3
		11.36	3.22×10^6			1

(*Note:* Failure is from 1.5-cps frequency shift and will be lower value than complete fracture.)

350 Describing Materials Behavior with Random Variables

and

$$s_x = 0.195$$

Then

$$(\overline{N}, s_N) = (96,120, 45,370) \text{ cycles}$$

where

$$C_N = \frac{s_N}{\overline{N}} = 0.472$$

8.15 STRENGTH, A RANDOM PROCESS

It has long been apparent that further development in modeling of materials behavior was necessary because of the growing need for optimum design. Strength has in the past been estimated deterministically or as a distribution from a single collection of test values, assuming that such a statistical description would be typical. This was not always true. Rather, the description of strength as a random process [34, p. 22] completes the task of modeling, with a complete behavioral profile.

The theory of random processes, discussed in Chapter 3, permits inclusion of time, cycle, or trend effects, as well as random variability in a dynamic materials strength behavior model. To illustrate, consider an experiment in determining the static ultimate strength of SAE-4340 cold-drawn–annealed

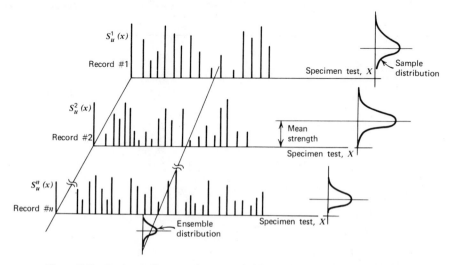

Figure 8.43 Static tensile strength as a probable ergodic random process [74].

Table 8.15 Ultimate Strength Data of 50 Wire Specimens of SAE 4340 Cold Drawn and Annealed Steel Wires[a] Obtained Using the Riehle Testing Machine

Ultimate Tensile Strength (psi)	
114,426	111,475
112,131	114,754
114,754	111,803
115,737	114,754
113,442	112,459
114,754	114,098
112,131	114,098
114,754	113,442
112,131	112,786
111,475	111,803
114,754	114,098
115,409	112,459
116,065	112,786
113,442	113,114
114,754	111,803
113,770	109,836
115,409	114,426
113,770	112,786
114,754	113,770
118,032	112,459
116,065	113,442
115,737	113,114
114,098	112,786
117,377	112,459
114,098	113,442

[a]Normal wire diameter = 0.0625 ± 0.0005 in.
Source: Mil = A = 8866 (ASG) [7, p. 191].

steel, say, 50 tests. The 50 values may be considered as one experiment or sample function, $S'_u(x)$ (see Table 8.15). If the experiment is repeated, the result will be somewhat different. A random process consists of an ensemble of (all) such possible experiments $S_u(x)$. From such an ensemble, the statistical characteristics of the random process may be estimated. It may be inferred that static ultimate strength of 4340 steel, for example, is probably an ergodic (see Chapter 3) random process of the discrete type, as represented in Figure 8.43 [74]. In the limit the distributional characteristics of all experiments (each a collection of strength values), will be the same, and also the same as the across the ensemble statistical characteristics, taken at any time or cycle or test of interest.

The concept of a random process is useful in developing the strength behavioral model of a specific material produced by a number of different

352 Describing Materials Behavior with Random Variables

companies. It is well known that the statistical characteristics of a given material produced to the same specification will vary from company to company. Therefore, if the sample of test results from each company is treated as a discrete sample function $S_u^i(x)$, the samples from a number of sources will make up a finite ensemble of finite length records. The sample distributions will vary randomly, but the ensemble distributions at any two x values will approach identity as the ensemble length increases, that is, reflect a discrete stationary process.

Now consider a sample of n specimens. The static strength from one specimen to another will vary in a random fashion. The strength of each specimen, however, may not vary over time. Thus for the n specimens, strength may be considered a continuous time function. Figure 8.44 shows static strength as a continuous time function. This random process is clearly stationary; that is, the across-the-ensemble distributions at times t_1 and t_2 are identical.

Ideally, random processes require both infinite length and number of records in an ensemble. In contrast, experimental research and testing records are always of finite length and of a finite number. From these collections, the characteristics of a process are estimated, as in the following example.

In accord with a suggestion by Bendat et al. [38], the data of Table 8.15 have been divided into seven subrecords, each containing seven test values. In Table 8.16 the resulting matrix of values is shown together with the mean values and standard deviation of each (row) record, $S_u^i(x)$, and also the column means and standard deviations across the ensemble, $S_u(x)$. The correspondence of results among the rows, among the columns, and between the two sets, is reasonable considering the small samples. It may be postulated that the discrete process is very likely ergodic and also stationary. For an ergodic random process, a single sample record is sufficient to define the

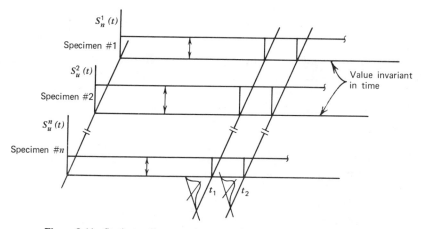

Figure 8.44 Static tensile strength as a stationary random process (in time).

Table 8.16 Ultimate Tensile Strength, AISI 4340 Cold-drawn–Annealed 0.0625-in.-diameter Wire Specimens, Plotted as a Random Process[a]

	Specimen Number								
		1	2	3	4	5	6	7	Mean	Std. Dev.
Test Number	1	114,426	112,131	114,754	115,737	113,442	14,754	112,131	113,911	1,389
	2	114,754	112,131	111,475	114,754	115,409	116,065	113,442	114,004	1,711
	3	114,754	113,770	115,409	113,770	114,754	118,032	116,065	115,222	1,489
	4	115,737	114,098	117,377	114,098	111,475	114,754	111,803	114,192	2,080
	5	114,754	112,459	114,098	114,098	113,442	112,786	111,803	113,349	1,049
	6	114,098	112,459	112,786	113,114	111,803	109,836	114,426	112,646	1,536
	7	112,786	113,770	112,459	113,442	113,114	112,786	112,459	112,974	1,496
Mean		114,473	112,974	114,051	114,145	113,348	114,145	113,161		
Standard Deviation		897	864	2,007	876	1,425	2,639	1,600		

[a]Population statistics (50 tests), estimators: mean = 113,000 psi; standard deviation = 1,600 psi
Source: Data from Table 8.15.

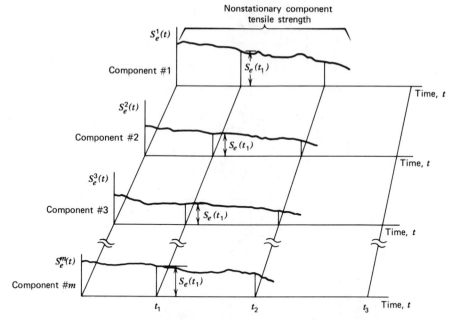

Figure 8.45 Component strength behavior as a nonstationary random process [$X(c)$], showing an ensemble of sample functions. Stress level required to produce the process shown here must correspond to a finite cycle life (i.e., $c < 10^6$ cycles).

process, thus often justifying the use of the distributional statistics determined from a single sample of strength data.

Both static loading and static strength of materials behavior are stationary random processes (see Figure 8.44). Strength from one specimen to another or load values from one application to another vary, but each remains constant in time.

Although strength of a material and external loading may be stationary random processes, with a collection of components subject, such as to corrosion, the strength may be a monotonically decaying random process in time (Figure 8.45). Local stress magnitude due to the development of more and more severe stress concentrations will be a monotonically increasing random process.

If the data from a cycles to failure experiment at a constant-amplitude stress level are grouped as was the previous tensile strength data in Table 8.16, it is found to suggest a discrete ergodic random process [45].

The strength of specimens subjected to constant-amplitude sinusoidal bending stresses above the endurance limit (overstressed) appears to decay monotonically with increasing cycle history, that is, a continuous nonstationary stochastic process $S_f(n)$ (see Figure 10.24). Monotonic decay of fatigue strength occurs with cyclic constant amplitude or completely random-loaded materials.

8.15.1 Random Processes in Design

Since the evidence indicates that fatigue strength is a continuous nonstationary random process, the question is how to develop the information needed for design. No method has existed for nondestructively determining fatigue strength $S_f(n)$ quantitatively at a given cycle n and then continuing the test.

Needed most often is the distribution of strength $S_f(t)$ across an ensemble after some time t or cycles n, for example, after t_2 cycles in Figure 8.45. It turns out that in the statistical $S-N$ envelope (e.g., Figure 8.29) across-the-ensemble statistics of material strength already exist, corresponding to fatigue strength after n cycles of loading. The statistical $S-N$ envelope provides an alternative representation of a stochastic materials behavior fatigue process (Figure 8.46).

Employing the concept of random process with probabilistic design methodology, certain classes of design problems for both constant-amplitude and narrow-band random loading and with nonstationary random materials behavior, can be solved (see Chapter 10).

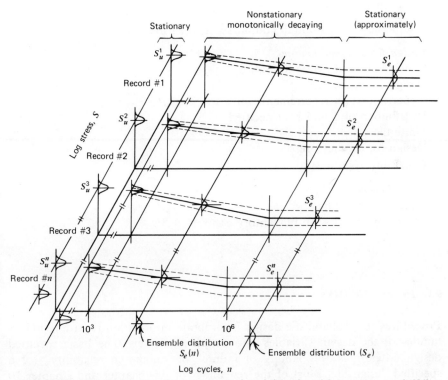

Figure 8.46 Statistical $S-N$ surfaces representative of those found for common steels [54].

CHAPTER 9

Design and Analysis

The following topics are considered in this chapter:

Classical and probabilistic approaches to design

General reliability Model

Static design and analysis
Normal stress and strength
Random static axial loads and bending loads
Nonnormal stress

Additional stress–strength combinations (distributions)

Design for dynamic loading
Consideration of cycles to failure
Design for endurance
Upper bound on requirements for a specified R
Distribution of S and s unspecified
Finite life
Design for contact stresses
Reliability assurance

Miscellaneous
Time to first failure
Pricing, warranties, and inventory control

9.1 INTRODUCTION

Procedures for probabilistic design incorporate information regarding uncertainties of the design variables into design algorithms. The basic specified design criterion is "probability of failure" or "minimum reliability" for a specified design life. Most of the remainder of this chapter and Chapter 10 are devoted to developing details of the mechanical probabilistic design process and in working out typical examples.

9.2 CLASSICAL APPROACH TO DESIGN

In Section 1.5 certain principal features of classical design are compared with similar points of theory in the probablistic approach. For classical design that has served engineers well within its limitations, the criterion is stated in equation (1.1) in a form indicated by experience. It is appealing because to its simplicity in form and application.

Custom in certain areas of design has dictated the use of allowable stress values as specified by recognized authorities or by codes. Safety factors have been assigned on the basis of experience, where possible, with similar problems to the one in hand.

In current classical engineering practice [90] stresses and strengths associated with dynamic loading are estimated by taking some cognizance of load fluctuations, surface finish, environmental factors, and so on, and safety factors are commonly assigned in the range 1.3 to 2. The tendency has been to conservatism resulting from the use of limit values of the design variables. In classical models, variables have been treated as single-valued phenomena.

Since safety factors are not performance-related measures there is no way by which an engineer can know whether his designs are near optimum or overconservative. In many instances this may not be important, but in others it can be critical [91]. Allowable stresses and safety factors are listed in Table 9.1.

Beginning with Section 9.3 probablistic design theory is developed in detail. The objective is to create near-optimum designs.

9.3 PROBABLISTIC APPROACH TO DESIGN

Figure 1.11 and 9.1 suggest that in design in addition to models describing the interactions of design variables, a second tier of modeling is necessary. The reason is that each design variable is seen as a multivalued random phenomenon. Thus both strength and load stress are functions of design variables, each of which is a random variable; that is, strength and stress are random variables. The design criterion utilized in the probablistic approach specifies

$$P\{\text{Strength} > \text{stress}\} \equiv R(t) = R \tag{9.1}$$

(where t denotes time or cycles) during the service life of a system or component. In other words, the probability that strength will exceed stress must equal or exceed a specified performance measure (R) called *reliability*.

A legitimate class of design problems also exists in which stress must exceed strength with a high probability. The relationship stated in equation (9.1) may be written

$$P\{S > s\} > R \tag{9.2}$$

Table 9.1 Allowable Stresses and Safety Factors (Deterministic)[a]

Application	Materials	Allowable stress, s_w	Safety factor, n
1. Machinery	Steel (shafts, etc.)	Steady tension, compression or bending; $s_w = s_y/n$; pure shear $s_w = (s_y/2n)$	1.5–2
		Tension s_t plus shear s_s; $s_t^2 + 4s_s^2 < s_y/n$	
		Alternating stress s_a plus mean stress s_m; Point representing an alternating stress nk_fs_a and a mean stress ns_m must lie below a Goodman diagram	
	Steel (SAE) 1095, leaf springs, thickness = $t < 0.10$ in.	Static leading $s_w = 230{,}000\text{–}1{,}000{,}000t$; variable loading, 10^7 cycles: $s_w = 200{,}000\text{–}800{,}000t$; dynamic loading, 10^7 cycles: $s_w = 155{,}000\text{–}600{,}000t$	
	Steel (wire ASTM-A228 helical springs)	$s_w = 100{,}000$ (for $d = 0.2$ in, 10^7 cycles repeated stress)	
2. Pressure vessels (unfired)	Carbon steel	Membrane stress $s_w = 0.211\,s_u$; membrane plus discontinuity stresses $s_w = 0.9\,s_y$ or $0.6\,s_u$	5
	Alloy steels	Membrane stress $s_w = 0.25\,s_u$	4
		Membrane plus discontinuity stresses $s_w = 0.95\,s_y$ or $0.6\,s_u$	
	Cast iron	Membrane stress $s_w = 0.1\,s_w$; bending stress $0.15\,s_u$	10, 6.67
	Nonferrous metals	Same rule as for alloy steels	4
3. Airplanes	Aluminum alloy and steel	Ultimate strength design	1.5 against ultimate 1.0 against yields
	Wood	Ultimate strength design	

4. Bridges	Structural metals and reinforced concrete	s_w about 0.9 s_w buildings	2.05
	Wood	s_w same as for buildings	6
5. Buildings and other structures	Structural steel	Direct tension, $s_w = 0.6\, s_y$; 0.45 s_y on net section at pin holes; bending, $s_w = 0.6\, s_y$; shear, $s_w = 0.45\, s_y$ on rivets; 0.40 s_y on girder webs; bearing, $s_w = 1.35\, s_y$ on rivets in double shear; 1.0 s_y in single shear	1.70 for beams 1.85 for continuous frames
	Structural aluminum 6061-T6, 6062-T6 $s_u = 38000, s_y = 35000$	Direct tension, $s_w = 17{,}000$; bending, structural shapes, $s_w = 17{,}000$; rectangular sections $s_w = 23{,}000$; shear, $s_w = 10{,}000$ on cold driven rivets; 11,000 on girder webs; bearing, $s_w = 30{,}000$ on rivets	1.8 for beams
	Reinforced concrete	Bending, compression in concrete, $s_w = 0.45\, s'_c$; bending, tension in steel, $s_w = 0.40\, s_y$; bending, tension in plain concrete footings, $s_w = 0.03\, s'_c$; shear, on concrete in unreinforced web, $s_w = 0.03\, s'_c$; compression in concrete column, $s_w = 0.225\, s'_c$	
	Wood	Bending, $s_w = \frac{1}{6}\, s'$; long compression, $s_w = \frac{1}{6}\, s'_c$; transverse compression, $s_w = \frac{1}{2}$ elastic limit (400 for Douglas fir); shear parallel to grain $s_w = 120$ for Douglas fir	6 in general

[a]Key: s_w = allowable or working stress; s_u = ultimate; tensile strength; s_y = tensile yield strength; s' = modulus of rupture in cross-binding of rectangular bar; s'_c = ultimate shear strength; s'_c = ultimate compressive strength; s_c = endurance limit or endurance strength for specified life; n = dividing factor applied to s_u, s_y, or s_c to obtain s_w (all stresses are in lb/in.^{2}).

Source: Roarke [91].

Figure 9.1 Histograms of fatigue stresses and strengths of structural components [26].

or

$$P\{S-s>0\} \geqslant R \tag{9.3}$$

Equation (9.2) predicts a finite probability of failure P_f, consistent with experience, since

$$P_f = 1 - R(t) = 1 - R; \quad R < 1$$

In equations (9.2) and (9.3), S and s may each be modeled by a functional combination of two or more independent design variables.

9.3.1 General Reliability Model

Consider a component of strength S, described by a probability density function $f(S)$, selected at random from a population of nominally identical components and subjected to random loading, developing applied load stress s described by a probability density function $f(s)$. It is desired to determine the probability R from strength of the component and the applied stress, as stated in equations (9.2) and (9.3).

The probability of a stress of value s_1 is numerically equivalent to the area of the element ds or of A_1 in Figure 9.2:

$$P\left\{s_1 - \frac{ds}{2} \leqslant s \leqslant s_1 + \frac{ds}{2}\right\} = f(s_1)\, ds = A_1$$

Since these are probability density functions, the probability that $S > s_1$ is numerically equivalent to the shaded area under the strength density curve indicated by A_2; thus [9]

$$P\{S > s_1\} = \int_{s_1}^{\infty} f(S)\, dS = A_2 \tag{9.4}$$

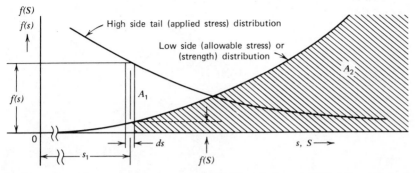

Figure 9.2 Allowable versus applied stress. (See D. Kececioglu and D. Cormier, "Designing a Specific Reliability into a Component," 3rd Annual Aerospace and Maintainability Conference, Washington, D.C., June 1964.)

The reliability R (i.e., the probability of no failure at s_1) is the product of probability equivalents A_1 and A_2:

$$P\left\{s_1 - \frac{ds}{2} \leqslant s \leqslant s_1 + \frac{ds}{2}\right\} \cdot p(S > s_1) \tag{9.5}$$

and

$$dR = f(s_1)\,ds \cdot \int_{s_1}^{\infty} f(S)\,dS$$

Reliability of the component is the probability of strength or allowable stress being greater than all possible values (over the range) of applied stress; thus

$$R = \int dR = \int_{-\infty}^{-\infty} f(s) \cdot \left[\int_{s}^{\infty} f(S)\,ds\right] ds \tag{9.6}$$

$$\int_{-\infty}^{\infty} f(S)\,dS = 1; \qquad \int_{-\infty}^{\infty} f(s)\,ds = 1$$

Alternatively, an expression for reliability R may be obtained by considering that a no-failure probability exists when applied stress remains less than some given value of allowable stress. Following the same reasoning as that just given, the probability of an allowable stress S_1 in an interval dS_1 is

$$P\left\{S_1 - \frac{dS_1}{2} \leqslant S \leqslant S_1 + \frac{dS_1}{2}\right\} = f(S_1)\,dS_1 = A_1'$$

The probability of applied stress less than S_1 is

$$P(s < S_1) = \int_{-\infty}^{S} f(s)\,ds = A_2'$$

Thus the no-failure probability is the product

$$(A'_1)(A'_2) = p\left(S_1 - \frac{dS_1}{2} \leqslant S \leqslant S_1 + \frac{dS_1}{2}\right) \cdot p(s < S_1)$$

and

$$dR = f(S_1) dS_1 \int_{-\infty}^{S} f(s) ds$$

Component reliability is the probability of no failure for all possible values of allowable stress (see Figure 9.3).

$$R = \int dR = \int_{-\infty}^{\infty} f(S) \cdot \left[\int_{-\infty}^{S} f(s) ds\right] dS \qquad (9.6a)$$

Probability of failure (p_f), the complement of the reliabilities just given, may be obtained by repeating the previous analysis, employing the fact that a failure probability exists when strength is less than a given stress:

$$p_f = \int_{-\infty}^{\infty} f(s) \left[\int_{-\infty}^{s} f(S) dS\right] \cdot ds$$

or

$$p_f = \int_{-\infty}^{\infty} f(S) \left[\int_{S}^{\infty} f(s) ds\right] \cdot dS \qquad (9.7)$$

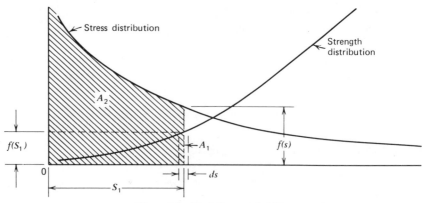

Figure 9.3 No failure probability.

STATIC DESIGN AND ANALYSIS

9.3.2 Normal Stress and Strength

Stress and strength density functions are given in Figure 9.4. Given a strength density function $f(S)$ (see Figure 2.1)

$$f(S) = \frac{1}{\sigma_S \sqrt{2\pi}} \exp\left[-\frac{(S-\mu_S)^2}{2\sigma_S^2}\right]$$

$$(-\infty < S < \infty) \qquad (9.8)$$

and an applied stress density function $f(s)$*

$$f(s) = \frac{1}{\sigma_s \sqrt{2\pi}} \exp\left[-\frac{(s-\mu_s)^2}{2\sigma_s^2}\right]$$

$$(-\infty < s < \infty) \qquad (9.9)$$

The difference z is written $z = S - s$ from which it follows that the mean [equation (2.40)] is given by

$$\mu_z = \mu_S - \mu_s$$

The standard deviation σ_z is [equation (2.41), with $\rho = 0$] given by

$$\sigma_z = \sqrt{\sigma_S^2 + \sigma_s^2}$$

With S and s normally distributed, Z is also normally distributed. It is noted that when $S > s, Z$ is positive. Thus the probability R may be equated with the positive area under the density curve $f(z)$ between the limits $(0, \infty)$ as

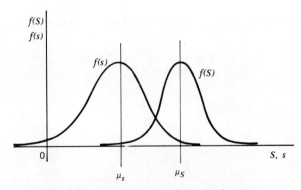

Figure 9.4 Stress and strength density functions.

*Figure 2.1.

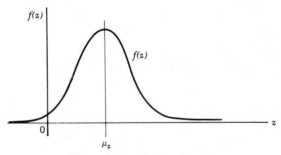

Figure 9.5 Density function of z.

shown in Figure 9.5. Hence

$$R = \frac{1}{\sigma_z \sqrt{2\pi}} \int_0^\infty \exp\left[-\frac{(z-\mu_z)^2}{2\sigma_z^2}\right] dz \qquad \text{(a)}$$

To evaluate equation (a) using standard normal area tables (Appendix 1), the transformation utilized is

$$t = \frac{z - \mu_z}{\sigma_z} \qquad \text{(b)}$$

The limits of the variable t are (Figure 9.6)
For $z = \infty$

$$t = \frac{z - \mu_z}{\sigma_z} = \frac{\infty - \mu_z}{\sigma_z} = \infty$$

For $z = 0$

$$t = \frac{z - \mu_z}{\sigma_z} = \frac{0 - \mu_z}{\sigma_z} = -\frac{\mu_z}{\sigma_z}$$

Substitution of equation (b) into equation (a), where $dz = \sigma_z \, dt$, yields

$$R = \frac{1}{\sqrt{2\pi}} \int_{-\frac{\mu_z}{\sigma_z}}^{\infty} e^{-t^2/2} dt \qquad (9.10)$$

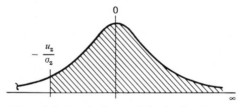

Figure 9.6 Standard normal density function.

where R is determined by integrating equation (9.10) from $t=(-\mu_z/\sigma_z)$ to $t=\infty$, with the aid of Appendix 1. The numerical value of t (lower limit) is obtained by utilizing equation (9.11), called the *coupling formula*.

$$t = -\frac{\mu_z}{\sigma_z} = \frac{\mu_S - \mu_s}{\sqrt{\sigma_S^2 + \sigma_s^2}} \qquad (9.11)$$

Equations (9.10) and (9.11) can be utilized to estimate an unknown cross-sectional area, or other variable, if μ_s and σ_s are expressions containing the unknown, and R is specified. The variable t is found in Appendix 1, corresponding to any specified R.

Provision has been made in probabilistic design theory to accommodate the statistical treatment of design variables. Cornell [92] and Ellingwood and Ang [93] suggest that uncertainties in prediction and modeling may be included explicitly in the formulation of design.

A second approach to contend with the effects of imponderables also addresses the problem of reliability assurance. This approach involves the employment of equation (9.11) in design, and a posteriori reliability assessment based on the results of one or more tests. In Example A, which follows, with the statistical estimators of static strength (S) of an eye bar and applied static load stress (s), equation (9.11) is used to estimate t. Reliability of the member is estimated from Appendix 1 as the probability associated with the range (t, ∞):

$$R = \frac{1}{\sqrt{2\pi}} \cdot \int_t^\infty e^{-t^2/2} dt$$

Probabilistic design synthesis of simple members subjected to random static loading is illustrated in Sections 9.3.3 and 9.3.4.

Example A Reliability Calculation (Analysis)

Given the statistics of the normal random variables describing strength and stress in an eye bar shown in Figure 9.7:

$$\text{Strength} = (\bar{S}, \sigma_S) = (27{,}000, 3200) \text{ psi}$$

$$\text{Stress} = (\bar{s}, \sigma_s) = (18{,}400, 1500) \text{ psi}$$

In this example S is static tensile strength and s is static load stress. Compute R and P_f, where $R = P[(S-s) > 0]$.

Solution. Reliability R is determined by integrating equation (9.10) over the range t to ∞ with the aid of Appendix 1. The lower limit of integration (t) is estimated by using equation (9.11).

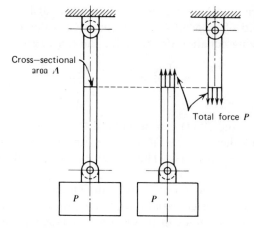

Figure 9.7 Eye bar loaded in tension [49] (courtesy of Spotts).

1. Assume that S and s are statistically independent, with $\rho=0$. Write the difference expression $Z=S-s$. From equations (2.40) and (2.41), the statistics of Z are

$$\bar{z} = 27{,}000 - 18{,}400 = 8600 \text{ psi} = \bar{S} - \bar{s}$$

$$\sigma_z = \sqrt{(3200)^2 + (1500)^2} = 3530 \text{ psi} = \sqrt{\sigma_S^2 + \sigma_s^2}$$

The transformation of the random variable z to the standardized normal (see Figure 9.8) is

$$t = \frac{z - \bar{z}}{\sigma_z} = \frac{0 - \bar{z}}{\sigma_z} = -\frac{\bar{z}}{\sigma_z}$$

Integration from 0 to ∞ is required; thus

$$t = \frac{8600}{3530} = -2.43$$

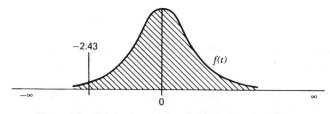

Figure 9.8 Standard normal probability density function.

Utilization of standard normal area tables (Appendix 1) and equation (9.10) yields

$$R = \int_{-2.43}^{\infty} \frac{1}{\sqrt{2\pi}} \cdot e^{-t^2/2} dt = 0.9925; \qquad P_f = 0.0075$$

The safety factor (SF) equivalent of $R = .9925$ is $\bar{S}/\bar{s} = 1.47$ for this problem.

2. Assume that S and s are correlated, with $\rho = +1$. Write the difference equation $Z = (S - s)$. From equations (2.40) and (2.41):

$$\bar{z} = -8600 \text{ psi}$$

$$\vartheta_z = \sqrt{(3200)^2 + (1500)^2 - 2(3200)(1500)} = 1610 \text{ psi}$$

$$= \sqrt{\vartheta_S^2 + \vartheta_s^2 - 2\rho \vartheta_S \vartheta_s}$$

$$t = \frac{-8{,}600}{1{,}610} = -5.34$$

Integrating as in step 1, see Appendix 1:

$$R = .9999995.$$

Positive correlation of stress and strength tends to improve reliability, and the converse holds for negative correlation.

Reliability a Performance Measure. If the probability of failure of a component selected at random is $1 - R$ and a population of components of size N is considered, then

$$\text{Expected failures} = N[P(\text{failure})] = N(1 - R) = N - NR$$

Thus there exists a linear relationship between the reliability and expected performance of a component population. With more than a single failure mode:

$$\text{Expected failures} = N[P_1\{\text{failure}\} + P_2\{\text{failure}\} + \cdots +$$

Subscripts identify the various failure modes.

Confidence in Design. The central limit theorem was introduced in Chapter 3, and in Chapter 12 bounds (u, l) associated with confidence δ and the confidence interval $u - l$ associated with δ are considered.

Emphasis in design is shared by analysis (reliability assessment, including estimation of bounds) and synthesis (designing in the required minimum reliability), as can be seen by examining Example A, illustrating analysis, and

Example B, illustrating design synthesis. In analysis reliability is estimated by a calculation involving random variables and is hence a probability and a random variable ($\bar{R}, s_R$). In design synthesis, reliability (R) is specified and a dimensional random variable ($\bar{d}, s_d$) for example is estimated by a calculation involving random variables.

Design Procedure with S and s Normal. Many mechanical components and systems are designed for conditions in which the loading will be constant in time. These may involve (1) various spacial distributions of applied loads, (2) mechanical loads due to thermal gradients in the system, and (3) inertial loads. Static design methodology may also be applied where a load is applied once and removed.

It may be advantageous, for a number of reasons, to make the following assumption of normality:

1. The mathematical form of the distribution is tractable; consequently, procedures based on normality assumptions are often conceptually and computationally simple.

2. It is one of the family of two-parameter distributions completely described by the mean and the standard deviation.

3. A large body of statistical methods based on the assumption of normality already exists, including tables necessary for implementation.

The normal assumption for strength and stress is often made in part because of indications that this model provides a reasonable representation of strength and stress behavior (see Table 9.5).

There may be some compelling reason to favor the use of distributional form other than the normal in a strength–stress analysis or design synthesis. Otherwise, it is recommended that the following design procedure be used [10]. The design procedure is as follows:

1. Carry out the necessary preliminary steps discussed in Section 1.4.

2. Write expressions for the mean and the standard deviation of the load-induced stress model.

3. Determine the mean and the standard deviation of strength from Chapter 8, Table 8.1, or other references.

4. Obtain the z value corresponding to specified reliability, R (Appendix 1 and *Tables of Normal Probability Functions* [40]).

5. Substitute statistics determined in the previous steps into the "coupling formula," [equation (9.11)], and solve for the unknowns.

6. Using the results from step 5, estimate dimensional tolerances consistent with the required reliability.

7. Utilize the algebra of normal functions in the previous steps.

The examples that follow retain the preceding numerical order in developing design solutions.

To illustrate probabilistic design synthesis in the initial examples, it is assumed that applied stress (s) and allowable stress (S) are statistically independent and normal and loading is static.

An approach to design synthesis not dependent on a normality assumption is presented in Section 9.3.9 (subsection on design with distribution of S and s unspecified) and is applied in the same section (subsection on finite life design with distributions of S and s unknown). The only restrictions on the design variables are that they be continuous and unimodal.

The examples in this chapter are kept simple, with emphasis on illustrating methods. All techniques used can be applied in more practical situations. Some of the detail in the examples, included to explain the process, could very likely be eliminated. Use of programming and a computer eliminate much tedious calculation.

9.3.3 Design for Random Static Axial Loads

Numerous mechanical components such as eye bars, bolts loaded in tension, cross and sway braces, reinforcing bars, and frame members are designed using the tension rod as a model.

Design of a Tension Bolt. The following are determined:

1. Design of the element for specified loading, material, and R. (This subsection).
2. The multiple roots of the design equations are interpreted. (Subsection on determination of the significant root).
3. Sensitivity of component reliability to cross-sectional variability is estimated. (Subsection on sensitivity of R to tolerance on r).

The required shank diameter of a bolt loaded in static tension is estimated by the procedure outlined in Section 9.3.2. (Subsection on design procedure with S and s normal).

Example B

1. Preliminary assumptions are as follows. There is no preloading of the bolt: stress is uniformly distributed over the element cross section, the axis is straight, and the random loading is concentrically applied, as indicated in Figure 9.9. Failure is defined as fracture, and the design criterion is

$$P(S_u - s > 0) \geqslant R$$

Problem

Estimate r and tolerance on r for specified static axial loading, reliability, and material.

370 Design and Analysis

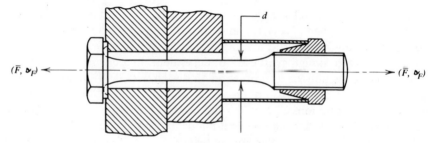

Figure 9.9 Tension bolt.

Given.

$$\text{Load} = (\bar{F}, s_F) = (6000, 90) \text{ lb.} \tag{a}$$

Tensile ultimate strength (4140 steel, 160-ksi heat treatment (Appendix 10.A):

$$(\bar{S}_u, s_{S_u}) = (156{,}000, 4300) \text{ psi} \tag{b}$$

Economic considerations indicate that fewer than 13 in 10,000 bolt replacements can be tolerated under warranty. The element is of circular cross section, and equation (5.1) is a sufficiently precise model for predicted stress.

2. The load-induced stress model is

$$s = \frac{F}{A} = \frac{F}{\pi r^2}$$

Since the load statistics [equation (a)] are given, only the statistics of A require determination, that is, the statistics of the product of a constant (π) and the square of a random variable (r):

$$A = \pi r^2$$

According to equations (2.59) and (2.8), the mean cross section is

$$\bar{A} \approx \pi \bar{r}^2 = f(\bar{r})$$

The approximate standard deviation on cross section, from equations (2.18) and (2.10), is (also see Appendix 5)

$$s_A \approx (2\pi \bar{r}) \cdot s_r = \left[\left(\frac{\partial A}{\partial r} \right)^2 s_r^2 \right]^{1/2} = \left[(2\pi \bar{r})^2 s_r^2 \right]^{1/2}$$

Predicted values for $\bar{s}$ and $\mathfrak{d}_s$ are

$$(\bar{s}, \mathfrak{d}_s) \approx \frac{(\bar{F}, \mathfrak{d}_F)}{(\bar{A}, \mathfrak{d}_A)} = \frac{(6000, 90)}{(\pi \bar{r}^2, 2\pi \bar{r} \mathfrak{d}_r)} \text{ psi} \qquad (c)$$

where $\bar{s}$ and $\mathfrak{d}_s$ are unknown because $\bar{r}$ and $\mathfrak{d}_r$ are unknown; thus another relationship is needed to make a unique solution for $\bar{r}$ and $\mathfrak{d}_r$ possible. One way to reduce the unknowns is by finding a relationship involving the variability on geometry, that is, manufacturing process variability or supplier process variability (see Chapter 4).

From considerations of manufacture's tolerance (Table 4.9), $1.5\% \cdot \bar{r} = \Delta$(99% interval). According to the rationale of Section 3.3.2, the standard deviation $\mathfrak{d}_r$ is estimated:

$$\mathfrak{d}_r \simeq \frac{0.015 \bar{r}}{3} = 0.005 \bar{r}$$

The mean value of applied stress, from equation (2.28) with values from equation (c), is

$$\bar{s} \approx \frac{\bar{F}}{\pi \bar{r}^2} = \frac{6000}{\pi \bar{r}^2} \text{ psi} \qquad (d)$$

The variance of stress, from equation (2.30) with values from equation (c), is

$$\mathfrak{d}_s^2 \approx \frac{\bar{F}^2 \mathfrak{d}_A^2 + \bar{A}^2 \mathfrak{d}_F^2}{\bar{A}^4}$$

$$\mathfrak{d}_s^2 \approx \frac{0.0081 \pi^2 \bar{r}^4 + 36[2\pi \bar{r}(0.005\bar{r})]^2}{(\pi \cdot \bar{r}^2)^4} \cdot (1000)^2$$

and

$$\mathfrak{d}_s^2 \approx \frac{0.0117}{\pi^2 \bar{r}^4} \cdot (1000)^2 (\text{psi})^2 \qquad (e)$$

The estimated statistics of applied stress are, from equations (d) and (e):

$$(\bar{s}, \mathfrak{d}_s^2) = \left[\frac{6000}{\pi \bar{r}^2} \text{ psi}, \frac{0.0117}{\pi^2 \bar{r}^4} \cdot 1000^2 \text{psi}^2 \right] \qquad (f)$$

3. Determine the statistics of static ultimate strength. In this case the statistics $(\bar{S}_u, \mathfrak{d}_{S_u})$ are given [see equation (b)] as

$$(\bar{S}, \mathfrak{d}_{S_u}) = (156{,}000, 4300) \text{ psi} \qquad (g)$$

4. The required reliability is

$$R = 1 - \frac{13}{10{,}000} = 0.9987 \leqslant \frac{1}{\sqrt{2\pi}} \int_z^\infty e^{-t^2/2} dt$$

from Appendix 1:

$$0.9987 \leqslant \frac{1}{\sqrt{2\pi}} \int_{-3.00}^\infty e^{-t^2/2} dt$$

Thus

$$z \approx -3.00$$

5. Estimate the mean radius by substituting equations (f) and (g) into equation (9.11):

$$z = -\frac{(\bar{S}_u - \bar{s})}{\sqrt{s_{S_u}^2 + s_s^2}} = -\frac{156 - (6/\pi \bar{r}^2)}{\sqrt{\dfrac{0.0117}{\pi^2 \bar{r}^4} + (4.3)^2}} = -3.00$$

Expand

$$9\left[\frac{0.0117}{\pi^2 \bar{r}^4} + 18.5\right] = 24{,}340 - \frac{1870}{\pi \bar{r}^2} + \frac{36}{\pi^2 \bar{r}^4}$$

and simplify

$$0 = \bar{r}^4 - 0.0246 \cdot \bar{r}^2 + 0.00015 \tag{h}$$

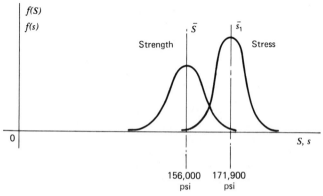

Figure 9.10 Case 1 ($\bar{r} = 0.1054$ in.).

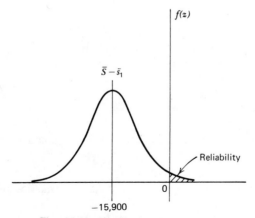

Figure 9.11 Distribution of $z = (S - s)$.

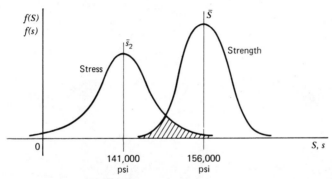

Figure 9.12 Case 2 ($\bar{r} = 0.116$ in.).

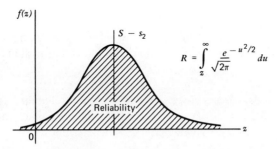

$$R = \int_z^\infty \frac{e^{-u^2/2}}{\sqrt{2\pi}}\, du$$

Figure 9.13 Distribution of $z = (S - s)$.

374 Design and Analysis

Solve equation (h) as a quadratic in $\bar{r}^2$:

Case I	Case II
$\bar{r}_1^2 = 0.0111$ in.^{2}	$\bar{r}_2^2 = 0.0135$ in.^{2}
$\bar{r}_1 = 0.105$ in.	$\bar{r}_2 = 0.116$ in.
$\bar{d}_1 = 0.211$ in.	$\bar{d}_2 = 0.233$ in.

6. The maximum tolerance is a direct outfall of the design calculations, as follows:

$$\text{Tolerance} \approx 1.5\%(\bar{d}_2) = \pm 0.0035 \text{ in. (maximum)}$$

Requirements in the threaded area must be determined separately.

Consistent with prudent engineering practice, designs such as the one just exemplified should be verified by test, regardless of the methodology used.

Design assuming determinism is considered in the following paragraph.

Given. Load $P = 6000$ lb., strength $S_u = 156{,}000$ psi., and $SF = 1.5$

From equation (1.1), the deterministic design criterion is

$$S_u = (SF)s = 1.5\frac{P}{A} \tag{i}$$

Substitution of assumed values into equation (i) yields

$$156{,}000 = 1.5\frac{6000}{\pi r^2}$$

from which $r = 0.136$ in. The predicted radius requirement by classical methods is significantly larger than the predicted radius in step 5, early in this section.

Determination of Significant Root. To estimate the reliability associated with radii r_1 and r_2, it is necessary to estimate the statistics of s_1 and s_2 and then utilize these statistics with given material tensile strenth.
Case 1 (Figures 9.10 and 9.11):

$$\bar{s}_1 = \frac{\bar{F}}{\bar{A}_1} = \frac{\bar{F}}{\pi \bar{r}_1^2} = \frac{6000}{\pi (0.105)^2} = 171{,}900 \text{ psi} \tag{j}$$

*Note: S_1 and S_2 are not principal stresses in this example.

Case 2 (Figures 9.12 and 9.13):

$$\bar{s}_2 = \frac{\bar{F}}{\bar{A}^2} = \frac{\bar{F}}{\pi \bar{r}_2^2} = \frac{6000}{\pi(0.116)^2} = 141{,}000 \text{ psi} \qquad (k)$$

First, estimate the z value of an element that has a radius $\bar{r}_1 = 0.1055$ in. Knowing strength $(156{,}000, 4300)$ psi and the mean value of stress $171{,}900$ psi, it remains to estimate the standard deviation of stress $\jmath_{s_1}$:

$$\jmath_A \approx 2\pi \bar{r} \jmath_r = 2\pi \bar{r}(0.005\bar{r}) = 2\pi \bar{r}^2(0.005)$$

$$\bar{s}_1 = \frac{\bar{F}}{\pi \bar{r}_1^2} = \frac{6000}{\pi \bar{r}_1^2}$$

$$\jmath_{s_1}^2 = \frac{1}{\bar{A}_1^4}\left(\bar{F}^2 \jmath_{A_1}^2 + \bar{A}_1^2 \jmath_F^2\right)$$

$$\jmath_{s_1}^2 = \frac{1}{\bar{A}_1^4} \cdot \left[\bar{F}^2(2\pi \bar{r}_1^2) \cdot (0.005)^2 + (\bar{A}_1^2 \cdot \jmath_F^2)\right] = 9{,}605{,}000 \text{ psi}^2$$

$$\jmath_{S_1}^2 = (4300 \text{ psi})^2 = (18{,}490{,}000 \text{ psi})^2 \qquad (l)$$

Substitution of equations (g), (j), and (l) into equation (9.11) yields

$$z = -\frac{156{,}000 - 171{,}900}{\sqrt{9{,}605{,}000 + 18{,}490{,}000}} = \frac{15{,}900}{5{,}300} = +3.00$$

from Appendix 1, case I reliability is $R \approx 0013$.

Second, consider case II, where $\bar{r}_2 = 0.116$ in. and mean stress $\bar{s}_2 = 141{,}000$. The variance $\jmath_{s_2}^2$ is

$$\jmath_{s_2}^2 = \frac{4(0.005)^2(6000)^2}{\pi^2(0.1164)^4} + \frac{(90)^2}{\pi^2(0.116)^4} = 6{,}462{,}000 \qquad (m)$$

Substitution of equations (g), (k), and (m) into equation (9.11) yields

$$z = -\frac{156{,}000 - 141{,}000}{\sqrt{6{,}462{,}000 + 18{,}490{,}000}} = -\frac{14{,}990}{4{,}990} = -3.00$$

Case II reliability is $R \approx 9987$. The value r_2 is associated with the required reliability (i.e., is significant).

Sensitivity of R to Tolerance on r. As indicated in Table 9.2, reliability (R) is sensitive to the level of geometric variability, if bilateral tolerances are used. This sensitivity can be avoided by using unilateral tolerances.

Table 9.2 Sensitivity of R to Geometric Variation ($\bar{r}=0.116$)

Tolerance on r	Reliability
0	.999
±0.00174 in.	.998
±0.00523 in.	.977
±0.0240 in.	.775

9.3.4 Design for Random Static Bending Loads

Design of a Simply Supported Beam. A mechanical system usually contains one or more components in whose design the model used was a beam. Furthermore, it has been shown [93] that within-member variability is insignificant compared to that among separate members. The implication is that failure will occur near the point of maximum load effect, with a high probability.

Example C

Estimate the outside radius and wall thickness of a simply supported tubular beam consistent with the conditions and assumptions given in the following paragraphs and utilizing the design procedure outlined in Section 9.3.2 (subsection on design procedure with S and s normal).

1. Preliminary assumptions are as follows. It is assumed that the effect of the beam weight relative to P may be considered negligible (see Figure 9.14). The consequences of a system malfunction indicate that fewer than 1 in 100,000 substandard production parts can be tolerated. Thus, the specified

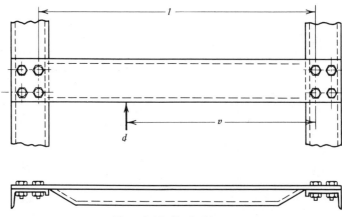

Figure 9.14 Typical beam.

reliability must be $R > .99999$. Failure is defined as yield, and the criteron is

$$P(S_y - s > 0) \geq R = .99999$$

Given. Estimated statistics of P and S_y (Table 8.1):

$$(\bar{P}, s_P) = (6070, 200) \text{ lb} \qquad (a)$$

$$(\bar{S}_y, s_{S_y}) = (117{,}300, 3200) \text{ psi (AISI 4140 steel)} \qquad (b)$$

Dimensional variables are as follows:

$$l = 120 \text{ in} \pm \tfrac{1}{8} \text{ in.}; \qquad \Delta l = \tfrac{1}{8} \text{ in.}$$

$$a = 72 \text{ in} \pm \tfrac{1}{8}.; \qquad \Delta a = \tfrac{1}{8} \text{ in.}$$

$$3 s_l = 3 s_a \approx 0.125.^*$$

$$s_l = s_a \approx \frac{0.125}{3} = 0.0416 \text{ in.}$$

$$(\bar{l}, s_l) \approx (120, 0.0416) \text{ in.}; \quad b = l - a$$

$$(\bar{a}, s_a) \approx (72, 0.0416) \text{ in.} \qquad (c)$$

2. Write expressions for the statistics of stress. In this case there is simple bending and no axial load; $P_a = 0$. Thus the model for maximum fiber stress is [equation (5.23)]:

$$s = \frac{Mc}{I} + \frac{P_a}{A} = \frac{M}{I/c}$$

where M is the bending moment (in. lb), c is the distance from neutral axis to outer fibers (inches), and I is the moment of inertia of beam cross section about the neutral axis (in.^{4}). To obtain expressions estimating the applied stress statistics, $(\bar{s}, s_s)$ psi, models for (M, s_M) and $(\bar{I}, s_I)$ must be written:

a. The moment statistics. [From Appendix 4, $M = p(ab/l)$]. Since (independent) P, a, and l are random variables,

$$(\bar{M}, s_M) = \frac{(\bar{P}, s_P)(\bar{a}, s_a)(\bar{b}, s_b)}{(\bar{l}, s_l)} = \frac{(6070, 200)(72, 0.0416)(48, 0.059)}{(120, 0.0416)}$$

*See Section 3.3.2.

Estimate the mean and standard deviation of u, where $b = (l - a)$ and

$$u = \left(\frac{ab}{l}\right) = \frac{al - a^2}{l} \tag{d}$$

Applying equation (2.62) to equation (d), the approximate standard deviation of u is

$$s_u \approx \left[\left(\frac{\partial u}{\partial a}\right)^2 \cdot s_a^2 + \left(\frac{\partial u}{\partial l}\right)^2 \cdot s_l^2\right]^{1/2}$$

The partial derivatives of u with respect to a and l are

$$\frac{\partial u}{\partial a} = \frac{l - 2a}{l}$$

$$\frac{\partial u}{\partial l} = \frac{a - (al - a^2)}{l^2} = \frac{a^2}{l^2}$$

If equation (c) and the partial derivatives are substituted into the equation for s_u, the standard deviation of u is

$$s_u \approx \left[\left(\frac{\bar{l} - 2\bar{a}}{\bar{l}}\right)^2 \cdot s_a^2 + \left(\frac{\bar{a}^2}{\bar{l}^2}\right)^2 \cdot s_l^2\right]^{1/2} = 0.0171 \text{ in.} \tag{d'}$$

From equation (2.59), the mean value estimator of u is

$$\bar{u} \approx \frac{\overline{ab}}{\bar{l}} = 28.8 = f(\bar{a}, \bar{b}, \bar{l}) \text{ in.} \tag{d''}$$

Since the moment is $M = Pu$,

$$(\overline{M}, s_M) = (6070, 200)(28.8, 0.0171) \text{ in. lb}$$

and from equations (2.42) and (2.43), the product of random variables P and u is

$$\overline{M} = \overline{Pu} \; ; \quad s_M \approx \left[\bar{P}^2 s_u^2 + \bar{u}^2 s_P^2 + s_P^2 s_P^2\right]^{1/2}$$

$$(\overline{M}, s_M) = (147{,}800, 5760) \text{ in. lb} \tag{e}$$

b. Estimate the moment of inertia statistics from the following model (see Appendix 5):

$$I \simeq \pi r^3 t \tag{f}$$

where r is the outside radius, t is the wall thickness, and $r=c$. On the basis of known requirements for stability of a tubular section [94]:

$$\frac{\bar{r}}{t} \geq 50; \qquad \bar{t} \geq \frac{\bar{r}}{50}$$

Thus

$$I = \pi r^3 \frac{r}{50} = \frac{\pi r^4}{50}$$

and

$$\left(\frac{I}{c}\right) = \frac{\pi r^3}{50} \qquad \text{(g)}$$

From equation (2.59) and (g):

$$\frac{\bar{I}}{\bar{c}} \approx \frac{\pi}{50} \bar{r}^3 = 0.0628 \bar{r}^3 = f(\bar{r}) \qquad \text{(g')}$$

From equations (2.62) and (g):

$$s_{I/c} \approx \frac{\pi}{50}\left[\left(\frac{\partial(I/c)}{\partial r}\right)^2 \cdot s_r^2\right]^{1/2}$$

$$\approx 0.0628\left[(3\bar{r}^2)^2 \cdot s_r^2\right]^{1/2} \qquad \text{(h)}$$

Based on manufacturers tolerances (see Table 4.4),

$$s_I \approx 0.015 \bar{r} \qquad \text{(i)}$$

Thus substitution of equation (i) into equation (h) yields

$$s_{I/c} \approx 0.00283 \bar{r}^3 \qquad \text{(g'')}$$

Applied stress is estimated from equations (e), (g'), and (g'') substituted into equation (5.23):

$$(\bar{s}, s_s) = \frac{(\bar{M}, s_M)}{(\bar{I}/\bar{c}, s_{I/c})} = \frac{(174{,}800, 5760)}{(0.0628\bar{r}^3, 0.002826\bar{r}^3)}.$$

From equations (2.45) and (2.47), statistics of the preceding quotient are

$$(\bar{s}, s_s) = \left[\frac{2.784}{\bar{r}^3} \cdot 10^6; \frac{15.67}{\bar{r}^3} \cdot 10^4\right] \text{psi} \qquad \text{(j)}$$

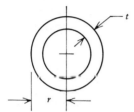

Figure 9.15 Tubular beam cross section.

3. In this exercise, yield strength statistics are given by equation (b).

4. $$R = 0.99999 = \frac{1}{\sqrt{2\pi}} \cdot \int_z^\infty e^{-t^2/2} dt = \frac{1}{\sqrt{2\pi}} \int_{-5.0}^\infty e^{-t^2/2} dt$$

Thus $z = -5.0$ (see Appendix 1).

5. Substitution of equations (b) and (j) into equation (9.11) yields

$$z \approx -\frac{\bar{S}_y - \bar{s}}{\sqrt{s_{S_y}^2 + s_s^2}} = -\frac{117{,}300 - \left(\dfrac{2.784}{\bar{r}^3}\right) \cdot 10^6}{\left[\left(\dfrac{15.67}{\bar{r}^3} \cdot 10^4\right)^2 + (3200)^2\right]^{1/2}} = -5.0$$

Expand and simplify:

$$\bar{r}^6 - 33.18\bar{r}^3 + 248.6 = 0$$

With solution of the quadratic in $\bar{r}^3$, the principal root is

$$\bar{r}^3 = 21.58 \text{ in.}^3$$

$$\bar{r} = 2.79 \text{ in. (outer radius)}$$

$$s_r \approx 0.015\bar{r} = 0.0419 \text{ in.}$$

$$\bar{t} = \frac{\bar{r}}{50} = 0.0558 \text{ in. (wall) (see figure 9.1.5)}$$

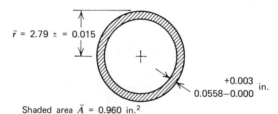

Shaded area $\bar{A} = 0.960$ in.^{2}

Figure 9.16 Minimum wall thickness for elastic stability.

6. Maximum tolerance on $r \approx \pm 0.126$ in. (compatible with $R = .99999$). From Table 4.10, a realistic tolerance is taken as ± 0.015 in. Since the minimum wall thickness required for elastic stability is 0.0558 in., manufacturing tolerances are imposed as shown in Figure 9.1.6.

Figure 9.16 gives dimensions of the model for 0.99999 reliability. The reliability of the component would correspond to R only if the model were valid and complete.

Design with Deterministic Assumption. The following calculations are presented for comparison: load $p = 6070$ lb, $l = 120$ in., $a = 72$ in., $b = 48$ in., strength $S_y = (17300$ psi$)$, and $SF = 1.5$. From equation (1.1):

$$S_y = (SF) \cdot s = 1.5s \qquad (k)$$

For a simple beam, maximum stress is $s = (Mc/I)$:

$$M = \frac{Pab}{l} = 6070\left(\frac{72 \cdot 48}{120}\right) = 174{,}800 \text{ in. lb}$$

$$I \approx \pi \cdot r^3 t, \text{ where } t \approx \frac{r}{50}$$

$$\frac{I}{c} \approx 0.0628 r^3, r = c$$

$$s = \frac{174{,}800}{I/c} = \frac{174{,}800}{0.0628 r^3} \qquad (l)$$

Substitution of equation (l) into equation (k) yields

$$117{,}300 = 1.5\left(\frac{174{,}800}{0.0628 r^3}\right)$$

$$r^3 = \frac{174{,}800}{117{,}300} \cdot \left(\frac{1.5}{0.0628}\right)$$

$$r = 2.901 \text{ in. (outer radius)}$$

$$t = \frac{r}{50} = 0.058 \text{ in. (wall)}$$

Nonconventional Applications. In numerous static engineering problems, analogues of stress and strength exist and solutions may be obtained by utilizing the procedures of Sections 9.3.3 and 9.3.4.

382 Design and Analysis

Example D

1. Estimate the probability that a link attached at point A in Figure 9.17, will not contact the stop after a random load is applied. In the solution, follow the design procedure in Section 9.3.2. In the solution of this problem $(\bar{x}, s_x)$ is interpreted as *allowable deflection (analogue of allowable stress)* and $(\bar{\delta}, s_\delta)$ as *applied deflection (analogue of applied stress)*.

2. Calculations indicate that deflection (δ) after the application of load $(\bar{P}, s_P) = (4000, 400)$ lb will be (see Chapter 6, Example H):

$$(\bar{\delta}, s_\delta) = (1.07 \cdot 10^2, 1.07 \cdot 10^3) \text{ in.}$$

3. Statistics of initial clearance (x) estimated from a series of measurements indicates that

$$(\bar{x}, s_x) = (0.0274, 0.005) \text{ in.}$$

4. The probability of the link not contacting the stop is

$$P(x > \delta) = \int_z^\infty \frac{e^{-t^2/2}}{\sqrt{2\pi}} dt = P(x - \delta > 0) \qquad (a)$$

5. From Figure 9.17, the difference distribution $z = x - \delta$ is utilized with equation (9.11):

$$z = -\frac{\bar{x} - \bar{\delta}}{\sqrt{s_x^2 + s_\delta^2}} = -\frac{0.0274 - 0.0107}{\sqrt{(0.00107)^2 + (0.005)^2}} = -\frac{0.0167}{0.00511} = -3.27$$

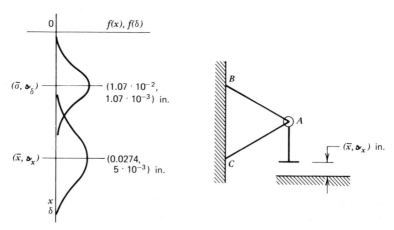

Figure 9.17 Probability of exceeding critical deflection.

6. Substitution into equation (a) yields

$$P(x>\delta) = \int_{-3.27}^{\infty} \frac{e^{-t^2/2} dt}{\sqrt{2\pi}}$$

From Appendix 1, $P(x>\delta) = 0.9995$, and the probability of contact P_f is

$$P_f = 1 - P(x>\delta) = 1 - .9995 = .0005$$

Example E

1. In this example available data are utilized to design a aluminum alloy compression element for a marine engine (Figure 9.18), such that a random selection from a population of such rods will not buckle due to an applied axial force F with probability R. Since this part can be quickly and economically replaced, a sufficient reliability goal is $R = .999$. The critical load P_{cr}, estimated by equation (6.9), may be loosely interpreted as a *strength analogue* and F as an *applied stress analogue*. It is required to estimate d such that $P(P_{cr} > F) \geqslant .999$.

2. The estimated statistics of applied loading are

$$(\bar{F}, s_F) = (593, 33) \text{ lb} \tag{a}$$

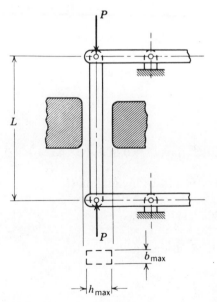

Figure 9.18 Compression element.

384 Design and Analysis

3. The model for P_{cr} is given by equation (6.9):

$$P_{cr} = \frac{\pi^2 EI}{l^2} \qquad (b)$$

In this example the moment of inertia I, is the unknown. More specifically, since the element has a solid circular cross section, the problem will be solved for d. The following are given:

$$(\bar{E}, s_E) = (10.3, 0.515) \cdot 10^6 \text{ psi (Table 5.1)}$$

$$(\bar{l}, s_l) = (22.0, 0.0625) \text{ in.}$$

From equations (2.16) and (2.18), the statistics of l^2 are

$$(\bar{l}^2, s_{l^2}) \approx (484, 2.75) \text{ in.}^2 \qquad (d)$$

The model for moment of inertia I is (from Appendix 5)

$$I = \frac{\pi}{64} \cdot d^4 \qquad (e)$$

From equations (2.59) and (e):

$$\bar{I} \approx \frac{\pi}{64} \cdot \bar{d}^4 = f(\bar{r}). \qquad (e')$$

From equations (2.62) and (e):

$$s_I \approx \frac{\pi}{64} (4\check{d}^3 s_d) = \left[\left(\frac{\partial I}{\partial r} \right)^2 s_r^2 \right]^{1/2} \qquad (e'')$$

Substitution of equations (c), (d), (e'), and (e'') into (b) yields

$$(\bar{P}_{cr}, s_{P_{cr}}) = \frac{\pi^3 (10.3, 0.515) 10^6}{64 (484, 2.75)} (\bar{d}^4, 4\bar{d}^3 s_d)$$

$$= 0.849 (2.072 \cdot 10^4, 1070)(\bar{d}^4, 4\bar{d}^3 s_d) = (17{,}980{,}900) \cdot (\bar{d}^4, 4\bar{d}^3 s_d) \qquad (f)$$

$s_d \approx (0.025\bar{d}/3)$, from example G, Chapter 6, and with equation (f):

$$(\bar{P}_{cr}, s_{P_{cr}}) \approx (17{,}980\bar{d}^4, 1110\bar{d}^4) \text{ lb} \qquad (f')$$

4. Obtain the z value corresponding to R, using equation (9.10):

$$R = 0.999 \leq \frac{1}{\sqrt{2\pi}} \cdot \int_z^\infty e^{-t^2/2} dt$$

From Appendix 1:

$$0.999 \leqslant \frac{1}{\sqrt{2\pi}} \int_{-3.10}^{\infty} e^{-t^2/2} dt$$

Thus $z = -3.10$.

5. Substitution of values from equations (f') and (a) into equation (9.11) yields

$$-3.10 = -\frac{\bar{P}_{cr} - \bar{F}}{\sqrt{s_{P_{cr}}^2 + s_F^2}} = -\frac{17,980\bar{d}^4 - 593}{\sqrt{(1110\bar{d}^4)^2 + (33)^2}}$$

$$0 = \bar{d}^8 - 0.0685\bar{d}^4 + 0.0011$$

$$\bar{d}^4 = 0.0409$$

$$\bar{d} = 0.450 \text{ in.}$$

$$s_d \approx (0.450) \cdot (0.025) = 0.011 \text{ in.}$$

6. The diametral tolerance is

$$\pm \Delta_d = \pm 0.033 \text{ in. (maximum)} \approx \pm 3 s_d$$

Design for Failure. The design problem of assuring mechanical failure at a single loading (with a high probability) occurs frequently. It may be necessary to completely separate the command module–service module complex of the Apollo from the Fourth Stage Saturn Booster. Shape charges have been used to effect shearing with a single controlled explosion. To eliminate a mechanical connection, such as fasteners joining the launch escape tower and the Apollo command module, explosive bolts have been utilized (Figure 9.19). The required relationship is

$$S_{su} \ll \tau$$

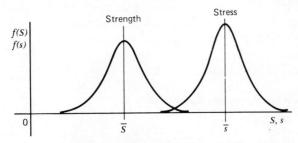

Figure 9.19 Failure with high probability.

or

$$P(S_{su} < \tau) \geq R \sim 1$$

where S_{su} is ultimate strength in shear and τ is shear stress. The shear key and shear pin are more common examples.

Design for failure is observed in metal working. Power shears and punches require complete shear failure at a single stroke. Punch presses, power brakes, stretch presses, and hydropresses require yielding (and permanent set) in material, forming

$$P(S_{sy} < s) = R \sim 1$$

9.3.5 Design for Nonnormal Stress

The question arises as to the amount of error in reliability calculations due to the assumption of normally distributed strength and stress when, in fact, one or both of these distributions may be nonnormal (i.e., γ), lognormal, or Weibull (see Tables 9.3 and 9.4). There are sometimes empirical and theoretical justifications for considering models of these general descriptions (see Figure 2.1).

For the preceding distributions, the degree of skewness is related to the size of the coefficient of variation C, as illustrated in Tables 9.5 and 9.6. Distributions with small values of C (≤ 0.1 for the lognormal) tend to be fairly symmetric and very similar in general shape to a normal distribution having the same mean and standard deviation. Differences between the normal and skewed distributions may exist insofar as tail probabilities are concerned. Since small probabilities are concerns in stress–strength analysis and design synthesis (the upper tail of the stress distribution and the lower tail of the strength distribution), significant errors could conceivably occur from substituting the symmetrical normal for skewed distributions. Substituting a normal model for a right-skewed* strength distribution, may tend to overestimate the calculated failure probability. This effect may be pronouned if the lower limit of the range of the skewed distribution is greater than zero. Conversely, substituting a normal model for a right-skewed stress distribution may tend to overestimate the predicted reliability, due to the relatively heavier right tail of the skewed distribution (note that not all skewed distributions have right tails heavier than the normal, i.e., a Weibull model with large shape parameter). When normal models are substituted for both right-skewed strength–stress models, these effects tend to cancel each other, and there is no general tendency for an increase or decrease in calculated reliability [95]. Each case requires individual consideration.

*Skewness, the third moment about the mean, is

$$\mu_3 = E(x - \mu_x)^3$$

Table 9.3 Comparison of Unreliabilities—Stress–Strength Analyses

Case No.	Strength						Stress						Unreliabilities		
	Weibull Parameters			Mean	Standard Deviation	Coefficient of Variation	Weibull Parameters			Mean	Standard Deviation	Coefficient of Variation	Weibull	Lognormal	Normal
	λ	θ	α^c				λ	θ	α^c						
1	25.25	10.42	2.0	34.48	4.8247	0.14	8.60	3.54	2.0	11.74	1.6422	0.14	$0.0_{12}6^b$	0.0_72	0.0_54
2	18.91	17.28	3.6	34.48	4.8049	0.14	6.44	5.88	3.6	11.74	1.6354	0.14	$0.0_{12}5$	0.0_72	0.0_54
3	9.38	16.07	5.0	24.14	3.3798	0.14	3.19	5.47	5.0	8.22	1.1505	0.14	0.0_61	0.0_72	0.0_54
4	15.35	2.68	2.0	17.72	1.2396	0.07	8.35	1.46	2.0	9.64	0.6740	0.07	$0.0_{12}5$	0.0_94	0.0_85
5	13.72	4.44	3.6	17.72	1.2347	0.07	7.47	2.41	3.6	9.64	0.6717	0.07	0.0_98	0.0_93	0.0_84
6	12.30	5.90	5.0	17.72	1.2407	0.07	6.70	3.21	5.0	9.64	0.6750	0.07	$0.0_{14}1$	0.0_94	0.0_85
7	12.30	5.90	5.0	17.72	1.2407	0.07	8.35	1.46	2.0	9.64	0.6745	0.07	0.0_87	0.0_94	0.0_85
8	15.35	2.68	2.0	17.72	1.2396	0.07	6.70	3.21	5.0	9.64	0.6750	0.07	$0.0_{33}1$	0.0_94	0.0_85

[a] Strength and stress distributions both Weibull, both lognormal and both normal.
[b] The notation utilized here indicates an unreliability value of 6×10^{-13} or a decimal value with 12 zeros followed by a 6.
[c] For a discussion of symbols λ, θ, and α see Ref. 13. *Source*: Ref. 95.

Table 9.4 Comparison of Unreliabilities—Stress–Strength[a] Analyses

Case No.	Standard Deviation of Stress	Coefficient of Variation of Stress	Unreliabilities Strength Normal, Stress γ	Strength Normal, Stress Lognormal	Strength Normal, Stress Normal
1	0.70	0.10	0.00286	0.00453	0.00025
2	1.22	0.17	0.02018	0.02165	0.01137
3	1.40	0.20	0.03199	0.04070	0.02180
4	1.75	0.25	0.95961	0.07679	0.04964

[a] Strength distribution normal, mean 10, standard deviation 0.5; stress distributions γ, lognormal, and normal, mean 7.
Source: Ref. 95.

Table 9.5 Comparison of Reliability for Normally Distributed Strength $N(\mu_s, \sigma_s) = (10, 1)$ and Stress Normal versus γ Distributed for $\mu_s = 7$ and σ_s for Values 0.10, 0.125, 0.150, 0.175, 0.20, and 0.250

Illustration	γ C_s	$\sqrt{\beta_1}$	R Stress Normal	Stress γ	σ_s
Figure 9.20	0.1000	0.2000	0.993007	0.994193	0.700
	0.1250	0.2500	0.988018	0.990647	0.875
	0.1508	0.3015	0.980451	0.982400	1.050
Figure 9.21	0.1741	0.3482	0.974178	0.978397	1.220
	0.2000	0.4000	0.959394	0.958488	1.400
	0.2500	0.5000	0.931678	0.925239	1.750

Source: Haugen and Eckert [22].

Table 9.6 Comparison of Reliability for Normally Distributed Strength $N(\mu_S, \sigma_S) = (10, 1)$ and Stress Normal versus Lognormally Distributed for $\mu_s = 7$ and σ_s for values of 0.700, 0.875, 1.050, 1.220, 1.400, and 1.750

Illustration	Lognormal C_s	$\sqrt{\beta_1}$	R Stress Normal	Stress Lognormall	σ_s
Figure 9.22	0.1000	0.3807	0.993007	0.961760	0.700
	0.1250	0.4805	0.988018	0.931812	0.875
Figure 9.23	0.1508	0.5867	0.980451	0.909100	1.050
	0.1741	0.6860	0.971478	0.887004	1.220
	0.2000	0.8007	0.959394	0.891722	1.400
	0.2500	1.0377	0.931678	0.846265	1.750

Source: Haugen and Eckert [22].

In Table 9.3 eight cases involving likelihood of failure estimation are studied. For each case the failure probability was calculated with both static strength and stress assumed to be distributed (1) in a Weibull pattern, (2) lognormally, and (3) normally. Corresponding mean values and standard deviations were equal, and in all cases the *normal–normal combination predicted the largest failure probability*. In Table 9.4 in all cases the static strength distribution was assumed normal, with mean of 10 and a standard deviation of 0.5, with static stress distributions: (1) γ, (2) lognormal, and (3) normal. The means of stresses were in all cases 7, and the standard deviations were varied from 0.7 to 1.75. Substituting a normal for the right-skewed stress distributions produced lower failure predictions (optimistic results). The degree of optimism relative to the size of the failure probability decreases as the coefficient of variation of stress increases.

Table 9.5 (compiled from data in Haugen and Eckert [22]) lists the reliabilities calculated for normally distributed strength, $N(10,1)$, and various parameter combinations of γ and normally distributed stress. Table 9.6 presents the reliabilities calculated for normally distributed strength $N(10,1)$ with various combinations of lognormally and normally distributed stress. The range of C_s (stress coefficient of variation) values employed was 0.10 to 0.25 for both nonnormal distributions, fairly representative of the range of applied stress often found in practice. Figures 9.20 and 9.21 [22] illustrate

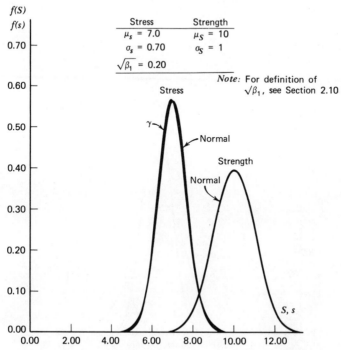

Figure 9.20 Normally distributed strength with normal and γ distributed stress [22].

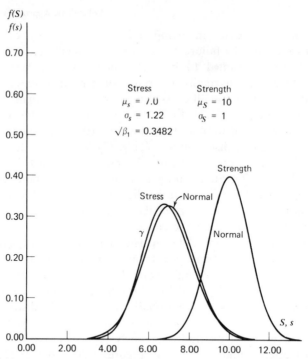

Figure 9.21 Normally distributed strength with normal and γ distributed stress.

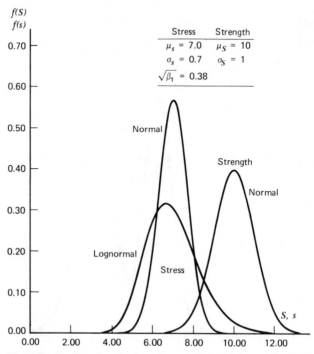

Figure 9.22 Normally distributed strength and normally and lognormally distributed stress.

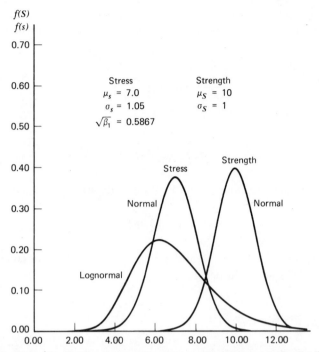

Figure 9.23 Normally distributed strength and normally and lognormally distributed stress.

plots of normally distributed stress compared with γ-distributed stress. Figures 9.22 and 9.23 [22] illustrate plots of normally distributed stress compared with lognormally distributed stress.

The practical question of whether it is safe to assume normality for strength and stress distributions, and if not, how appropriate distributions are to be determined, is further considered in Section 9.3.9.

Substitution of normal for theoretical skewed stress and/or strength distributions often leads to conservative results. However, in practice it is often not known which of the theoretical distributions are the appropriate models of true strength or stress distributions (see Mischke [68] and Whittaker [82] in case of fatigue). Most theoretical distributions either extend to infinity in both directions or to plus infinity, whereas expirical distributions are sometimes bounded. Depending on the point of truncation, the difference between the true truncated distribution and a theoretical approximation may be critical for the prediction of high reliabilities. It would be risky to use a stress distribution with an upper bound in a design synthesis unless it were certain that the bound exceeded the true truncation point of the distribution *The normal distribution has an element of conservatism in being unbounded.*

Design for Exponential Stress and Normal Strength. Given equation (9.6a):

$$R = \int dR = \int_{-\infty}^{\infty} f(S) \cdot \int_{-\infty}^{S} f(s)\,ds\,dS$$

392 Design and Analysis

If strength is normally distributed $N(\mu,\sigma)$ and stress is exponentially distributed with parameter λ, the probability density functions are (see Figure 2.1).
Strength:

$$f(S) = \frac{1}{\sigma\sqrt{2\pi}} \exp-\frac{1}{2}\left(\frac{S-\mu}{\sigma}\right)^2 (-\infty, \infty) \qquad (a)$$

Stress:

$$f(s) = \lambda e^{-\lambda s} (0, \infty) \qquad (b)$$

Substitute equations (a) and (b) into equation (9.6a) and integrate:

$$R = 1 - \exp-\tfrac{1}{2}\left[2\mu\lambda - \lambda^2\sigma^2\right] \qquad (9.12)$$

In this instance $\mu \approx \bar{S}$, $\lambda \approx (1/\bar{s})$, and $\sigma = \sigma_S$. Rearrangement of equation (9.12) into a form more useful in design yields

$$R = 1 - \exp-\frac{1}{2}\left[\frac{2\bar{S}}{\bar{s}} - \frac{\sigma_S^2}{(\bar{s})^2}\right] \qquad (9.12a)$$

$$P_f = 1 - R\text{-}\exp-\frac{1}{2}\left[\frac{2\bar{S}}{\bar{s}} - \frac{\sigma_S^2}{(\bar{s})^2}\right] \qquad (9.13)$$

where $\bar{s}$ is the mean value estimator of exponentially distributed stress, $\bar{S}$ is the mean value estimator of normally distributed strength, and σ_S is the standard deviation estimator of normally distributed strength.

Example G   Normal Strength and Exponential Stress (Analysis)

With strength normally distributed and stress exponentially distributed, estimate the probability R of no shear failure. Assume:
Strength (static shear):

$$N(\bar{S}, \sigma_S) = (27{,}000;\ 3200)\ \text{psi} \qquad (a)$$

Stress (shear):

$$\text{Exponential } \bar{s} = \frac{1}{\lambda} = 18{,}400\ \text{psi}$$

$$\sigma_s = \frac{1}{\lambda} = 18{,}400\ \text{psi} \qquad (b)$$

Solution. Utilizing equations (9.12a):

$$R = 1 - \exp-\tfrac{1}{2}\left(\frac{2\bar{S}}{\bar{s}} - \frac{\sigma_S^2}{(\bar{s})^2}\right)$$

substitute from equations (a) and (b) into equation (9.12a):

$$R = 1 - \exp-\tfrac{1}{2}\left(\frac{54{,}000}{18{,}400} - \frac{(3200)^2}{(18{,}400)^2}\right)$$

$$R = 1 - \exp-\tfrac{1}{2}(2.940 - 0.030) = 1 - e^{-1.455}$$

From tables of exponentials:

$$e^{-1.455} = 0.233,$$

and

$$R = 1 - .233 = .767.$$

For static shear strength, $N(27{,}000, 3200)$ psi and shear stress, $N(18{,}400; 18{,}400)$ psi, the reliability would be $R = .677$.

Example H Design with Normal Strength and Static Load Stress Exponentially Distributed

1. Consider an AISI-1117 cold-drawn steel bolt loaded in single shear (Figure 9.24), with strength normally distributed. What cross-sectional area (A) is required to avoid yielding in shear? Since this bolt will not be acting as a single fastener, $R = .999$ is sufficient, and the criterion is $R = .999 \leqslant P(S_{sy} > s)$.

2. A random load is applied once. It is exponentially distributed with

$$\bar{L} = 5500 \text{ lb}; \qquad \lambda = \frac{1}{5500} \text{ lb} \qquad \text{(a)}$$

Frictional resistance due to bolt tensile preload is not considered. The model for shear stress is

$$\tau = \frac{L}{A} \qquad \text{(b)}$$

and from equations (2.59) and (b):

$$\bar{\tau} \approx \frac{\bar{L}}{A} = \bar{s}$$

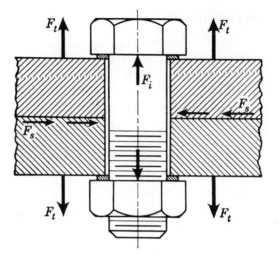

Figure 9.24 Typical bolted connection.

3. Given $S_{sy} \approx 0.577 S_y$; shear yield strength (S_{sy}) may be estimated as follows. From data in Table 8.1; the tensile yield strength is

$$\left(\bar{S}_y, {}_{\vartriangleright S_y}\right) = (81.4, 4.71) \text{ ksi}$$

$$\left(\bar{S}_{sy}, {}_{\vartriangleright S_{sy}}\right) \approx 0.577(81.4, 4.71) \text{ ksi} \approx (47.0, 2.72) \text{ ksi} \tag{c}$$

4. Equation (9.13) provides the following expression:

$$P_f = e - \frac{1}{2}\left[\frac{2\bar{S}}{\bar{s}} - \frac{\sigma_S^2}{(\bar{s})^2}\right]; \quad \sigma_S = {}_{\vartriangleright S_{sy}}$$

Then

$$2\ln P_f \cdot (\bar{s})^2 + 2 \cdot \bar{S}(\bar{s}) - \sigma_S^2 = 0 \tag{d}$$

Substitution of equation (b) into equation (d) yields

$$2\left(\frac{\bar{L}}{\bar{A}}\right)^2 \ln(P_f) + 2\bar{S} \cdot \left(\frac{\bar{L}}{\bar{A}}\right) - \sigma_S^2 = 0$$

$$(2\bar{L}^2)\ln(P_f) + 2 \cdot \bar{S} \cdot \bar{L} \cdot \bar{A} - \sigma_S^2 \cdot \bar{A}^2 = 0 \tag{e}$$

Since $R = .999$, $P_f = 1 - R = 0.001$, and $\ln P_f^* = 2.30 \cdot L_{10}(0.001) = -6.91$.
5. Solve the quadratic equation (e) for A:

$$-2(6.918) \cdot (5.5)^2 + 2(47.0) \cdot (5.5)\bar{A} - (2.72)^2 \cdot \bar{A}^2 = 0$$

$$\bar{A}^2 - 68.84\bar{A} + 56.49 = 0$$

$$\bar{A} = 0.818 \text{ in.}^2$$

9.3.6 Additional Stress–Strength Combinations (Distributions)

1. γ *Stress and normal strength.* Utilization of equation (9.6a) yields

$$R = \int_{-\infty}^{\infty} f(S) \left[\int_{0}^{S} f(s) ds \right] dS \qquad (a)$$

The probability density functions (from Figure 2.1) are as follows. Strength:

$$f(S) = \frac{1}{\sqrt{2\pi}\sigma} \exp -\tfrac{1}{2}\left(\frac{S-\mu}{\sigma}\right)^2 (-\infty, \infty) \qquad (b)$$

Stress:

$$f(x) = \frac{1}{\beta^{\delta+1}\Gamma(\delta+1)} x^\delta e^{-x/\beta} (0, \infty) \qquad (c)$$

Substitution of equations (b) and (c) into equation (a) and integration yields

$$R = 1 - \frac{1}{\sqrt{2\pi}\sigma} \sum_{i=0}^{\delta} \frac{1}{\beta^i (i)!} \int_{-\infty}^{\infty} S^i \exp -\tfrac{1}{2}\left(\frac{S-u}{\sigma}\right)^2 - \frac{S}{\beta} \, dS \qquad (9.14)$$

For $\delta = 1$ and $i = 1$, utilization of equation (9.14) yields

$$R = 1 - \exp -\tfrac{1}{2}\left[\frac{2\mu}{\beta} - \left(\frac{\sigma}{\beta}\right)^2\right] - \frac{1}{\sqrt{2\pi}\sigma}$$

$$\cdot \frac{1}{\beta^1(1)!} \int_{-\infty}^{\infty} S \exp\left[-\tfrac{1}{2}\left(\frac{S-\mu}{\sigma}\right)^2 - \frac{S}{\beta} \right] dS \qquad (d)$$

Integration of equation (d) yields

$$R = 1 - \left(1 + \frac{\mu}{\beta} - \frac{\sigma^2}{\beta^2}\right) \cdot \exp\left[\left(\frac{\sigma^2}{2\beta^2}\right) - \left(\frac{\mu}{\beta}\right)\right] \qquad (9.15)$$

*Natural log of $x = \ln(x)$; log to base 10 of $x = L_{10}(x)$.

where [9]

$$\mu = \beta(1+\delta); \quad \sigma = \beta\sqrt{1+\delta}; \quad \delta = 1.$$

2. *Strength and stress—lognormally distributed.*

$$R = \int_{Z_1}^{\infty} \phi(Z)\,dZ$$

$$Z_1 = -\frac{\ln \check{S} - \ln \check{s}}{\sqrt{\sigma_{\ln S}^2 + \sigma_{\ln s}^2}}$$

where $\check{S}$ and $\check{s}$ are the medians of the strength and stress distributions, respectively, and $\sigma_{\ln S}$, and $\sigma_{\ln s}$ are the standard deviations for $\ln(S)$ and $\ln(s)$ respectively.

3. *Strength—Weibull Distribution with Parameters (β, Θ, S_0) and Stress—Normally Distributed $N(\mu, \sigma)$**

$$R = 1 - \int_{Z_1}^{\infty} \phi(Z)\,dZ + \frac{1}{\sqrt{2\pi}} \left[\frac{\Theta - S_0}{\sigma}\right] \int_0^{\infty}$$

$$\cdot \exp\left\{-y^\beta - \frac{1}{2}\left[\left(\frac{\Theta - S_0}{\sigma}\right)y + \frac{S_0 - \mu}{\sigma}\right]\right\}^2 dy$$

where

$$Z_1 = \frac{S_0 - \mu}{\sigma} \qquad y = \frac{s - S_0}{\Theta - S_0}$$

4. *Strength—Weibull distributed with parameters (β_S, Θ_S, S_0) and stress with Weibull distribution (β_s, Θ_s, s_0).*

$$R = 1 - \int_0^{\infty} e^{-y} \exp\left\{-\left[\frac{\Theta'_S}{\Theta'_s}(y)^{1/\beta_s}S + \frac{S_0 - s_0}{\Theta'_s}\right]^{\beta/s}\right\} dy$$

where

$$\Theta'_S = \Theta_S - S_0, \Theta'_s = \Theta_s - s_0; \qquad y = \left[\frac{S - S_0}{\Theta'_S}\right]^{\beta/S}$$

*K. Kapur and L. Lamberson, *Reliability in Engineering Design*, Wiley, New York 1977.

5. Strength—normal $N(\bar{S}, \hat{s}_S)$ and stress—Rayleigh distributed, σ_s.

$$\mu_s = E(s) = \frac{(\sigma_s^2 \pi)^{1/2}}{\sqrt{2}}$$

$$V(s) = E(s - \mu_s)^2 = 0.429\sigma_s^2$$

$$R = 1 - \frac{\sigma_s}{\sqrt{\hat{s}_S^2 + \sigma_s^2}} \cdot \exp -\tfrac{1}{2} \frac{\bar{S}}{\hat{s}_S^2 + \sigma_s^2}$$

In some cases reliability can be computed from a closed-form expression. In other cases reliability is expressed in integral form, and the integral can be evaluated in terms of other well-known and tabulated functions. For cases 3 and 4, the integration is by numerical methods, that is, according to Simpson's rule [96].

Although the engineer encounters a number of mechanical analysis and design problems in which loading is constant in time, the mechanical problems involving one or another type of dynamic loading described in Chapter 3 are more numerous. The following sections in this chapter are concerned with loading phenomena that are dynamic in nature.

DESIGN FOR DYNAMIC LOADING

9.3.7 Consideration of Cycles to Failure

The distribution of cycles to failure for a given type of steel specimen or component at a fixed stress amplitude (in fully reversed bending) has been found to be modeled by a lognormal distribution. The probability density function of cycles to failure is [97]

$$f(N') = \frac{1}{\hat{s}_{N'}\sqrt{2\pi}} \exp \tfrac{1}{2} \left(\frac{N' - \hat{\bar{N}}'}{\hat{s}_{N'}} \right)^2 \tag{a}$$

where N represents cycles to failure, $N' = L_{10}(N)$, $\hat{\bar{N}}$ is the mean of N' values, and $\hat{s}_{N'}$ is standard deviation of N' values. Integration of equation (a) from $-\infty$ to a specified value N_1 (Appendix 1) yields the probability of failure $P_f(N_1)$ in less than N_1 cycles

$$P_f(N_1) = P_f(N_1') = \int_{-\infty}^{N_1'} \frac{1}{\hat{s}_{N'}\sqrt{2\pi}} \exp -\tfrac{1}{2} \left(\frac{N' - \hat{\bar{N}}'}{\hat{s}_{N'}} \right)^2 dN' \tag{b}$$

Design and Analysis

Reliability is given by

$$R(N_1) = R(N_1') = 1 - \int_{-\infty}^{N_1'} \frac{1}{\hat{s}_{N'}\sqrt{2\pi}} \exp\left[-\frac{1}{2}\left(\frac{N' - \overline{N'}}{\hat{s}_{N'}}\right)^2\right] dN'$$

or

$$R(N_1) = \int_{N_1'}^{\infty} \frac{1}{\hat{s}_{N'}\sqrt{2\pi}} \exp\left[-\frac{1}{2}\left(\frac{N' - \overline{N'}}{\hat{s}_{N'}}\right)^2\right] dN' = \int_{z_1}^{\infty} \phi(z)\,dz$$

Since the lognormal distribution of cycles to failure (N') plots as a normal distribution, it is feasible to integrate under the probability density function curve by transforming to the standard normal probability density function, given by

$$\phi(z) = \frac{1}{\sqrt{2\pi}} e^{-z^2/2}; \qquad z = \frac{N' - \overline{N'}}{\hat{s}_{N'}}$$

Example I

The probability of exceeding a specified cycle life is discussed in this example. Assume a spindle of AISI 4340 cold-drawn steel that is to function as a flexible shaft, subjected to a constant amplitude fully reversed stress of 89,000 psi.

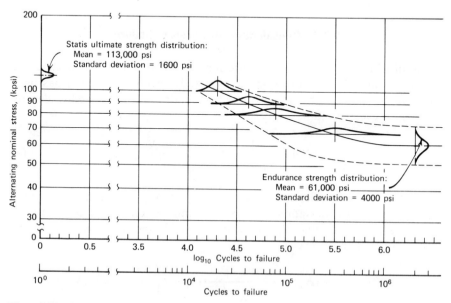

Figure 9.25 Cycles to failure and endurance strength for SAE 4340 wire specimens of 0.0625-in diameter [97].

Table 9.7 Wire Fatigue Machine Results for 0.0625-in.-diameter[a] SAE 4340 Steel Wire

Maximum Alternating Stress Level (psi)	Cycles to Failure			
	Normal Distribution		Lognormal Distribution	
	Mean	Standard Deviation	Mean	Standard Deviation
68,000	356,900	193,400	5.497	0.218
80,800	84,900	34,000	4.895	0.174
89,000	43,900	13,100	4.623	0.130
101,000	21,100	3,700	4.317	0.076

[a] Endurance strength distribution (normal) parameters at 2×10^6 cycles; mean; 61,000 psi; standard deviation; 4000 psi.
Source: Kececioglu and Haugen [97].

What is the probability, $R(N_1') = R(15,800)$, of surviving 15,800 cycles of loading? See Figure 9.25 and Table 9.7.

Solution.

$$N_1' = L_{10}(15,800) = 4.200$$

From Table 9.7, corresponding to a stress amplitude of 89,000 psi, the log statistics are

$$\hat{\overline{N'}} = 4.623$$

$$\hat{s}_{N'} = 0.130$$

The normalizing transformation is

$$z = \frac{N_1' - \hat{\overline{N'}}}{\hat{s}_{N'}} = \frac{4.200 - 4.623}{0.130} = -3.254$$

If Appendix 1 is utilized with equation (9.10), the estimated probability of survival is

$$R(N_1') = R(15,800) = \frac{1}{\sqrt{2\pi}} \int_{-3.254}^{\infty} e^{-z^2/2} dz = 0.9994$$

9.3.8 Design for Endurance

In the design of mechanical components to survive an indefinite number of direct loading or bending cycles, the critical stress is the endurance-limit distribution (ferrous materials) or the fatigue-limit distribution (nonferrous materials). In this section the loadings are limited to constant-amplitude

400 Design and Analysis

repeated and constant-amplitude fully reversed sinusoidal regimes. The materials behavior relevant in the examples are the endurance limit S_e in bending and the endurance limit in shear S_{se} of C1035 steel. More involved examples of fatigue are presented in subsequent sections and in Chapter 10.

Wire-rope Sheave Design. To illustrate design for unlimited life, consider wire rope used in hoisting and conveying equipment. In the following example the primary design variables are assumed to be normally distributed. Thus the design procedure for S and s normal decribed in Section 9.3.2 is used here. The following is a preliminary design exercise.

Example J

1. Flexibility in wire rope is secured by using a large number of small-diameter wires. The 6×19 and 6×37 configurations shown in Figure 9.26 are most widely used. Failure of a wire rope occurs as the result of
   (a) fatigue and
   (b) wear in passing over the sheave. Only design to avoid fatigue failure caused by flexing and unflexing of the rope in passing over a sheave is addressed in this example. Each passage is counted as one stress cycle.

The pitch diameter d_s of a sheave is estimated such that the probability of fatigue failure of the rope will be 0.005; that is, the design criterion will be $R = 0.995 \leqslant P(S_e > s)$:
For plow steel:

$$\left(\bar{S}_e, {}_{^3S_e}\right) \approx 0.5\left(\bar{S}_u, {}_{^3S_u}\right) = (87.5, 6.0) \text{ ksi} \tag{a}$$

where ${}_{^3S_u} \approx 12$ ksi is conservative (Table 8.1).

2. The model from which the stress statistics is estimated is given in Spotts [49]. Equation (c) is a crude estimator of s. From equation (6.1):

$$\frac{M}{EI} = \frac{1}{r} = \frac{2^*}{d_s} \tag{b}$$

Figure 9.26 Cross section of typical wire rope.

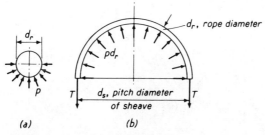

Figure 9.27 Forces in wire rope passing over a sheave.

where r is the radius of curvature. From equation (5.23):

$$s = \frac{Mc}{I} = \frac{Md_w}{2I} \tag{c}$$

where $c = (d_w/2)$. Combination of equations (b) and (c) yields

$$s = \frac{Ed_w}{d_s} \tag{d}$$

where

d_w = Wire diameter = (0.040, 0.003) in. (from Chapter 4)

d_s = Sheave pitch diameter (Figure 9.27)

$(\bar{E}, s_E) = (12 \cdot 10^6, 3.33 \cdot 10^5)$ psi (Tables 9.8 and 5.1)

Table 9.8

Type	Metallic Cross-Sectional Area (d_r = diam rope) in.²	Diameter of Outer Wire, d_w in.	Minimum Diameter d_s for Sheaves for Steel Rope in.	Modulus of Elasticity of Steel Rope psi	Breaking Stress Type of Material	psi
6 × 7	$0.380 d_r^2$	$1/9 d_r$	$42 d_r$	14,000,000	Improved plow steel	200,000
6 × 19	$0.404 d_r^2$	$1/16 d_r$*	$24 d_r$	12,000,000	Plow steel	175,000
6 × 37	$0.404 d_r^2$	$1/22 d_r$	$18 d_r$	11,000,000	Extra-strong cast steel	160,000
8 × 19	$0.352 d_r^2$	$1/19 d_r$	$20 d_r$	10,000,000	Cast steel	140,000
					Iron	65,000

*Filler wire type. For Warrington type large outer wires, $d_w = 1/14$.

Application of equations (2.25) and (2.27) to the product $E d_w$ in equation (d) yields

$$\left(\bar{d}_w, s_{d_w}\right) \cdot \left(\bar{E}, s_E\right) \approx (4.8 \cdot 10^5, 3.65 \cdot 10^4) \tag{e}$$

Stress amplitude in the wires passing over the sheave is estimated by substituting equation (e) into equation (d):

$$\left(\bar{s}, s_s\right) = \frac{(4.8 \cdot 10^5, 3.65 \cdot 10^4)}{\left(\bar{d}_s, s_{d_s}\right)} \tag{f}$$

In equation (f) applicable machining tolerances indicate that

$$s_{d_s} \approx 0.05 \, \bar{d}_s \text{ in.}$$

Application of equations (2.28) and (2.30) to equation (f) yields

$$\left(\bar{s}, s_s\right) = \left[\frac{4.8 \cdot 10^5}{\bar{d}_s}, \frac{4.329 \cdot 10^4}{\bar{d}_s}\right] \text{ psi} \tag{g}$$

3. The models of endurance limit (S_e) are the statistics of stress at failure associated with experiencing and surviving an unlimited number of cycles of loading [equation (a)]:*

$$\left(\bar{S}_e, s_{S_e}\right) \approx (87.5, 6.0) \text{ ksi}$$

4. From the design criterion and equation (9.10):

$$R = 0.995 \leq \frac{1}{\sqrt{2\pi}} \int_z^\infty e^{-t^2/2} dt = \frac{1}{\sqrt{2\pi}} \int_{-2.57}^\infty e^{-t^2/2} dt$$

Thus z taken from Appendix 1 is $z = -2.57$.

5. Substitution of equations (g) and (a) into equation (9.11) yields

$$z = -\frac{\bar{S}_e - \bar{s}}{\sqrt{(s_{S_e})^2 + s_s^2}} = -2.57 = -\frac{87{,}500 - (4.8 \cdot 10^5 / \bar{d}_s)}{\sqrt{(6000)^2 + \left(\frac{4.327 \cdot 10^4}{\bar{d}_s}\right)^2}}$$

Expansion and simplification yields

$$0 - \bar{d}_s^2 - 11.5 \bar{d}_s + 29.8 \tag{h}$$

*Since a rough estimate of S_e will be used in the preliminary design, some testing is required to validate the endurance limit statistics prior to finalizing.

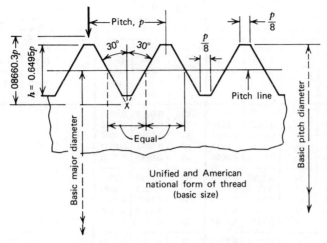

Figure 9.28 Screw-thread geometry.

The solution of equation (h) for sheave pitch diameter is

$$(\bar{d}_s)_1 = 7.46 \text{ in.}$$

$$(\bar{d}_s)_2 = 3.91 \text{ in.}$$

$$s_{d_s} \approx 0.05(\bar{d}_s) = 0.12 \text{ in.}$$

$$\pm \Delta'_{d_s} = \pm 3(0.12) = \pm 0.36 \text{ in.}$$

Fastener Design for Endurance. A second example of design for endurance will involve a bolted connection.

A knowledge of the elementary methods of fastening is assumed. Typical methods of fastening and joining parts include the use of bolts, nuts, cap screws, set screws, rivets, spring retainers, locking devices, and keys.

Details and tables of dimensions and tolerances of the various thread forms used in mechanical design are found in engineering and machinery handbooks and industrial publications [98]. The geometry of common thread forms is shown in Figure 9.28.

In practice, tables such as Table 9.9 are utilized in selecting and designing threaded connectors, illustrated in Figures 9.30 through 9.33.

Example K Rigid Bolted Joint Design for Endurance [75]

1. In this part of Example K we determine bolt shank requirements for a rigid joint. Stress is dynamic shear (τ) fully reversed. The design procedure for S and s normal outlined in Section 9.3.2 is utilized. In design for (shear)

---

*A tolerance $\Delta_{d_s} = 0.05$ in. is more reasonable.

Table 9.9

Size	Outside or Major Diameter, in.	Coarse-Thread Series			Fine-Thread Series			Extra-Fine-Thread Series			Hex. Nut Width Across Flats
		Threads per in.	Basic Pitch Diameter, in.	Stress Area, in.^{2}	Threads per in.	Basic Pitch Diameter, in.	Stress Area, in.^{2}	Threads per in.	Basic Pitch Diameter, in.	Stress Area, in.^{2}	
0	0.0600				80	0.0519	0.0018				
1	0.0730	64	0.0629	0.0026	72	0.0640	0.0028				
2	0.0860	56	0.0744	0.0037	64	0.0759	0.0039				
3	0.0990	48	0.0855	0.0049	56	0.0874	0.0052				
4	0.1120	40	0.0958	0.0060	48	0.0985	0.0066				
5	0.1250	40	0.1088	0.0080	44	0.1102	0.0083				
6	0.1380	32	0.1177	0.0091	40	0.1218	0.0102				
8	0.1640	32	0.1437	0.0140	36	0.1460	0.0147				
10	0.1900	24	0.1629	0.0175	32	0.1697	0.0200				
12	0.2160	24	0.1889	0.0242	28	0.1928	0.0258	32	0.1957	0.0270	
1/4	0.2500	20	0.2175	0.0318	28	0.2268	0.0364	32	0.2297	0.0379	7/16
5/16	0.3125	18	0.2764	0.0524	24	0.2854	0.0580	32	0.2922	0.0625	1/2
3/8	0.3750	16	0.3344	0.0775	24	0.3479	0.0878	32	0.3547	0.0932	9/16
7/16	0.4375	14	0.3911	0.1063	20	0.4050	0.1187	28	0.4143	0.1274	5/8
1/2	0.5000	13	0.4500	0.1419	20	0.4675	0.1599	28	0.4768	0.170	3/4
9/16	0.5625	12	0.5084	0.182	18	0.5264	0.203	24	0.5354	0.214	13/16
5/8	0.6250	11	0.5660	0.226	18	0.5889	0.256	24	0.5979	0.268	15/16
3/4	0.7500	10	0.6850	0.334	16	0.7094	0.373	20	0.7175	0.386	1-1/8
7/8	0.8750	9	0.8028	0.462	14	0.8286	0.509	20	0.8425	0.536	1-5/16
1	1.0000	8	0.9188	0.606	12	0.9459	0.663	20	0.9675	0.711	1-1/2
1-1/8	1.1250	7	1.0322	0.763	12	1.0709	0.856	18	1.0889	0.901	1-11/16
1-1/4	1.2500	7	1.1572	0.969	12	1.1959	1.073	18	1.2139	1.123	1-7/8
1-3/8	1.3750	6	1.2667	1.155	12	1.3209	1.315	18	1.3389	1.370	2-1/16
1-1/2	1.5000	6	1.3917	1.405	12	1.4459	1.581	18	1.4639	1.64	2-1/4
1-3/4	1.7500	5	1.6201	1.90				16	1.7094	2.24	2-5/8
2	2.0000	4½	1.8557	2.50				16	1.9594	2.95	3
2-1/4	2.2500	4½	2.1057	3.25							3-3/8
2-1/2	2.5000	4	2.3376	4.00							3-3/4
2-3/4	2.7500	4	2.5876	4.93							4-1/8
3	3.0000	4	2.8376	5.97							4-1/2
3-1/4	3.2500	4	3.0876	7.10							4-7/8
3-1/2	3.5000	4	3.3376	8.33							5-1/2
3-3/4	3.7500	4	3.5876	9.66							5-5/8
4	4.0000	4	3.8376	11.08							6

ASA B 1.1—1960, N. Y.

endurance the criterion is (see Figure 8.34)

$$P(S_{se} = 0.577 \cdot S_e > \tau) \geq R$$

Assumptions.

a. Existing functional models for stress and strength provide sufficiently precise predictions.

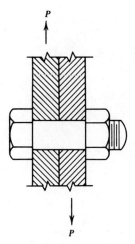

Figure 9.29 Connection with one shear plane.

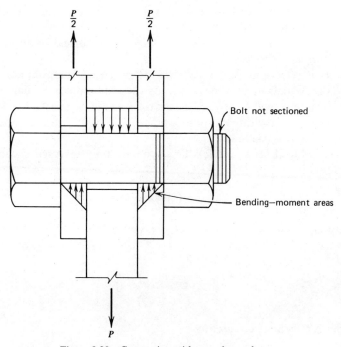

Figure 9.30 Connection with two shear planes.

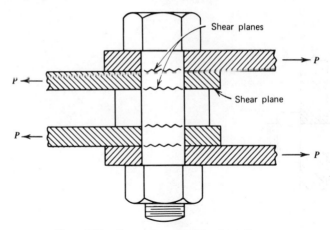

Figure 9.31 Connection with four shear planes.

b. The strength of connecting members, not entering the shank design, is assumed to be sufficient.
c. The design variables are considered to be normal or approximately normally distributed.
d. Bolt selection (Table 9.9) will be based on the common fastener that equals or that least exceeds the required shank diameter.
e. The joint will have $n = 2$ shear planes (Figure 9.30).
f. Consistent with good design practice, only the unthreaded shank of the bolt is loaded.

2. Load-induced stress model. The dynamic fully reversed shear loading (amplitude) L is

$$(\bar{L}, s_L) = (2000, 330) \text{ lb}$$

Since loading L and cross section of the bolt A are independent (random) variables, dependent (random) variable τ is estimated by using equation (5.1):

$$(\bar{\tau}, s_\tau) = \frac{(\bar{L}, s_L)}{(\bar{A}, s_A)} \text{ psi} \qquad (a)$$

The model for shear stress area is

$$A = \frac{\pi d^2}{4} \text{ in.}^2$$

where d is the shank diameter (inches). From Appendix 5 the shear area statistics are

$$(\bar{A}, s_A) = \left(\frac{\pi \bar{d}^2}{4} ; \frac{\pi \bar{d} s_d}{2} \right) \text{ in.}^2$$

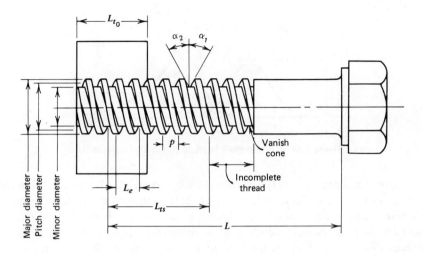

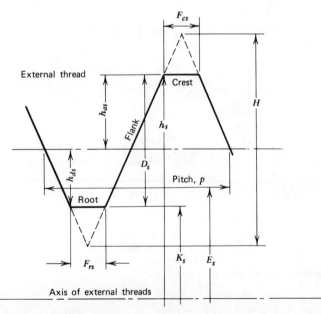

Figure 9.32 Thread terminology and symbols: $K_s = D_s - 2h_s = D_s - 1.22687p = D_s - \dfrac{1.22687}{N}$; D_s = major diameter of external thread; $E_s = D_s - 2h_{as} - D_s - 0.64952p = D_s - \dfrac{0.64952}{N}$; h_s = height of external thread = $0.61343p = \dfrac{0.61343}{N}$; H = height of V thread = $0.86603p = \dfrac{0.86603}{N}$; $\dfrac{3}{16}H = \dfrac{h_{as}}{2} = 0.32476 = 0.16238p = \dfrac{0.162}{N}$.

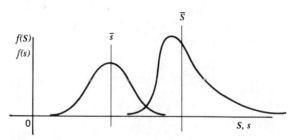

Figure 9.33 Right-skewed strength (reliability increased).

To solve for area uniquely, a relationship involving the shank diameter variability is needed. Since tolerances on bolt geometry are usually based on diameter, pitch, and length of thread engagement, there is considerable variation in tolerances, dependent on nominal size, class of fit, and series of thread. For standard series (unified and American), screw geometry values are available in Table 9.9, from which mean outside diameter d may be estimated.

Based on UNC thread series and class 1A fit, the average tolerance is $\Delta \approx 0.0056\bar{d}$. According to the rationale in Section 3.3.2:

$$s_d \approx \frac{0.0056\bar{d}}{3} \approx 0.002\bar{d}$$

In single shear:

$$(\bar{\tau}, s_\tau) = \frac{4(\bar{L}, s_L)}{\pi(\bar{d}^2, 0.004\bar{d}^2)} \tag{b}$$

From equations (2.59) and (b), in double shear:

$$\bar{\tau} \approx \frac{4}{n\pi} \cdot \left(\frac{\bar{L}}{\bar{d}^2}\right) \text{ psi} = \left(\frac{8000}{n\pi\bar{d}^2}\right) \text{ psi} \tag{b'}$$

and from equations (2.30) and (b):

$$s_\tau \approx \frac{1}{\pi n} \left[\frac{\bar{L}^2 \cdot (0.004\bar{d}^2) + (\bar{d}^2)^2 \cdot s_L^2}{\bar{d}^8}\right]^{1/2} \approx \left(\frac{330}{n\pi\bar{d}^2}\right) \text{ psi} \tag{b''}$$

Equations (b') and (b'') provide estimators of the mean value and standard deviation of applied stress amplitude τ in terms of shank diameter.

3. In this problem the shear endurance statistics of C1035 cold-drawn steel are required. Assume that the only available data relate to tensile ultimate strength from Appendix 10.A:

$$(\bar{S}_u, s_{S_u}) \approx (92.1, 6.1) \text{ ksi}$$

Equation (8.6b) can be used to estimate tensile endurance limit, for purposes of preliminary design:

$$(\bar{S}_e, {}^s S_e) \approx (0.498, 0.051)(\bar{S}_u, {}^s S_u)$$
$$= (46.0, 5.6) \text{ ksi}$$

The empirical relationship in equation (8.7) can be used for a preliminary estimate of shear endurance limit $S_{se} \approx 0.577(S_e)$:

$$(\bar{S}_{se}, {}^s S_{se}) \approx 0.577(46.0, 5.6) = (26.5, 3.23) \text{ ksi}$$

4. Due to safety considerations, it is known that for public acceptance, fewer than 2 in 10^4 fastener breakdowns caused by this bolt can be tolerated. Thus specified reliability must be $R = .9998$. From equation (9.10) and Appendix 1:

$$R = 0.9998 \leqslant \frac{1}{\sqrt{2\pi}} \int_z^\infty e^{-t^2/2} dt = \frac{1}{\sqrt{2\pi}} \int_{-3.50}^\infty e^{-t^2/2} dt$$

5. Substitution of equations (b'), (b''), and (c) into equation (9.11), where $z = -3.50$, yields

$$z = -\frac{\bar{S}_{se} - \bar{\tau}}{\sqrt{{}^s S_{se}^2 + {}^s \tau^2}} = -3.50 = -\frac{\bar{S}_{se} - \dfrac{4\bar{L}}{n\pi \bar{d}^2}}{\left[{}^s S_{se}^2 + \left(\dfrac{660}{\pi \bar{d}^2}\right)^2\right]^{1/2}}$$

$$-3.5 = -\frac{26{,}500 - \dfrac{8000}{n\pi \bar{d}^2}}{\sqrt{(3230)^2 + \left(\dfrac{660}{\pi \bar{d}^2}\right)^2}} \tag{d}$$

Expand and solve equation (d) as a quadratic in $(\bar{d})^2$:

$$\bar{d}^2 = 0.0676 \text{ in.}^2$$
$$\bar{d} = 0.26 \text{ in.}$$
$${}^s d \approx 0.005 \text{ in.} = 0.002 \bar{d} \text{ in.}$$
$$\Delta_d = 3 {}^s d = 0.0017 \text{ in.}$$

410 Design and Analysis

According to Table 9.9, the closest common fastener configuration satisfying $\bar{d}$ is

$$\tfrac{5}{16}\text{ dia}-18-\text{UNC}-1\Lambda$$

(*Note:* Had the bolt been subjected to dynamic tensile loading, the minor diameter would have been critical as would stress concentration due to thread geometry.) The shear area is

$$(\bar{A}, \triangleright_A) = (\pi \bar{r}^2, 2\pi(0.004)\bar{r}^2) = (0.0531, 0.00042) \text{ in.}^2$$

The shear stress amplitude is

$$(\bar{\tau}, \triangleright_\tau) = \frac{(1000, 165)}{(0.0531, 0.00042)} = (18{,}330, 3110) \text{ psi} \qquad (e)$$

In this second part of Example K consider the consequences of superimposing the fully reversed dynamic shear stressing, [equation (e)] on a range of levels of tensile prestress (see Table 9.10).

1. Based on dynamic shear loading $(\bar{L}, \triangleright_L) = (2000, 330)$ lb acting on two shear planes, to satisfy $R = .9998$, the diametral requirements were $(\bar{d}, \triangleright_d) = (0.26, 0.0005)$ in. As a consequence, the dynamic shear stress amplitude in the shank of the bolt was as given in equation (e). Utilizing the shear area estimated in part 1, assume the following tensile preloads with which $(\bar{\tau}, \triangleright_\tau)$ will be combined:

    1. (212, 21) lb
    2. (531, 53) lb
    3. (1060, 106) lb   (f)

A coefficient of variation $(\triangleright_s / \bar{s}) = 0.10$ is assumed.

Table 9.10 Tensile Prestress and Dynamic Shear Stress Amplitude (in psi)

Tensile Prestress		Dynamic Shear-stress Amplitude	
$(\bar{s}/2) = \bar{x}$	$\triangleright_{s/2} = \triangleright_x$	$\bar{\tau}$	$\triangleright_\tau$
1. 2,000	200	18,830	3110
2. 5,000	500	18,830	3110
3. 10,000	1000	18,830	3100

2. The model for combined tensile prestress with dynamic shear stress is given by equation (5.20):

$$\tau_{max} = \sqrt{x^2 + \tau^2} \tag{g}$$

From equation (2.63), $\bar{\tau}_{max}$ is

$$\bar{\tau}_{max} \cong \sqrt{\bar{x}^2 + \bar{\tau}^2} \tag{g'}$$

and from equation (2.64), $s_{\tau_{max}}$ is

$$s_{\tau_{max}} \approx \sqrt{\frac{\bar{x}^2 \cdot s_x^2 + \bar{\tau}^2 s_\tau^2}{\bar{x}^2 + \bar{\tau}^2}} \tag{g''}$$

The combined stress statistics are estimated by utilizing equations (g') and (g'') with values from Table 9.10:

1. $(\bar{\tau}_{max}, s_{\tau_{max}}) \approx (18{,}940, 3040)$ psi

2. $(\bar{\tau}_{max}, s_{\tau_{max}}) \approx (19{,}480, 3010)$ psi

3. $(\bar{\tau}_{max}, s_{\tau_{max}}) \approx (21{,}320, 2750)$ psi $\quad$ (h)

3. From part 1 of this example, the shear endurance limit S_{se} of C1035 steel was estimated as

$$(\bar{S}_{se}, s_{S_{se}}) \approx (26{,}500, 3230) \text{ psi.} \tag{i}$$

4. Combined stress reliabilities were estimated by utilizing values from equations (h) and (i) with equation (9.11):

$$z = -\frac{\bar{S}_{se} - \bar{\tau}_{max}}{\sqrt{s_{S_{se}}^2 + s_{\tau_{max}}^2}}$$

Table 9.11 Summary—Impact of Combined Stress on R (in psi)

Tensile Prestress		Dynamic Shear		Reliability
$\bar{s}$	s_s	$\bar{\tau}$	s_τ	R
0	0	18,830	3,110	0.9998
2,000	200	18,830	3,110	0.9834
5,000	500	18,830	3,110	0.9783
10,000	1,000	18,830	3,110	0.9152

and from equation (9.10) (with Appendix 1):

$$R = \frac{1}{\sqrt{2\pi}} \int_z^\infty e^{-t^2/2} dt$$

Table 9.11 summarizes the impact on reliability R due to adding varying levels of tensile prestress to the dynamic shear.

9.3.9 Upper Bound on Requirements for a Specified R

In design situations exact distributional forms are sometimes in doubt. However, it is conservative to assume normally distributed strength that is unbounded on the high side (except in the rare case of negatively skewed strength) and lognormally distributed applied stress. These conclusions follow from (1) the central limit theorem for product strings of random variables, such as stress functions, and (2) the argument that follows.* If the distribution of strength were in fact right skewed (Figure 9.33) rather than normal and symmetrical, for a given mean value ($\bar{S}$) and standard deviation (σ_S), the actual reliability of the designed component would (on the average) exceed the required target reliability R. The larger actual reliability would be due to the larger than assumed density in the right tail of the strength distribution. If the stress distribution were in fact skewed right, (Figure 9.34) rather than symmetrical and normal as assumed, for a given mean value $\bar{s}$ and standard deviation σ_s, the actual reliability of the designed component would be less than the required target reliability (Table 9.6). The unsatisfactory reliability would be due to the larger than expected density in the right tail of the stress distribution.

In summary, assumptions leading to an adequate design appear to be associated with strength (assumed to be) normally distributed and stress (assumed to be) lognormally distributed.

Reliability for Stress Normally versus Lognormally Distributed. Consider two possible design situations in each of which the statistics ($\bar{S}, \sigma_S$) and ($\bar{s}, \sigma_s$) are identical. Strength of the material in both cases is normally distributed. However, in the first case stress is normally distributed and in the second case lognormally. Tables 9.12 and 9.13 illustrate the fact that reliability with stress normally distributed invariably exceeds reliability when stress is lognormally distributed.

In actual instances where the statistics of strength and stress can be estimated but the correct distributional models are in doubt, we may assume normal strength and lognormal stress and solve the design problem by numerical methods employing a computer (which may become costly). There is, however, a simpler approach to completing a design. Assume that the

---

*Also see Bayes decision procedures in Section 3.2.2.

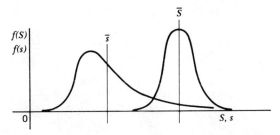

Figure 9.34 Right-skewed stress (reliability reduced).

following are given:

1. Specified reliability R.
2. Strength statistics $(\bar{S}, \sigma_S)$; hence $C_S = (\sigma_S / \bar{S})$.
3. Stress statistics $(\bar{s}, \sigma_s)$; hence $C_s = \sigma_s / \bar{s}$ stated in terms of an unknown.

Now consider the information provided in Figures 9.35 through 9.42. For each figure, strength of the material is of a specific coefficient of variation (C_S^1), and the plotted curves are for reliability versus stress coefficient of variation (C_s^1). Two curves are plotted on each figure (1) the upper curve applies for normally distributed stress, and (2) the lower curve applies for lognormally distributed stress.

When these curves are to be utilized in the design process in a specific instance, a figure must be selected in which $C_S^1 \geq C_S$ (C_S applies to the selected material). Furthermore, referring to Figure 9.37, for example, at the

Table 9.12 Data for $(\mu_S/\mu_s) = 1.25$

σ_s/μ_s	R_{normal}	$R_{lognormal}$
Strength = $(\mu_S, \sigma_S) = (10, 1)$; Stress, $\mu_s = 8$		
.1000	.940824	.856378
.1250	.921349	.825063
.1508	.899083	.800589
.1741	.878282	.785017
.2000	.855426	.762006
.2500	.814452	.679250
σ_s/μ_s	R_{normal}	$R_{lognormal}$
Strength = $(\mu_S, \sigma_S) = (10, 0.5)$; Stress, $\mu_s = 8$		
.1000	.982996	.885652
.1250	.963180	.846207
.1508	.937176	.808406
.1741	.911733	.778269
.2000	.883584	.650165
.2500	.834011	.586355

Source: Haugen and Eckert [22].

Table 9.13 Data for $(\mu_S/\mu_s) = 1.43$

σ_s/μ_s	R_{normal}	$R_{\text{lognormal}}$
\multicolumn{3}{l}{Strength $= (\mu_S, \sigma_S) = (10, 1)$; Stress $= \mu_s = 7$}
.1000	.993007	.961760
.1250	.988018	.931812
.1508	.980451	.909100
.1741	.971478	.887004
.2000	.959394	.891722
.2500	.931678	.846265
σ_s/μ_s	R_{normal}	$R_{\text{lognormal}}$
\multicolumn{3}{l}{Strength $= (\mu_S, \sigma_S) = (10, 0.5)$; Stress, $\mu_s = 7$}
.1000	.999755	.975584
.1250	.998543	.942551
.1508	.994891	.911774
.1741	.988618	.906797
.2000	.978294	.891660
.2500	.950356	

Source: Haugen and Eckert [22].

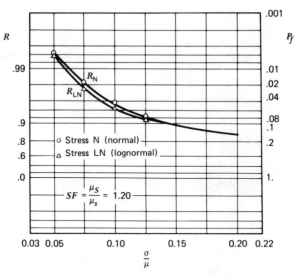

Figure 9.35 Reliability versus stress coefficients of variation. Normal strength $= (\mu_S, \sigma_S) = (10.0000, 0.5000)$; normal and lognormal stress $= \mu_s = 8.3330$; $C_S = 0.05$.

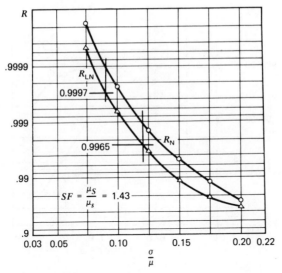

Figure 9.36 Reliability versus stress coefficients of variation. Normal strength $=(\mu_S, \sigma_S)=$ (10.0000, 0.5000); normal and lognormal stress $=\mu_s=7.0000$; $C'_S=0.05$.

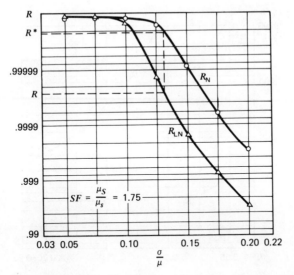

Figure 9.37 Reliability versus stress coefficients of variation. Normal strength $=(\mu_S, \sigma_S)=$ (10.0000, 0.50000); normal and lognormal stress $=\mu_s=5.7140$; $C_S=0.50$.

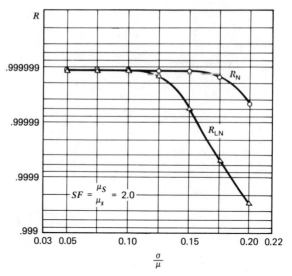

Figure 9.38 Reliability versus stress coefficients of variation. Normal strength $=(\mu_S, \sigma_S) = 10.0000, 0.5000$; normal and lognormal stress $= \mu_s = 5.0000$; $C'_S = 0.05$.

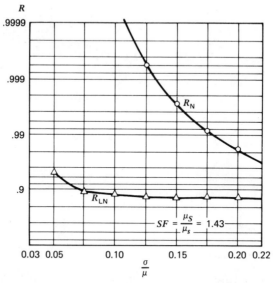

Figure 9.39 Reliability versus stress coefficients of variation. Normal strength $=(\mu_S, \sigma_S) = (10.0000; 0.3000)$; normal and lognormal stress $= \mu_s = 7.0000$; $C'_S = 0.03$.

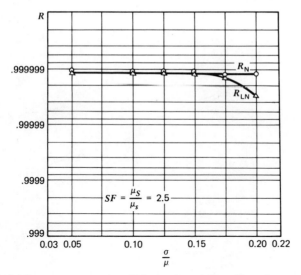

Figure 9.40 Reliability versus stress coefficients of variation. Normal strength $=(\mu_S, \sigma_S)=$ (10.0000; 0.5000); normal and lognormal stress $= \mu_s = 4.0000$; $C'_S = 0.05$.

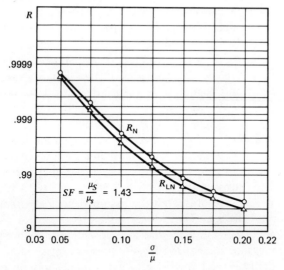

Figure 9.41 Reliability versus stress coefficients of variation. Normal strength $=(\mu_S, \sigma_S)=$ (10.0000, 0.75); normal and lognormal stress $= \mu_s = 7.0000$; $C'_S = 0.075$.

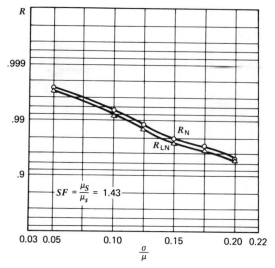

Figure 9.42 Reliability versus stress coefficients of variation. Normal strength $=(\mu_S, \sigma_S)=$ (10.0000, 1.0000); normal and lognormal stress $= \mu_s = 7.0000$; $C'_S = 0.100$.

specified value of R, C_s^1 (on the curve for lognormal stress) must equal or exceed C_s of the design stress expression.

It can now be seen that if the design is carried out by using theory applying to the case of normally distributed strength and normally distributed stress to satisfy R^* (Section 9.3.2, subsection on design procedure with S and s normal) the result will be the same as if the design were for normally distributed strength and lognormally distributed stress to satisfy the reliability R.

Design with Distribution of S and s Unspecified. Distributional strength and stress models that result in designs that equal or exceed specified target reliability (for given means and standard deviations) were discussed in Section 9.3.9. In the subsection in Section 9.3.9 in on fixed strength statistics, reliability–C'_s curves were presented for assumed normal strength in combination with normal stress and with lognormal stress.

The discussions in Section 9.3.9 and the subsection therein on reliability as a performance measure suggest the feasibility of a probabilistic design procedure not dependent on the knowledge of exact distributional forms of S and s. At a given stress (s) coefficient of variation C'_s (see Figure 9.35 through 9.42), lognormally distributed stress results in a value of reliability on the LN curve. At the same C'_s, normally distributed stress results in a value of reliability on the N curve. Usually $R_N > R_{LN}$. Strength is assumed normally distributed in both cases.

In a design situation, reliability is specified and C_s for stress can be estimated. Assuming that stress is lognormally distributed, a point is located on the appropriate LN curve at the specified R. At the same stress coefficient of variation, the point directly above on the N curve identifies the value R.* Estimated C_s must equal or be less than the abscissa value C_s' in the figure (i.e., indicate less scatter).

For a given set of stress and strength statistics, a design satisfying a specified reliability R can be achieved by treating both strength and stress as normally distributed and designing for a higher than specified fictitious R,* with the advantage of utilizing well-developed probabilistic design theory and methodology of the normal strength–normal stress case. The design procedure is as follows:

1. Treat strength as normally distributed and stress as lognormally distributed, although the actual distributions may be unknown.
2. Write the expressions for the mean values, $\bar{S}$ and $\bar{s}$ and the standard deviations, σ_S and σ_s, of strength and stress by utilizing the algebra of normal functions (Section 2.6). If loading is dynamic, estimate strength (stress at failure) from the appropriate S–N envelope, by interpolation, and then estimate C_S needed to select a graph in Section 9.3.9 (subsection on reliability as a performance measure).
3. Estimate the coefficient of variation of stress C_s from the stress function statistics.
4. At specified reliability R locate a point on the appropriate LN curve and verify that on the graph, $C_s' \geqslant C_s$.
5. At the value C_s' on the graph, locate the point on the N curve directly above LN and then determine the value R^* associated with it, noting that $R^* \geqslant R$.
6. Utilize the fictitious value R^* with the expressions for stress and strength determined in step 2 to carry out the design synthesis by the normal theory and methodology procedure for S and s normal in Section 9.3.2.

In the following example the design procedure outlined in the preceding list is used. The component must satisfy or exceed a specified finite cycle life, that is, $n = 31{,}600$ cycles in this case, with a specified probability R.

Finite Life Design with Distributions of S and s Unknown. In Example L loading is dynamic and the distributions of S and s are only known to be continuous and unimodal.

Example L

1. Preliminary assumptions are as follows. Assume that line AC in Figure 9.43 is straight, deflection and strain are negligible, and angular relationships do not change due to loads.

Only the tensile case is considered since it is assumed that columnar buckling is not a likely mode of failure. A probability of .994 is sufficient for surviving a required $n = 31{,}600$ cycles of loading since safety is not a consideration. The

420 Design and Analysis

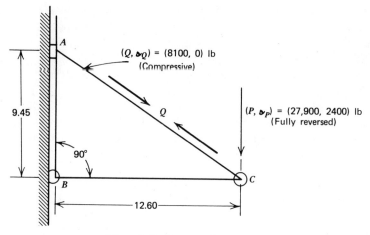

Figure 9.43 Two-member frame.

material selected for the frame members is C1045 steel, water quenched at 1520°F and tempered at 1210°F. The design criterion is

$$R = 0.994 \leqslant P(S_f > s)$$

From the given information and that in Figure 9.43, estimate the required mean value and the standard deviation of the cross section in member AC.

2. Estimate the load-induced stress. Since the distribution of stress being unknown it is treated as approximately lognormal. Load P is essentially fully reversed with a small element of force peak random variation, due to transients, startups, shutdowns, and so on. Load Q is in phase with P. Only the tensile stresses are assumed to contribute to fatigue.

a. Loading in member AC is as follows. The pinned frame is a right triangle. From equation (2.8):

$$\bar{P}_{AC} = \frac{5}{3} \cdot (27,900) = 46,300 \text{ lb}$$

From equation (2.10):

$$s_{P_{AC}} = \frac{5}{3} \cdot (2400) = 3840 \text{ lb} \qquad (a)$$

From Figure 9.43:

$$(\bar{Q}, s_Q) = -(8100, 0) \text{ lb}$$

Then

$$(\bar{L}, s_L) = (\bar{P}_{AC}, s_{P_{AC}}) - (\bar{Q}, 0) \tag{b}$$

Tensile peak load amplitude (L) is estimated from equations (2.40) and (2.41):

$$\bar{L} = \bar{P}_{AC} - \bar{Q} = 46{,}300 - 8100 = 38{,}200 \text{ lb}$$

$$s_L = \sqrt{s_{P_{AC}}^2 + s_Q^2} = 3840 \text{ lb.}$$

and, in couple notation, the loading amplitude is

$$(\bar{L}, s_L) = (38{,}200, 3840) \text{ lbs} \tag{b'}$$

b. Estimate the applied stress. Applied stress in member AC is modeled by Equation (5.1):

$$s = \frac{L}{A} \tag{c}$$

Substitution of equation (b′) into equation (c) yields

$$(\bar{s}, s_s) = \frac{(38{,}200, 3840)}{(\bar{A}, s_A)} \text{ psi}$$

From manufacturing tolerancers, assume that tolerance on A is $\Delta \approx \pm 0.05\bar{A}$. The estimated standard deviation s_A is (see Section 3.3.2)

$$s_A \approx \frac{0.05\bar{A}}{3} = 0.0167\bar{A}$$

$$(\bar{s}, s_s) = \frac{(38{,}200, 3840)}{(\bar{A}, 0.0167\bar{A})} \tag{c'}$$

From equations (2.28) and (2.30), with equation (c′), the statistics of the quotient are

$$(\bar{s}, s_s) = \left[\frac{38{,}200}{\bar{A}}, \frac{3890}{\bar{A}} \right] \text{ psi} \tag{c''}$$

and

$$C_s = \frac{3890}{38{,}200} = 0.102$$

422 Design and Analysis

3. Estimate the statistics of fatigue strength (after experiencing and surviving $n=31{,}600$ load cycles). For the AISI 1045 steel to be used, the needed information regarding the mechanical behavior is found in Appendix 10.B.

Estimate the statistics of S_f at 31,600 cycles as follows:

   a. Using Table 9.14 and equation (8.6a), estimate the S_f at 10^4 and 10^5 cycles.
   b. Estimate $\bar{S}_f$ and $\char"0290 S_f$ at 31,600 cycles by interpolation.

Equations (8.6a) provide the transformation from Weibull parameters to sample statistics:

$$\bar{S}_f \approx x_0 + (\theta - x_0) \cdot \Gamma\left(1 + \frac{1}{b}\right)$$

$$C_{S_f} \approx \frac{\char"0290 S_f}{\bar{S}_f} = \frac{\bar{S}_f - x_0}{\bar{S}_f} b^{-0.926}$$

Estimators of $\bar{S}_f$ are as follows:

   At 10^4 cycles: $79.0 + (86.2 - 79.0) \cdot 0.888 = 85.39$ ksi
   At 10^5 cycles: $67.0(73.0 - 67.0) \cdot 0.89 = 72.34$ ksi

Estimators of C_{S_f} are as follows:

   At 10^4 cycles: $C_{S_f} = 0.0748(0.888^{-0.926}) = 0.0835$
   At 10^5 cycles: $C_{S_f} = 0.0738(0.80^{-0.926}) = 0.0822$
   At 10^4 c_y: $\char"0290 S_f \approx 7.13$ ksi
   At 10^5 c_y: $\char"0290 S_f \approx 5.95$ ksi

At 10^4 c_y: $L_{10}10^4 = 4.0$	$L_{10}S_f$
$\bar{S}_f = 85.34$ ksi	1.93115
$\char"0290 S_f = 7.13$ ksi	
$S_f + \char"0290 S_f = 92.47$ ksi	1.96600
At 10^5 c_y: ($L_{10}10^5 = 5.0$)	
$\bar{S}_f = 72.34$ ksi	1.85938
$\char"0290 S_f = 5.95$ ksi	
$S_f + \char"0290 S_f = 78.29$	1.89368

Probabilistic Approach to Design

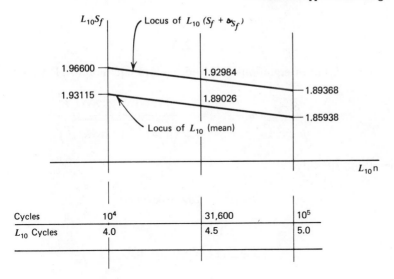

$$\bar{S}_{31,600} = \text{antilog}\,(1.89026) = 77.76 \text{ ksi}$$

$$\bar{S}_{31,600} + {}_{\gg S_f} = \text{antilog}\,(1.92984) = 85.08 \text{ ksi}$$

$$({}_{\gg S_f})_{31,600} = 85.08 - 77.67 = 7.41 \text{ ksi}$$

Thus the estimated statistics of AISI 1045 steel at 31,600 cycles are

$$\left(\bar{S}_f, {}_{\gg S_f}\right) = (77.67, 7.41) \text{ ksi} \tag{d}$$

4. Following the design procedure in for distribution of S and s unspecified described Section 9.3.9 and consulting Figure 9.42 (since for AISI C1045 Steel, $C_S = 0.095$, it is seen that by designing for $R^* \simeq .996$, using normal theory, the result will be a component satisfying $R = .992$. if stress is lognormal. Note for load stress that $C_s = 0.102 \sim$ equals $C_s' = 0.100$ the abscissa value from the figure associated with $R = .992$.

Utilizing equation (9.10) and Appendix 1, we obtain

$$R^* = .996 \leqslant \frac{1}{\sqrt{2\pi}} \cdot \int_z^\infty e^{-t^2/2}\,dt = \frac{1}{\sqrt{2\pi}} \cdot \int_{-2.70}^\infty e^{-t^2/2}\,dt$$

Table 9.14 Fatigue Strength

Cycles	Weibull Parameters			$\Gamma[1 + (1/b)]$(Figure 8.30)
	x_0 ksi	b	θ ksi	
10^4	79.0	2.6	86.2	~ 0.888
10^5	67.0	2.75	73.0	~ 0.890

424 Design and Analysis

Thus

$$z \approx -2.70$$

5. Substitution of equations (c″) and (d) into equation (9.11) yields

$$z = -\frac{\bar{S}-\bar{s}}{\sqrt{\mathring{s}_S^2 + \mathring{s}_s^2}} \approx -2.70 = -\frac{77{,}600 - (38{,}200/\bar{A})}{\sqrt{(7410)^2 + (3890/\bar{A})^2}}$$

Expand and simplify:

$$0 = \bar{A}^2 - 1.28\bar{A} + 0.021$$

$$\bar{A} = 1.30 \text{ in.}^2$$

$$\Delta \approx 0.05\bar{A} = \pm 0.065 \text{ in.}^2$$

9.3.10 Gears and Bearings

Example M Probabilistic Gear-tooth Design

1. In Example G in Chapter 5 the statistics of dynamic bending stresses in a spur gear (tooth) that has a pitch diameter $d = 12.00$ in. and $N = 24$ teeth were estimated. The force transmitted to the interacting gear was

$$\left(\bar{P}_t, \mathring{s}_{P_t}\right) \simeq (10{,}000, 500) \text{ lbf}$$

In this following example (a) loading is assumed to be of constant amplitude, with no shock effects, (b) lubrication is adequate, (c) the component is protected against dirt intrusion, and (d) adequate cooling is provided. The spur gear is designed to survive an indefinite number of bending load cycles, with a probability of $R > .997$.

Since the distributions of stress and endurance limit are very likely in doubt, in Examples M and N the distribution free design procedure outlined in Section 9.3.9 (subsection on design with distribution of S and s unspecified) is utilized.

2. Bending stress s stated as a function of gear face width F was determined to be (Chapter 5, Example G):

$$(\bar{s}, \mathring{s}_s) \simeq \left[\frac{272{,}300}{\bar{F}}; \frac{14{,}230}{\bar{F}}\right] \text{ psi} \qquad (a)$$

and

$$C_s = \frac{14{,}230}{272{,}300} = 0.052.$$

3. Consider a hypothetical preliminary design where the spur gear material will be specified as AISI 1045 steel heat treated to 420 BHn. If this were the only information available, static tensile strength S_u would be estimated by the empirical formula in equation (8.5):

$$\left(\bar{S}_u, \vphantom{S}_{S_u}\right) \simeq \text{BHn} \cdot (495, 20.5) \text{ psi} \approx (198{,}000, 8110) \text{ psi}$$

and by using the empirical formula in equation (8.6b):

$$\left(\bar{S}_e, \vphantom{S}_{S_e}\right)^* \simeq (0.498, 0.051)\left(\bar{S}_u, \vphantom{S}_{S_u}\right)$$

$$\approx (98{,}600, 8260) \text{ psi}$$

4. Since the required reliability is $R > .997$, we utilize normal design theory and a fictitious reliability $R^* = .999$ (see Figure 9.36) and note that $C_s = (\vphantom{s}_s / \bar{s}) = 0.052$. Associated with fictitious $R^* = .999$ is $z \simeq -3.0$ (see Appendix 1).

5. Face width F of the spur gear is now estimated by utilizing equations (a) and (b) with equation (9.11):

$$-3.0 = -\frac{98{,}600 - 272{,}300/\bar{F}}{\sqrt{(8260)^2 + \left(\dfrac{14{,}230}{\bar{F}}\right)^2}}$$

Expand and simplify:

$$\bar{F}^2 - 6.33 \cdot \bar{F} + 8.220 = 0$$

$$\bar{F} = 4.50 \text{ in.}$$

$$\vphantom{s}_F \simeq 0.01 \bar{F} = 0.045 \text{ in.}$$

The spur gear designed by the foregoing calculations should perform for an indefinite number of bending load cycles with a probability of .997 if the models used were valid and realistic. However, since the materials behavior model was based on empirical estimators rather than tests, this design must be verified by test.

Example N

1. As noted in Section 5.19, gears are subject to failure in two modes: (1) bending fatigue and (2) surface failure. In this example F is estimated such that the probability of surface failure due to repeated contact stresses will be:

$$P_f \leqslant 0.003$$

*In an actual situation the statistics would be (b) validated by test.

2. In Example H in Chapter 5 the statistics of contact stress for the spur gear under study were estimated. The mean value and standard deviation of s_c were predicted as

$$(\bar{s}, \mathord{\vphantom{s}}_{s_c}) \simeq \left[\frac{2.22 \cdot 10^5}{\bar{F}^{1/2}} ; \frac{1.14 \cdot 10^4}{\bar{F}^{1/2}} \right] \text{ psi} \qquad (a)$$

3. As in Example M, the material of the spur gear will be AISI 1045 steel heat treated to 420 BHn. The surface endurance limit S_c must be estimated from the following (conservative) empirical formulas, [1, p. 514]:

$$\bar{S}_c \simeq 400 \text{ BHn} - 10{,}000 = 158{,}000 \text{ psi}$$

$$\mathord{\vphantom{s}}_{S_c} \simeq 20.5 \text{ BHn} = 8200 \text{ psi } [\text{equation (8.5)}]$$

$$(\bar{S}_c, \mathord{\vphantom{s}}_{S_c}) \simeq (158{,}000, 8200) \text{ psi.} \qquad (b)$$

4. The design objective now is to size F such that the spur gear will survive 10^8 contact stress cycles with a probability of $R > .997$ (for which we utilize fictitious $R^* = .999$; see Figure 9.36). Associated with $R^* = .999$ is $z \simeq -3.0$.

5. Utilize equation (9.11) with equation (a) and (b):

$$z = -3.0 = -\frac{158{,}000 - \dfrac{2.22 \cdot 10^5}{\bar{F}^{1/2}}}{\sqrt{(8200)^2 + \left(\dfrac{1.14 \cdot 10^4}{\bar{F}^{1/2}}\right)^2}}$$

Expand and simplify:

$$0 = \bar{F} - 2.43 \bar{F}^{1/2} + 2.027$$

$$\bar{F}^{1/2} = 1.81 \sqrt{\text{in.}}$$

$$\bar{F} = (1.81)^2 \text{ in.} = 3.29 \text{ in.}$$

$$\mathord{\vphantom{s}}_F \simeq 0.01 \bar{F} = 0.033 \text{ in.}$$

Compare the value of face width $\bar{F}$ with that estimated in Example M.

*See Appendix 1.

Probabilistic Ball-bearing Analysis.

Example O

1. In the following analysis the allowable loading of a #207 bearing for survival of 10^8 load cycles (with a probability of .9997) is estimated.

2. The contact stresses in a #207 Conrad type single-row radial ball bearing were estimated in Section 5.20.2 (Table 5.11). The contact stress s_c statistics (in the balls and the race) stated as functions of mean dynamic loading $\bar{P}$ were

$$(\bar{s}_c, {}_{\backprime}s_c) \simeq \left[(17{,}800) \cdot \sqrt[3]{\bar{P}}, 610 \cdot \sqrt[3]{\bar{P}} \right] \text{ psi} \tag{a}$$

3. Almost all manufacturers of ball bearings now use steels of the same type (E52100) and subject them to the same heat treatment. Experimental measurements at the General Motors Research Laboratories disclose that a range of threshold contact stress intensities exists below which no gross structural change appears to occur, even after exposure to 10^8 or more cycles of loading.* For E52100 steel, S_c ranges from 422,000 to 480,000 psi. Thus utilizing the rationale of Section 3.3.2 and assuming an approximate normal model

$$\bar{S}_c \simeq \frac{480{,}000 + 422{,}000}{2} = 451{,}000 \text{ psi}$$

and

$${}_{\backprime}S_c \simeq \frac{480{,}000 - 422{,}000}{6} = 9670 \text{ psi}$$

Thus

$$(\bar{S}_c, {}_{\backprime}S_c) \simeq (451{,}000;\ 9{,}670) \text{ psi} \tag{b}$$

4. Load-carrying capacity P of the bearing may now be estimated utilizing equations (a) and (b) with equation (9.11). The value of z associated with required ($R = .9997$) reliability (Appendix 1) is -3.62. The design procedure for S and s normal discussed in Section 9.3.2 applies in this example because a goodness-of-fit test of contact stress using Monte Carlo simulation and a plot of data on probability paper suggested a normal distribution model.

5. Thus from equation (9.11), with equations (a) and (b):

$$-3.62 = -\frac{451{,}000 - 17{,}800\bar{P}^{1/3}}{\sqrt{(9670)^2 + (610\bar{P}^{1/3})^2}} \tag{c}$$

---

*According to Mischke, these figures may be subject to question.

428 Design and Analysis

Table 9.15 Loads and Associated Reliabilities

Reliability	Load (lb)	Derating
.90	13,740	3.09
.99	12,460	2.81
.999	11,250	2.53
.9999	10,490	2.36

Expand and simplify equation (c):

$$\bar{P}^{2/3} - 51.5\bar{P}^{1/3} + 648. = 0$$

$$\bar{P}^{1/3} = 21.9$$

$$\bar{P} \approx 10{,}500 \text{ lb}$$

Comparing $\bar{P}$ with the B-10 load rating of 4440 lb for bearing #207 in Table 5.11, the published allowable loading appears to be severely derated. Table 9.15 provides the calculated dynamic loadings supportable by bearing #207 for 10^8 cycles at a variety of probabilities.

For $R = .90$ of surviving 10^8 contact stress cycles, the value $\bar{P} = 13,400$ lb compares closely with results published by Bush et al. in "Microstructural and Residual Stress Changes in Hardened Steel Due to Rolling Contact," Proceedings of G. M. Symposium 1962 (see Figure 9.44).

9.3.11 Reliability Assurance

Tests of actual hardware are rarely sufficient to assure reliability by relative frequency calculations. Thus there is a need for methods that will give meaningful checks based on a few test results. Furthermore, these methods

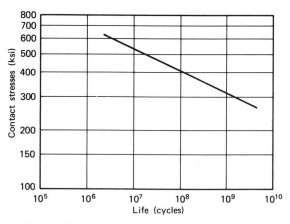

Figure 9.44 Average S–N curves for contact stresses.

should be such that reliability estimates can be amended as more performance information becomes available [92].

The primary purpose of probabilistic design procedures is the production of better designs whose intrinsic and target reliability are in close agreement. Primarily engineering decisions and secondarily scientific descriptions are the goals of design efforts. Furthermore, the endeavor of designing probabilistically must eventually satisfy the legalistic problem of demonstrating that a sufficient degree of reliability has been attained while satisfying the technical problem of producing sound designs that represent a consistent economical balance between costs and risks.

In efforts where the emphasis is on design rather than description, a number of unresolved arguments within science about the fundamental nature of probability become unimportant. It no longer matters whether nature is basically deterministic or whether probability is just a tool to describe phenomena too complex to treat with the present level of knowledge. It only matters that from an engineering design point of view it is useful to represent particular quantities as random variables. If useful in design, it may be reasonable to treat all unknown factors as random, including the errors implicit in approximate stress analysis, the errors in professional simplifications, and in load distributions, or the quality of an unknown producer. Uncertainties may not be solely probabilistic (in the relative frequency sense) but may also be statistical (i.e., due to practical limitations on the data available to estimate the parameters) and professional (i.e., incomplete information or knowledge about the underlying model). At the same time, our assurance method should provide a mechanism for incorporating the information contained in professional judgment (i.e., indirect related experience with similar problems).

Bayes's formula may be adapted for the purpose of coupling judgmental information and data based on tests.

Suppose that A is an event that can occur only if one of the mutually exclusive events $B_1, B_2, \ldots, B_n$ occurs. That is $A \subset \bigcup_{i=1}^{n} B_i$; $B_i \cdot B_j = 0$; $i = 1, 2, \ldots, n$. Then

$$P(B_1|A) = \frac{P(B_i) \cdot P(A|B_i)^*}{\sum_{i=1}^{n} P(B_i) \cdot P(A|B_i)} \qquad (9.16)$$

In the form of equation (9.16), Bayes's formula is often regarded as a device for adjusting the a priori probability $P(B_i)$ as a result of observing outcome A. Before utilizing equation (9.16) in reliability assurance, its use is illustrated by a simple engineering example.

---

*In words, $P(A|B_1)$ states the probability of A given that B_1 has been observed, called *conditional probability*.

Example P

Estimate the likelihood of column failure in a given mode, employing Bayes's formula, where B_1 represents Euler buckling and B_2 represents fatigue failure. The probability of Euler buckling (B_1) is

$$P(A|B_1) = P_{f_1} \leqslant .00100 = 100 \cdot 10^{-5}$$

$$P(P_{cr} > P) \geqslant .99900$$

The probability of axial fatigue failure (B_2) is

$$P(A|B_2) = P_{f_2} \leqslant .00005 = 5 \cdot 10^{-5}$$

$$P(S_e > s) \approx .99995$$

where A is the event "a failure is observed." Use of these data yields

$$P(B_1) = \frac{.99900}{.99900 + .99995} = .4998$$

$$P(B_2) = \frac{.99995}{.99900 + .99995} = .5002$$

From equation (9.16), if a random failure is observed, the probability that it is due to Euler buckling (B_1) is

$$P(B_1|A) = \frac{.4998(100 \cdot 10^{-5})}{.4998(100 \cdot 10^{-5}) + .5002(5 \cdot 10^{-5})} = .9523$$

and that the failure due to axial fatigue (B_2) is

$$P(B_2|A) = \frac{.5002(5 \cdot 10^{-5})}{.4998(100 \cdot 10^{-5}) + .5002(5 \cdot 10^{-5})} = .0477$$

Bayes's Theorem in Design. In and during design development, direct observational data often becomes available, in addition to a priori data and professional opinion. A contribution of Bayesian statistical decision theory is a procedure for coupling professional information (a priori and a posteriori data) and the information contained in this statistical data. The vehicle is Bayes's formula.

In words, the theorem says that the probability $P(B_i|A)$ of hypothesis i, given observation A, is proportional to the prior probability $P(B_i)$ times the likelihood of A, given B_i, that is, $P(A|Bi)$. Thus equation (9.16)

$$P(B_i|A) = \frac{P(A|B_i)P(B_i)}{\Sigma P(A|B_i)P(B_i)}$$

may be applied in making predictions of inherent reliability.

Example O

Consider the problem of the reliability verification of a new, expensive mechanical component never before designed, built, or tested. If designed properly to meet the performance criteria, the time to failure has a near-normal distribution, with a mean value of $\mu_1 = 1000$ hr. If improperly designed, the mean time to failure may be as low as $\mu_2 = 100$ hours (Figure 9.45). The engineer knows the character of this design with respect to the state of the design art. He may express this by saying (based on design calculations, etc.) that his prior probability that the design is satisfactory is .9. (In this example it is assumed that the variables are normally distributed);

$$P(\mu = \mu_1) = 0.9 \quad \text{(1000 hr average life)}$$
$$P(\mu = \mu_2) = 0.1 \quad \text{(100 hr average life)}$$

(*Note:* Estimated $\sigma = (0.10) \cdot 1000 \text{ hr} = 100 \text{ hr}$.) A single component is now tested in a simulated (or actual) environment for 400 hr, at which point the test is stopped.

The engineer can combine his prior and test information into a single statement as follows. The likelihood of observing a lifetime in excess of 400 hr

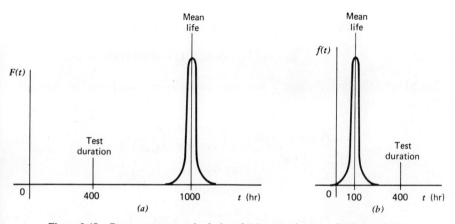

Figure 9.45 Component properly designed (*a*) versus improperly designed (*b*).

is from equation (9.11):

$$z = \frac{400 - 1000}{100} = -6.00$$

and from equation (9.10):

$$P(A|\mu_1 = \mu) + \frac{1}{\sqrt{2\pi}} \int_{-6.00}^{\infty} e^{-z^2/z} \, dz$$

Thus

$$P(A|\mu = \mu_1) = .999999998$$

Calculating the conditional probability:

$$P[\mu = \mu_1|A] \approx P[A|\mu = \mu_1] \cdot P(\mu = \mu_1)$$

$$\approx (.999999998)(.9) = .899999998.$$

From equation (9.11):

$$z = \frac{400 - 100}{100} = +3.000$$

and from equation (9.10):

$$P(A|\mu = \mu_2) = \frac{1}{\sqrt{2\pi}} \int_{+3.00}^{\infty} e^{-z^2/2} \, dz$$

$$P(A|\mu = \mu_2) = .00269980 \text{ (from Appendix 1)}$$

Thus

$$P[\mu = \mu_2|A] \approx P[A|\mu = \mu_2] P(\mu = \mu_2)$$

$$\approx (.00269980)(.1) = .00026998$$

The absolute values of these *a posteriori* probabilities are found by normalizing:

$$P(\mu = \mu_1|A) = \frac{.899999998}{.899999998 + .00026998}$$

$$= \frac{.899999998}{.900269978} = .99970000$$

$$P(\mu = \mu_2|A) = \frac{.00026998}{.900269978} = .00029980$$

Mean Life	Probability Based on Calculation that $\mu_1 = 1000$ hr (Before Test)	Probability Based on Calculation + Test that $\mu_1 = 1000$
$\mu_1 = 1000$	.90	.99970000
$\mu_2 = 100$	.10	.0002998
	(Prior information based on best information prior to design)	(After one performance test)

The results of the preceding calculations can now be utilized. For example, if the component is asked to perform for 500 hr, the probability that one selected at random will meet this requirement is found by utilizing equation (9.11); thus

$$z = \frac{500 - 1000}{100} = -5.000; \quad z = \frac{500 - 100}{100} = +4.000$$

and from the *Tables of Normal Probability Functions:*

$$R = .9999997 \quad R = .00004$$

The engineer can now state, on the basis of his design calculations and the limited available performance data, that the system reliability (for a 500-hr mission) is .9999997 (i.e., nearly certain), with a probability of .9997.

Note that this statement is comparable to a classical statistical confidence statement. The unknown parameter μ_1 has been treated as e random value with a distribution. This distribution is conditional on the amount of information and changes with the information. Prior to testing, the engineer could state that the R was .9999997, with a probability of only .9.

ADDENDA

Time to First Failure.* Ordinarily the goal in a dynamic design problem is in efficiently designing such that a component (produced from the design), selected at random from among many, will survive service in its intended use environment for at least a specified cycle life and with satisfactory probability.

A meaningful question that may arise is in regard to when (in cycle count) the first failure can be expected to occur in a population of nominally identical mechanical components or devices. An analogous situation is described in Hahn and Shapiro [13, p. 116], that is, if in a circuit nominally identical components are connected in series and if their time to failure distribution is γ, the circuit time to failure follows a type III asymptotic (or

---
*Also see Crandall and Mark [45].

434 Design and Analysis

Weibull) distribution. A series circuit failure is equivalent to a first component failure in a population. For mechanical components, however, time or cycles to failure may be better described by a lognormal model. Thus determination of time to first failure for populations of mechanical elements or devices may pose a similar although not an identical problem to that of a series circuit failure.

Pricing, Warranties, and Inventory Control. In an on-going probabilistic design it is often necessary to develop the statistics of time or cycles to failure of various components. This is a simple task once the design equations have been written, the data assembled, and the component dimensions estimated.

When coupled with information such as projected sales figures, developmental and manufacturing costs, and so on usually available within a company, the potential information available from time or cycle life diagrams can be extremely valuable in making management decisions.

For example, if unit cost and projected total sales figures are known, Fig. 9.46 can be utilized to estimate the cost of a performance guarantee:

$$\text{Total cost} = (\text{unit cost})(\text{total sales}) \int_0^{N_1} f(N)\,dN$$

Alternatively, Figure 9.46 could be utilized to establish a guarantee period N_1, acceptable to management.

Beyond the warranty period, normal parts replacements become necessary. Figure 9.46 can be utilized to estimate the replacement demand in any time period, hence the anticipated inventory requirements. For instance, in the time period $N_3 - N_2$:

$$\text{Replacement parts} = (\text{total sales}) \left[\int_0^{N_3} f(N)\,dN - \int_0^{N_2} f(N)\,dN \right]$$

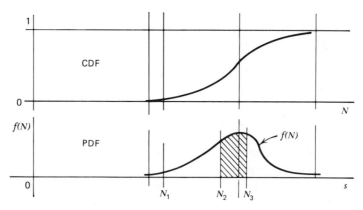

N_1 = Guarantee period

Figure 9.46 Time or cycles to failure.

Thus replacement parts manufacturing schedules can be rationally developed as to quantity and time of manufacture. With such information, inventory costs can be optimized (i.e., minimized).

Some parts or components must be serviced or replaced periodically over the life of a device or a system. Figure 9.46 can be utilized to establish a service or replacement policy such that the probability of a breakdown before service or replacement is scheduled will be acceptably low, that is, optimize (maximize) customer satisfaction.

CHAPTER 10

Strength of Mechanical Components

The following topics are considered in this chapter:

Theories of static failure

Maximum normal stress theory
Maximum shear stress theory
Distortion energy theory
Significant stress–strength models

Design factors

Strength–stress modifiers (K_a, k_b, k_d, K_t, q)

Design to avoid fatigue failure (zero mean stress)

Endurance and finite life (constant amplitude)
Fatigue due to narrow-band random stress

Design for fluctuating stress (nonzero mean)

Constant-amplitude dynamic stresses
Random dynamic stresses
Design for Multiaxial dynamic loading
Multiple-member systems
Variability in fracture mechanics

10.1 INTRODUCTION

The strength of mechanical components is directly identified with the way in which they can fail; thus a failure mode analysis is appropriate in the design process. The engineer intuitively attempts to adjust the reliability of his designs to reflect the consequences of failure, as illustrated in Figure 10.1.

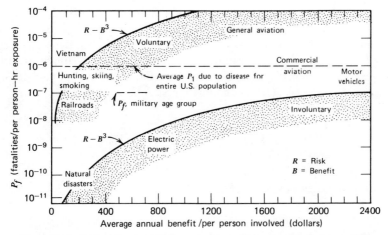

Figure 10.1 Risk versus benefit for voluntary and involuntary exposure [99].

After a probabilistic base has been established in stress, strain, deflection, materials behavior, and the theory of design, the strength of mechanical components is considered in this chapter.

The recent trend in design has been to develop new criteria from an objective point of view so that reasonable agreement with experiment is obtained. Elaboration to embrace additional parameters such as cycles to failure, mean stress, and phenomena of variability is included. Further considerations involve comparison of the strength of a material with the strength of a mechanical component.

Yield strength and ultimate strength (Sections 8.6 and 8.7) or endurance limit (Section 8.14.1), bending strength, axial strength, and so on come to mind when the strength of a material is involved. However, with a machine element, the strength is a limiting value of the applied stress at a particular point. The endurance limit S_e', for example, of a mechanical component is often very different than the endurance limit S_e of the material.

Formerly, nominal applied stress s was often utilized as determined from simple equations such as $s=(P/A)$, $s=(Mc/I)+(P/A)$ and $\tau=(Tc/J)$. More realistically, significant applied stress s' is approximated by an expression such as equation (10.15), where s is nominal applied stress and the K values are stress-modifying factors. "Significant stress" is that experienced at a particular point by a mechanical component in its use environment. The preceding observations suggest that mechanical design may be viewed as (among other things) being involved with (1) modeling each individual variable and (2) mathematically modeling stress and strength in functional forms to realistically reflect subsequent real-world behavior of mechanical systems.

To assess materials behavior, the results of many tests of a material should be available, as discussed in Chapter 8. The specimens should be of the same heat treatment, surface finish, and size as a component to be designed, with test loading conditions similar to those anticipated in the subsequent use environment. As such information was rarely available in the past, a long-standing fundamental problem in design has been to utilize simple (deterministic) tension test data, attempting to correlate them with component strengths regardless of the stress state or the loading situation.

THEORIES OF FAILURE

10.2 MAXIMUM NORMAL STRESS THEORY

In the 19th century it was generally accepted in engineering that (1) stresses were deterministic and (2) principal stresses were orderable. These were unstated beliefs when Rankline (1802 to 1872) postulated the "maximum normal stress" theory. However, probabilistic consideration has shown such hypotheses to be questionable.

Classical maximum normal stress theory postulated failure whenever $s_1 > S_y$ or $s_1 > S_u$, whichever was applicable [27]. Loading was assumed to be deterministic static, and s_1 was the largest of the (deterministic) principal stresses.* However, random variables described by two parameter models are not orderable, and engineering variables behave like random variables.

Further reason for questioning general use of this failure theory for the biaxial stress state is apparent from consideration of Figure 10.2, in which $\tau = s_1$. The maximum normal stress theory postulates failure when shear stress equals to strength in tension or compression, not supported by experiment.

In the uniaxial stress case, $s_2 = s_3 = 0$, and s_1 takes on nonzero values. The maximum normal stress theory can be effectively employed in probablistic design of many uniaxially loaded components involving:

1. A probablistic requirement to avoid yield:

$$R = p \cdot (S_y > s_1) \qquad (10.1)$$

2. A probablistic requirement to avoid tensile fracture:

$$R = p \cdot (S_u > s_1) \qquad (10.2)$$

3. A probablistic requirement to avoid shear yielding:

$$R = p(S_{sy} = 0.577 S_y > s_1)^\dagger \qquad (10.3)$$

---
*For definition of principal stress, see Section 5.7.
†See Section 10.4.

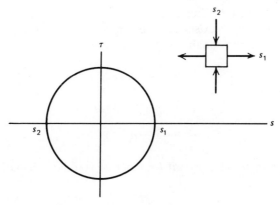

Figure 10.2 Postulated stress at failure.

In the usual situation in design, s_1, s_2, and s_3 must be treated as random variables. Thus probablistically, the inequality $s_1 > s_2 > s_3$ is in a strict sense, without meaning.

The only possible ordering is usually that of the mean values

$$\bar{s}_1 > \bar{s}_2 > \bar{s}_3.$$

Examine Figure 10.3 and consider only the probability density functions of s_1 and s_2. It is seen that if s_1 and s_2 are statistically independent, there will be a nonzero probability that

$$s_2 > s_1$$

Next, consider the case where s_1 and s_2 are correlated with $\rho = +1$; thus

$$\bar{s}_2 = c\bar{s}_1$$

$$\sigma_{s_2} = c\sigma_{s_1}$$

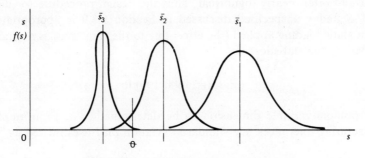

Figure 10.3 Ordering of principal stresses.

where $c<1$ is a constant. For any random value $\bar{s}_1+\Delta$, $s_1>s_2$. Alternatively, since $\sigma_{s_2}<\sigma_{s_1}$, a value θ exists such that $s_2=s_1$ and further for a value $\theta-\Delta$, $s_2>s_1$ with a nonzero probability. For example,

$$\theta = \bar{s}_2 - a_i\cdot(\sigma_{s_2})$$

$$\theta = c\bar{s}_1 - a_i(c\sigma_{s_1}) \tag{a'}$$

$$\theta = \bar{s}_1 - a_i(\sigma_{s_1}), \tag{b'}$$

where a_i is a numerical constant (number of standard deviations). Equating (a') and (b'):

$$c\bar{s}_1 - a_i\cdot(c\sigma_{s_1}) = \bar{s}_1 - a_i\cdot(\sigma_{s_1})$$

$$a_i(\sigma_{s_1} - c\sigma_{s_1}) = \bar{s}_1 - c\bar{s}_1$$

$$a_i = \frac{\bar{s}_1 - c\bar{s}_1}{\sigma_{s_1} - c\sigma_{s_1}} = \frac{\bar{s}_1}{\sigma_{s_1}}$$

Example A

1. Preliminary assumptions are as follows. An example in which the maximum normal stress theory is appropriate is the following design of an element of rectangular cross section. The stress is essentially uniaxial bending and the objective is to avoid fracture. Thus the design criterion, [equation (10.2)] is $R=P(S_u>s_1)$. In this example the structural requirements for a static loaded sling are estimated. Geometry of the sling is shown in Figures 10.4 and 10.5. It is to be manufactured from AISI 4340 alloy steel, whose tensile strength S_u is given in Figure 8.41. The loading as applied at the ends acts downward. The specified probability of surviving a random loading is to be $R=.9965$ since replacement cost is modest and safety will not be a consideration. The distribution followed by static stress is unknown.

If loading is normally distributed, load-induced stress is nonetheless nonnormal and often nearly lognormal. Thus the design procedure for distribution of S and s unspecified discussed in Section 9.3.9 is appropriate. The random static loading applied (the effect due to the sling mass is negligible) is described by the statistics:

$$(\bar{P}, \sigma_p) = (1800, 100) \text{ lb}$$

The random-variable dimension to be determined is h. From applicable manufacturing tolerances, the standard deviation on l is estimated as

$$\sigma_h \simeq 0.015\bar{h}$$

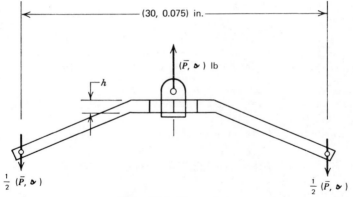

Figure 10.4 Cargo sling.

2. Stress statistics are estimated as follows. Since the cross section will be rectangular and $\bar{l} \gg \bar{b}$, the formula for extreme fiber stress [equation (5.23)] is considered an acceptable model for estimating stress:

$$s = \frac{Mc}{I}, \tag{a}$$

where $c = (h/2)$ and $I = (bh^3/12)$ (Appendix 5). Thus the model for applied load stress is

$$s = \frac{6M}{bh^2} \text{ psi} \tag{b}$$

The statistics of $(\overline{M}, s_M)$ must be estimated, that is, the product of random variables $P/2$ and $l/2$. According to equations (2.8) and (2.10), the statistics of the product of a random variable and a constant are

$$\tfrac{1}{2}(\bar{p}, s_p) \approx \tfrac{1}{2}(1800, 100) = (900, 50) \text{ lb}$$

and

$$\tfrac{1}{2} \cdot (\bar{l}, s_l) = (15.00; 0.0375) \text{ in.}$$

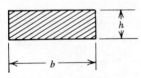

$(\bar{b}, s_b) = (2.00, 0.035)$ in.   **Figure 10.5** Cargo sling cross section.

Thus the moment M is

$$(\overline{M}, s_M) = (900, 50) \cdot (15.00; 0.0375) \text{ in. lb}$$

From equations (2.25) and (2.27), the statistics are $M = (P/2)(\bar{l}/2)$:

$$(\overline{M}, s_M) = (13{,}500, 750) \text{ in. lb}$$

and the product $6M$, from equations (2.8) and (2.10), is

$$6(\overline{M}, s_M) = 6(13{,}500, 750) = (81{,}000, 4500) \text{ in. lb}$$

Let $u = (6M/b)$; then

$$(\bar{u}, s_u) = \frac{(81{,}000; 4500)}{(2.00; 0.035)}$$

From equations (2.24) and (2.26), the statistics of u are

$$\bar{u} \approx \frac{81{,}000}{2.00} = 40{,}500$$

and

$$s_u \approx \sqrt{\frac{(81{,}000)^2 \cdot (0.035)^2 + (4500)^2 \cdot (2.00)^2}{(2.00)^4}} = 2440$$

The statistic estimators for a random variable squared in terms of the statistics of the variable are, from equations (2.16) and (2.18)

$$(\overline{h^2}, s_{h^2}) \approx (\bar{h}^2, 2\bar{h} \cdot s_h)$$

and

$$(\bar{s}, s_s) = \frac{(40{,}500, 2440)}{(\bar{h}^2, 2\bar{h} \cdot s_h)} \text{ psi} \qquad (c)$$

where s_h was given as $s_h \approx 0.015\bar{h}$. Thus from equations (2.24) and (2.26) with equation (c), the stress statistics are

$$(\bar{s}, s_s) = \frac{(40{,}500, 2440)}{(\bar{h}^2, 0.03\bar{h}^2)} = \left[\frac{40{,}500}{\bar{h}^2}, \frac{2730}{\bar{h}^2}\right] \text{ psi} \qquad (d)$$

$$C_s = \frac{s_s}{\bar{s}} = \frac{2730}{\bar{h}^2} \bigg/ \frac{40{,}500}{\bar{h}^2} = 0.067$$

3. The allowable stress of the material is (from Figure 8.41)

$$\left(\bar{S}_u, \mathrm{\mathchar'26\mkern-9mu s}_{S_u}\right) = (140{,}000; 5600) \text{ psi} \qquad (e)$$

and

$$C_{S_u} = \frac{\mathrm{\mathchar'26\mkern-9mu s}_{S_u}}{\bar{S}_u} = \frac{5600}{140{,}000} \approx 0.04$$

4. Actual strength is normal and stress follows an unknown distribution. The value $\bar{h}$ must be determined such that $R = .9965$ is satisfied. Solve the design problem assuming that strength and stress are normal, with a ficticious requirement that $R' = .9990$ (see Figure 9.36 since C_{S_u} for the material is 0.04). Use the design procedure for distribution of S and s unspecified in Section 9.3.9:

$$R' = .9990 \leqslant \frac{1}{\sqrt{2\pi}} \cdot \int_z^\infty e^{-t^2/2}\,dt = \frac{1}{\sqrt{2\pi}} \cdot \int_{-3.0}^\infty e^{-t^2/2}\,dt \qquad (f)$$

5. Substitution of z, equation (d), and equation (e) into equation (9.11) yields

$$z = -\frac{\bar{S}_u - \bar{s}}{\sqrt{\mathrm{\mathchar'26\mkern-9mu s}_{S_u}^2 + \mathrm{\mathchar'26\mkern-9mu s}_s^2}} \approx -3.0 = -\frac{140{,}000 - \dfrac{40{,}500}{\bar{h}^2}}{\sqrt{(5600)^2 + \left(\dfrac{2730}{\bar{h}^2}\right)^2}} \qquad (g)$$

The principal root of equation (g) satisfies ficticious $R' = .9990$, hence actual $R = .9965$ is

$$\bar{h} = 0.602 \text{ in.}$$

$$\mathrm{\mathchar'26\mkern-9mu s}_h \approx 0.015\bar{h} = 0.0090 \text{ in.}$$

Total $\Delta \approx 3\mathrm{\mathchar'26\mkern-9mu s}_h = 0.027$ in.

(*Note:* Compare the value of $\bar{h}$ in this example with that in Example G, where the loading is dynamic.)

10.3 MAXIMUM SHEAR STRESS THEORY

The opening remarks in Section 10.2 also apply to the maximum shear stress theory. Predictions based on this failure theory (Coulomb, 1736 to 1806) have been shown to be on the safe side by test results. It has been

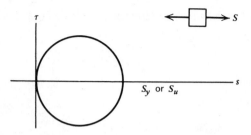

Figure 10.6 Maximum shear stress theory.

useful (within the deterministic framework) as a failure theory for ductile materials. With variations, it has provided a conservative theory for brittle materials, although this application may soon be taken over by fracture mechanics analysis.

For the triaxial state of stress, three mean principal shear stresses are postulated:

$$\bar{\tau}_{12} = \frac{\bar{s}_1 - \bar{s}_2}{2}; \quad \bar{\tau}_{23} = \frac{\bar{s}_2 - \bar{s}_3}{2}; \quad \bar{\tau}_{13} = \frac{\bar{s}_1 - \bar{s}_3}{2} \quad (10.4)$$

Yielding is predicted (deterministic) whenever the largest of the three shear stresses equals $S_y/2$ from a simple tension test (see Figure 10.6). According to this theory, shearing yield strength S_{sy} of a material is half of the uniaxial tensile yield strength S_y:

$$\left(\bar{S}_{sy}, \circ_{S_{sy}}\right) \approx 0.5\left(\bar{S}_y, \circ_{S_y}\right) \quad (10.5)$$

It follows that S_{sy} will be linearly related to S_y and correlated $\rho = +1$. The distribution of S_{sy} should be of the same general type as S_y, that is, approximately normal. The graph of expected behavior is shown in Figure 10.7. If both principal stresses have the same sign, the maximum shear stress and maximum normal stress theories are identical.

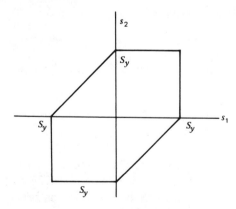

Figure 10.7 Graph of maximum shear stress theory.

Maximum Shear Stress Theory

For the biaxial stress state, the mean maximum shear stress in the xy plane is

$$\bar{\tau}_{max} = \frac{\bar{s}_1 - \bar{s}_2}{2} \tag{10.6}$$

Statistically, two cases for the determination of τ_{max} can be identified: (1) s_1 and s_2 statistically independent and (2) s_1 and s_2 correlated with $\rho = +1$. It is seen that for $(\bar{\tau}_{max}, s_{\tau_{max}}) = (0,0)$, s_1 and s_2 must be equal, in phase, and correlated $\rho = +1$, because from equation (2.41):

$$s_{\tau_{max}} = s_{s_1 - s_2} = \sqrt{s_{s_1}^2 + s_{s_2}^2 - 2\rho s_{s_1} \cdot s_{s_2}} = 0$$

and

$$\bar{\tau}_{max} = \frac{\bar{s}_1 - \bar{s}_2}{2} = 0$$

Consider the case where s_1 and s_2 are equal, normally distributed, and statistically independent, that is,

$$\bar{s}_1 = \bar{s}_2; \qquad s_{s_1} = s_{s_2}; \qquad \rho = 0$$

Then the consequences are seen to be

$$\bar{\tau}_{max} = \frac{\bar{s}_1 - \bar{s}_2}{2} = 0$$

$$s_\tau = \tfrac{1}{2} \cdot \sqrt{s_{s_1}^2 + s_{s_2}^2} = \frac{1}{\sqrt{2}}(s_{s_1})$$

Example B

The following calculations indicate that with a biaxially loaded component with s_1 and s_2 equal and independent, a nonzero probability of a shear failure will nevertheless exist. Given

$$(\bar{S}_{sy}, s_{S_{sy}}) \tag{a}$$

and assuming

$$(\bar{\tau}_{max}, s_{\tau_{max}}) = \left(0, \frac{s_{s_1}}{\sqrt{2}}\right) \tag{b}$$

the difference equation (a) − equation (b) = z is $z = S_{sy} - \tau_{max}$, and since $\bar{\tau}_{max} = 0$, $\bar{z} = \bar{S}_{sy}$ and

$$s_z = \sqrt{s_{\tau_{max}}^2 + \left(\frac{s_{y'}}{\sqrt{2}}\right)^2}$$

From equation (9.10):

$$R = \frac{1}{s_z \sqrt{2\pi}} \cdot \int_0^\infty \exp\left[-\frac{(z-\bar{z})^2}{2 s_z^2}\right] dz$$

Making the transformation to the unit random variable, t:

$$t = \frac{z - \mu_z}{s_z}$$

At $z = 0$, $t = (-\mu_z / s_z)$. Thus substitution into equation (9.11) yields

$$t = -\frac{\bar{S}_{sy}}{\sqrt{s_{\tau_{max}}^2 + \frac{s_{s'}^2}{2}}}$$

where t is finite; thus $R < 1$. It follows that $P_f = 1 - R > 0$.

10.4 DISTORTION ENERGY THEORY (VON MISES–HENCKY)

The distortion energy theory appears to have been the best predictor of yielding among past available failure theories for use with ductile materials. It has been utilized to predict the initiation of yielding, assumed to begin when the distortion energy produced in a unit element (resisting combined stresses) becomes equal to the distortion energy at yield for a unit element subjected to simple tension [100].

The distortion energy theory does not require ordering of the principal stresses, an important consideration in any failure theory utilized in probabilistic design. In the postulation of this theory it has been argued that yielding was more than a simple tensile or compressive phenomenon. Rather, yielding appeared to be in some manner related to the angular distortion of a component under stress. There appeared to be logic in considering the difference between total strain energy and the portion that produced volume change. The difference would presumably account for the energy producing angular distortion. In Figure 10.8 an element is shown subjected to stresses s_1, s_2, and s_3 acting on the zero shear planes.

Distortion Energy Theory (Von Mises–Hencky)

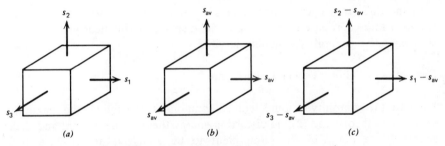

Figure 10.8 Element subjected to stresses s_1, s_2, and s_3.

For a unit cubic element, the work done in any principal direction is

$$U_i = \frac{s_i \varepsilon_i}{2}, \quad i = 1, 2, 3 \tag{a}$$

where ε_i is the strain associated with the ith principal stress. Total energy of strain is (Section 6.8):

$$U = U_1 + U_2 + U_3, \tag{b}$$

and

$$U = \frac{1}{2E}\left[s_1^2 + s_2^2 + s_3^2 - 2\mu(s_1 s_2 + s_2 s_3 + s_1 s_3)\right] \tag{10.7}$$

The distortion energy criterion is

$$2S_y^2 = (s_1 - s_2)^2 + (s_2 - s_3)^2 + (s_3 - s_1)^2 \tag{10.8}$$

Equation (10.8) defines the stress state at which yielding is postulated for the triaxial case. Equation (10.9) defines the stress state at which yielding is postulated for the biaxial case:

$$S_y^2 = s_1^2 - s_1 s_2 + s_2^2 = s_x^2 + s_y^2 - s_x \cdot s_y + 3\tau_{xy}^2 = s_{eq}^2. \tag{10.9}$$

In the case of pure torsion, $s_2 = -s_1$, and s_1 and s_2 are correlated, $\rho = -1$; thus

$$\tau = s_1 \tag{c}$$

Substitution of equation (c) into equation (10.9) yields

$$S_y^2 = 3\tau^2$$

$$0.577 \cdot S_y \approx S_{sy} \tag{10.10}$$

448 Strength of Mechanical Components

In the distortion energy theory, tensile yield and shear yield strength are assumed to be correlated, $\rho = +1$, and the distribution of shear yield strength is of the same general type as tensile yield.

Example C Failure Theory Comparison

Estimate the reliability of a mechanical component utilizing (1) the maximum shear stress theory and (2) the distortion energy theory. The normal and shear stresses at a point in the component to be analyzed are (Figure 10.9) described as follows:

$$(\bar{s}_x, \vartheta_{s_x}) = (13{,}000, 650) \text{ psi} \tag{a}$$

$$(\bar{s}_y, \vartheta_{s_y}) = (3000, 200) \text{ psi} \tag{b}$$

$$(\bar{s}_z, \vartheta_{s_z}) = (0, 0)$$

and

$$(\bar{\tau}_{xy}, \vartheta_{\tau_{xy}}) = (12{,}000, 600) \text{ psi} \tag{c}$$

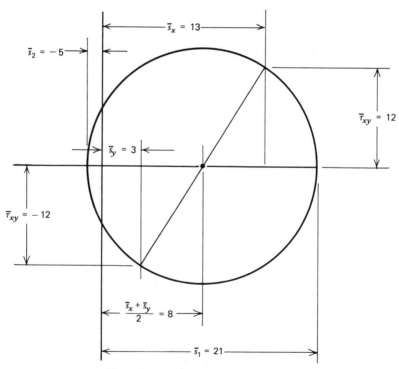

Figure 10.9 Circle of stress (units in ksi).

Distortion Energy Theory (Von Mises–Hencky) 449

The material of the component to be analyzed is ASTM-A7, for which (from Table 8.1)

$$(\bar{S}_y, \,_{s_y}) \approx (40{,}000;\, 4{,}000)\ \text{psi} \tag{d}$$

Assume that all variables are normally distributed and that s_x and s_y are independent.

1. Analysis employing the maximum shear stress theory
    a. The model for predicting shear stress as a function of s_x, s_y, and τ_{xy} is [equation (5.20)]

$$\tau_{max} = \sqrt{\left(\frac{s_x - s_y}{2}\right)^2 + \tau_{xy}^2} \tag{e}$$

From equations (2.59), (a), (b), (c), the mean of τ_{max} is

$$\bar{\tau}_{max} \approx \sqrt{\left(\frac{\bar{s}_x - \bar{s}_y}{2}\right)^2 + \bar{\tau}_{xy}^2} = 13{,}000\ \text{psi}$$

To estimate $\,_{\tau_{max}}$, first estimate the statistics of $z = (s_x - s_y/2)$. From equations (2.8) and (2.22):

$$\bar{z} = \frac{\bar{s}_x - \bar{s}_y}{2} = 5000\ \text{psi}.$$

From equations (2.10) and (2.24), with equations (a) and (b):

$$\,_z = \tfrac{1}{2}\sqrt{(650)^2 + (200)^2} = 340\ \text{psi}$$

From equations (2.62), (2.10), and (c), $\,_{\tau_{max}}$ is

$$\,_{\tau_{max}} \approx \left[\frac{\bar{z}^2\,_z^2 + \bar{\tau}_{xy}^2 \sigma_{\tau_{xy}}^2}{\bar{z}^2 + \bar{\tau}_{xy}^2}\right]^{1/2} \approx 610\ \text{psi}$$

Thus

$$(\bar{\tau}_{max},\,_{\tau_{max}}) = (13{,}000,\, 610)\ \text{psi} \tag{f}$$

b. Shear strength estimated by the maximum shear stress theory, applied to equation (d), is

$$(\bar{S}_{sy},\,_{S_{sy}}) \approx 0.5(\bar{S}_y,\,_{S_y}) \approx (20{,}000,\, 2000)\ \text{psi} \tag{g}$$

450 Strength of Mechanical Components

c. Estimate t by using equation (9.11) with equations (f) and (g):

$$t = -\frac{20{,}000 - 13{,}000}{\sqrt{(2000)^2 + (610)^2}} = -3.35$$

d. If equation (9.10) and Appendix 1 are utilized, the reliability estimate is

$$R = \frac{1}{\sqrt{2\pi}} \cdot \int_t^\infty e^{-t^2/2} dt = \int_{-3.35}^\infty \frac{1}{\sqrt{2\pi}} \cdot e^{-u^2/2} du = 0.9996$$

2. Solution employing the distortion energy theory

a. From equation (10.9) the equivalent stress s_{eq} as a function of s_x, s_y, and τ_{xy} is

$$s_{eq} = \sqrt{s_x^2 + s_y^2 - s_x s_y + 3\tau_{xy}^2} \tag{h}$$

From equation (2.59), the mean-value estimator of equivalent stress is

$$\bar{s}_{eq} \approx 23{,}900 \text{ psi} \tag{h'}$$

From equations (2.62) and (h) the estimator of $\mathfrak{s}_{s_{eq}}$ is

$$\mathfrak{s}_{s_{eq}} \approx \left[\left(\frac{\partial s}{\partial s_x}\right)^2 \cdot \mathfrak{s}_{s_x}^2 + \left(\frac{\partial s}{\partial s_y}\right)^2 \cdot \mathfrak{s}_{s_y}^2 + \left(\frac{\partial s}{\partial \tau_{xy}}\right)^2 \cdot \mathfrak{s}_{\tau_{xy}}^2 \right]^{1/2} \tag{i}$$

Next, taking partial derivatives of s_x, s_y, and τ_{xy} with respect to s and evaluating at the mean values

$$\frac{\partial s}{\partial s_x} = \frac{2s_x - s_y}{2\sqrt{s_x^2 + s_y^2 - s_x s_y + 3\tau_{xy}^2}} = \frac{23.0}{47.8} = 0.48$$

$$\frac{\partial s}{\partial s_y} = \frac{2s_y - s_x}{2\sqrt{s_x^2 + s_y^2 - s_x s_y + 3\tau_{xy}^2}} = \frac{7.0}{47.8} = 0.15 \tag{j}$$

$$\frac{\partial s}{\partial \tau_{xy}} = \frac{6\tau_{xy}}{2\sqrt{s_x^2 + s_y^2 - s_x s_y + 3\tau_{xy}^2}} = \frac{72.0}{47.8} = 1.5$$

Substitute equation (j) into equation (i) and evaluate:

$$\mathfrak{s}_{s_{eq}} \approx 945 \text{ psi} \tag{h''}$$

Thus

$$(\bar{s}_{eq}, {}^s s_{eq}) \approx (23{,}900, 945) \text{ psi} \tag{k}$$

b. Yield strength is [equation (d)]

$$(\bar{S}_y, {}^s S_y) \approx (40{,}000, 4000) \text{ psi} \tag{l}$$

c. Estimate z from equation (9.11) by using values from equations (k) and (l):

$$z = -\frac{40{,}000 - 23{,}900}{\sqrt{(4000)^2 + (945)^2}} = -3.92$$

d. According to the Tables of Normal Probability Functions [40] or Appendix 1, predicted reliability is [equation (9.10)]

$$R = \frac{1}{\sqrt{2\pi}} \cdot \int_z^\infty e^{-u^2/2} du = \int_{-3.92}^\infty \frac{1}{\sqrt{2\pi}} e^{-u^2/2} du = 0.99995$$

Example C is summarized in Table 10.1 and Figure 10.10. For biaxially stressed components such as tubes, pressure vessels, diaphragms, and many others, the criterion for avoiding tensile yield, according to the classical distortion energy theory [equation (10.9)], is

$$S_Y^2 = s_1^2 - s_1 s_2 + s_2^2 \tag{10.11}$$

The probabilistic criterion for avoiding tensile yield is

$$R = P\left[S_Y > \sqrt{s_1^2 - s_1 s_2 + s_2^2} \right] \tag{10.12}$$

Table 10.1 Summary of Example C

Failure Theory Employed	Safety Factor, $\bar{S}/\bar{s}$	Reliability for Normal Strength and Normal Stress (Predicted)	Reliability for Normal Strength and Lognormal Stress (Predicted)
Maximum shear strength theory	1.54	.9998	$R \approx .9994$
Distortion energy theory	1.67	.99995	$R \approx .99985$

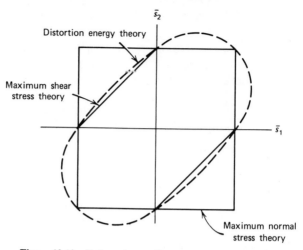

Figure 10.10 Failure theory diagrams of mean values.

Example D Thin-walled Cylinder

In a thin-walled cylinder the model for tangential stress is [from equation (5.10)]

$$s_1 = \frac{Pd}{2t}$$

The longitudinal stress model is [from equation (5.11)]

$$s_2 = \frac{Pd}{4t}$$

As P is a random variable, s_1 and s_2 are random variables correlated $\rho = +1$, that have the type of distribution determined by the (random) function pd/t. In this case $s_1 = 2s_2$;

$$S_y^2 = 4s_2^2 - (2s_2)s_2 + s_2^2 = 3s_2^2$$

and

$$S_y > \sqrt{3} \cdot s_2 = \frac{\sqrt{3}}{2} \cdot s_1$$

According to the probabilistic distortion energy theory of failure:

$$R = P\left(S_y > \sqrt{3s_2} = \frac{\sqrt{3}}{2} \cdot s_1\right)$$

Example E Thin-walled Pressure Vessel

Consider a thin-walled vessel assumed to be a surface of revolution. The vessel is subjected to internal pressure of magnitude p symmetrically distributed with respect to the axis of revolution, with no sharp bends or other discontinuities:

$$\frac{s_1}{r_1} + \frac{s_2}{r_2} = \frac{p}{t}$$

1. In the case of a spherical vessel, $r_1 = r_2$ and $s_1 = s_2$ (correlated $\rho = +1$):

$$s_1 = s_2 = \frac{pr_1}{2t} = \frac{pd}{4t}$$

Then from equation (10.11):

$$S_y^2 > s_1^2 - s_1 s_2 + s_2^2$$

$$S_y^2 > s_1^2 = s_2^2$$

$$S_y > s_1 = s_2$$

From the probabilistic distortion energy theory of failure, the design criterion is

$$R \leqslant p\{S_y > s_1 = s_2\}$$

2. Assume that

$$S_y^2 = s_1^2 - s_1 s_2 + s_2^2$$

and s_1 and s_2 are statistically independent random variables:

$$s_{eq} = \sqrt{s_1^2 - s_1 s_2 + s_2^2} \tag{a}$$

Taking partial derivatives of the variables in equation (a) with respect to s_{eq} yields

$$\frac{\partial s_{eq}}{\partial s_1} = \frac{2s_1 - s_2}{\sqrt{s_1^2 - s_1 s_2 + s_2^2}} \,;\quad \frac{\partial s_{eq}}{\partial s_2} = \frac{2s_2 - s_1}{\sqrt{s_1^2 - s_1 s_2 + s_2^2}}$$

From equation (2.62):

$$\mathring{s}_{eq} \approx \left[\frac{(2\bar{s}_1 - \bar{s}_2)^2}{\bar{s}_1^2 - \bar{s}_1 \bar{s}_2 + \bar{s}_2^2} \cdot (\mathring{s}_{s_1}^2) + \frac{(2\bar{s}_2 - \bar{s}_1)^2}{\bar{s}_1^2 - \bar{s}_1 \bar{s}_2 + \bar{s}_2^2} \cdot (\mathring{s}_{s_2}^2) \right]^{1/2}$$

10.5 SIGNIFICANT STRESS–STRENGTH MODELS

10.5.1 Significant Strength Models [75]

The endurance limit of an engineering material (steel) in a conventional S–N envelope is denoted S_e for a rotating-beam specimen (see Section 8.14.1). The usual model for the significant endurance limit S'_e of a component is

$$S'_e = S_e \cdot k_b \cdot k_d \cdot \ldots . \tag{10.13}$$

$$\left(\bar{S}'_e, {}_{\scriptscriptstyle \circ S'_e}\right) = \left(\bar{S}_e, {}_{\scriptscriptstyle \circ S_e}\right) \cdot \left(\bar{k}_b, {}_{\scriptscriptstyle \circ k_b}\right) \cdot \left(\bar{k}, {}_{\scriptscriptstyle \circ kd}\right) \cdot \ldots . \tag{10.13a}$$

where k_b is the size factor (see Figure 10.16), k_c is the reliability factor, k_d is the temperature factor (see Figure 10.17), and k_f is the miscellaneous effects factor.

Strength design factors are discussed in Section 10.6 and employed in design examples in later sections. The value k_c should be omitted from the product string as its influence is restored by treating all other factors as random variables. If each factor is a random variable, then regardless of the distributional form of each above-mentioned random variable, the distribution of S'_e will be skewed right (see Section 2.9.1).

The S–N diagram is usually modified to reflect the effects of strength design factors. For example, the value of point D in Figure 10.11 is the product $S_e \cdot k_d$, and the value of point B is the product $S_e \cdot k_d \cdot k_b$. According to sources in the literature [27], there may also be a modification of the S–N curve at 10^3 cycles.

The way in which the k values are applied must be carefully considered, to satisfy probabilistic requirements. Each design factor is a random variable and thus affects both the mean value and the distributional form of the product distribution. According to a central limit theorem for products of

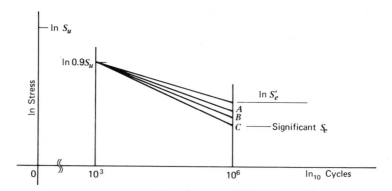

Figure 10.11 Deterministic S–N curve modified for design factors.

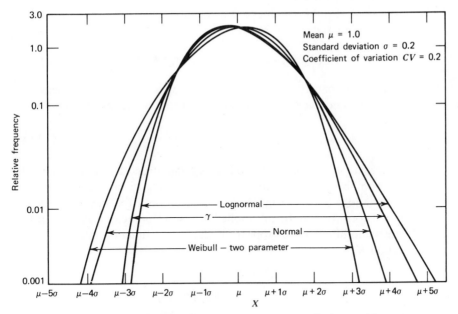

Figure 10.12 Relative frequencies for common distribution models.

random variables, the *distribution of a product series approaches the lognormal form as an asymptotic limit* [20] as the number of random variables increases.

For a given mean value and standard deviation, significant material strength (if lognormal) tends to produce a higher reliability (component or system) population, on the average, than does normally distributed strength. Conversely, the assumption that strength is normally distributed tends to produce conservative designs (see Figures 9.33 and 9.34) (strength negatively skewed may require special attention [68]; see Figure 10.12).

Since significant strength is that displayed by the component in its use environment, only those design factors that actually add realism to the materials behavior model should appear in a significant strength expression. Clearly, temperature is an applicable strength design factor [44]. Increasing temperature tends to reduce tensile ultimate and yield strength and modulus of elasticity, whereas lowering temperature produces the reverse effect. The notch sensitivity effect may also modify strength, as it is a property of a specific material (see Section 10.6 and Figure 10.21). Also, variations in metallurgic effects within the body of a material may reflect on the mechanical behavior.

10.5.2 Significant Stress Models

If simple applied stress (s) is considered, an equation often given [1] is

$$s = cf(x_1, x_2, \ldots, x_n) \cdot F(F_1, F_2, \ldots, F_r) \qquad (10.14)$$

where c is a constant, the x values are geometric values, and the F values are forces. A model for significant stress is

$$s' = F(F_1, F_2, \ldots, F_r) f(x_1, x_2, \ldots, x_n) \cdot K_a \cdot K_t \cdots . \qquad (10.15)$$

where each F_i is a load random variable, each x_i is a geometric dimensional random variable, and each K_i is a random-value stress design factor. Thus

$$(\bar{s}', \circ_{s'}) = (\bar{s}, \circ_s) \cdot (\overline{K}_a, \circ_{K_a}) \cdot (\overline{K}_t, \circ_{K_t}) \cdots . \qquad (10.15a)$$

The lognormal may usually be considered the stress distributional model associated with the lower bound on reliability for a given mean value and standard deviation. Stress design factors modify both the mean values and distributional form of significant applied stress. Thus factors that do in fact model stress modifiers must appear in the stress function, such as stress concentration K_t.

DESIGN FACTORS

10.6 STRENGTH AND STRESS MODIFIERS

Among the most dramatic modifiers encountered in design are those due to thermal effects on strength in general [78, 79] and stress concentration effects on local stress magnitudes. Chapter 7 was devoted to a discussion of geometric stress concentration, which has been studied by computer simulation for a number of geometric configurations and size ranges, and the general conclusions are that stress concentration is a random phenomenon whose effects are aggravated as notch and fillet radii become smaller. Goodness-of-fit tests usually favor the normal distribution as a model of K_t [15].

Several modifiers of strength and of stress are considered in the following discussion. Their influences on mean values, standard deviations, and distributions of significant strength and stress are shown in subsequent examples.

10.6.1 Strength Modifiers

Since endurance limit and fatigue strength are basic materials behavior properties, care must be exercised in applying only factors that in reality modify these materials behavior properties in design. Some of the factors that can be expected to modify the endurance limit and fatigue strength and the shape of their distributions are (1) temperature (see Figures 10.16–10.18), (2) size factor (metallurgy and basic material structure are to some extent functions of size), (3) directional characteristics of the material, (4) notch sensitivity, (5) corrosion, and (6) plating.

10.6.2 Stress Modifiers

Some of the factors that can be expected to modify applied stress and the shape of its distribution are (1) geometric stress concentration (Chapter 7) K_t, (2) fatigue stress concentration factor, (3) surface finish (Figure 10.13) K_a, (4) internal defects, (5) residual stresses, and (6) assembly stresses. Following the suggestion of Marin that modifying factors can be used to account separately for these effects, equations (10.13) and (10.15) are utilized.

Surface Factor. With dynamic bending and torsional loading, dynamic stress amplitude tends to be increased locally with increasing surface roughness of a component (see Figure 10.13), as follows:

Ground:
$$\mu_{k_a} = 1.006278 - 0.71493\,(10^{-6})\mu_{S_u}$$
$$\sigma_{k_a} = 0.103$$

Machined:
$$\mu_{k_a} = 0.9472 - 0.159\mu_{S_u}$$
$$\sigma_{k_a} = 0.0406$$

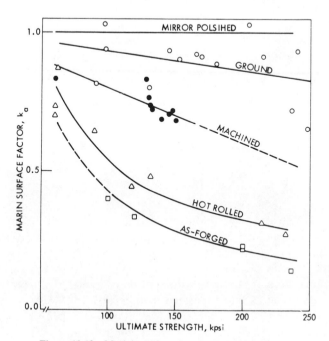

Figure 10.13 Marin's surface factors [101, Mischke].

Hot rolled:

$$\mu_{k_a} = \frac{20919 + 0.0545\,\mu_{S_u}}{\mu_{S_{u/2}}}$$

$$\sigma_{k_a} = \frac{2780.5}{\mu_{S_{u/2}}}$$

As forged:

$$\mu_{k_a} = \frac{20955 - 0.00266\,\mu_{S_u}}{\mu_{S_{u/2}}}$$

$$\sigma_{k_a} = \frac{2763}{\mu_{S_{u/2}}}$$

Effect of Size. With dynamic bending and torsional loading, endurance limit and fatigue strength tend to decrease as size increases, with a considerable amount of scatter observed in test values (see Figures 10.14 and 10.15). Small elements usually exhibit higher endurance strength values than do standard 0.30-in.-diameter specimens of the same materials. Some filaments display dramatic increases in strength [102]. As component diameters increase above 0.30 in., endurance strength tends to decay, reaching mean values (accompanied by a considerable amount of scatter) considerably less than 1.00, on the basis of $k_b = 1.00$ for a 0.3-in.-diameter specimen. The data for Figure 10.15 are as follows:

$$\mu_{k_b} = 1 \qquad\qquad 0 < x \leqslant 0.3$$
$$\mu_{k_b} = 1 - 0.08696(x - 0.3) \qquad 0.3 < x \leqslant 1$$
$$\mu_{k_b} = 0.97 - 0.029787x \qquad 1 < x \leqslant 6$$
$$\sigma_{k_b} = 0.051282(x - 0.3) \qquad 0 < x \leqslant 2.3$$
$$\sigma_{k_b} = 0.13 - 0.00833x \qquad 2.3 < x \leqslant 6$$

where μ_{k_b} is the mean value of Marin's fatigue modification factor k_b and σ_{k_b} is the standard deviation of Marin's fatigue modification factor k_b.

Figure 10.14 gives a summary of test results obtained by various investigations, using steels in the ultimate strength range of 50 to 165 ksi. Based on results such as these, it may be postulated that the mean value of size effect follows a curve such as that suggested in Figure 10.14. Moore found for the materials in Figure 10.14 that the mean deviation was 4.2% of the mean value [103]. Thus $\sigma_{k_b} \simeq$ (mean deviation)/0.7979 if the random variable is Gaussian

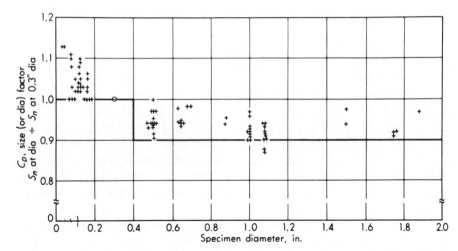

Figure 10.14 Size effect [27]: bending and torsional loads (data are for steel, $S_y = 50$ to 165 ksi). (*Note*: Test points were compiled from data reported by many investigators.)

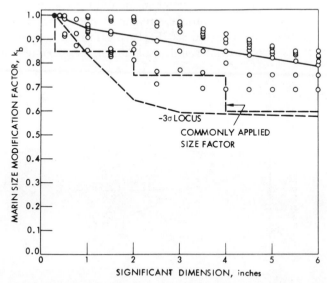

Figure 10.15 Marin's fatigue modification factor k_b (size) for carbon steels of diameter 0.3 to 6 in. [101].

459

distributed:

$$s_{k_b} \simeq \frac{0.042\bar{k}_b}{0.7979} = 0.053\bar{k}_b \tag{10.16}$$

At diameters of 4 to 12 in. the apparent mean size factors are in the range 0.65 to 0.75. However, these may not be true size effects but may result from metallurgical factors related to component size [27, p. 232]. Generally, uniform microstructure is not obtained throughout very large heat-treated components. Additionally, poorer surface finishes and undetected residual stress fields may contribute to apparent lower fatigue strength.

In the absence of standard deviations estimated from data, a minimum value of 6% of the mean is suggested.

The effect of size on static and low cycle fatigue strength is not pronounced and is commonly neglected, except for large components.

Temperature Factor. Elevated temperature characteristics, static and dynamic, of a number of commonly used metal alloys are described by means of Weibull parameters (as illustrated in Figures 10.19 and 10.20) by Lipson et al. [78, 79]. Thus for use in elevated temperature design, the mean value and standard deviation estimators of materials behavior may be readily obtained by applying the equations for $\bar{S}_f$ and C_S in Section 8.14.1.

Curves in Figure 10.16 provide for the mean and the standard deviation of the temperature factor k_d for fatigue strength modification of carbon and low-alloy steels over the range 0 to 800°F, [101]. Effects of temperature on the fatigue limit of gray iron are deterministically illustrated in Figure 10.18.

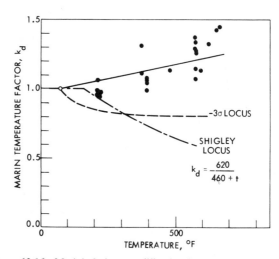

Figure 10.16 Marin's fatigue modification factor k_d (temperature).

Strength and Stress Modifiers   461

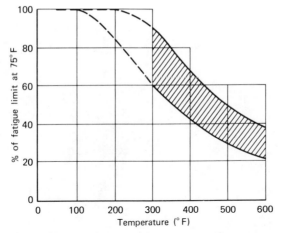

Figure 10.17 Generalized effect of elevated temperature on fatigue strength at 5×10^5 cycles (smooth, rotating, bending); stress ratio $= -1$. [3, p. 52].

Mechanical components operating in an elevated-temperature (Figure 10.17 and 10.18) environment may be prone to failure due to creep and/or fatigue. Materials become much more susceptible to corrosion at elevated temperature, thus reducing the load-carrying capacity of the component. Stress concentrations are developed and metallurgical changes in the material result, particularly at the surfaces.

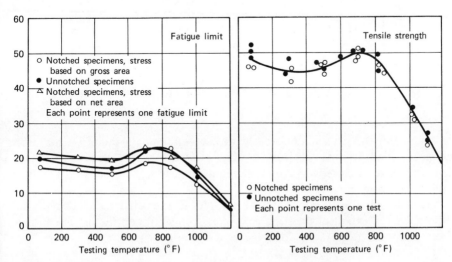

Figure 10.18 Effects of temperature on fatigue limit of gray iron; composition: 2.84 c, 1.52 Si, 1.05 Mn, 0.07 P, 0.12 S, 0.31 Cr, 0.20 Ni, 0.37 Cu [48, p. 356].

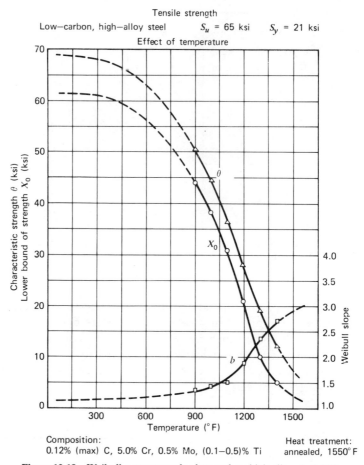

Figure 10.19 Weibull parameters for low-carbon high-alloy steel [78].

NOTCH SENSITIVITY

In Section 7.2 it was noted that with ductile materials, modifiers for stress concentration are not necessary in cases of static loading. Studies of fatigue failures have disclosed that some materials appear to be more sensitive than others to geometric discontinuities.

Thus the modification of nominal stress for less sensitive materials need not be to the full extent of the geometric stress concentration factor. The fatigue stress concentration factor for a material is defined as

$$K_f = \frac{S_e \text{ (notch free)}}{S_e \text{ (notched)}} = \frac{S_e}{S_e'} \qquad (10.17)$$

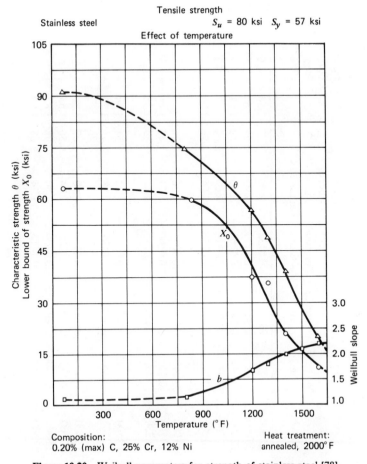

Figure 10.20 Weibull parameters for strength of stainless steel [78].

Since S_e (notch free) and S_e' (notched) are random variables, the quotient K_f is also a random variable; S_e' and S_e are correlated $\rho = +1$. Notch sensitivity q is defined as

$$q = \frac{K_f - 1}{K_t - 1} \tag{10.18}$$

where K_f has been determined experimentally from fatigue tests [59, p. 13]. See Table 10.2 for typical deviations in q for carbon and low-alloy steels.

Example F

For a rough estimation of the statistics of q for SAE 1020 as rolled steel, see Table 10.2. Assume the mean values $\overline{S}_e$ unnotched $\approx 34{,}000$ psi,

Table 10.2 Check of Computed Strength Reduction Factor (k') with Strength Reduction Factor Determined Directly by Fatigue Tests

(1) Metal	(2) Diameter of Specimen, d, in.	(3) Endurance Limit from Tests S'_e, m psi. Un-notched	(4) Endurance Limit from Tests S'_e, m psi. Notched	(5) Radius at Root of Notch, p, (in.)	(6) Depth of Notch, t, (in.)	(7) p' from Formula (in.)	(8) "Theoretical" Stress Concentration Factor, $k_t{}^b$	(9) Strength Reduction Factor, k'	(10) Strength Reduction Factor from Tests, (3)/(4)	(11) Deviation of (10^c) from (9)%
SAE 1020 Steel, as rolled (series II)	0.125	34000	22000	0.01	0.01	0.0133	2.00	1.47	1.54	−4.6
	0.250	32000	20000	0.02	0.02	0.0177	2.00	1.51	1.60	−5.6
	0.500	28900	17800	0.04	0.04	0.0200	2.00	1.59	1.62	−1.9
	1.000	28200	17200	0.08	0.08	0.0210	2.00	1.66	1.67	−0.6
	1.875	28200	16000	0.14	0.14	0.0216	2.06	1.76	1.76	0
SAE 1020 steel, strain relieved (series I)	0.125	—	23000	0.0625	0.0625	0.0087	1.26	1.26	—	—
	0.160	29000	20500	0.0200	0.0200	0.0100	1.74	1.43	1.41	+1.4
	0.250	29000	21250	0.0625	0.1250	0.0116	1.47	1.33	1.36	−2.2
	0.500	28000	19500	0.0625	0.0625	0.0131	1.74	1.51	1.44	+4.9
	0.500	28000	16500	0.0200	0.0200	0.0131	2.38	1.76	1.70	+3.5
	1.000	28000	15500	0.0200	0.0200	0.0138	2.60	1.87	1.81	+3.3
	1.000	28000	18000	0.125	0.125	0.0138	1.74	1.56	1.55	+0.6
	2.000	28000^d	15000	0.0200	0.0200	0.0200	2.80	1.98	1.87	+5.9
	2.000	28000^d	17500	0.2500	0.2500	0.1415	1.74	1.60	1.60	0
SAE 1035 steel, as rolled (series II)	0.125	39000	27000	0.01	0.01	0.0117	2.00	1.47	1.44	+2.1
	0.250	39000	25000	0.02	0.02	0.0156	2.00	1.55	1.56	−0.6
	0.500	35000	22000	0.04	0.04	0.0175	2.00	1.60	1.59	+0.6
	1.000	34400	19300	0.08	0.08	0.0185	2.00	1.68	1.78	−5.6
	1.875	34400	19300	0.14	0.14	0.0189	2.06	1.78	1.78	0

Material	d	S_u			ρ'	k_t	k'	k	Dev. %
SAE 1035 steel, bright anneal (series II)[e]	0.250	39000	0.02	0.02	0.0136	2.00	1.55	1.66	−7.1
SAE 2345 steel, heat treated (series I)	0.125	70250	0.0625	0.0625	0.00107	1.26	1.20	1.24	−3.2
	0.160	70750	0.0200	0.0200	0.00123	1.75	1.60	1.59	+0.7
	0.250	66750	0.0625	0.1250	0.00143	1.46	1.43	1.34	+6.7
	0.300	70000	0.0625	0.0625	0.00148	1.54	1.47	1.45	+1.4
	0.300	70000	0.0200	0.0200	0.00148	2.10	1.87	1.89	−1.1
	0.300	70000	0.0375	0.0375	0.00148	1.75	1.65	1.61	+2.5
	0.500	66500	0.0200	0.0200	0.00160	2.38	2.07	1.99	+3.0
	0.500	66500	0.0625	0.0625	0.00160	1.75	1.64	1.59	+3.1
	0.875	64000	0.0625	0.0625	0.00169	2.06	1.92	1.83	+4.9
	1.500	66500	0.0625	0.1250	0.00173	2.54	2.32	2.25	+3.1
SAE X4130 steel (series II)	0.125	75400	0.01	0.01	0.00072	2.00	1.79	1.95	−8.2
	0.250	69800	0.02	0.02	0.00096	2.00	1.82	1.87	−2.7
	0.500	65000	0.04	0.04	0.00108	2.00	1.86	1.82	+2.2
	1.000	63600	0.08	0.08	0.00114	2.00	1.87	1.84	+1.6
	1.750	63600	0.14	0.14	0.00116	2.06	1.91	1.83	+4.4

[a] The Neuber constant ρ' is computed from the formula $\rho' = 0.2[1 - (S_y/S_u)]^3 \times (1 - 0.05/d)$, in which S_y is the yield strength of the metal, S_u the tensile strength, and d is the diameter of the specimen. Then the "theoretical" stress concentration factor k_t may be determined by the use of the Neuber diagram and using the value of ρ' obtained by the formula and knowing the angle between the straight sides of the notch (in semicircular notches, and in notches cut "straight in" perpendicular to the surface, the flank angle ω is zero) the computed strength reduction factor k' may be found by the use of the second Neuber diagram. In both Neuber diagrams ρ denotes the radius at the bottom of the notch.
[b] From Neuber diagrams.
[c] Mean deviation of predicted values of strength reduction factor k' from test values (disregarding sign) = 2.92%, σ = 3.7%; maximum deviations = +6.7% and −8.2%.
[d] Unnotched specimen 1.875 in. in diameter.
[e] These specimens were notched and polished *after* annealing.

Source: *ASTM Proceedings*, Vol. 42, 1942.

$\bar{S}'_e$ notched $\approx$ 22,000 psi, and theoretical $\bar{k}_t \approx 2.00$. From equation (10.17):

$$\bar{k}_f \approx \frac{34{,}000}{22{,}000} = 1.55$$

Utilization of equation (2.46) yields

$$\bar{q} \approx \frac{\bar{K}_f - 1}{\bar{K}_t - 1} = \frac{1.55 - 1}{2.00 - 1} \approx 0.55 \qquad (10.18a)$$

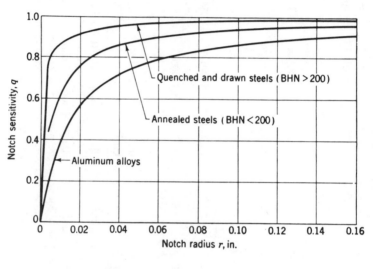

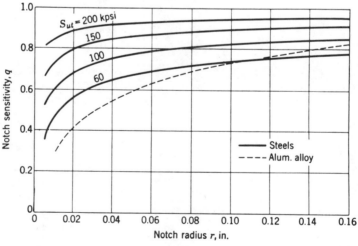

Figure 10.21 Notch sensitivity for steels and aluminum alloys Shigley. [1]

Assuming:

$$s_{K_t} \approx (0.06)(\bar{K}_t) = 0.12$$

$$s_{K_f} \approx (0.06)(\bar{K}_f) = 0.093$$

Utilization of equation (2.49) with equation (10.18) yields

$$s_q \approx \sqrt{\frac{(\bar{K}_t-1)^2 s_{K_f}^2 + (\bar{K}_f-1)^2 s_{K_t}^2}{(\bar{K}_t-1)^4}} = 0.114 \qquad (10.18b)$$

where $(\bar{q}, s_q) = (0.55, 0.114)$. Further,

$$q = \frac{K_{tf}-1}{K_t-1} \qquad (10.19)$$

where K_{tf} is a calculated factor based on average values of q obtained from Figure 10.21 or other such curves. Figure 10.21 indicates the fit of limited experimental results to q curves for steels and aluminum alloys. Normally, average values of q can be expected to fall within the range 0 to 1. Material lacking sensitivity to notches $q=0$ has a value $K_f=1$ or $K_{tf}=1$. In instances where a material is completely sensitive, $q=1$ would involve important statistical considerations. Both limit values of q imply that K_t and K_{tf} are correlated, $\rho = +1$.

For large-notch radii and high-strength materials $q \to 1$ and $K_{tf} \to K_f$. It should be noted that the behavior of q as a random variable is in need of study and experimental development.

10.7 DESIGN TO AVOID FATIGUE FAILURE

In Chapter 9 we examined the problem of design to avoid fatigue failure with a stated probability as it applied to geometrically very simple elements. The models used involved only basic mechanical materials strength and nominal load stress statistics. Next, the more realistic problem of designing typical mechanical components to satisfy specified cycle lives with stated probabilities R is examined. In Chapter 5 basic stress (attributable to various types of loading) was modeled as a random variable. However, in the design of typical mechanical components to avoid fatigue failure, the critical stress is s^1. Significant stress as discussed in Section 10.5.2. is that attributable to direct loading or bending modified by the appropriate stress design factors. In

468 Strength of Mechanical Components

Chapter 8 the various types of basic mechanical behavior of materials, also modeled as random variables, were studied.

From test series results it was found that realistic models of basic mechanical behavior are obtained because effects on basic material behavior arising from existing cracks, inclusions, and so on (in new materials) are considerations inherent in statistical models of materials strength. These are prime concerns in probablistic fracture mechanics. A multiplicity of test results, when reduced, reflect overall representative behavior of realistically imperfect materials. The tendency in a large number of specimens is to find a representative pattern of defects. The law of large numbers is operational if typical processing and machining practices are used and if test pieces are randomly selected.

In instances where cracks are detected early in components in service, probablistic fracture mechanics, outlined in Section 10.16, should be employed to estimate the residual lives (statistics) of damaged components [104]. It is notable that there are devices, such as large rifled gun tubes, in which cracking normally begins almost immediately. On the other hand, in ordinary fatigue a substantial fraction of a components life may be devoted to the crack initiation phase.

The realistic design of typical mechanical components to avoid fatigue failure involves modeling significant strength (s^1) statistically as discussed in Section 10.5.1. The significant strength of a component is estimated employing an expression of basic material fatigue strength modified by appropriate strength design factors.

Characteristics of $S-N$ envelopes are described now. As we have seen in Chapter 8, in the development of a specific $S-N$ envelope, a number n of specimens are exposed to dynamic loading at each of N constant amplitude levels. At each of the N levels, n cycle lives to failure are determined, each associated with a different specimen. These data are reduced to statistics for each of the selected levels (see Section 8.14.1). The following observations apply in the use of statistical $S-N$ envelopes in design:

1. The statistical $S-N$ envelope for a given material and condition appears to be unique.

2. For fixed cycle life, the distribution of stress at failure may be approximately normal (or two-parameter Weibull) (see Figure 8.26).

3. For a fixed level of stress, the distribution of cycles to failure is skewed to the right and appears to be approximately described by a lognormal distribution (see Figure 10.17).

4. At any specified cycle life n_i the distribution of stress at failure reflects *experiencing and surviving* $n_i - 1$ cycles of exposure to the sinusoidal constant amplitude-loading, *including any transient peaks*.

On the basis of these observations, the engineer can directly interpolate values from statistical $S-N$ envelopes for use in the design of components and systems expected to experience and survive specified cycles of loading, with a predicted R.

Utilization of the $S-N$ envelope is as follows:

1. *Dynamic behavior of ferrous materials.*
   a. In design for *endurance*, $\bar{S}_e$ and s_{S_e} are estimated directly from $S-N$ envelopes such as those in Appendix 10.B.
   b. In *finite-life* design, $\bar{S}_f$ and s_{S_f} are estimated from an appropriate figure, such as Figure 8.29a by interpolation to the applicable cycle life, or from Appendix 10.B.
   c. In design for *low-cycle fatigue*, the low-cycle life statistics of S_f are estimated from an appropriate figure, such as Figure 8.33, by interpolation to the applicable cycle life. To estimate the needed mean and $\pm 3s$ values it is sometimes necessary to estimate the mean value as $3s$ loci, as indicated by the dashed lines in Figure 8.33.
2. *Dynamic behavior of aluminum alloys.* Estimations of the mean and standard deviations of S_f for required cycle lives can be obtained from composite plots of test results such as Figure 8.32, or better using data from Appendix 10.B.

10.7.1 Constant-amplitude Stresses (Zero Mean)

A large class of design problems involving dynamic loading (hence dynamic stressing) of mechanical components can be approached in a simple way. Consider rotating shafts, axles, crank shafts, spindles, and so on that may be subjected to sinusoidal fully reversed essentially bending loads. Another group of components within this class is subjected to sinusoidal fully reversed essentially axial loads. $S-N$ Envelopes describe stress at failure of materials subjected to dynamic loading, and these are modified by applying the appropriate fatigue strength design factors for use in design.

Design for constant-amplitude loading. The stress or strength model is developed by isolating and describing the essential design variables. Sensitivity studies (see Section 2.7.3) disclose the variables of prime importance. In the example (Section 9.3.3) the sensitivity of reliability to variations in tolerances was examined. Example G, which follows, illustrates finite-life design of a simple element, uniaxially stresses, at constant amplitude. The following task is the design of a flat spring whose anticipated loading will be constant-amplitude sinusoidal with zero mean. Since the distributions of applied load stress and strength are unknown, the design procedure for distribution of S and s unspecified (in Section 9.3.9) is utilized, and the design criterion is $R \leqslant P(S_f > s)$.

Example G

1. *Flat-spring design.*

Shown in Figure 10.22 is a model of the flat spring (cross section shown in Figure 10.23) to be manufactured from AISI 4340 alloy steel whose constant-amplitude fatigue strength is given in Figure 8.41. The loading is applied at

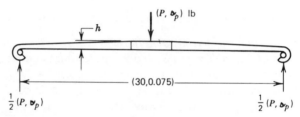

Figure 10.22 Flat spring.

the midpoint and perpendicular to the axis of symmetry of the spring. The finite life design is for $n = 100{,}000$ cycles of loading. The specified probability of survival is $R = .9965$ since replacement cost is modest and public safety is not a consideration. If loading is a stationary Gaussian process, load-induced stress is nonetheless non-Gaussian and often nearly lognormal. The loading considered (effect due to spring mass is negligible) is constant-amplitude sinusoidal:

$$\left(\bar{P}, s_p\right) = (1800, 0) \text{ lb}. \tag{a}$$

The random-variable dimension to be determined is h. From applicable manufacturing tolerances the standard deviation on h is estimated as

$$s_h \approx 0.015\bar{h}$$

2. Design the spring to sustain constant-amplitude stressing [see equation (10.15a)]. In this example it is assumed that $K_t \approx 1$, $K_a \approx 1$ since the component geometry involves no stress raisers and the surface will approach a ground finish; thus $(\bar{s}, s_s) \approx (\bar{s}', s_{s'})$. As in Example A, the formula used to estimate surface fiber stress is [equation (5.23)]

$$s = \frac{Mc}{I} = \frac{6M}{bh^2} \text{ psi}$$

$$\left(\bar{M}, s_M\right) = \frac{\left(\bar{P}, s_P\right)}{2} \cdot \frac{\left(\bar{l}, s_l\right)}{2}$$

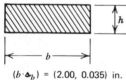

$(b \cdot s_b) = (2.00, 0.035)$ in.   **Figure 10.23** Flat spring cross section.

From equations (2.25) and (2.10):

$$\overline{M} = \tfrac{1}{4}(1800)(30.00) = 13{,}500 \text{ in. lb}$$

From equations (2.27) and (2.10):

$$\sigma_M = \tfrac{1}{4}\sqrt{(1800)^2(0.075)^2 + (0)^2(30.00)^2 + (0)^2(0.075)^2} = 34 \text{ in. lb}$$

Thus

$$6(\overline{M}, \sigma_M) = (81{,}000, 200) \text{ in. lb}$$

Let $u = (6M/b)$; then from equations (2.8) and (2.10):

$$(\overline{u}, \sigma_u) = \frac{(81{,}000, 200)}{(2.00, 0.035)}$$

Application of formulas (2.28) and (2.30) yields

$$\overline{u} \approx 40{,}500$$

and

$$\sigma_u \approx \sqrt{\frac{(81{,}000)^2(0.035)^2 + (200)^2 \cdot (2.00)^2}{(2.00)^4}} = 710$$

$$(\overline{u}, \sigma_u) \approx (40{,}500, 710)$$

From equations (2.16) and (2.18):

$$(\overline{h^2}, \sigma_{h^2}) \approx (\overline{h}^2, 2\overline{h} \cdot \sigma_h)$$

From equations (2.28) and (2.30), dynamic applied stress is

$$(\overline{s}, \sigma_s) = \frac{(40{,}500, 710)}{(\overline{h}^2, 0.03\overline{h}^2)} = \left[\frac{40{,}500}{\overline{h}^2}, \frac{1400}{\overline{h}^2} \right] \text{psi}$$

$$C_s = \frac{1400}{40{,}500} = 0.035. \tag{b}$$

3. The strength at 100,000 cycles of constant-amplitude loading is estimated as (Figure 8.41)

$$(\overline{S}_f, \sigma_{S_f}) \text{ psi} \approx (85{,}000, 4100) \text{ psi} \tag{c}$$

(*Note*: Temperature will be ambient, that is, $K_d = 1$, and $K_b \approx 1$ is used in the first design iteration.)

4. Substitute equations (b) and (c) into equation (9.11) and apply a fictitious reliability of $R' = .999$,* for which $z = -3.0$:

$$z = -3.0 = -\frac{85{,}900 - (40{,}500/\bar{h}^2)}{\sqrt{(4100)^2 + (1400/\bar{h}^2)^2}} \quad \text{(d)}$$

and solve equation (d) for $\bar{h}$:

$$\bar{h} = 0.752 \text{ in.}$$

$$s_h \approx 0.011 \text{ in.}$$

$$\Delta_h = \pm 0.033 \text{ in.}$$

FINITE-LIFE DESIGN WITH STRESS CONCENTRATION

Example H

1. Initial considerations are as follows. The problem here is to design a tubular rotating shaft (see Figure 10.25) supported by self-aligning bearings. The loading is sinusoidal fully reversed constant amplitude. A shaft, selected at random from among many, is intended to survive 65,000 cycles of loading (see Section 8.14.1). High performance is required, avoiding undue conservatism; thus $R = .99999$ is specified. All variables are treated as normally distributed, and normal design theory is used, based on the procedure with S and s normal described in Section 9.3.2.

The presence of a geometric stress concentration is unavoidable because of the manner of load support. The following rough estimation of K_t is

$$(K_t, s_{K_t}) = (1.20, 0.06) \quad \text{(a)}$$

Loading is

$$(\bar{p}, s_p) = (6070; 200) \text{ lb} \quad \text{(b)}$$

2. The stress model.

    a. Estimate nominal load stress. The stressing is essentially constant-amplitude sinusoidal bending, as a result of P acting perpendicular to the shaft axis of symmetry. The value $\bar{P}$ is the mean amplitude

---

*See Section 9.3.9 (subsection on reliability of normal vs. lognormal distributed stress) for the estimation of R'.

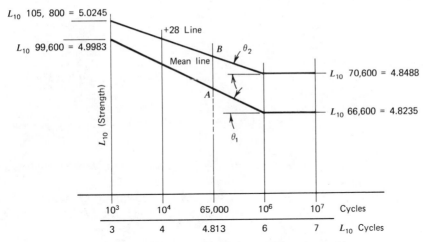

Figure 10.24 Estimated statistical $S-N$ envelope (no scale).

of the loading, and s_p takes cognizance of a small element of peak load variability (startup, etc.). It is assumed that the response of the component is linear, frequency is not near resonance, and strain rate is moderate and does not effect material behavior nor involve a stress multiplying factor much greater than unity. A further assumption is that phase is not important. Also see equation (10.15a), the general model for significant stress.

b. Geometry and nominal stress are as follows. The geometry of the shaft and point of load application are (see Figure 10.11)

$$(\bar{l}, s_l) = (120, 0.0416) \text{ in.}$$

$$(\bar{a}, s_a) = (72, 0.0416) \text{ in.}$$

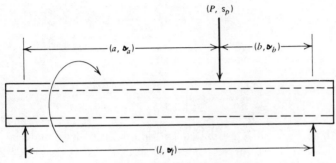

Figure 10.25 Rotating tubular shaft.

Maximum moment is (from Example C, Chapter 9):

$$(\overline{M}, \gt_M) = (174{,}800, 5760) \text{ in. lb} \quad (c)$$

Maximum fiber stress in the shaft is [from equation (5.23)]

$$(\bar{s}, \gt_s) = \frac{(\overline{M}, \gt_M)}{(\bar{I}/\bar{c}), (\gt_I/c)} \text{ psi}$$

As in Example C, Chapter 9, the approximate expression for the moment of inertia of a tubular section is $I \approx \pi r^3 t$, and $\bar{r}/\bar{t} \approx 50$ or $\bar{t} \approx (\bar{r}/50)$, is required for stability (see beginning of Section 9.3.4). From equation (2.59), applied to the model for I/c (where $c=r$):

$$\left(\frac{\bar{I}}{\bar{c}}\right) = \frac{\pi \bar{r}^3 \bar{t}}{\bar{c}} = (\pi \bar{r}^2) \cdot \left(\frac{\bar{r}}{50}\right) = 0.0628 \bar{r}^3$$

and

$$\gt_{I/c} \approx 0.063 \cdot (0.045 \bar{r}^3) = 0.00282 \bar{r}^3 \quad (d)$$

The expression for nominal applied stress Mc/I, utilizing equations (c) and (d), is

$$(\bar{s}, \gt_s) = \frac{(174{,}820, 5760)}{(0.06283 \bar{r}^3, 0.002827 \bar{r}^3)} \text{ psi}$$

$$(\bar{s}, \gt_s) = \left[\frac{2.782 \cdot 10^6}{\bar{r}^3}, \frac{15.52 \cdot 10^4}{\bar{r}^3}\right] \text{ psi} \quad (e)$$

c. The statistics of significant stress are obtained as follows. If nominal stress is modified for stress concentration K_t, the product of equations (e) and (a), and if equation (10.15a), is applied, then

$$(\bar{s}', \gt_{s'}) = \left[\frac{2.782}{\bar{r}^3} \cdot 10^6, \frac{15.52}{\bar{r}^3} \cdot 10^4\right] \cdot (1.20, 0.06) \text{ psi}$$

The statistics of significant stress are

$$(\bar{s}', \gt_{s'}) \approx \left[\frac{3.339}{\bar{r}^3} \cdot 10^6, \frac{2.500}{\bar{r}^3} \cdot 10^5\right] \text{ psi} \quad (f)$$

3. Fatigue strength is estimated as follows. The shaft material is a low-alloy steel, AISI 4130, for which the S_f statistics at 10^3 and 10^6 cycles are known (see Table 10.3). Since the statistics of stress at failure for a life of

Design to Avoid Fatigue Failure   475

Table 10.3   AISI 3140 Steel[a] (see Appendix 10.B)[a]

Specification Life, Cycles	x_0 (ksi)				b				θ (ksi)			
	10^3	10^4	10^5	10^6	10^3	10^4	10^5	10^6	10^3	10^4	10^5	10^6
No-Notch	89	74	66	57	3.7	5.2	5	5.5	100.4	87.7	76.7	67.2

[a]With $S_u = 108$ ksi, 109 ksi; $S_y = 87$ ksi, 75 ksi; rotary bending.

65,000 cycles are not given, these must be estimated by interpolation (see Figure 10.24). Also, since the values in Table 10.3 are Weibull parameters x_0, b, and θ, the means and standard deviations for S_f at 10^3 and 10^6 cycles of life must be estimated by using equations (8.6a) [equations (g) and (h), which follow]:

$$\bar{S}_f = x_0 + (\theta - x_0)\Gamma\left(1 + \frac{1}{b}\right) \tag{g}$$

$$C_{S_f} = \frac{\sigma_{S_f}}{\bar{S}_f} = \frac{\bar{S}_f - x_0}{\bar{S}_f} \cdot b^{-0.926} \tag{h}$$

The statistics of fatigue strength and endurance limit of AISI 3140 steel are estimated from the data in Table 10.4 with the use of equations (g) and (h). Values of $\Gamma[1+(1/b)]$ are estimated from Figure 8.30 for specific values of b. The results are

Fatigue strength $(10^3 \text{ cycles}) \simeq (99{,}600; 3{,}150)$ psi

Endurance limit $(10^6 \text{ cycles}) \simeq (66{,}600; 2{,}000)$ psi

From the S–N envelope (Figure 10.24), we estimate the mean and standard deviation at 65,000 cycles by interpolation. Only the upper half of the envelope is shown; thus:

$$\tan \theta_1 = 0.0582; \qquad \tan \theta_2 = 0.0586$$

Table 10.4   Data Summarized from Table 10.3

Weibull[a] Parameter	Fatigue Strength, 10^3 Cycles	Endurance Limit 10^6 Cycles
x_0	89,000 psi	57,000 psi
b	3.7	5.5
θ	100,400 psi	67,200 psi

[a]See Ref. 13.

476 Strength of Mechanical Components

At 65,000 cycles, the mean is

$$\bar{S}_f = 78,000 \text{ psi}$$

$$L_{10}(A) = 4.8235 + 0.0582 \cdot (6.000 - 4.813) = 4.8926$$

At L_{10} 65,000 cycles, B on the $2\mathrm{s}$ line is

$$L_{10}(B) = 4.8488 + 0.0586 \cdot (6.000 - 4.813) = 4.9179$$

$$\bar{S}_f + 2\mathrm{s}_{S_f} = 82,800 \text{ psi}$$

$$\mathrm{s}_{S_f} \approx \frac{82,800 - 78,000}{2} \text{ psi} = 2500 \text{ psi}$$

Thus an estimate of fatigue strength at 65,000 cycles is

$$(\bar{S}_f, \mathrm{s}_{S_f}) \simeq (78,000, 2500) \text{ psi} \tag{i}$$

Further modification of fatigue strength is not required in this problem.

4. Estimate z associated with $R = .99999$. Use of equation (9.10) yields

$$R = .99999 \leqslant \frac{1}{\sqrt{2\pi}} \cdot \int_{z}^{\infty} e^{-z^2/2} dz = \frac{1}{\sqrt{2\pi}} \cdot \int_{-5.0}^{\infty} e^{-z^2/2} dz$$

Thus $z \approx -5.0$ (see Appendix 1).

5. Substitution of equations (f) and (i) into equation (9.11) yields

$$z = -\frac{\bar{S} - \bar{s}'}{\sqrt{\mathrm{s}_S^2 + \mathrm{s}_{s'}^2}} = -5 = -\frac{78,000 - (3.339/\bar{r}^3) \cdot 10^6}{\sqrt{(2500)^2 + [(2.501/\bar{r}^3) \cdot 10^5]^2}}$$

Expand and then simplify:

$$0 = \bar{r}^6 - 86.9\bar{r}^3 + 1600$$

$$\bar{r} = 3.38 \text{ in.}$$

The wall thickness is $\bar{t} \approx (\bar{r}/50) = 0.079$ in. After the first design iteration is completed, a check of K_t is made to determine the advisability of modification and a second iteration.

Sensitivity of Design to Mean K_t. Since the interest is in comparison, the stress concentration is first set equal to $\bar{K}_t = 1$ and then increased to $(\bar{K}_t, \mathrm{s}_{K_t}) = (1.50, 0.06)$. The effects on radius requirements are estimated for $R = .99999$.

The stress concentration effect on radius requirements is as follows:

$(\overline{K_t}, {}_{^sK_t})$	Radius, $\bar{r}$ (in.)
(1, 0)	3.00
(1.20, 0.06)	3.38
(1.50, 0.06)	3.62

Because stress concentration is a first-order influence on design, efforts to minimize or eliminate it are certainly in the interest of efficiency. In design for endurance, ignoring the effects of stress concentration could very well result in stress levels in the finite-life region.

Endurance versus Finite Life, Weight Penalty. It has sometimes been the practice to design for endurance any component expected to experience dynamic loading. In such cases the endurance limit (rather than fatigue strength) of the material was used.

To assess the penalty for failure to design for finite life, the geometric requirements for endurance of the shaft previously designed for 65,000 cycles of life are calculated. The two cross sections are compared. For the AISI 4130 steel, the endurance limit was previously estimated as

$$\left(\overline{S_e}, {}_{^sS_e}\right) = (66{,}600, 2000) \text{ psi} \tag{j}$$

All other conditions in the previous design remain the same. Note that the endurance limit $\overline{S_e}$ is less than $\overline{S_f}$ and ${}_{^sS_e}$ is less than ${}_{^sS_f}$. Substitution of equations (f) and (j) into equation (9.11) yields

$$z = -5 = -\frac{66{,}600 - (3.339/\bar{r}^3) \cdot 10^6}{\sqrt{(2000)^2 + \left[(2.501/\bar{r}^3)\right] \cdot 10^5)^2}} \tag{k}$$

The principal root in equation (k) is

$$\bar{r} \approx 4.15 \text{ in.}$$

$${}_{^sr} \approx 0.015\bar{r} = (0.015) \cdot 4.15 = 0.062 \text{ in.}$$

The estimated tolerance is $\Delta \approx 3 s_r = 0.186$ in., the estimated wall thickness is

$$\bar{t} = \frac{\bar{r}}{50} = \frac{4.15}{50} = 0.083 \text{ in.}$$

478 Strength of Mechanical Components

The weight penalty is

$$\frac{\text{Weight(endurance design)}}{\text{Weight(65,000 cycles)}} \cdot 10^2 = \frac{2\pi r_1 t_1}{2\pi r_2 t_2} = \frac{0.344}{0.310} = 1.11$$

The percent excess weight is 11%.

10.7.2 Bearing Selection

Estimations of contact stress statistics for ball bearings were presented in Section 5.20.2, and bearing analysis was treated in Section 9.3.10.

Antifriction bearings represent high levels of industrial accomplishment in terms of design and manufacture to satisfy life, accuracy, load-carrying capacity, and dependability requirements. Thus the problem of the mechanical design engineer is often not how to design a ball or roller bearing, but how to select one.

The B-10 (or rating) life of a sample of nominally identical ball bearings is defined as the number of revolutions (or hours) at some given speed, that 90% of the members in the sample will complete or exceed before evidence of fatigue develops (Figure 10.26).

The AFBMA [106] has established a standard load rating C for ball bearings in which speed is not a consideration. The variable C is defined as the radial load that a sample of nominally identical bearings can support for a rating life of 10^6 revolutions of the inner race (stationary outer race and

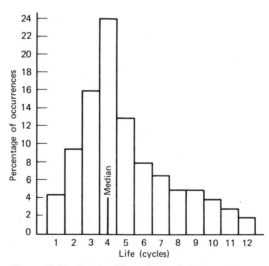

Figure 10.26 Bearing life versus probability of survival.

load). For any other load intensity, the life L is estimated as

$$L = \left(\frac{C}{F}\right)^3 \qquad (10.21)$$

where L is millions of revolutions, and F is any arbitrary loading.

Example I

If a life of $42.875 \cdot 10^6$ revolutions is required, then

$$C \approx (42.875)^{1/3} \cdot F = 3.5F$$

A rating C of $3.5F$ is needed.

Studies indicate that the Weibull model often closely approximates the distribution of bearing life (Figures 10.27 and 10.28).

10.7.3 Fatigue Due to Narrow-band Random Loading (Zero Mean)

Many components experience repeated loads of random amplitude. For instance, machines that are started and stopped experience overloads, transient vibrations, and natural random force phenomena. As discussed by Swanson [29], automotive systems are normally subject to loads of random amplitudes. The random-amplitude loading of ship hulls and structure has

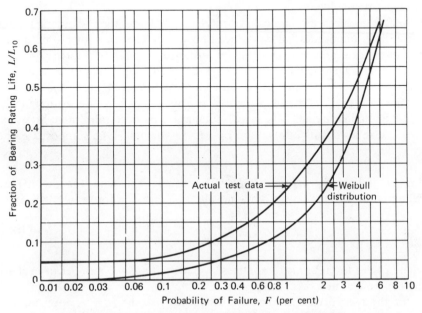

Figure 10.27 Reduction in bearing life for increased reliability.

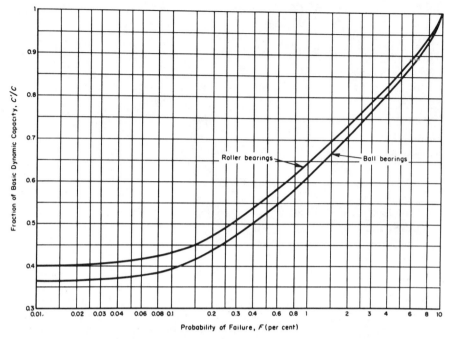

Figure 10.28 Reduction in basic dynamic capacity C [107].

been discussed by R. R. Little. Also see Taylor [28] for detailed discussion of random aircraft loads.

Since high-speed mechanical systems came into operation, dynamic loading of multiple amplitudes has often been observed. The need for life prediction was immediate and critical with, for instance, aircraft structures subject to gust loading. Deterministic approaches for predicting fatigue life of components in random dynamic environments were first attempted. An empirical rule was sought that might provide a relatively simple guide in design [27]. One postulation, namely, the linear-cumulative damage rule, came into common use following its publication by M. A. Miner in 1945 (proposed earlier by Palmgren).

Analysis Employing Miner's Rule. The usual expression for the Miner–Palmgren linear cumulative damage rule, as a criterion for failure [64], is

$$\sum_{i=1}^{r} \frac{n_i}{N_i} = 1 \qquad (10.22)$$

where cycles to failure is N_i at stress level s_i and n_i is the number of cycles of the component at stress level s_i.

For the special case of constant-amplitude loading ($n=1$), application of the concept poses no problems, and equation (10.22) predicts total damage (failure) when $n = N$. The constant-amplitude problem was solved by probabilistic methods in Example H.

For the case of $r > 1$, typical of multiple-amplitude random loading, equation (10.22) predicts total damage when

$$\sum_{i=1}^{r} \frac{n_i}{N_i} = \frac{n_1}{N_1} + \frac{n_2}{N_2} + \cdots + \frac{n_r}{N_r} = 1$$

and n_i/N_i indicates the predicted partial damage due to n_i cycles of stress at level s_i. The classical $S-N$ curve (such as Figure 8.23) has been utilized in analysis to estimate the N_i of the various s_i (hence partial damages) at observed stress amplitudes.

Miner's rule postulated that the total life of a component could be estimated by adding up the percentage of life consumed by each overstress cycle. *By overstress is meant a level above the endurance limit of the original material.*

Example J

For steel, approximately 1/75,000 of life might be used by each 136,000-psi stress cycle, 1/40,000 by each 160,000 psi cycle, and so on. If the total overstress cycles consume a total of 1/50 of the life of a component during one hour of normal operation, then by Miner's rule total life of the part should be approximately 50 hr if the mix of stress amplitudes remains the same.

Original tests yielded failures at a range of values, $\Sigma(n/N) = 0.61$ to $\Sigma(n/N) = 1.45$. Under conditions of progressively increasing stress levels or at progressively decreasing stress levels failures have been obtained from $\Sigma(n/N) = 0.18$ to $\Sigma(n/N) = 23$. Recently the behavior of engineering materials, such as AISI 4130 and 4340 steel and 7075-T6 aluminum alloy, 2024-T3, T4, and T6, 6061-T6, magnesium alloy, Rene 41, and maraging steel have been studied in narrow-band random loads environments (see Section 8.14.3) [29]. The published results of these studies provide data from which statistical $S-N$ envelopes for the materials may be estimated [54]. Figure 8.37 includes the statistical $S-N$ envelopes for AISI 4340 steel subject to constant-amplitude sinusoidal and to narrow-band random loading.

With present computer-controlled fatigue testing, including on-line loading and concurrent damage analysis, realistic life assessment is now feasible. One important reason for improved results in analysis is the step-by-step walk through which the random sequence of loading is preserved, in contrast to earlier block testing, at fixed amplitudes.

Cycle Counting with Random Loading. The application of Miner's linear cumulative damage rule using the results of block testing (to estimate damage) produced questionable correlations with actual specimen cycle lives. Thus efforts have been made to improve correlation between predicted and actual life by applying sophisticated cycle counting methods. Two such schemes are the "range pair" and the "rain flow" methods discussed by Swanson [41]. For most practical situations, the two methods are equivalent. If cycles are to be counted until failure occurs for a random dynamic load sequence, one complete cycle should be counted between the most positive and the most negative peaks in the sequences and other smaller complete cycles that are interruptions of the largest cycle should also be counted.

With the advent of computer control in fatigue testing, the on-line loading and concurrent damage analysis possible with the preceding counting methods make tedious life assessment much more palatable.

According to Swanson, the "law of linear cumulative damage" appears to work quite well (*in analysis*) when the cycles have been properly determined, the appropriate strain-life or stress-life curves used, and the damage $(1/N_i)$ at each cycle (*i*) estimated on the basis of the level of each Δs_i. Reasonable correlation of observed with estimated life might be expected, since the use life-environment sequence of loading amplitudes is preserved. However, more work needs to be done to confirm these observations.

Hillberry [88] has obtained experimental results that suggest that broad-band random loading may, on the average, provide a less severe operational environment (a slightly longer life at the same RMS loading value) than narrow-band random loading for the same number of peaks. Such results may be due to relatively fewer zero crossings and smaller stress ranges [77, 108, 109] (see Figure 8.40). Thus the design requirements for narrow-band random loads may often establish an upper bound on structural requirements for dynamic loading.

For *design synthesis*, a different approach (discussed next) is required. In contrast to the *step-by-step, point-by-point* walk through discussed previously here and in Swanson [41, Section 2.1], one of the key parameters utilized in design is the measure of amplitude of the random loading process, namely, the RMS level.

It is now feasible to attempt empirical probabilistic design for narrow-band random loading because of currently available materials behavior data relevant to random loading cases [54, 67, 68, 110].

10.7.4 Design for Narrow-band Random Loading (Zero Mean)

Referring to Figure 8.41, it is seen that the $S-N$ envelope for the fatigue-strength behavior of a material exposed to narrow-band random loading (zero mean) is similar to the envelope for constant-amplitude sinusoidal loading except that the stress at failure values are depressed.

Thus for a given cycle life n, the model for stress at failure is [equation (10.13a)]

$$\left(\overline{S}'_f, {}_{\triangleright S'_f}\right)^* \text{ psi}$$

*Determined in a narrow-band random loading environment.

The model for narrow-band dynamic stress is [equation (10.15a)]

$$(\bar{s}', \text{\char'40}_{s'}) = (0, RMS) \ldots \text{psi}$$

The criterion for an adequate design is

$$R \leqslant P(s' < S_f')$$

and

$$R \leqslant \frac{1}{\sqrt{2\pi}} \cdot \int_z^\infty e^{-t^2/2} dt$$

where (Appendix 1)

$$z = -\frac{\bar{S}_f' - 0}{\sqrt{\text{\char'40}_{S_f'}^2 + \text{\char'40}_{s'}^2}}$$

Example K Design of a Narrow-band Random Loaded Spring

1. In this example structural requirements to survive narrow-band random loading are estimated. Shown in Figure 10.22 is a model of a flat spring* to be manufactured from AISI 4340 alloy steel, whose tensile ultimate strength S_u, constant-amplitude fatigue strength, and narrow-band random fatigue strength are given in Figure 8.41. The loading will be applied at the midpoint and perpendicular to the axis of symmetry of the spring. The dynamic finite-life design is for $n = 100{,}000$ cycles. The specified probability of survival is $R = .9965$ since replacement cost is modest and public safety is not a consideration.

If loading is normally distributed, load-induced stress is nonetheless non-normal and often nearly lognormal. Thus the design procedure for distribution of S and s unspecified, discussed in Section 9.3.9 is appropriate. The loading considered (the effect due to spring mass is negligible) is described as follows:

$$(\bar{P}, \text{\char'40}_p) = (0, 1240) \text{ lb} = 0.707(0, 1800) \text{ lb}$$

The random-variable dimension to be determined is h. From applicable manufacturing tolerances, the standard deviation on h is estimated as

$$\text{\char'40}_h \approx 0.015 \bar{h}$$

For purposes of comparisons, the RMS level of random loading in this

---

*For helical spring stress equations, see Section 6.1.

484 Strength of Mechanical Components

example was chosen to be the same as the RMS level of the constant amplitude case (see Example G).

2. As in Example G, the model for stress is equation (5.23):

$$s = \frac{Mc}{I}; \qquad s = \frac{6M}{bh^2} \text{ psi}$$

The maximum bending moment is

$$(\bar{M}, s_M) = \frac{(\bar{P}, s_P)}{2} \cdot \frac{(\bar{l}, s_l)}{2} \text{ in. lb}$$

$$= \tfrac{1}{4}(0, 1240)(30.0, 0.075)$$

$$\bar{M} = \tfrac{1}{4} \cdot (0)(30) = 0 \text{ in. lb}$$

From equations (2.27) and (2.10):

$$s_M = \tfrac{1}{4}\sqrt{\bar{P}^2 \cdot s_l^2 + \bar{l}^2 s_p^2 + s_p^2 \cdot s_l^2} \approx 9275 \text{ in. lb}$$

$$6(\bar{M}, s_M) = (0, 55{,}650) \text{ in. lb}$$

Letting $u = (6M/b)$,

$$(\bar{u}, s_u) = \frac{(0, 55{,}650)}{(2.00, 0.035)}; \qquad \bar{u} = 0$$

$$s_u = 27{,}800; \qquad (\bar{u}, s_u) = (0, 27{,}800)$$

From equations (2.28) and (2.30), the statistics of stress are

$$(\bar{s}, s_s) = \frac{(0, 27{,}800)}{(\bar{h}^2, 0.03\bar{h}^2)} = \left[0; \frac{27{,}800}{h^2}\right] \text{ psi} \qquad \text{(a)}$$

3. Random fatigue strength at 10^5 cycles (from Figure 8.41) is estimated as

$$(\bar{S}_f, s_{S_f}) \approx (60{,}500, 2670) \text{ psi} \qquad \text{(b)}$$

4. Substitution of equations (a) and (b) into equation (9.11), with $z = -3.0$ (associated with $R' = .999$; see Figure 9.36) yields

$$-3.0 = -\frac{60500 - 0}{\sqrt{(2670)^2 + (27{,}800/\bar{h}^2)^2}}. \qquad \text{(c)}$$

Solve equation (c) for $\bar{h}$:

$$\bar{h} = 1.18 \text{ in. (required to satisfy } R = .9965)$$

Compare the preceding value of $\bar{h}$ with that in Examples A and G.

10.8 FLUCTUATING STRESS (NONZERO MEAN)

10.8.1 Constant-amplitude Dynamic Stresses

It is common in design to find that stresses fluctuate about a value other than zero (see Section 8.14.2). Figure 3.21 illustrates several such relationships. The stress components are commonly identified as follows:

s_{min} = Lower extreme of stress excursion

s_{max} = Upper extreme of stress excursion

s_m = Middle value of stress excursion

s_a = Alternating component

s_{static} = Static stress

The presence of a static stress is traceable to an unvarying or steady fixed load. The static stress is often statistically independent of the dynamic components. Both mean and alternating dynamic stress components are random variables, defined as follows:

$$s_m = \frac{s_{max} + s_{min}}{2} \tag{10.23}$$

$$s_a = \frac{s_{max} - s_{min}}{2} \tag{10.24}$$

The alternating component is often modeled as a function of time:

$$s_a = A \sin(\omega t)$$

However, according to Anderson [57] and Shigley [1], the exact wave shape does not appear to be of great significance. It is, however, necessary to consider the strength characteristics of materials subject to various combinations of mean and alternating loading. The results given in Section 8.14.2 illustrate the behavior of two annealed steels from tests employing mean and alternating tensile load components. For constant amplitude s_a, probabilistic design is carried out by utilizing radius vectors of allowable stress random variables and of applied load stress random variables as shown in Figure

486 Strength of Mechanical Components

10.30b. This empirical approach is motivated by observations of plotted test results of materials behavior (see Figure 8.35).

In classical design, utilizing deterministic values for strength and stress, the modified Goodman diagram or the Gerber parabola (as shown in Figure 8.36) led to design decisions; that is, if the mean and alternating components of stress in a component plotted as a point P below either the modified Goodman line or the Gerber parabola, presumably the design would be sufficient.

With the mean and alternating stress components treated as random variables s_a and s_m, effective stress s_R is modeled as a radius vector. Equation (2.63) yields

$$\bar{s}_R = \sqrt{\bar{s}_a^2 + \bar{s}_m^2} \tag{a}$$

and the standard deviation is given by equation (2.64):

$$s_{s_R} \approx \sqrt{\frac{\bar{s}_a^2 s_{s_a}^2 + \bar{s}_m^2 s_{s_m}^2}{\bar{s}_a^2 + \bar{s}_m^2}} \tag{b}$$

Equations (a) and (b) are utilized with equation (10.15a) to yield an expression for significant stress. The stress at failure (endurance) for the material is interpreted from a figure such as Figure 8.37 or Figure 10.30b, corresponding to the applicable stress ratio $\bar{s}_a/\bar{s}_m$, where s_a is a significant stress random variable.

It may be observed that the mean stress at failure radius vector exceeds the mean applied stress radius vector (Figure 10.30b):

$$\left(\bar{S}_a^2 + \bar{S}_m^2\right)^{1/2} > \left(\bar{s}_a^2 + \bar{s}_m^2\right)^{1/2}$$

Design for Constant-amplitude Loading (Nonzero Mean). From Section 5.17.1, the model for fluctuating stress is the radius vector s_R [equation (5.45)]:

$$s_R = \sqrt{s_a^2 + s_m^2}$$

The statistics of s_R in terms of the statistics of s_a and s_m are given by equations (5.45a) and (5.45b). The direction of s_R is indicated by θ in Figure 5.25.

From Section 8.14.2, the model for stress at failure is the radius vector S_R:

$$S_R = \sqrt{S_a^2 + S_m^2} \tag{c}$$

The statistics of S_R in terms of the statistics of S_a and S_m are given by

Fluctuating Stress (Nonzero Mean)

equations (2.63) and (2.64) with equation (c). The design criterion is

$$R \leqslant P(S_R > s_R) = P\left[\sqrt{S_a^2 + S_m^2} > \sqrt{s_a^2 + s_m^2}\,\right]$$

Also

$$R \leqslant P(S_R > s_R) = P(S_R \sin\theta > s_R \sin\theta) = P(S_a > s_a)$$

Since the distributions of S_R and s_R are usually unknown, the design procedure for distribution of S and s unspecified, described in Section 9.3.9 is used, and

$$R \leqslant \frac{1}{\sqrt{2\pi}} \cdot \int_z^\infty e^{-t^2/2}\,dt$$

where

$$z = -\frac{\bar{S}_R - \bar{s}_R}{\sqrt{{}_{\scriptscriptstyle{}^{2}\!s_R} + {}_{\scriptscriptstyle{}^{2}\!s_R}}}$$

Example L

1. The preliminary specifications are as follows. A link forged from AISI 4340 steel bar stock, (Figure 10.29) is to withstand (indefinitely) repeated tensile axial loading varying from 0 to $(13{,}000, 1000) = (\bar{F}, {}_{\scriptscriptstyle{}^{s}\!F})$ lb. Because of a snap ring groove, there will be a geometric stress concentration effect $(\bar{K}_t, {}_{\scriptscriptstyle{}^{s}\!K_t})$ $\simeq (1.57, 0.0785)$. The criterion of failure is fracture. The component is not expected to experience compressive loads. Estimate the required net diameter such that reliability is no less than .9993. Assume that the distributions describing strength and applied stress are unknown.

2. Determine the following parameters for the load stress model.

   a. Estimate mean and alternating load statistics. The first task is to estimate the loading component statistics. From these, expressions for significant mean and alternating stress statistics can be developed. The mean and alternating stress expressions are also utilized

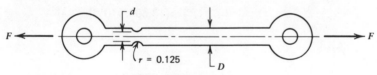

Figure 10.29 Forged link.

to estimate the stress ratio s_a/s_m. When a component or test specimen is loaded by a combination consisting of mean and dynamic load components, the strain response and internal stresses developed are functions of the mean applied stress and the alternating applied stress.

In this problem

$$\bar{F}_m = \bar{F}_a = 13{,}000/2 \text{ lb}. \tag{a}$$

The two force components are correlated and $(s_a/s_m) = (F_a/F_m)$, where F_a and F_m are significant loads. From Figure 10.31:

$$(\bar{F}_a, {}^{\scriptscriptstyle \gt}{F}_a) = (\bar{F}_m, {}^{\scriptscriptstyle \gt}{F}_m) = (6500, 500) \text{ lb} \tag{b}$$

Since the material of the link is ductile AISI 4340 steel, the stress design factors effect only the dynamic stress components s_a. Applicable design factors (including stress concentration) are determined in steps b and c, which follow.

b. Estimate the surface-finish factor (Figure 10.13). Since $\bar{S}_u \approx 110{,}000$ psi (see Figure 10.30a) and the link will be machined,

$$(\bar{K}_a, {}^{\scriptscriptstyle \gt}{K}_a) = (0.75, 0.041) \tag{c}$$

c. Determine the size factor (Figure 10.15). Since an estimate of link diameter is $d \approx 1$ in.,

$$(\bar{k}_b, {}^{\scriptscriptstyle \gt}{k}_b) = (0.95, 0.036) \tag{d}$$

The model for significant F'_a is

$$F'_a = \frac{F_a \cdot K_t}{K_a} \tag{e}$$

$$(\bar{F}'_a, {}^{\scriptscriptstyle \gt}{F}'_a) = \frac{(6500, 500) \cdot (1.57, 0.0785)}{(0.75, 0.041)}$$

and

$$(\bar{F}'_a, {}^{\scriptscriptstyle \gt}{F}'_a) \approx (13{,}600, 970) \text{ lb} \tag{f}$$

Also

$$(\bar{F}_m, {}^{\scriptscriptstyle \gt}{F}_m) \approx (6500, 500) \text{ lb} \tag{g}$$

Estimate applied load radius vector statistics. Apply equation (2.63) with equations (f) and (g):

$$\bar{F} \approx \sqrt{\bar{F}_a^2 + \bar{F}_m^2} = 15{,}080 \text{ lb}$$

Apply equation 2.64 with equations (f) and (g):

$$\, _sF \approx \sqrt{\frac{\bar{F}_a^2 \, _sF_a^2 + \bar{F}_m^2 \cdot \, _sF_m^2}{\bar{F}_a^2 + \bar{F}_m^2}} = 900 \text{ lb}$$

$$(\bar{F}, \, _sF) = (15{,}080, 900) \text{ lb} \tag{h}$$

Estimate applied stress radius vector statistics. The expression for cross-sectional area in terms of the unknown diameter d, is, from equations (2.59) and (2.62):

$$(\bar{A}, \, _sA) = \left(\frac{\pi \bar{d}^2}{4} \, ; \, \frac{\pi}{2} \bar{d} \, _sd \right)$$

From tolerance tables, Chapter 4:

$$\Delta = 0.015 \bar{d} \text{ tolerance equated with range (Section 3.3.2)}$$

$$\, _sd \approx \frac{\Delta}{3} = \frac{0.015 \bar{d}}{3} = 0.005 \bar{d}$$

$$(\bar{A}, \, _sA) = \left(\frac{\pi \bar{d}^2}{4} \, , \, \frac{\pi}{2} 0.005 \bar{d}^2 \right) \tag{i}$$

The applied stress vector s, combining equations (h) and (i), is

$$\bar{s} \approx \frac{15{,}080}{\pi \bar{d}^2 / 4} = \frac{19{,}200}{\bar{d}^2} = \frac{\bar{F}}{\bar{A}} \text{ psi}$$

$$\, _ss \approx \sqrt{\frac{\bar{F}^2 \cdot \, _sA^2 + \bar{A}^2 \, _sF^2}{\bar{A}^4}}$$

$$\approx \left(\frac{4}{\pi \bar{d}^2} \right)^2 \cdot \sqrt{(15{,}080)^2 \left(\frac{\pi}{2} \cdot 0.005 \bar{d}^2 \right)^2 + \left(\frac{\bar{d}^2 \cdot \pi}{4} \right)^2 \cdot (900)^2}$$

$$= \frac{1565}{\bar{d}^2} \text{ psi}$$

and

$$(\bar{s}, \mathit{v}_s) = \left(\frac{19{,}200}{d^2}, \frac{1565}{\bar{d}^2}\right) \text{ psi} \qquad \text{(j)}$$

The applied stress ratio, s_a/s_m, with equations (f) and (g), is

$$\tan\theta = \frac{\bar{s}_a'}{\bar{s}_m} = \frac{13{,}600}{A} \bigg/ \frac{6500}{A} = 2.09 \qquad \text{(k)}$$

3. Estimate significant stress at failure statistics. The tensile strength (see Figure 10.37 or Figure 8.37) is

$$\bar{S}_u = 110{,}600 \text{ psi}; \qquad \mathit{v}_{S_u} \approx 1300 \text{ psi}$$

The axial endurance limit (S_e') is

$$S_e' \approx 48{,}700 \text{ psi}; \qquad \mathit{v}_{S_e'} \approx 4{,}440 \text{ psi} \qquad \text{(k')}$$

The significant endurance limit [equation (10.13)] is

$$S_e' = k_b k_d S_e \qquad \text{(l)}$$

Figure 8.37 is utilized in this calculation. It is observed that because of the way the testing was carried out and the modified Gerber envelope developed, the definition of strength as stress at failure should be utilized. Thus, strength is the vector sum of a mean component and an alternating component. The mean allowable stress component and the alternating allowable stress component (because of the method of test loading) are correlated random variables (see Figure 3.19).

3. *Strength radius vector statistics.* Since the loading is axial eliminating the effects of a stress gradient, the notch sensitivity is considered as due entirely to material characteristics. From Figure 10.21, $q \simeq 0.85$. The factor modifying the dynamic strength component is

$$\frac{1}{q} = \frac{1}{0.85} = 1.18.$$

Application of equation (k) yields

$$S_a = 2.09 S_m \qquad \text{(k'')}$$

The Gerber parabola [equation (8.9)] is

$$\frac{S_a}{S_e} + \left(\frac{S_m}{S_u}\right)^2 = 1$$

Substitution from equations (k') and (k") into equation (8.9) yields

$$\frac{2.09\bar{S}_m}{48,700} + \left(\frac{\bar{S}_m}{110,600}\right)^2 = 1$$

$$\frac{\bar{S}_m^2}{1.223 \cdot 10^{10}} + \frac{2.09\bar{S}_m}{48,700} = 1$$

Thus

$$\bar{S}_m \approx 19,420 \text{ psi}$$

and

$$\bar{S}_a \approx 2.09(19.420) = 44,670 \text{ psi}$$

The mean radius vector is

$$\sqrt{\bar{S}_m^2 + \bar{S}_a^2} \approx [(4.467)^2 + (1.942)^2]^{1/2} \cdot 10^4 = 48,710 \text{ psi}$$

Taking cognizance of notch sensitivity, q, and k_b:

$$\bar{S}_R \approx K_b \cdot \frac{1}{q}\sqrt{\bar{S}_m^2 + \bar{S}_a^2} = 1.18(48710) \cdot (0.95) = 54,140. \qquad (m)$$

After $\bar{S}_R$ has been estimated, it is necessary to estimate the coordinates of $_{3S_R}$, which are S_a and S_m of the 3_s point (see Figure 10.30b). Again utilize equation (8.9):

$$\frac{S_a}{S_e} + \left(\frac{S_m}{S_u}\right)^2 = 1$$

Substitute the -3_s values of $\bar{S}_e$ and S_u:

$$\frac{2.09\bar{S}_m}{35,380} + \left(\frac{\bar{S}_m}{106,600}\right)^2 = 1$$

$$\frac{2.09\bar{S}_m}{3.538 \cdot 10^4} + \frac{\bar{S}_m^2}{1.136 \cdot 10^{10}} = 1$$

Solve for $\bar{S}_m$:

$$\bar{S}_m = 14,300 \text{ psi}$$

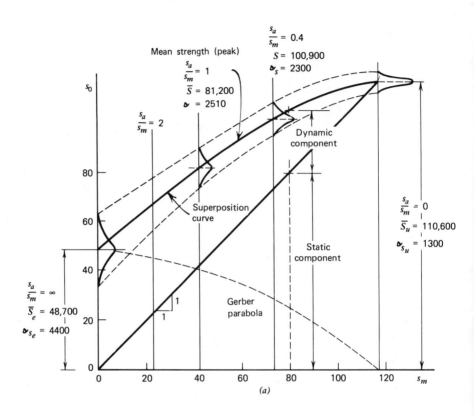
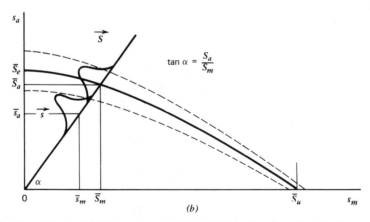

Figure 10.30 Signoidal relationship between da/dN and ΔK as plotted on logarithmic coordinates. Stress–strength distribution radius vectors plotted on a Gerber parabola (b).

and since $S_a = 2.09 S_m$,

$$S_a = 2.09(14{,}300) = 34{,}750 \text{ psi}$$

The radius vector of the $3\bar{s}$ point is

$$[(1.43)^2 + (3.475)^2]^{1/2} \cdot 10^4 = 37{,}580 \text{ psi}$$

The standard deviation estimate of the strength radius vector $\hat{s}_{S_R}$ is

$$\hat{s}_{S_R} \approx (1.18)(54{,}140 - 37{,}580) \cdot \frac{0.95}{3} \approx 6140 \text{ psi} \qquad (n)$$

Strength radius vector statistics [equations (m) and (n)] are

$$\left(\bar{S}'_R, \hat{s}_{S'_R}\right) = (54{,}140, 6140) \text{ psi}$$

$$C_{S_R} = \frac{6140}{54140} = 0.113 \qquad (o)$$

4. Since the required reliability of the design is $R = .9993$, design for $R' = 0.9995$, use the normal probabilistic design theory (see Figure 9.41) as explained in Section 9.3.9 (subsection on design with distribution of S and s unspecified). For reliability $R' = .9995$ (from Appendix 1):

$$z = -3.3$$

5. Substitution of equations (j) and (o) into equation (9.11) yields

$$-3.3 = -\frac{54140 - (19200/\bar{d}^2)}{\sqrt{(6140)^2 + (1565/\bar{d}^2)^2}}$$

$$\bar{d}^2 = 0.846 \text{ in.}^2$$

$$\bar{d} = 0.921 \text{ in.}$$

10.8.2 Random Dynamic Stresses (Nonzero Mean)

Stress models are discussed in Section 5.17 and strength models are discussed in Section 8.14.3 for the design situation where loading is narrow-band random associated with nonzero mean loads. It may be postulated, in the absence of test results, that stress at failure can be modeled as shown in the sketch (A) that follows, which indicates stress at failure for a life of N cycles.

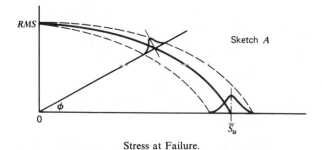

Stress at Failure.

10.9 DESIGN FOR MULTIAXIAL DYNAMIC LOADING

If the stress is known to be uniaxial (such as with flat springs, rotating shafts, etc.), no problem develops in relating life to stress (see Figure 10.24). The *S–N* envelopes and appropriate equations are directly applicable. If, however, the stressing of a component is multiaxial, the problem may be to utilize normal or principal stress components to calculate uniaxial equivalent stresses that can be related to uniaxial materials behavior.

Although exceptions have been identified, most fatigue failures are surface originated. Since stress normal to an unloaded surface is zero, it is not often necessary to consider the possible fatigue failure of a component subjected to a triaxial state of stress [110]. The proposed model for biaxial equivalent dynamic stress (s_1 and s_2 are dynamic components) is [1, p. 187]:

$$s_{eq} = \sqrt{s_1^2 - s_1 \cdot s_2 + s_2^2}$$

for use with the dynamic probabilistic design criterion, in cases of biaxial (dynamic) stress fields:

$$P(S_f > s_{eq}) \geq R$$

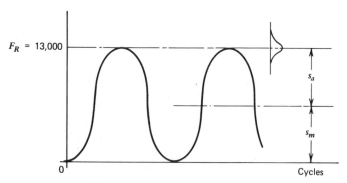

Figure 10.31 Repeated tensile loading.

Expressions for the mean values and standard deviations of s_{eq} are given in Section 5.17 for cases in which s_1 and s_2 are (1) statistically independent and (2) correlated $\rho = +1$.

Descriptions of fatigue strength S_f are obtained from uniaxial S–N envelopes, such as Figure 10.24, for endurance and for finite-life design (including low-cycle fatigue). Models that may be useful in estimating equivalent stress are described in the sections that follow.

10.10 MULTIPLE-MEMBER SYSTEMS

In a typical machine the various components perform tasks and develop loading histories in independent manners. Thus anticipated loading histories can be estimated a priori and components designed for required service conditions (Figure 10.32).

When components are joined in such a manner as to remove all possible degrees of freedom, such a combination must be considered a rigid body. With structures that have zero degrees of freedom, special techniques may be necessary to treat multiple degrees of redundancy [112]. In addition, where structural shapes are complicated, finite-element methods may be necessary to estimate stress statistics (see Section 5.20).

Fortunately, many systems of elements encountered in mechanical design can be considered determinate. For example, the elements in the power-delivery system of a ground vehicle from piston to rear wheel amount to a determinate system. Element strengths are statistically independent, being of different materials and from different processing.

10.10.1 Pin-jointed Systems

The probability of failure of a pin-jointed system depends on (1) the number (n) of elements, (2) the degree of static indeterminacy, and (3) the configuration of the system itself. In statically determinate systems the forces in the elements are determined by means of the equations of statics alone.

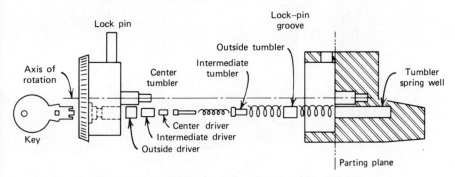

Figure 10.32 Multiple-member system of components [111].

In a statically determinate system of n elements, the failure of any one element constitutes failure of the system. Thus the strength of the system is determined by the minimum strength of one of the n elements. Reliability of the system is less than that of component R_i having the lowest probability of survivial $R_{\text{Sys}} < R_i$. Also, since strength of the elements are statistically independent [9, p. 22]:

$$R_{\text{sys}} = R_1 \cdot R_2 \cdot \ \cdots \ \cdot R_n \qquad (10.25)$$

If a system is to be designed to have elements of equal individual reliabilities, and if the system reliability is $R_{\text{sys}} = .9900$, then (Figure 10.33):

$$(R)^n = .9900 = R_1 \cdot R_2, \ldots, R_n$$

For $n = 3$:

$$(R_i)^3 = 0.9900$$

$$R_i = \sqrt[3]{0.9900} = 0.9967.$$

Assume that the stresses in a set of elements have been determined:

$$S_1, S_2, \ldots, S_n$$

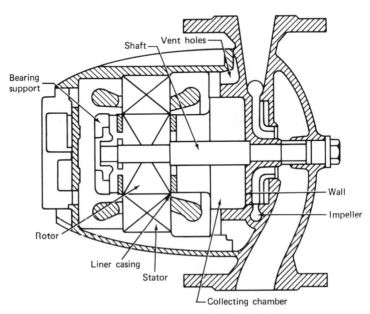

Figure 10.33 System of equal reliability elements.

If S' is the significant strength, then

$$P_f = P\{s > S'\}$$

The probability of failure (P_{f_i}) of the ith element is

$$P_{f_i} = P\{s_i > S\} = 1 - R_i \qquad (10.26)$$

and system reliability will be

$$R_{sys} = \prod_{i=1}^{n} (1 - P_{f_i})$$

For equal probability of failure in each of the n elements:

$$R_{sys} = (1 - P_{f_i})^n$$

Example M

1. Preliminary specifications are as follows. Consider the simple pinned frame ABC in Figure 10.34. Members AC and BC are subject to constant-amplitude reversed tension–compression loading. Member AB is loaded similarly but is 180° out of phase with AC and BC. Design is for an extended life of 10^8 cycles of loading, with a system reliability of $R = .9900$.

$$F_x = F_1 \cos 35° + F_2 \cos 35° = 0$$

$$F_z = F_1 \sin 35° + F_2 \sin 35° - P = 0$$

$$(F_1 + F_2) = \frac{P}{2 \sin 35°} = 2F_1 = 2F_2$$

$$F_1 = F_2 = \frac{P}{2 \sin 35°} = \frac{P}{1.147} = 0.872 P$$

The tensile amplitude in member AB is

$$= \frac{P}{2 \sin 35°} (\cos 35°) = 0.714 P$$

The geometry is

$$(\bar{l}, s_l) = (48, 0.6) \text{ in.}$$

$a = AC = BC:$ $\qquad (\bar{a}, s_a) = \frac{(48, 0.6)}{2} \frac{1}{\cos 35°} = (29.3, 0.366) \text{ in.}$

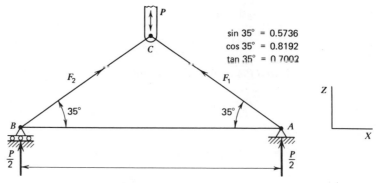

Figure 10.34 Pin-jointed frame.

System loading is essentially constant-amplitude sinusoidal (axial), with a small element of peak load variability:

$$(\bar{P}, s_P) = (230{,}000, 18{,}000) \text{ lb} \tag{a}$$

The material is aluminum alloy S_f estimated from Figure 8.32c:

$$(\bar{S}_u, s_{S_u}) = (76{,}000, 1600) \text{ psi}$$

and

$$(\bar{S}_f, s_{S_f}) = (27{,}000, 1600) \text{ psi} \tag{b}$$

$$(S_f)_{\text{axial}} \approx 0.9 (S_f)_{\text{bending}}$$

Since the loading is constant-amplitude sinusoidal, examination of the members for fatigue will be in the tension mode, with a check for stability in compression. Examination for column strength will be for static sufficiency. Also, note that since the loading is dynamic and member sizes will be large, the size-effect factor must be applied to properly reduce the endurance strength (see Section 10.6). It is assumed that (1) all members are straight, (2) loading is applied at the axial center of symmetry, and (3) loading is applied in the direction of the centerline of symmetry. It is also assumed that the Euler formula for P_{cr} applies. Thus for AB acting as a column, equation (6.9) is

$$P_{\text{cr}} = \frac{\pi^2 EI}{l^2} \tag{c}$$

$$(\bar{E}, s_E) = (10.3 \cdot 10^6, 15{,}400) \text{ psi}$$

(see Table 5.1). For a circular cross section (Appendix 5):

$$I = \frac{\pi d^4}{64} = 0.0491 d^4$$

A deterministic approximation (for the size-factor estimate) is as follows. First approximation of d for element AB, substituting mean values into (c), yields

$$163{,}200 = \frac{9.9(10{,}300{,}000)(0.0491 d^4)}{(48)^2}$$

$$d^4 \approx \frac{163{,}200}{2{,}170} = 71.5 \text{ in.}^4$$

$$d \approx 2.95 \text{ in.}$$

Based on the preceding estimate of member diameter, a size factor is estimated (see Figure 10.15):

$$(k_b, \mathit{s}_{k_b}) \simeq (0.88, 0.05)$$

It is apparent from the component details that $K_t \approx 1$.
The probabilistic design of element AB is as follows.

2. The significant applied stress:
Member AB load amplitude is:

$$(230{,}000, 18{,}000) \cdot (0.714) = (163{,}200, 12{,}900) \text{ lb}$$

From Table 4.5:

$$\Delta \approx 1.5\% \, \bar{r} \text{ and } \mathit{s}_r \approx \frac{0.015 \bar{r}}{3} = 0.005 \bar{r}$$

From Appendix 5, the statistics of A are

$$\mathit{s}_A \approx 2\pi \bar{r} \mathit{s}_r = 0.0314 \bar{r}^2$$

and

$$\bar{A} \approx \pi \bar{r}^2$$

$$(\bar{s}, \mathit{s}_s) = \frac{P}{A} = \frac{(163{,}200, 12{,}900)}{(\pi \bar{r}^2, 0.0314 \bar{r}^2)} = \left[\frac{52{,}000}{\bar{r}^2}, \frac{4140}{\bar{r}^2}\right] \text{ psi} \qquad (d)$$

$$C_s = \frac{\mathit{s}_s}{\bar{s}} = \frac{4140/\bar{r}^2}{52{,}000/\bar{r}^2} = 0.079$$

3. The significant allowable stress is $(b)\cdot k_b$, the product of two random variables [equation (10.13a)]:

$$\left(\bar{S}'_f, {}_{\backslash S'_f}\right) = (27{,}000, 1600)(0.88, 0.05)$$

$$= (23{,}800, 1950) \text{ psi} \qquad (e)$$

4. The specified system reliability for the frame is $R_{\text{sys}} = .9900$. The necessary reliability for each element is

$$R = [.9900]^{1/3} = .9967$$

From equation (9.10):

$$R = .9967 = \frac{1}{\sqrt{2\pi}} \cdot \int_z^\infty e^{-z^2/2} dz = \frac{1}{\sqrt{2\pi}} \cdot \int_{-2.73}^\infty e^{-z^2/2} dz$$

Thus (Appendix 1) $z = -2.73$.

5. Substitution of equations (d) and (e) into equation (9.11) yields

$$-2.73 = -\frac{23{,}760 - (52{,}400/\bar{r}^2)}{\sqrt{(1950)^2 + (4140/\bar{r}^2)^2}} \qquad (f)$$

Solve equation (f) for $\bar{r}$:

$$\bar{r} = 1.76 \text{ in.; tolerance} = \pm 0.026 \text{ in.}$$

$$\bar{d} = 3.52 \text{ in.; tolerance} = \pm 0.052 \text{ in.}$$

Members BC and AC are similarly designed.

Example N

Assume now that the structural element designed in Example M is placed in altered circumstances such that the loading is narrow-band random and the instantaneous amplitudes are normally distributed. Assume also that the peak stresses are Rayleigh distributed (see Sections 2.3 and 3.12), with an expected value $E(s) \approx \bar{s}$. Calculate the probability R of adequate performance of an element (designed in example M) selected at random and operating in the altered loading environment.

On the basis of results in Example M, the mean of the stress peaks $(\bar{s})$ estimator is calculated. If it is given that $\bar{F} = 163{,}200$ lb and, according to the preceding calculation, $(\bar{d}, {}_{\backslash d}) = (3.52, 0.0173)$ in., then from equation (2.59):

$$\bar{s} \approx \frac{4\bar{F}}{\pi \bar{d}^2} = 16{,}880 \text{ psi}$$

Utilizing expressions in Section 9.3.6, the Rayleigh parameter σ_s is estimated based on the mean of the peaks of the narrow-band random stress $E(s) = \mu_s$, as follows:

$$E(s) = \mu_s \approx \bar{s} = \left(\frac{\pi}{2}\right)^{1/2} \cdot \sigma_s$$

from which $\sigma_s \approx \dfrac{\bar{s}}{\sqrt{\pi/2}} = \dfrac{16{,}880}{\sqrt{\pi/2}} = 13{,}400$ psi.

As in Example J, fatigue strength is

$$\left(\bar{S}'_f, \,_{\!\!\!\!>S'_f}\right) = (23{,}760, 1950) \text{ psi}$$

From Section 9.3.6:

$$R = 1 - \frac{\sigma_s}{\left[\,_{\!\!\!\!>S'_f}^2 + \sigma_s^2\right]^{1/2}} \cdot \exp\left[-\frac{1}{2}\frac{\bar{S}'^2_f}{\,_{\!\!\!\!>S'_f}^2 + \sigma_s^2}\right]$$

$$R = 1 - \frac{13{,}400}{\sqrt{(1950)^2 + (13{,}400)^2}} \cdot \exp\left[-\frac{1}{2}\frac{(23{,}760)^2}{(1950)^2 + (13{,}400)^2}\right]$$

$$R = 0.7877$$

(in the altered random dynamic loading environment).

10.11 PROBABILISTIC FRACTURE MECHANICS

10.11.1 Monotonic Loading

Fracture toughness, often considered a fracture mechanics analogue of static strength (in design assuming elastic materials response), is considered in Section 8.11. Table 8.4 provides typical examples of the statistics describing both plane-strain (K_{1c}) and plane-stress (K_c) fracture toughness of a number of metallic alloys in common use.

The analogue of static load stress is called the *stress intensity factor* (K), defined as follows:

$$K = s\sqrt{\pi a \alpha} \quad \text{ksi} \cdot \sqrt{\text{in.}} \tag{10.27}$$

where K relates the stress state at the crack tip to the applied loads, s is the gross stress (ksi), a is the $\frac{1}{2}$ crack length (inches), and α is a dimensionless term depending on crack shape, configuration, and loading [71, 113, 114]* (see Table 10.5). To convert from ksi $\cdot \sqrt{\text{in}}$ to MPa $\cdot \sqrt{\text{m}}$, multiply by 1.099.

*Also see: D. P. Rooke and D. J. Cartwright, *Stress Intensity Factors*, Her Majesty's Stationary Office, London, 1976.

Table 10.5 Stress Intensity Factors for Tensile Loading
Stress Intensity Form, $K = \sigma\sqrt{\pi a \alpha}$

Crack Configuration	α
Center through crack in an infinite plate	1.0
Edge through crack in an infinite plate	1.25
Internal circular crack of radius a in an infinite body	$\dfrac{4}{\pi^2}$
Center through crack in a finite plate of width W	$\sec\dfrac{\pi a}{W}$

Since s and a are random variables, K is also a random variable. In design to satisfy a specified reliability, K is modeled in terms of applied loads and area statistics together with the statistics of the other relevant parameters [77, 104, 115–117].

The role of fracture mechanics in design and analysis has yet to be fully defined but is now being utilized in the analysis of plate and sheet structural elements and numerous others, categorized as follows:

1. Subject to monotonic loading in situations that may lead to unstable materials behavior, that is, high-strength brittle materials or normally ductile materials in cold environments (below their transition temperatures).
2. In prediction and control of subcritical crack grown in components operating in dynamic-load environments.
3. In prediction of residual component life after detection of cracks developed during service.

In this subsection the coupling of fracture mechanics and probabilistic concepts is explained by way of example. Probabilistic considerations are necessary in safety and weight sensitive structures. The essence of fracture mechanics is to relate original crack length a_i (detected or assumed to exist), to failure, that is, probability of failure under either static or dynamic loading conditions.

In probabilistic fracture mechanics it will be noted that utilization of stress concentration K_t factors to modify load stress and size (k_b) factors to modify strength are not applicable. Fracture toughness is determined by employing notched or cracked specimens (see Figure 10.35), and in specific designs fracture toughness values that correspond to the element thickness (estimated requirements) are utilized.

In cases where the variability in a may be inconsequential, the distribution of K will approximate that of s, often being normal; or lognormally distributed. Otherwise, the limiting distribution of K is assumed (i.e., lognormal) and the design procedure for distribution of S and $\bar{s}$ unspecified, outlined in Section 9.3.9 is applicable.

The following example is typical of quantitative estimations of the behavioral statistics of K.

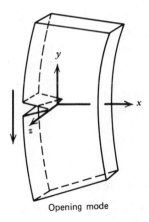

Opening mode   **Figure 10.35** Fracture toughness with notched specimens.

Example O

Assume that the element loaded in tension in Example A, Chapter 5, is a wide plate (treated as infinite), having a through-crack at one edge. Then (1) estimate the mean and the standard deviation of the stress intensity factor K, and (2) examine the sensitivity of s_K to the parameters in equation (10.27). Given:

Stress: $\quad (\bar{s}, s_s) = (10{,}160, 1{,}160)$ psi.

$\frac{1}{2}$ Crack length: $\quad (\bar{a}, s_a) = (0.06, 0.006)$ in.

$$\alpha = 1.25 \text{ (constant) (see Table 10.3)}$$

where s and a are statistically independent.

1. Application of equation (2.59) with equation (10.27) yields

$$\bar{K} = 10{,}160 \sqrt{\pi \cdot 0.06 \cdot 1.25} = 4930 \text{ psi} \cdot \sqrt{\text{in.}} \qquad (a)$$

Taking partial derivatives of the variables s and a in equation (10.27) with respect to K and evaluating the partials at $\bar{s}$ and $\bar{a}$ yields

$$\frac{\partial K}{\partial s} = \sqrt{\pi \alpha a} = 0.485; \quad \frac{\partial K}{\partial a} = \frac{s\sqrt{\pi \alpha}}{2} \left(\frac{1}{\sqrt{a}}\right) = 41{,}100 \qquad (b)$$

Substitution of results in equation (b) into equation (2.62) yields

$$s_K \simeq \left[\left(\frac{\partial K}{\partial s}\right)^2 \cdot (s_s)^2 + \left(\frac{\partial K}{\partial a}\right)^2 \cdot s_a^2 \right]^{1/2}$$

$$\simeq [(0.485)^2 \cdot (1{,}160)^2 + (41{,}000)^2 \cdot (0.006)^2]^{1/2}$$

$$s_K \simeq [314{,}000 + 60{,}800]^{1/2} = 610 \text{ psi } \sqrt{\text{in.}} \qquad (c)$$

2. Examination of the terms in equation (c) reveals that variability in stress is, in this case, the largest contributor to variability in K.

In the design process, loading statistics are estimated together with those of dimensional width w and length l of the plate, bar, or sheet element. Thickness $(B \pm \Delta B)$ is sometimes treated as the unknown to be determined. If $s = (P/A) = (P/wB)$, substitution into equation (10.27) yields

$$K = \frac{P}{wB} \cdot \sqrt{\pi \alpha a} \quad \text{psi} \cdot \sqrt{\text{in.}}$$

10.11.2 Design to Avoid Running Cracks (Monotonic Loading)

Since fracture toughness K_c (or K_{1c}) and the stress intensity factor K are random variables, design to avoid a running crack (with a specified probability R) can be accomplished. We adapt the design procedure outlined in Section 9.3.2 for normally distributed K and K_c or K_{1c} or that in Section 9.3.9 when the distributions of K^c and/or K_c or K_{1c} cannot be specified.

The criteria for adequate designs are

$$P(K_{1c} > K) \geq R \quad \text{(for thick sections)} \tag{10.28}$$

or

$$P(K_c > K) \geq R \quad \text{(for thin sections)} \tag{10.29}$$

The following example illustrates design to avoid a running crack (with a specified probability R) due to a random application of monotonic loading P.

Example P

1. Preliminary specifications are as follows. Assume a simply supported flat plate of radius $(\bar{r}, s_r) = (15.00, 0.0625)$ in. The thickness $(B \pm \Delta B)$ must be determined such that the probability of unstable crack growth will be no greater than $P_f = .0001$, due to the existence of a (undetected) flaw.

Monotonic loading P is applied uniformly over the surface of the plate. Total load is:

$$(\bar{P}, s_P) = (2400, 320) \text{ lb} \tag{a}$$

From Figure 10.36:

$$W = \frac{1}{2} w \pi r^2 = \frac{P}{2}; \quad P = 2W$$

where w is average loading per square inch:

$$w = \frac{P}{\pi r^2} \text{ lb/in.}^2 \tag{b}$$

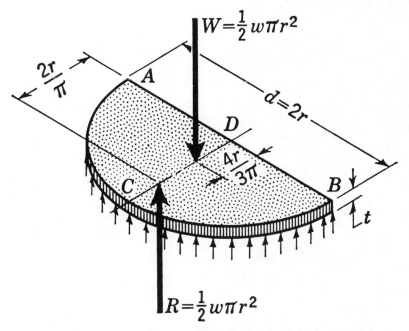

Figure 10.36 A flat circular plate uniformly loaded and simply supported.

The model to be utilized for expressing stress intensity in the plate (along a diameter) is, [1, p. 571]

$$s = \frac{wr^2}{B^2} = \frac{P}{\pi r^2} \cdot \left(\frac{r^2}{B^2}\right) = \frac{P}{\pi B^2} \text{ psi} \qquad (c)$$

The stress intensity factor K [equation (10.27)] is

$$K = s\sqrt{\pi a \alpha} = \frac{P}{\pi B^2} \cdot \sqrt{\pi a \alpha} \text{ psi } \sqrt{\text{in.}} \qquad (d)$$

In this exercise (1) loading and materials behavior are normally distributed and statistically independent, and (2) a center through-crack in the plate is suspected; thus $\alpha = 1.0$ (see Table 10.5). The (undetected) crack will be treated as having somewhat less than standard minimum detectable dimensions [104]. Assume that a will be a random variable described as

$$(\bar{a}, s_a) = (0.05, 0.01) \text{ in.} \qquad (e)$$

2. Next, the statistics of the stress intensity factor K [equation (d)] must be estimated as functions of sheet thickness B. The statistics of the root of half crack length a are [from equations (2.50) and (2.51)]

$$(\sqrt{\bar{a}}, s_{\sqrt{a}}) \approx (0.224, 0.100)$$

Thus use of equation (d) yields

$$K = \frac{P}{\pi B^2}\sqrt{\pi a \alpha} = 0.5642 \frac{P\sqrt{a}}{B^2}$$

and

$$(\bar{k}, s_k) = 0.5642 \frac{(2,400,320)}{(\bar{B}^2, 2\bar{B} s_B)}(0.224, 0.100).$$

$s_B \approx 0.001\bar{B}$ is estimated from tables in Chapter 4.

$$(\bar{k}, s_k) = 0.5642 \frac{(2400, 320)}{\bar{B}^2, 0.002\bar{B}^2}(0.2236, 0.100)$$

$$(\bar{k}, s_k) = \frac{(302.8, 141.3)}{(\bar{B}^2, 0.002\bar{B}^2)} \tag{f}$$

The statistics of K are

$$(\bar{k}, s_k) = \left(\frac{302.8}{\bar{B}^2}, \frac{141.3}{\bar{B}^2}\right) \tag{g}$$

3. To select material having K_c statistics (from Table 8.4) compatible with this design, a deterministic rough estimate of K is made for several values of B, utilizing equations (a), (d), and (e) (see Table 10.6):

$$K = \frac{2W}{\pi B^2}\sqrt{\pi a \alpha} \simeq \frac{302.8}{B^2} \text{ psi } \sqrt{\text{in.}}$$

For initial trial, the material selected from Table 8.4 is 2024-T3 aluminum alloy sheet, 0.080 thick, having the following K_c statistics:

$$(\bar{K}_c, s_{K_c}) = (83.40, 2.07) \text{ ksi } \sqrt{\text{in.}} \tag{h}$$

Note that K_{1c} (or K_c) is a function of B.

Table 10.6 Deterministic Rough Estimate of K

B (in.)	$K \cdot$ ksi $\sqrt{\text{in.}}$
0.10	30.25
0.075	54.00
0.05	121.00

4. Use of equation (9.10) yields

$$R = 1 - 0.0001 \approx \frac{1}{\sqrt{2\pi}} \int_z^\infty e^{-t^2/2} dt = \frac{1}{\sqrt{2\pi}} \int_{-3.62}^\infty e^{-t^2/2} dt$$

From Appendix 1, $z = -3.62$.

5. Solve for B, utilizing equations (g) and (h) with equation (9.11):

$$-3.62 = -\frac{83{,}400 - (302.8/\bar{B}^2)}{\sqrt{(2070)^2 + (141.3/\bar{B}^2)^2}}.$$

Expand and simplify:

$$0 = \bar{B}^4 - 0.00731\bar{B}^2 - 0.000024$$

$$\bar{B}^2 = 0.82 \cdot 10^3$$

The required thickness (B) of the plate in Figure 10.36 is

$$\bar{B} = 0.099 \text{ in.}$$

Note that $\bar{B} = 0.099$ in. exceeds the thickness of the sheet used to establish the statistics of K_c in equation (h). Thus a second iteration of the calculations may be necessary, using modified (smaller) values of K_c.

Example Q

There is frequently a need to estimate the statistics of a critical dimension (i.e., one that will result in unstable crack growth) a_{cr} of a flaw or notch in a mechanical component of known geometry, material, and loading. For the purpose of estimating a_{cr}, equation (10.30) may be used, where a is estimated for a specified material and component geometry. Assume a narrow thin sheet with an edge through-crack transverse to the direction in which tensile monotonic loading is applied. Then

$$K_c = s \cdot \sqrt{\pi a \alpha} \qquad (10.30)$$

where K_c is fracture toughness, in this case corresponding to the stress intensity necessary for unstable crack growth (Table 8.4) of 300M steel (70°F) strip, 0.125 in. thick and 5.00 inches wide:

$$(\bar{K}_c, s_{K_c}) = (141, 8.7) \text{ ksi } \sqrt{\text{in.}} \qquad (a)$$

Given. Maximum monotonic applied load stress intensity:

$$(\bar{s}, s_s) = (132.0, 7.50) \text{ ksi} \qquad (b)$$

508   Strength of Mechanical Components

For this example, $\alpha \simeq 1.25$ will be appropriate (see Table 10.5). Use of equation (10.30) yields

$$(\bar{a}_{cr}, {}_{s}a_{cr}) = \left(\frac{1}{\pi\alpha}\right) \cdot \frac{\left(\bar{K}_c, {}_{s}K_c\right)^2}{\left(\bar{s}, {}_{s}s\right)^2} \tag{c}$$

Application of equations (2.16) and (2.18) for squaring random variables K_c and s in equation (c) yields

$$(\bar{a}_{cr}, {}_{s}a_{cr}) \simeq 0.255 \cdot \left(\frac{(19{,}400, 2450)}{(17{,}400, 1480)}\right) \tag{d}$$

Critical crack length is, from use of equations (2.28) and (2.30) with equation (d),

$$(\bar{a}_{cr}, {}_{s}a_{cr}) \simeq (0.29, 0.050) \text{ in.}$$

The strength of a cracked material body is a function of geometry, fracture toughness, (K_{1c} or K_c) and size, shape, location, and orientation of the crack. An example of the relationship among these variables and applied stress is described next, in which allowable stress statistics are estimated.

Example R

A large D6AC steel sheet having an edge crack is loaded in tension. The crack length a through the entire thickness is

$$(\bar{a}, {}_{s}a) \simeq (1.00, 0.05) \text{ in.} \tag{a}$$

Also, from Table 10.5:

$$(\bar{\alpha}, {}_{s}\alpha) = (1.25, 0)$$

The value K_{1c} is given as follows (Table 8.4):

$$\left(\bar{K}_{1c}, {}_{s}K_{1c}\right) \simeq (65.50, 12.50) \cdot \text{ksi } \sqrt{\text{in.}} \tag{b}$$

It is assumed that plane-strain conditions dominate. The allowable stress S, utilizing equation (10.27), is

$$S = \frac{K_{1c}}{\sqrt{\pi a \alpha}} = \frac{K_{1c}}{\sqrt{3.14 \cdot 1.25 a}} = \frac{K_{1c}}{1.96 \sqrt{a}} \tag{c}$$

From equations (2.59) and (2.62):

$$\left(\overline{\sqrt{a}}, {}_{s}\sqrt{a}\right) \simeq \left(\bar{a}^{1/2}, \frac{{}_{s}a}{2\bar{a}}\right) = (1.00, 0.025) \sqrt{\text{in.}} \tag{d}$$

Allowable stress is estimated as

$$(\bar{S}, s_S) \simeq \frac{(65.50, 12.50)}{(1.96, 0.0500)} \qquad (e)$$

and from equation (2.28) and (2.30) used with equation (e):

$$(\bar{S}, s_S) \simeq (32.5, 6.33) \text{ ksi.}$$

Example S

Consider the situation where the plate in Example N is found to have a tool mark of depth

$$(\bar{a}, s_a) \simeq (0.210, 0.04) \text{ in.} \qquad (a)$$

after fabrication across one edge and approximately perpendicular to the direction of tensile loading. The plate is subjected to a random application of monotonic tensile loading. Determine the probability P_f of a running crack if the peak load stress is

$$(\bar{s}, s_s) = (132, 7.5) \text{ ksi.}$$

The critical crack depth determined in Example N is

$$(\bar{a}_{cr}, s_{a_{cr}}) \simeq (0.290, 0.050) \text{ in.} \qquad (b)$$

Application of equation (9.11) and substitution of values from equations (a) and (b) yield

$$z = -\frac{\bar{a}_{cr} - \bar{a}}{\sqrt{s_{a_{cr}}^2 + s_a^2}}$$

$$z = -\frac{0.291 - 0.210}{\sqrt{(0.049)^2 + (0.040)^2}} = -\frac{0.081}{0.063} = -1.29$$

According to standard normal area tables (Appendix 1):

$$R = \int_{-1.29}^{\infty} \frac{e^{-u^2/2} du}{\sqrt{2\pi}} = 0.8497$$

The probability P_f of a running crack is

$$P_f = 1 - R = 1.0000 - 0.850 = 0.150$$

This P_f value is unacceptable for most applications, and rework of the plate will be required to remove the tool mark and suitably reduce the failure probability.

10.11.3 Design to Avoid Fracture (Dynamic Loading)

With dynamic loading where crack growth may lead to failure, probabilistic fracture mechanics provides a means of estimating the requirements to limit propagation. Since even the most carefully processed material may not be flaw free, the probability of failure arising from crack propagation is a real concern in manufactured components. The parameter that promotes the growth of fatigue cracks in metals is called the *stress intensity factor range* (Δk). The value of ΔK as a function of a stress range is given by [113]

$$\Delta k = \Delta s \sqrt{\pi a \alpha} \tag{10.31}$$

where

$$\Delta k = k_{max} - k_{min}$$

and

$$\Delta s = s_{max} - s_{min}$$

Fatigue-crack propagation characteristics are described by plots (Figure 10.37) relating Δk and the crack growth rate, da/dN. The shape of the curve is sigmoidal and can be partitioned into three regions. Region I indicates the threshold value for Δk. In this region crack growth is extremely low or zero. In the second region the relationship is essentially linear, described by (see Figure 10.38):

$$\frac{da}{dN} = c(\Delta k)^n \tag{10.32}$$

where c and n are numerical values of materials characteristics. In region 3 crack growth is accelerated leading to fracture. Life here is extremely short. Regions 1 and 2 are of greatest concern to the designer. It may be desirable to design for region 1, that is, for conditions of nonpropagating fatigue cracks, with a specified probability.

In the case where the variability on half crack length a is inconsequential, the distribution of Δk will be approximately that of Δs. To design for a specified probability of endurance R, equation (9.11) can be used, in application of the design procedure for distribution of S and s unspecified, described in Section 9.3.9:

$$z = -\frac{\overline{\Delta k}_{th} - \overline{\Delta k}}{\sqrt{s_{\Delta k_{th}}^2 - s_{\Delta k}^2}} \tag{10.33}$$

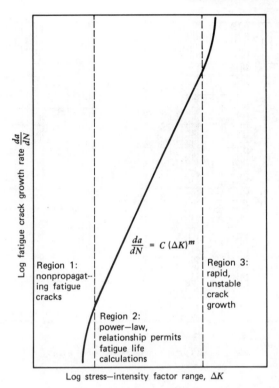

Figure 10.37 Signoidal relationship between da/dN and ΔK as plotted on logarithmic coordinates.

where Δk_{th} is threshold fracture toughness (analogous to endurance limit). The value of z is found in Appendix 1, corresponding to R, where

$$R \leqslant P(\Delta K < \Delta K_{th})$$

Typical values for threshold conditions of fatigue crack growth [118] are

$$\begin{aligned} \text{Steels:} & \quad 6\text{--}8 \text{ ksi } \sqrt{\text{in.}} \\ \text{Aluminum alloys:} & \quad 1\text{--}3 \text{ ksi } \sqrt{\text{in.}} \\ \text{Titanium alloys:} & \quad 2\text{--}6 \text{ ksi } \sqrt{\text{in.}} \end{aligned}$$

Also, see Figure 10.38 and similar figures.

Example T

1. To illustrate design for region 1 of Figure 10.37, consider a simply supported flat plate of radius r, as in Example P. Thickness (B) must be determined such that the probability of fracture (with indefinite load cycles) will be no greater than $P_f = .0001$.

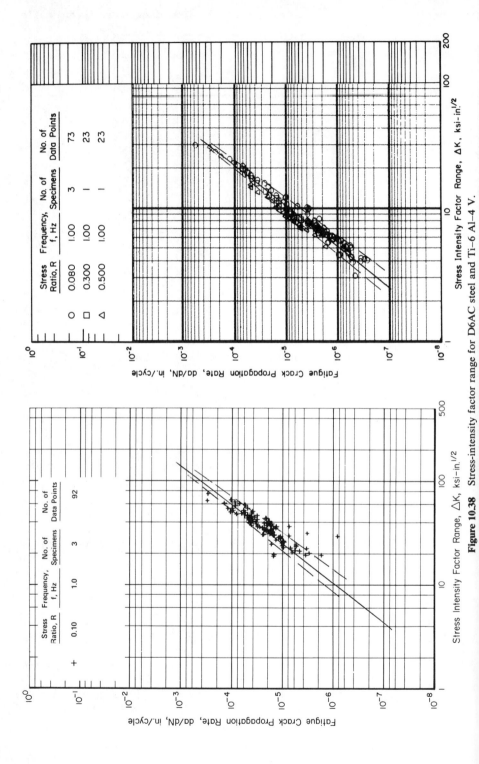

Figure 10.38 Stress-intensity factor range for D6AC steel and Ti-6 Al-4 V.

In this problem consider that the range of constant-amplitude cyclic loading applied uniformly over the surface of the plate is

$$(\Delta \bar{p}, s_{\Delta p}) = (1200, 200) \text{ lb} \qquad (a)$$

From Figure 10.36 and Example M:

$$W = \frac{1}{2} w \pi r^2 = \frac{P}{2}$$

$$w = \frac{P}{\pi r^2} \text{ lb/in}^2$$

The model for stress amplitude [1, p. 571] is

$$s = \frac{wr^2}{B^2} = \frac{P}{\pi r^2} \frac{r^2}{B^2} = \frac{P}{\pi B^2} \qquad (b)$$

A center through-crack is suspected; thus $\alpha = 1.0$ (Table 10.5). The undetected crack will be treated as having somewhat less than standard minimum detectable dimensions, 0.05 in. deep by 0.10 in. long [104], and a will be described as

$$(\bar{a}, s_a) = (0.05, 0.01) \text{ in.} \qquad (c)$$

Use of equations (2.50) and (2.51) with equation (b) yields

$$(\overline{\sqrt{a}}, s_{\sqrt{a}}) = (0.224, 0.10) \qquad (d)$$

2. From equation (10.31), the stress-intensity factor range, Δk, is:

$$\Delta k = \Delta s \sqrt{\pi a \alpha}$$

and

$$\Delta s = \frac{\Delta p}{\pi B^2}$$

Thus

$$\Delta k = \frac{\Delta P}{\pi B^2} \cdot \sqrt{\pi a \alpha} \; ; \quad \frac{\sqrt{\pi \alpha}}{\pi} = 0.5642$$

$$(\overline{\Delta k}, s_{\Delta k}) = 0.5642 \frac{(\overline{\Delta P}, s_{\Delta P})(\overline{\sqrt{a}}, s_{\sqrt{a}})}{(\bar{B}^2, 2\bar{B} s_B)}$$

and since (see Example M)

$$s_R \approx 0.001 \bar{B} \qquad (\overline{\Delta k}, s_{\Delta k}) = \frac{(\Delta P, s_{\Delta P})(\overline{\sqrt{a}}, s_{\sqrt{a}})}{(\bar{B}^2, 0.002\bar{B}^2)} \cdot 0.5642 \qquad (e)$$

substitution of equation (a) into equation (d) into equation (e) yields

$$(\overline{\Delta k}, s_{\Delta k}) = \frac{(1200, 200)(0.224, 0.100)}{(\bar{B}^2, 0.002\bar{B}^2)} 0.5642$$

$$= \frac{(268.8, 128.1)}{(\bar{B}^2, 0.002\bar{B}^2)} 0.5642 = \frac{(151.1, 72.3)}{(\bar{B}^2, 0.002\bar{B}^2)}$$

$$(\overline{\Delta k}, s_{\Delta k}) = \left[\frac{151.1}{\bar{B}^2}, \frac{73.0}{\bar{B}^2} \right] \qquad (f)$$

3. The plate material is C1050 carbon steel, for which the threshold Δk_{th} ranges from 6 to 8 ksi $\sqrt{in}$. The estimated mean value $\overline{\Delta k_{th}}$ is 7 ksi $\sqrt{in}$. Dividing the range by 6 provides a rough estimate of the standard deviation (i.e., 0.333 ksi $\sqrt{in}$.):

$$(\overline{\Delta k}_{th}, s_{\Delta k_{th}}) \approx (7000, 333) \text{ psi } \sqrt{in.} \qquad (g)$$

4. Use of equation (9.10) yields

$$R = 1 - 0.0001 = \frac{1}{\sqrt{2\pi}} \int_z^\infty e^{-t^2/2} dt = \frac{1}{\sqrt{2\pi}} \int_{-3.62}^\infty e^{-t^2/2} dt$$

From Appendix 1, $z = -3.62$.

5. Substitution from equations (f) and (g) into equation (9.11) yields

$$-3.62 = \frac{7000 - (151/\bar{B}^2)}{\sqrt{(333)^2 + (73/\bar{B}^2)^2}}. \qquad (h)$$

Expansion and simplification of equation (h) yields

$$1{,}456{,}000 B^4 - 7000 B^2 + 70{,}200 = 0$$

The required thickness, B, of the plate is

$$\bar{B} = 0.47 \text{ in.}$$

In the absence of more specific information it was assumed that Δk and Δk_{th} were approximately normally distributed in this example.

CHAPTER 11

Mechanical Element Optimization

11.1 INTRODUCTION

All design problems are actually optimization problems [85]. The imperative is to design a component, system, or process that will perform a specified task optimally, subject to certain solution constraints. Loosely, optimization can be defined as finding the best way to produce a design result. More accurately, the objective is to design a system, identify its variables and the conditions they must satisfy, define a measure of effectiveness (reliability), and then seek the state of the system (values of the variables) that give the most desirable measure of effectiveness [27].

Classical optimization procedures require application of calculus to the basic problem of finding the maximum or minimum of a continuous function (see Figure 11.1). An important characteristic of certain functions of one variable is related to the existance of maxima or minima. This property is called *convexity* and its opposite, *concavity*. Whereas elegant theory is available for treating convex and concave functions, most real-life functions in engineering design are neither [119, p. 134].

The concern in mechanical engineering design is often to minimize (optimize) weight to be lifted or moved, inertia of components in a dynamic system, and/or physical size, subject to a likelihood of failure constraint. An acceptable solution is required at the least weight, inertia, and size. Typical classical geometric models of constrained optimization problems are illustrated in Figures 11.2 through 11.7.

To verify engineering designs, experimentation (testing) is often cited as being essential. However, experimentation in the narrow sense, such as physical manipulation of variables, is frequently neither practical nor possible. Thus an approach that does not involve experimentation (entailing physical manipulation of the subject under study) on the total system must be used in the majority of cases.

We construct models of systems and their operations with which to conduct studies. The models often take the form of equations, which although

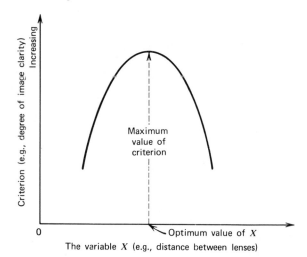

Figure 11.1 Graphic representation of the concept of optimum [2]. Distribution envelope of allowable combined stresses for AISI 4340 steel subject to fluctuating loading [84], endurance (a).

mathematically complicated, possess a simple underlying structure:

$$U = f(\mathbf{x}_n, \mathbf{y}_m); \qquad U^* = f(\mathbf{x}_n^*, \mathbf{y}_m^*)$$

where U is the value of system performance (figure of merit, utility, criterion or objective function), $\mathbf{x}_n$ are variables that can be controlled, $\mathbf{x}_n^*$ is the set of design variables corresponding to optimal U, $\mathbf{y}_n$ are variables that affect U

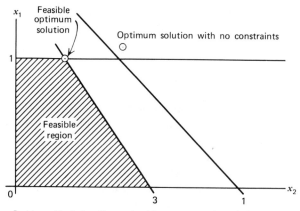

Problem: Maximize $f(x_1, x_2)$ subject to constraints, 1, 2, and 3.
1. $a_1 x_1 + b_1 x_2 \leq k_1$
2. $a_2 x_1 + b_2 x_2 \leq k_2$
3. $x_1 \leq k_3$

$a_1, b_1, a_2, b_2, k_1, k_2, k_3$ are same constraints.

Figure 11.2 A simplified but general optimization problem involving linear constraints. (*Note:* The range of possible solutions consists of all points such that $x_1, x_2 \leq 0$.)

Introduction 517

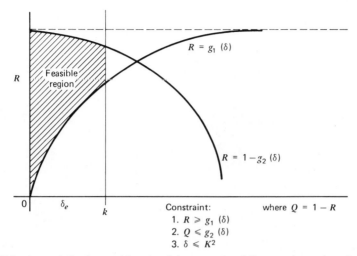

Figure 11.3 An optimization problem involving opposing failure modes and a single linear constraint.

but are not controlled, f is the relationship between U and x and y, and U^* is the value of the objective function at optimum.

Once the model is constructed, it can be used to find optimal values of the design variables that produce the best system performance for specified values of the uncontrolled variables (Figure 11.2). Regardless of the procedure used, an optimal solution maximizes or minimizes (as appropriate) the objective function $f(x,y)$, subject to constraints represented in that model.

Only constrained stochastic optimization problems that are functions of one variable are considered. In the examples attempts are made to find the

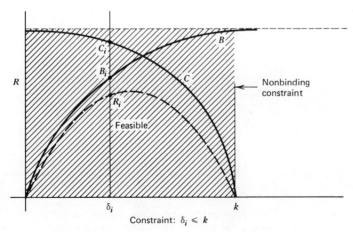

Figure 11.4 An optimization problem involving opposing failure modes and a nonbinding constraint.

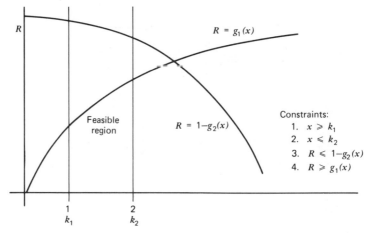

Figure 11.5 An optimization problem involving opposing failure modes and two linear constraints.

optimum value of a dimensional variable subject to sets of decisions (performance, life, and cost requirements, etc.) and constraints (reliability, physical size and configuration envelopes, strength behavior of materials envelopes, environmental factor envelopes, etc.).

Since a model is never a perfect representation of the problem, the formal optimal solution is never the ultimate solution. At the optimal value of the objective function U^*, the basic variables of the solution have specified a component that should have the intended impact on the system performance, which happens only if the model is a good representation of what is needed.

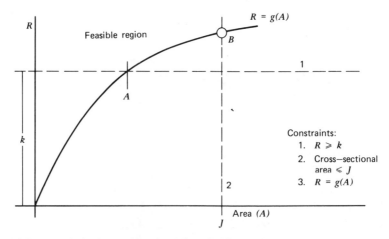

Figure 11.6 An optimization problem involving reliability and area constraints. (*Note*: Depending on the functional relation between cost and cross-section area, the optimum solution based on cost might be at either A or B.)

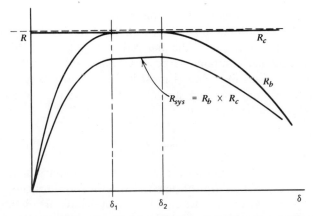

Figure 11.7 An optimization problem involving opposing failure modes, no constraints, and the existence of alternate optimum solutions. (*Note*: It can be seen in the figure that, from the macroscopic viewpoint of the engineer, there is essentially no change in the system reliability between δ_1 and δ_2. This implies that any value of δ, such that $\delta_1 < \delta < \delta_2$, will yield an optimum solution. Any δ satisfying this inequality is usually called an *alternate optimum solution*.)

The utilization of statistical models is clearly necessary because these provide complete behavioral profiles. Attempts to optimize employing deterministic assumptions may produce a tendency to conservatism.

11.1.1 Optimum Design of Mechanical Components

A study of mechanical elements suggests that the following three characteristics are common to most:

1. The mechanical components must satisfy certain requirements of the structure or machine in which they perform. For example, in the design of a helical spring (Figure 11.8) the functional requirements of the machine might

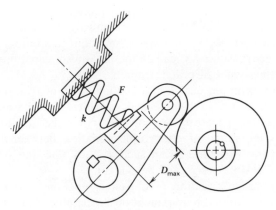

Figure 11.8 Cam load spring.

520 Mechanical Element Optimization

require resisting a specified distribution of force $(\bar{F}, s_F)$, possessing spring constant $(\bar{k}, s_k)$, and functioning at a stress level having low probability of exceeding the endurance limit of the material in shear [120, p. 205].

2. Most mechanical components possess undesirable attributes, (cost, inertia, weight, etc.), and their values must be limited to tolerable ranges. For example, the maximum shearing stress in a helical spring must be less than shearing strength of the material:

$$P\{\tau_{max} < S_{sy}\} \geq R$$

3. Real mechanical components are results of geometry specification including tolerances, surface finish, material specification, process or forming, and heat or other treatments. Mechanical components are defined by their geometry, material conditions, and dispersions. However, certain important properties of materials cannot be well defined mathematically. For example, machining characteristics, availability, and restriction to standard sizes and shapes are considerations that are often only vaguely defined. Generally, the selection of a material for a mechanical component will be based primarily on the distribution parameters of materials mechanical behavior.

Design Equations. Design equations generally express requirements in terms of design variables that can be classified in three basic categories: (1) functional requirement variables (Chapter 9 and 10), (2) materials behavior variables (Chapter 8), and (3) geometric variables (Chapter 4). These variables are generally random variables, estimated from analysis of the preliminary design form, that may often be influenced by factors external to the component. Random geometric variables can be varied far more continuously than materials behavior variables.

Optimum Design. The particular basis of optimization depends on the component under consideration and its application in a structural or machine setting. The procedure might require minimizing deflection or maximizing any one of such values as power-transmission capability, energy-storage capability, force-transmission capability, or speed capability, often complicated by the existence of practical limits.

A problem of optimum design may require study of relationships between functions that normally involve several variables. To determine an explicit solution, it may be necessary to consider not only individual functional variations, but the range of compatible variation between several limited functions of many variables. From such a study, the domains of variation with respect to feasible design will be evident.

The primary design equation or objective function is determined by the most significant functional requirement or undesirable effect. For example, in optimizing the weight or inertia (minimizing) of the offset link in Figure

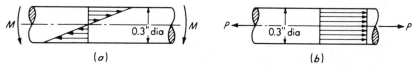

Figure 11.9 Stress gradients for bending and axial loads: (*a*) bending load; (*b*) axial load [27, p. 227].

11.18, cross-sectional area must be minimized, subject to the criterion [equation (9.3)]

$$P(S>s) = P(S-s>0) \geq R$$

Satisfactory ranges of values of certain variables often exist, which are expressed as simple *limit equations*. Stress limitations in a mechanical element are imposed by material behavioral (strength) properties. Geometric limitations may result from functional requirements of a mechanical or structural system, (due to space restrictions, manufacturing process limitations, etc.).

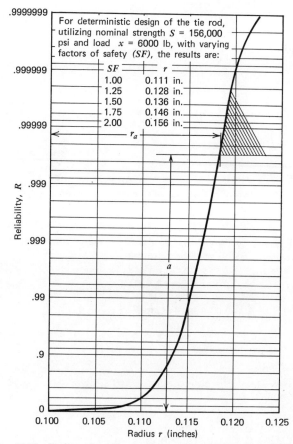

For deterministic design of the tie rod, utilizing nominal strength $S = 156{,}000$ psi and load $x = 6000$ lb, with varying factors of safety (SF), the results are:

SF	r
1.00	0.111 in.
1.25	0.128 in.
1.50	0.136 in.
1.75	0.146 in.
2.00	0.156 in.

Figure 11.10 Radius of tensile element versus reliability (data from Table 11.1).

522 Mechanical Element Optimization

Initial Steps in Mechanical Optimization [121]. The essential steps in design development are stated in Section 1.4 and formalized in Section 9.3.2 (subsection on design procedure with S and s normal). The practical task of optimization may be materially assisted by:

1. Utilization of probabilistic methods in design synthesis and analysis.
2. Stress gradient (see Figure 11.9) reduction or elimination, since minimizing geometric dimensions implies maximizing stress intensity everywhere (subject to material strength constraints) (see Figure 9.11).
3. Dealing locally with necessary stress raisers. Steps to minimize their potential effects should not be carried to the entire component.
4. Since design factors tend to intensify stress and reduce strength, attention should be given to minimizing their potential effects.

In design synthesis, the concern may be with minimizing the significant effect of failure. Therefore, an initial step is to determine the ways in which failure can occur. For example, with a tie rod required to resist a static tensile load, the effect to be avoided may be fracture.

Reliability of zero is associated with zero cross-sectional area of the tie rod. With increasing cross-sectional area, reliability improves. Reliability is a monotonically increasing function of the design variable. Optimization of a tie-rod design is illustrated in Figure 11.10, with the area constraint removed.

11.2 OPTIMUM DESIGN: R SPECIFIED, STRESS GRADIENT ZERO

Example A Tie Rod Optimization

Sensitivity of reliability to mean radius change is examined in this example. The component is a solid right circular cylinder loaded in tension (static) by $(\bar{P}, s_P)$.

Reliability is calculated for various mean radii over the range 0.09 to 0.125 in. The results are tabulated in Table 11.1 and plotted in Figure 11.10. The variability on the radius was determined as follows:

$$\text{Tolerance} = \Delta \approx 0.015 \bar{r}$$

from which the standard deviation is estimated as (Section 3.3.2):

$$s_r \approx \frac{\Delta}{3} = \frac{0.015 \bar{r}}{3} = 0.005 \bar{r}$$

Data for constraints and limit equations are as follows:

1. For AISI 4130 steel, the ultimate tensile strength is $(\bar{S}_u, s_{S_u}) = (156{,}000, 4300)$ psi (Appendix 10.A).
2. Static loading is $(\bar{P}, s_p) = (6000, 90)$ lb.
3. Failure is defined as fracture.

Table 11.1 A single Independent Variable versus Reliability, in Tension Element Design

(1) $\bar{r}$ (in.)	(2) $\bar{r}^2$ (in.^{2})	(3) $\pi\bar{r}^2$	(4) $\bar{r}^4$ (in.^{4})	(5) $\pi^2\bar{r}^4$	(6) $\dfrac{6.00}{\pi\bar{r}^2}$	(7) $156-6$	(8) $\dfrac{0.0117}{\pi^2\bar{r}^4}$	(9) $8+18.49$	(10) $\dfrac{7}{9}=z$	(11) R
0.09	0.0081	0.02545	6.561·10^{-5}	6.475·10^{-4}	−235.75	−79.75	18.11	6.050	+13.18	0.000...
0.095	0.009025	0.02835	8.136·10^{-5}	8.029·10^{-4}	−211.65	−55.65	14.505	5.744	+9.041	0.000...
0.100	0.0100	0.031416	1.0·10^{-4}	9.868·10^{-4}	−190.95	−34.95	11.86	5.509	+6.344	0.00000000012
0.105	0.011025	0.034636	1.2155·10^{-4}	1.1995·10^{-3}	−173.20	−17.20	9.9772	5.2255	+3.223	0.00064
0.1055										0.001
0.110	0.0121	0.03801	1.4641·10^{-4}	1.445·10^{-3}	−157.85	−1.85	8.097	5.156	+0.3588	0.35987
0.1125	0.012656	0.03976	1.60176·10^{-4}	1.5807·10^{-3}	−150.90	+5.10	7.4017	5.0884	−1.0023	0.841900
0.115	0.01323	0.04156	1.7503·10^{-4}	1.727·10^{-3}	−144.33	+11.67	6.775	5.0264	−2.322	0.98988
0.1163										0.999
0.1175	0.013806	0.04337	1.90608·10^{-4}	1.881·10^{-3}	−138.34	+17.66	6.2200	4.9709	−3.5526	0.999809
0.120	0.0144	0.045239	2.0736·10^{-4}	2.046·10^{-3}	−132.63	+23.37	5.7185	4.920	−4.750	0.9999989
0.1225	0.015006	0.04714	2.2518·10^{-4}	2.222·10^{-3}	−127.28	+28.72	5.2655	4.8739	−5.8926	0.99999998
0.125	0.01563	0.049103	2.4430·10^{-4}	2.4105·10^{-3}	−122.20	+33.80	4.8538	4.831	−6.996	1.000...

524 Mechanical Element Optimization

The functional relationships in this problem are similar to those of Example B, Chapter 9, and are solved for R values corresponding to a range of r values (see Table 11.1). This is a stochastic constrained optimization problem in one variable (r) with a load-stress constraint due to material strength and a process tolerance constraint $\pm \Delta$ on the radius. By utilizing equation (h), Example B, Chapter 9, values of z are calculated corresponding to a range of r values. Appendix 1 is utilized to estimate the values of R.

A notable result of these calculations is revealed in Figure 11.10. Reliability is shown to be extremely sensitive to the mean radius of the rod's cross section. For a required reliability $R > a$, the region of acceptable solutions is indicated by the slanted lines. Example:

$$z = -\frac{156 - (6.00/\pi \bar{r}^2)}{\sqrt{(4.3)^2 + 0.0117/\pi^2 \bar{r}^4}},$$

and with $\bar{r} = 0.09$, and $z = +13.18$. Thus

$$R = \frac{1}{\sqrt{2\pi}} \int_z^\infty e^{-z^2/2} dz = \frac{1}{\sqrt{2\pi}} \cdot \int_{+13.18}^\infty e^{-z^2/2} dz.$$

From Appendix 1, $R \approx 0.00 \cdots$.

In the preceding example the following are established:

1. Minimum radius r_a associated with a required reliability $R \geq a$.
2. Fully stressed design is realized, since all material over the cross section is loaded to capacity.

11.3 OPTIMUM DESIGN: R SPECIFIED, SURFACE STRESS GRADIENT

Example B Optimization by Eliminating Surface Stress Gradients

To optimize the profile of a solid shaft, we must maximize the outer fiber stress intensity at all points along the profile from support to support, subject to a reliability and a material strength constraint. Outer fiber stress gradients must be eliminated to achieve "fully stressed design."

Utilizing probabilistic design methods to estimate $\bar{r}_{min}$ at all points, we assure equal maximum bending stresses in the outer fibers over the length of the shaft. Three conditions of shaft support are examined:

Case 1. Self-aligning bearings at both ends (no capability for developing end moments).

Case 2. Self-aligning bearing at one end and a journal at the other end (end moment developed at the journal end).

Case 3. Journals at both ends (end moments developed at both ends of shaft).

Optimum Design: R Specified, Surface Stress Gradient

To simplify this multipart example without reducing its utility in illustrating optimization, only bending stresses are considered. The nominal stress for a cylindrical shaft, is given by equation (5.23):

$$s_x = \frac{Mc}{I}$$

where (from Appendix 5)

$$I = \frac{\pi d^4}{64} \text{ in.}^4$$

$$d = 2r = \text{Shaft diameter} = 2c$$

Thus

$$s_x = \frac{Mc}{I} = \frac{Md}{2\left(\frac{\pi d^4}{64}\right)} = \frac{32M}{\pi d^3} \quad \text{(a)}$$

Assume that all random variables in the probabilistic portion of this example are normally distributed and statistically independent.

Constraints and limit equations are as follows. The shaft material is AISI 4130 cold-rolled and annealed steel (Table 8.1):

$$\bar{S}_u = 98{,}000 \text{ psi}$$

Assume that

$$(\bar{S}_e, s_{S_e}) \approx (49{,}000, 3300) \text{ psi} \quad \text{(b)}$$

Loading (fully reversed sinusoidal) is

$$(\bar{P}, s_P) = (6070, 200) \text{ lb.} \quad \text{(c)}$$

For the probabilistic design $R = .9987$. Safety is not a consideration, and replacement is economical.

$$(\bar{l}, s_l) = (72.0, 0.0416) \text{ in.}$$

Stress concentration K_t (see Figure 7.1) is $s_{K_t} \approx 0.05$ ($\bar{K}_t$) $= 0.06$ (estimated), and $\bar{K}_t = 1.20$.

Design with deterministic assumptions is as follows. The working stress (s_x) is [from equation (a)]

$$s_x = \frac{S_e}{K_t(SF)} = \frac{32M}{\pi d^3} = \frac{4M}{\pi r^3} \quad \text{(d)}$$

Thus

$$r = \left[\frac{4K_t(SF)}{\pi S_e} M\right]^{1/3} = [4.677 \cdot 10^{-5} \cdot M]^{1/3} \text{ in.} \quad (e)$$

For the estimation of M and r, see Figure 11.11 through 11.13.

1. The probabilistic optimum design is as follows. The bending moment diagrams in Figures 11.11 through 11.13 indicate that varying moments, hence a stress gradient, will exist over the length of each shaft from the left to the right bearing support. Each shaft profile can be optimized by utilizing probabilistic methods to calculate the minimum radii such that uniform maximum surface fiber stress is developed at every point between the supporting bearings. The outer fiber stress, given by equation (5.23), will be estimated to yield $(\bar{r}, s_r)$ as a function of the distance $(\bar{x}, s_x)$ along the shaft from the left support point (see Figure 11.14). Stress concentrations were detected in localized areas of the shafts, at hub faces, and at bearing faces. The shaft fillet radius adjacent to the hub and bearing faces was determined in each case for $\overline{K}_t = 1.20$ (Figure 7.1b). The diameter D was set equal to twice the largest of $\bar{r}_{34}$, $\bar{r}_{38}$, or $\bar{r}_{36}$ (see Figure 11.14), for which:

$$\left(\overline{K}_t, s_{K_t}\right) \approx (1.20, 0.06)$$

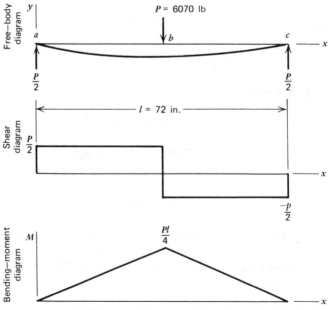

Figure 11.11 Shaft with self-aligning bearings, both ends. Classical approach—case 1.

Optimum Design: R Specified, Surface Stress Gradient 527

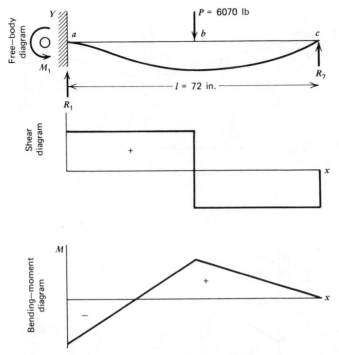

Figure 11.12 Shaft with self-aligning bearing and pillow block. Classical approach—case 2: shaft supported by self-aligning bearing and pillow block as shown; approximated by a beam with one fixed and one simply supported, with a center load.

Significant applied stress is [equation (10.15a)]

$$(\bar{s}', \circ_{s'}) = (\bar{s}, \circ_s)(\bar{K}_t, \circ_{K_t}) = (\bar{s}, \circ_s) \cdot (1.20, 0.06) \text{ psi}$$

2. The probabilistic solution, case 1 (Figure 11.11), is as follows. Estimate the applied stress statistics. The bending moment for a simply supported centrally loaded beam (Appendix 4) is

$$(\bar{M}_{ab}, \circ_{M_{ab}}) = \frac{(\bar{P}, \circ_P)(\bar{x}, 0)}{2}$$

$$= \frac{(6070, 200)(\bar{x}, 0)}{2} = (3035\bar{x}, 100\bar{x}) \qquad (a')$$

Equation (a') provides the moment M_{ab} statistics as a function of $\bar{x}$, the distance from a. For a round shaft, the moment of inertia (Appendix 5) about the neutral axis is

$$I = \frac{\pi \bar{d}^4}{64} \text{ in.}^4 = \frac{\pi \bar{r}^4}{4} \text{ in.}^4; \quad \frac{I}{r} = \frac{\pi r^3}{4} \qquad (b')$$

528 Mechanical Element Optimization

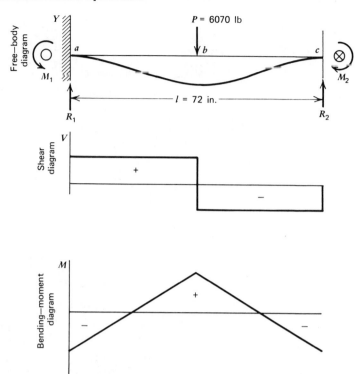

Figure 11.13 Shaft with pillow block, both ends. Classical approach—case 3: shaft supported by pillow blocks at each end; approximated by a beam with fixed supports at both ends, with a center load.

Hence, [from equation (2.59)]:

$$\frac{\bar{I}}{\bar{r}} = \frac{\bar{r}^4}{4\bar{r}} = \frac{1}{4} \cdot \bar{r}^3 = 0.785\bar{r}^3 \tag{b''}$$

on the basis of manufacturing tolerances (Chapter 4 and Section 3.3.2), the standard deviation estimate is $\mathfrak{s}_r \approx 0.015\bar{r}$. From equation (2.62) and $f(r) = r^3$:

$$\mathfrak{s}_r 3 = \frac{\partial}{\partial r}(\bar{r}^3)\mathfrak{s}_r = 3\bar{r}^2 \mathfrak{s}_r = 3(0.015)\bar{r}^3 = 0.045\bar{r}^3$$

Table 11.2 Summary, Deterministic Solution with Safety Factor of 1.5

	Shaft Radius	Self-aligning Bearings
Case 1	$r = 1.72$ in.	2
Case 2	$r = 1.56$ in.	1
Case 3	$r = 1.37$ in.	0

The design solutions summarized in Table 11.2 are not optimum because the shaft outer fibers are fully stressed only at the points of maximum moment.

Optimum Design: R Specified, Surface Stress Gradient

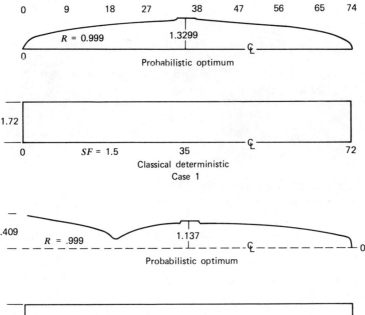

Figure 11.14 Optimum shaft profiles.

and

$$s_{I/r} = \frac{\partial}{\partial r}\left(\frac{\pi}{4}\bar{r}^3\right) \cdot s_r = 0.785 \cdot \frac{\partial}{\partial r}(\bar{r}^3) \cdot 0.015\bar{r} = 0.0353\bar{r}^3 \qquad (b''')$$

The applied stress is, from equations (5.23), (a'), (b''), and (b'''):

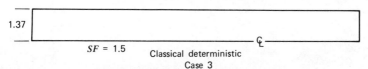

(c')

530 Mechanical Element Optimization

3. The endurance limit has been specified as [equation (b)]

$$\left(\bar{S}_e, {}_{{}^s S_e}\right) = (4.9 \cdot 10^4, 3.3 \cdot 10^3) \text{ psi} \tag{d'}$$

4. Utilization of equation (9.11) and substitution of values of $\bar{x}$ from 0 to 34 in. from the left support, yields

$$z = -\frac{\bar{S}_{ab} - \bar{S}_e}{\sqrt{{}^2_{S_{ab}} + {}^2_{S_e}}}; \quad \bar{S}^2_{ab} - z^2 {}^2_{S_{ab}} - 2\bar{S}_{ab} \cdot \bar{S}_e + \bar{S}^2_e - z^2 {}^2_{S_e} = 0 \tag{e'}$$

where $z = -3$ correspond to a probability of .9987 (Appendix 1). Substitution of values from equations (c') and (d') into equation (e') yields

$$\frac{1}{\bar{r}^6}(1.452 \cdot 10^7 \bar{x}^2) + \frac{1}{\bar{r}^3}(3.79 \cdot 10^8 \bar{x}) + (2.303 \cdot 10^9) = 0$$

5. Solve the quadratic in terms of $1/\bar{r}^3$, and take the root of the reciprocal, as follows:

$$\bar{r}_{ab} = 0.393 \cdot (\bar{x})^{1/3} \quad (0 < x_i < 34) \text{ in.}$$

and by symmetry:

$$r_i = r_{72-i}; \quad (38 < x_{72-i} < 72) \text{ in.}$$

The profile is shown in Figure 11.14 (also see Table 11.3).

Table 11.3 Radii Yielding Equal Maximum Outer Fiber Stresses

r Location (See Figure 11.14)		Case 1 1–33 in. and 71–38 in.	Case 2		Case 3 1–33 in. 71–38 in.
			1–33 in.	71–38 in.	
1	71	0.393	1.384	0.335	1.207
3	69	0.567	1.333	0.484	1.158
6	66	0.714	1.247	0.610	1.075
9	63	0.818	1.148	0.699	0.976
12	60	0.900	1.028	0.769	0.853
15	57	0.969	0.870	0.829	0.677
18	54	1.030	0.614	0.881	0.151
21	51	1.085	0.484	0.927	0.567
24	48	1.134	0.714	0.969	0.714
27	45	1.179	0.850	1.008	0.818
30	42	1.222	0.953	1.044	0.900
33	39	1.261	1.037	1.078	0.969

6. The probabilistic solution, case 2 (Figure 11.12), is as follows. The bending moment for a beam with one end fixed and one end simply supported (with center load) is given by (Appendix 4)

$$\left(\overline{M}_{ab}, {}^{s}M_{ab}\right) = \frac{\left(\overline{P}, {}^{s}{}_{P}\right)}{16}\left[11(\overline{x},0) - 8(l, {}^{s}{}_{l})\right] \text{ in. lb}$$

and

$$\left(\overline{M}_{bc}, {}^{s}M_{bc}\right) = \frac{5}{16}\left(\overline{P}, {}^{s}{}_{P}\right)\left[(l, {}^{s}{}_{l}) - (\overline{x},0)\right] \text{ in. lb}$$

Following the general procedure used in case 1:

$$\bar{r}_{ab} = \left[\frac{5.49 \cdot 10^7 \bar{x}^2 - 2.156 \cdot 10^9 \bar{x} + 2.116 \cdot 10^{10}}{5.207 \cdot 10^8 \bar{x} - 1.022 \cdot 10^{10} + \left[1.83 \cdot 10^{16} \bar{x}^2 - 7.199 \cdot 10^{17} \bar{x} + 7.068 \cdot 10^{18}\right]^{1/2}}\right]^{1/3} \text{ in.}$$

and

$$\bar{r}_{bc} = \left[\frac{1.134 \cdot 10^7 \bar{x}^2 - 1.633 \cdot 10^7 \bar{x} + 5.879 \cdot 10^{10}}{-2.367 \cdot 10^8 \bar{x} + 1.704 \cdot 10^{10} + \left[3.788 \cdot 10^{15} \bar{x}^2 - 5.454 \cdot 10^{17} \bar{x} + 1.963 \cdot 10^{19}\right]^{1/2}}\right]^{1/3} \text{ in.}$$

The optimized shaft profile for case 2 is shown in Figure 11.14 (also see Table 11.3).

7. The probabilistic approach, case 3 (Figure 11.13), is as follows. The bending moment for a shaft with fixed supports and a center load is given by Appendix 4:

$$\left(\overline{M}_{ab}, {}^{s}M_{ab}\right) = \frac{\left(\overline{P}, {}^{s}{}_{p}\right)\left[4(\overline{x},0) - (l, {}^{s}{}_{l})\right]}{8}$$

$$\left(\overline{M}_{ab}, {}^{s}M_{ab}\right) = \left[(3.035 \cdot 10^3 \bar{x} - 5.463 \cdot 10^4);\ (10^4 \bar{x}^2 - 3.6 \cdot 10^5 \bar{x} + 3.241 \cdot 10^6)^{1/2}\right]$$

Following the mathematical procedures used previously:

$$\bar{r}_{ab} = \left[\frac{2.90 \cdot 10^7 \bar{x}^2 - 1.045 \cdot 10^9 \bar{x} + 9.406 \cdot 10^9}{3.787 \cdot 10^8 \bar{x} - 6.817 \cdot 10^9 + \left[9.697 \cdot 10^{15} \bar{x}^2 - 3.491 \cdot 10^{17} \bar{x} + 3.142 \cdot 10^{18}\right]^{1/2}}\right]^{1/3} \text{ in.}$$

The optimized shaft profiles are shown in Figure 11.14 (also see Table 11.3).

11.4 OPTIMUM DESIGN: R SPECIFIED, CROSS-SECTIONAL STRESS GRADIENT

Example C Optimizing by Eliminating Cross-sectional Stress Gradients

1. In this example the cross section in the shank of an offset link (Figure 11.15) is designed. Case 1 assumes a rectangular cross section. In case 2, weight and inertia are minimized, using the idea of fully stressed design. Case 3, which deterministic, is included for comparison. The limit equations are as

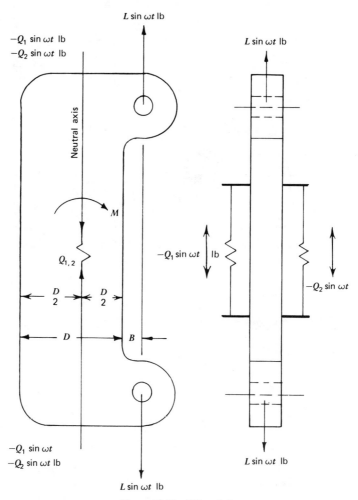

Figure 11.15 Offset link.

follows

Reliability: $R \geqslant 0.9987$
Cycle life: $N \geqslant 235{,}000$
Material: titanium alloy, Ti 6Al-4V
Loading: constant-amplitude sinusoidal: $(\bar{L}, s_L) = (12{,}500, 980)$ lb

$$(-\bar{Q}_1, s_{Q_1}) = (-\bar{Q}_2, s_{Q_2}) = (-6250, 490) \text{ lb} \tag{b}$$

The geometry is

$$(\bar{B}, s_B) = (0.125, 0.004) \text{ in.}$$

$$(\bar{D}, s_D) = (2.00, 0.060) \text{ in.} \tag{c}$$

Stress and strength are assumed to be normal or approximately normally distributed, thus the design procedure for S and s normal, described in Section 9.3.2 is appropriate. Assume that the links will be machined from plate or bar stock. They will be free from geometric discontinuities ($K_t \approx 1$) and surface roughness ($K_a \approx 1$).

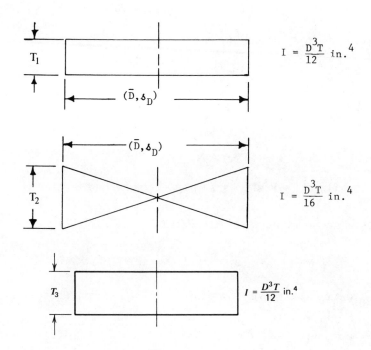

534 Mechanical Element Optimization

The following cases are optimized:

a. *Case 1.* The cross section of the offset link is rectangular by specification, $R = 0.9987$.
b. *Case 2.* The cross section of the offset link is of unspecified shape. Any stress gradients are exploited in optimizing the design, $R = 0.9987$.
c. *Case 3.* Deterministic design with rectangular cross-section and Safety Factor of 1.5.

In case 1 a partial optimization of the link cross section is achieved by designing the rectangular cross section probabilistically. In case 2 optimization is carried further by designing the cross section such that stress is everywhere uniformly maximum (subject to material endurance strength and reliability constraints). Case 3 assumes determinism and is included for comparison.

2. Stress determination is as follows. The loading will be identical for cases 1 and 2. From Figure 11.15, the correlated sinusoidal fully reversed forces Q_1 and Q_2 are applied on opposite sides of the link and act along the centerline of symmetry. From equations (2.36) and (2.37):

$$-\bar{Q} = -\bar{Q}_1 - \bar{Q}_2 = -6250 - 6250 = -12{,}500 \text{ lb}$$

and

$$s_Q = \sqrt{s_{Q_1}^2 + s_{Q_2}^2 + 2 s_{Q_1} s_{Q_2}} = 980 \text{ lb}$$

The load transferred through the pins is

$$\text{Load}_{\text{pin}} = L \sin wt$$

The shank load is

$$\text{Load}_{\text{shank}} = Q \sin wt$$

The combined direct load effect P is

$$p = \text{Load}_{\text{pin}} + \text{load}_{\text{shank}} = L \sin wt - 2Q \sin wt = (L - Q) \sin wt.$$

The total direct load amplitude is $L - Q$:

$$\bar{P} = (\bar{L} - \bar{Q}) \sin wt = (12{,}500 - 12{,}500) \sin wt = 0$$

Since L and Q are positively correlated with $\rho = +1$, from equation 2.41:

$$s_P = \sqrt{s_L^2 + s_Q^2 - 2\rho s_L s_Q} = \sqrt{2(980)^2 - 2(+1)(980)^2} = 0$$

Optimum Design: R Specified, Cross-sectional Stress Gradient 535

Thus the total direct load sums to zero. There is, however, a dynamic moment. From Figure 11.15, it is seen that the link must resist a moment M:

$$M = L\left(B + \frac{D}{2}\right) \text{ in. lb}$$

The maximum bending stress in the link is given by

$$s = \frac{Mc}{I} = \frac{M(D/2)}{I} = \frac{MD}{2I} \tag{d}$$

Case 1, bending stress (rectangular cross-sectional link), varies from 0 at neutral axis to s_{max} at outer fibers:

$$s = \frac{L[B+(D/2)]D}{2\left(\frac{T_1 D^3}{12}\right)} = \frac{L}{T_1}\left(\frac{6B}{D^2} + \frac{3}{D}\right) \tag{e}$$

Case 2, bending stress—link cross-section uniformly stressed from neutral axis to outer fibers:

$$s = \frac{MC}{I} = \frac{L[B+(D/2)]D}{2(D^3 T_2/16)} = \frac{L}{T_2}\left(\frac{8B}{D^2} + \frac{4}{D}\right) \tag{e'}$$

Case 1 design calculations are as follows.

3. Mean and standard deviation of tensile applied stress, from equations (2.59) and (e), is

$$\bar{s} \approx \frac{\bar{L}}{\bar{T}_1}\left(\frac{6\bar{B}}{\bar{D}^2} + \frac{3}{\bar{D}}\right) \approx \frac{12{,}500}{\bar{T}_1}\left(\frac{6(0.125)}{4} + \frac{3}{2}\right) = \frac{21{,}090}{\bar{T}_1} \text{ psi} \tag{f}$$

The standard deviation of s is estimated by utilizing equations (2.62) and (e). Taking partial derivatives of s with respect to the variables in (e) and evaluating at the mean values of each yields

$$\frac{\partial s}{\partial L} = \frac{1}{\bar{T}_1}\left(\frac{6\bar{B}}{\bar{D}^2} + \frac{3}{\bar{D}}\right) = \frac{1}{\bar{T}_1}\left(\frac{6(0.125)}{4} + \frac{3}{2}\right) = \frac{1.68}{\bar{T}_1}$$

$$\frac{\partial s}{\partial B} = \frac{6\bar{L}}{\bar{D}^2 \bar{T}_1} = \frac{6(12{,}500)}{4\bar{T}_1} = \frac{18{,}740}{\bar{T}_1}$$

$$\frac{\partial s}{\partial T} = -\frac{\bar{L}}{\bar{T}_1^2}\cdot\left(\frac{6\bar{B}}{\bar{D}^2} + \frac{3}{\bar{D}}\right) = -\frac{21{,}100}{\bar{T}_1^2}$$

$$\frac{\partial s}{\partial D} = -\frac{12\bar{B}\bar{L}}{\bar{D}^3 \bar{T}_1} - \frac{3\bar{L}}{\bar{D}^2 \bar{T}_1} = -\frac{11{,}700}{\bar{T}_1}$$

Utilizing the given data, the terms of equation (2.62) are computed:

$$\left(\frac{\partial s}{\partial L}\right)^2 \cdot s_L^2 = \frac{2,740,000}{\bar{T}_1^2} \qquad s_L = 980 \text{ lb}$$

$$\left(\frac{\partial s}{\partial B}\right)^2 \cdot s_B^2 = \frac{5620}{\bar{T}_1^2} \qquad s_B = 0.004 \text{ in.}$$

$$\left(\frac{\partial s}{\partial T}\right)^2 \cdot s_T^2 = \frac{999}{\bar{T}_1^4} \qquad s_T = 0.0015 \text{ in.}$$

$$\left(\frac{\partial s}{\partial D}\right)^2 \cdot s_D^2 = \frac{4.93 \cdot 10^5}{\bar{T}_1^2} \qquad s_D = 0.060 \text{ in.}$$

The standard deviation estimator of applied stress s is

$$s_s \approx \left[\left(\frac{\partial s}{\partial L}\right)^2 \cdot s_L^2 + \left(\frac{\partial s}{\partial B}\right)^2 \cdot s_B^2 + \left(\frac{\partial s}{\partial T}\right)^2 \cdot s_T^2 + \left(\frac{\partial s}{\partial D}\right)^2 \cdot s_D^2\right]^{1/2}$$

$$\approx \left[\frac{2.737 \cdot 10^6}{\bar{T}_1^2} + \frac{5620}{\bar{T}_1^2} + \frac{10^3}{\bar{T}_1^4} + \frac{4.93 \cdot 10^5}{\bar{T}_1^2}\right]^{1/2} = \left[\frac{3.236 \cdot 10^6}{\bar{T}_1^2} + \frac{10^3}{\bar{T}_1^4}\right]^{1/2} \quad \text{(g)}$$

Clearly, loading and dimension D are the largest potential contributors to variability in stress. Equations (f) and (g) provide the stress statistics:

$$(\bar{s}, s_s) \approx \left[\frac{21,090}{\bar{T}_1}, \left(\frac{3.24 \cdot 10^6}{\bar{T}_1^2} + \frac{10^3}{\bar{T}_1^4}\right)^{1/2}\right] \text{ psi}$$

It is seen that the first term in s_s dominates:

$$\frac{3.2a \cdot 10^6}{\bar{T}_1^2} \gg \frac{10^3}{\bar{T}_1^4}$$

Thus

$$(\bar{s}, s_s) \approx \left[\frac{21,100}{\bar{T}_1}, \left(\frac{3.24 \cdot 10^6}{\bar{T}_1^2}\right)^{1/2}\right] \text{ psi} \quad \text{(h)}$$

4. Estimate the material fatigue strength at the required 235,000 cycle life:

Titanium alloy bar: Ti 6Al-4V,

Ultimate strength: $(\bar{S}_u, s_{S_u}) = (135.5, 6.7)^*$ ksi (i)

Endurance limit: $(\bar{S}_e, s_{S_e}) = (95.3, 5.31)^\dagger$ ksi (i')

*See Table 8.1.
†See Appendix 10.B.

Optimum Design: R Specified, Cross-sectional Stress Gradient 537

Assume a strength value of $0.9S_u$ at 10^3 cycles, that is,

$$(122, 6.03) \text{ ksi} \approx 0.90(135.5, 6.7) \text{ ksi} \tag{j}$$

The $S-N$ envelope (Figure 11.16) is developed by the technique explained in Section 10.10.1. The fatigue strength of the 6Al-4V titanium alloy after 235,000 cycles of loading is estimated by interpolation, using equations (i′) and (j) and the figure

$$\left(\overline{S}_f, {}_{\circ S_f}\right) \approx (101.2, 5.78) \text{ ksi} \tag{k}$$

5.

$$R = 0.9987 \leqslant \frac{1}{\sqrt{2\pi}} \int_z^\infty e^{-t^2/2} dt = \frac{1}{\sqrt{2\pi}} \int_{-3.08}^\infty e^{-t^2/2} dt$$

6. Substitution of values from equations (h) and (k) into the primary design equation (9.11) yields

$$-3.08 = -\frac{101.2 - \left(21.100/\overline{T}_1\right)}{\sqrt{(5.78)^2 + \left(3.240/\overline{T}_1^2\right)}}$$

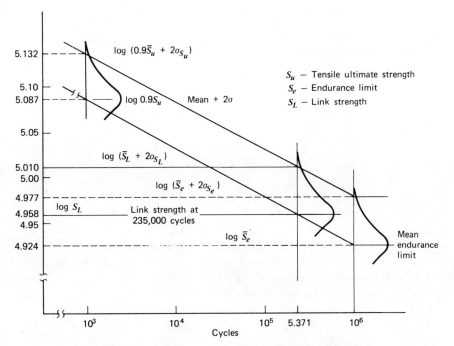

Figure 11.16 $S-N$ Envelope. For Ti 6 Al–4V alloy see Table 8.1.

538 Mechanical Element Optimization

Expand and simplify:

$$0 = \bar{T}_1^2 - 0.431\bar{T}_1 + 0.0542$$

For case 1 with a link of rectangular cross section:

$$\bar{T}_1 = \frac{0.441 \pm 0.140}{2} = 0.280, 0.151 \text{ in.}$$

The significant root is $\bar{T}_1 = 0.280$ in. The preceding solution falls short of optimum because only the outer surface fibers are fully stressed.

7. Case 2 design calculations are as follows, beginning with mean and standard deviation of tensile applied stress. From equation (2.59), and the procedure followed in case 1, use of equation (e′) yields

$$\bar{s} = \frac{\bar{L}}{\bar{T}_2}\left(\frac{8\bar{B}}{\bar{D}^2} + \frac{4}{\bar{D}}\right)$$

from equation (2.62):

$$\mathfrak{d}_s \approx \left[\frac{5.51 \cdot 10^6}{\bar{T}_2^2} + \frac{7120}{\bar{T}_2^4}\right]^{1/2} \approx \left[\frac{5.75 \cdot 10^6}{\bar{T}_2^2}\right]^{1/2}$$

The expression for applied stress in case 2 is

$$(\bar{s}, \mathfrak{d}_s) \approx \left[\frac{28,100}{\bar{T}_2}, \left(\frac{5.75 \cdot 10^6}{\bar{T}_2^2}\right)^{1/2}\right] \text{ psi} \qquad (a')$$

8. The fatigue strength of the titanium alloy bar Ti 6Al-4V at 235,000 cycles of loading in case 1 for fully reversed bending was estimated as [see equation (a)]

$$(\bar{S}_f, \mathfrak{d}_{S_f}) \approx (101.2, 5.78) \text{ ksi}$$

Since in case 2 the cross section of the offset link is being designed to eliminate any stress gradient by fully loading the entire cross section of the element, the loading will be analogous to axial [27, p. 227] rather than fully reversed bending. Juvinall [27] discusses a ratio of endurance limit for axial and bending loading called a *load constant* (C_L) for axial loading. From experimental evidence it is suggested that a value $C_L = 0.9$ be used. Thus if values from equation (i) and C_L are used, strength at 10^6 cycles is

$$(\bar{S}_f, \mathfrak{d}_{S_f}) \approx 0.9(95.3, 5.31) \text{ ksi} \approx (85.8, 4.78) \text{ ksi} \qquad (b')$$

Proceeding as in case 1: $0.9(\bar{S}_u, {}_{\flat S_u}) = (122.1, 6.12)$ ksi is the estimate of strength at 10^3 cycles. Interpolating between $(122.1, 6.12)$ and $(85.8, 4.78)$, fatigue strength at 235,000 cycles is (Figure 11.17):

$$\left(\bar{S}_f, {}_{\flat S_f}\right)_2 = (94.4, 4.2) \text{ ksi} \tag{c'}$$

9. As in case 1, $R = 0.9987$, from standard normal area tables (Appendix 1):

$$z \approx -3.10 \text{ (limit equation)}$$

In this example equation (9.11), with the values calculated in equations (a') and (c'), yields

$$z = -\frac{\bar{S}_f - \bar{s}}{\sqrt{{}_{\flat S_f}^2 + {}_{\flat s}^2}} = -3.1 = -\frac{94.41 - (28.10/\bar{T}_2)}{\sqrt{(4.2)^2 + \frac{5.51}{\bar{T}_2^2}}}$$

Expansion and simplification yields

$$0 = \bar{T}_2^2 - 0.70\bar{T}_2 + 0.110$$

Solving of the quadratic yields

$$\bar{T}_2 = 0.395 \text{ in.}$$

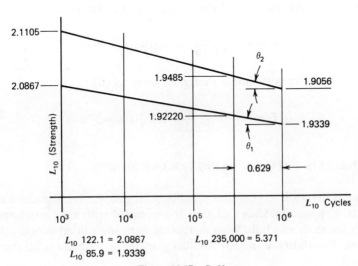

Figure 11.17 S–N.

Probabilistic optimization is summarized as follows. Case 1:

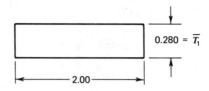

$$A_1 = \text{Cross section} = 2.00(0.28) = 0.56 \text{ in.}^2$$

Case 2:

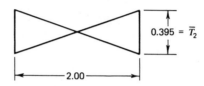

$$A_2 = \text{Cross section} = 1.00 \cdot \frac{0.395}{2} \cdot 2 = 0.395 \text{ in.}^2$$

Case 3: The deterministic design of the offset link, utilizing a safety factor $SF = 1.5$, yields

$$T_3 = 0.31 \text{ in.}$$

$$A_3 = 2.00 \ (0.31) = 0.62 \text{ in.}^2$$

The weight penalty with case 1 design (relative to case 2) is

$$\text{Percent} = \left(\frac{0.56}{0.395} - 1.00\right) \cdot 100 = 39.6\%$$

The weight penalty with case 3 design (relative to case 2) is

$$\text{Percent} = \left(\frac{0.62}{0.395} - 1.00\right) \cdot 100 = 57\%$$

11.5 ANALYSIS WITH OPPOSING FAILURE MODES

In pump design it is common practice to shrink fit the impeller hub onto the shaft. Slippage reliability of zero is associated with zero interference (δ) between the shaft and hub. With increasing hub–shaft interference, reliability improves. Reliability is a monotonically increasing function of the design variable δ.

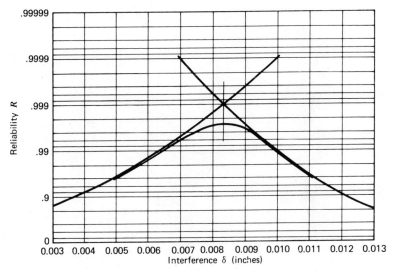

Figure 11.18 Interference versus reliability in a pump shaft–impeller assembly.

Consider the impeller hub–shaft assembly in a second failure mode, tensile yielding or fracture. Reliability diminishes (from $R = 1$ at $\delta = 0$) with increasing interference and, therefore, is a monotonically decreasing function of geometrical interference δ.

Optimization of an impeller hub–shaft assembly involves opposing failure modes, illustrated in Figure 11.18 with geometrical interference δ plotted on the abscissa and reliability R on the ordinate. Curve B represents probability of no slip and curve C, probability of no tensile failure. Since slip and tensile failure represent different failure modes, reliability R_i of the assembly is $R_i = B_i C_i$. Maximum reliability R_{max} occurs in this case at the value of δ for which $B_i = C_i$, although it may not be true in general.

Example D Optimization With Opposing Failure Modes

Some interference between a pump shaft and the impeller hub is necessary for the elimination of relative slippage. However, with increasing interference, tensile stresses in the impeller hub increase and may result in tensile failure due to yielding or fracture if they become excessive. Both the shaft outer diameter and the hole diameter in the impeller hub are random variables, and stresses are functions of the dimensions and fit of the members (Figure 11.18).

Tangential tension and radial compression develop as a result of hub interference with the shaft. Holding ability of the joint must be considered for the transmission of axial forces, torque, and the effect of centrifugal forces. The rotor shaft is in equilibrium axially because of equal fluid pressure on each side of the rotor. Shaft and impeller interference fit are shown in the following diagram.

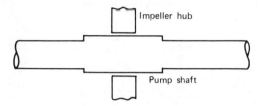

Deformation due to centrifugal forces tends to reduce the original static interference. The decrease in interference is given by

$$\delta = \frac{U^2 \gamma}{4g} \cdot \frac{3+u}{E} \tag{a}$$

where γ is the material specific weight, or the Poisson ratio, g is the gravitational constant, U is the linear velocity at the outer diameter of the rotor, and E is the modulus of elasticity. Torsional holding power T_h is given by

$$T_h = \frac{\pi}{2} \cdot f_1 (pLd^2) \tag{b}$$

where f_1 is the coefficient of friction, L is the width of rotor, d is the shaft outer diameter, δ is the necessary interference (inches), p is the unit pressure between shaft and rotor (lb/in.²); thus

$$p = \frac{E\delta}{2d} \frac{D_v^2 - d^2}{D_v^2} \tag{c}$$

where D_v is the distance across roots of vane slots.

By equating applied torque and torsional holding ability and then substituting for p, δ can be calculated.

Assumptions.

1. Shaft and rotor are of same material.
2. Tolerances are approximately equal to 3 standard deviations (see Section 3.3.2), $\Delta \approx 3_s$.
3. A design factor, (starting) K_d, accounts for increased torque due to shock loading.
4. For friction coefficients, the maximum and minimum listed in the literature are taken as $+3_s$ and -3_s values.

Solution.

1. Determine the interference necessary to prevent slippage caused by torque on the shaft, using equations (b) and (c). Applied torque to the shaft is (from pump performance curves):

$$T = f(hp, RPM, K_d)$$

$$hp = f(RPM) = A_0 + B_0 \cdot RPM, \qquad (d)$$

2. Torque is calculated as a function of pump speed.
3. Interference is calculated for the centrifugal force component $\delta = f[(RPM)^2]$.
4. The two components are summed and then differentiated with respect to pump speed, to yield the minimum value of interference:

$$(\delta)_{min} = (\delta_t \text{ to prevent slippage at any torsional load})_{@RPM}$$

$$+ (\delta_c \text{ separation due to centrifugal force})$$

5. The tensile stress in the rotor (s_{t_r}) is calculated as a function of interference.
6. Hub reliability (probability of no tensile failure) is calculated as a function of interference.

The tensile stress in the rotor (s_{t_r}) as a function of the interference between shaft and rotor bore is

$$s_{t_r} = \frac{E\delta_f}{2d}\left[1 + \left(\frac{d}{D_v}\right)^2\right]$$

where d is the shaft diameter and D_v is the distance across roots of vane slots. To relate interference to reliability, equation (9.11) is used to estimate z:

$$z = -\frac{\bar{S}_y - \bar{s}_{t_r}}{\left[{}_*s_{S_y}^2 + {}_*s_{S_{t_r}}^2\right]^{1/2}} \qquad (e)$$

Then, from equation (9.10) and Appendix 1:

$$R \leqslant \frac{1}{\sqrt{2\pi}} \int_z^\infty e^{-t^2/2} dt$$

where $\bar{S}_y$ is the mean yield strength of rotor, $\bar{S}_{t_r}$ is the mean tensile stress in rotor, ${}_*s_{S_y}$ is the yield strength standard deviation, and ${}^*{}_*s_{S_{t_r}}$ is the standard deviation of rotor tensile stress. Assembly no-slip reliability is calculated (this would step 7 in the preceding "solution" list) as a function of interference. The expression for the probability of no rotor slip on the shaft versus interference between shaft and rotor is [from equation (9.11)]

$$z = -\frac{\bar{\delta}_f - \bar{\delta}_s}{\left[s_{\delta_f}^2 + s_{\delta_s}^2\right]^{1/2}} \tag{f}$$

Then from equation (9.10) and Appendix 1:

$$R \leqslant \frac{1}{\sqrt{2\pi}} \int_z^\infty e^{-t^2/2} \, dt$$

where $\bar{\delta}_f$ is the mean interference fit between the rotor and the shaft, $\bar{\delta}_s$ is the mean minimum interference required such that the rotor is on the verge of

Table 11.4 Interference (inches) versus Reliability in a Pump Shaft—Impeller Assembly, with Opposing Failure Modes[a]

Assembly Interference, δ in.	Assembly No-Slip Reliability		Hub Reliability (Tensile)	
	z, No–slip	R_s, No–slip	z, No Tensile Failure	R_t, No Tensile Failure
0.0002	−0.120	.452	6.9976	.99999999999
0.0010	0.200	.579	6.6147	.9999999999
0.0020	0.590	.722	6.1345	.999999999
0.0030	0.983	.837	5.6532	.99999999
0.0040	1.376	.915	5.1712	.9999998
0.0050	1.770	.962	4.6888	.9999976
0.0060	2.163	.985	4.2064	.999987
0.0070	2.556	.995	3.7245	.99990
0.0080	2.949	.9984	3.2434	.9994
0.0090	3.3425	.9996	2.7635	.997
0.0100	3.7357	.99990	2.2853	.989
0.0110	4.1290	.99998	1.8091	.964
0.0120	4.5222	.9999969	1.3352	.909
0.0130	4.9155	.9999995	0.8641	.806
0.0140	5.3087	.9999999	0.3959	.653
0.0150	5.7019	.999999994	−0.0688	.527
0.0160	6.0952	.999999999	−0.6299	.298

[a]Display in this table was accomplished by using increments of δ of 0.001 in., for plotting and tabulation, with the same number of function evaluations (viz., 17) of $R = R_s R_t$ in the interval $0 < \delta < 0.0160$.

slipping, $s_{\delta_s}^2$ is the standard deviation of δ_s, and

$$\delta_s = \delta_t + \delta_c$$
$$s_{\delta_s}^2 = s_{\delta_t}^{2*} + s_{\delta_c}^{2*}$$

The assembly can fail functionally as a result of rotor hub failure in tension due to excessive interference or due to insufficient interference (slippage). The probability of no slip increases with increasing interference (δ_f) and the probability of rotor hub nonfailure due to tensile stresses decreases.

Equation (e) is utilized in calculating reliability versus tensile stress in the rotor hub as a function of δ, a monotonically decreasing function. Equation (f) is utilized in calculating reliability or probability of no slippage versus interference δ, a monotonically increasing function.

A typical result in an opposing mode failure problem (reliability versus assembly interference) is illustrated in Table 11.4 and Figure 11.18.

---

*Obtained by the method of partial derivatives, equation (2.62).

CHAPTER 12

Reliability Confidence Intervals

12.1 INTRODUCTION

In Section 3.3.1 the central limit theorem is utilized in several types of estimation problems. In this chapter confidence intervals on reliability, which involve functions of random variables, are considered.

A *point estimate* is a single value that is utilized to estimate a parameter of interest. Point estimates, however, are sometimes inadequate as estimates of parameters since they rarely coincide with the parameters. *Interval estimates* are often of value where they take the form (u,l), where u is the upper bound and l, the lower bound. Such interval estimates are functions of the sample, or in words, random variables with the property that probability statements concerning the intervals and the parameters can be made before testing. Thus if the parameter to be estimated is designated by θ, an interval estimate of θ is (u,l) such that there is a probability $1-\delta$ that $l \leq \theta \leq u$; δ is the probability associated with the level of significance [15, p. 98; 11, p. 167], that is, such as the 5% significance level. The variables l and u are usually called the $100(1-\delta)\%$ *confidence limits* for the parameter of interest. The interval between these values is called the $100(1-\delta)\%$ *confidence interval*. If a 95% confidence interval is to be computed, this implies that in the long run 95% of such limits can be expected to include the true average, μ. In words, the statement that the true average lies within the computed range has a probability of .95 of being correct.

12.1.1 Confidence Interval of a Normal Random Variable of Known Standard Deviation

Assume that the normally distributed random variable X has a mean value μ_x and known standard deviation σ_x. Let $\bar{x}$ be the sample mean of a random sample of size n. The $100(1-\delta)\%$ confidence interval for μ_x is

$$\bar{x} \pm k_{(\delta/2)}\left[\sigma_x/\sqrt{n}\,\right]$$

where $k_{(\delta/2)}$ is the $100(\delta/2)$ percentage point of the normal distribution (Appendix 6). This result arises from considering the distribution of

$[(\bar{x} - \mu_x)/(\sqrt{n}/\sigma_x)]$, which is distributed $N(0,1)$. Thus

$$P\left\{-(k)_{(\delta/2)} \leq \frac{\bar{x} - \mu_x}{\sigma_x/\sqrt{n}} \leq (k)_{(\delta/2)}\right\} = 1 - \delta$$

Multiplication through the inequality by $\sigma_x/\sqrt{n}$ yields

$$P\left\{-(k)_{\delta/2} \cdot \frac{\sigma_x}{\sqrt{n}} \leq \bar{x} - \mu_x \leq (k)_{(\delta/2)} \cdot \frac{\sigma_x}{\sqrt{n}}\right\} = 1 - \delta$$

Addition through the inequality by $\bar{x}$ yields

$$P\left\{\bar{x} - (k)_{(\delta/2)} \cdot \frac{\sigma_x}{\sqrt{n}} \leq \mu_x \leq \bar{x} + (k)_{(\delta/2)} \cdot \frac{\sigma_x}{\sqrt{n}}\right\} = 1 - \delta$$

Consequently:

$$l = \bar{x} - (k)_{(\delta/2)} \cdot \frac{\sigma_x}{\sqrt{n}}$$

and

$$u = \bar{x} + (k)_{(\delta/2)} \cdot \frac{\sigma_x}{\sqrt{n}}$$

12.1.2 Confidence Interval of a Normal Random Variable of Unknown Standard Deviation

Assume that the normally distributed random variable X that has a mean value μ_x and unknown standard deviation σ_x, and let $\bar{x}$ be the sample mean value of a random sample of size n; the $100(1-\alpha)\%$ confidence interval of μ_x is

$$\bar{x} \pm (t_{\delta/2;\,n-1}) \cdot \frac{\sigma_x}{\sqrt{n}}$$

where $t_{\delta/2;\,n-1}$ is the $100(\delta/2)$ percentage point of the t distribution with $n-1$ degrees of freedom and $\bar{x}$ is the unbiased estimator of μ_x from a sample of size n. The preceding result arose from considering the distribution of $(\bar{x} - \mu_x)/(\sigma_x/\sqrt{n})$, which can be shown to have a t distribution with $n-1$ degrees of freedom. Thus

$$P\left\{-t_{\delta/2;\,n-1} \leq \frac{\bar{x} - \mu_x}{\sigma_x/\sqrt{n}} \leq t_{\delta/2;\,n-1}\right\} = 1 - \delta$$

548 Reliability Confidence Intervals

which by multiplication through the inequality by $s_x/\sqrt{n}$ and then addition of $\bar{x}$, yields

$$l = \bar{x} - t_{\delta/2; n-1}\frac{s_x}{\sqrt{n}}$$

and

$$u = \bar{x} + t_{\delta/2; n-1}\frac{s_x}{\sqrt{n}}$$

Example A

In the manufacture of a connecting rod, the average distance $\bar{x}$ between hole centers, based on a sample of $n = 6$, was 8.376 in. and the sample standard deviation, $s_x = 0.0124$ in. Assume that the dimensions are normally distributed; then with the stated values of $\bar{x}$ and s_x, a 95% confidence interval for μ_x is

$$\bar{x} \pm 2.571 \frac{s_x}{\sqrt{n}}$$

where 2.571 is the 2.50% point of the t distribution (see Appendix 7)* for 5 degrees of freedom. Thus the confidence interval is

$$(l, u) = \left(8.376 - 2.571 \frac{0.0124}{\sqrt{6}}, 8.376 + 2.571 \frac{0.0124}{\sqrt{6}}\right) = (8.363, 8.389) \text{ in.}$$

12.1.3 Confidence Interval of the Standard Deviation of a Normally Distributed Random Variable

Consider a normally distributed random variable X with unknown mean value μ_x and unknown standard deviation σ_x. Let $\bar{x}$ be the sample mean value determined from a sample of size n, and [equation (2.2)]:

$$s_x = \left[\frac{1}{n-1} \sum_{i=1}^{n} (x_i - \bar{x})^2\right]^{1/2}$$

The $100(1 - \delta)\%$ confidence interval for σ_x is

$$\left[s_x \cdot \sqrt{\frac{n-1}{\chi^2_{\delta/2; n-1}}} \; ; \; s_x \cdot \sqrt{\frac{n-1}{\chi^2_{1-\delta/2; n-1}}}\right]$$

*Note: $\delta = \alpha$.

where $\chi^2_{\delta/2;n-1}$ relates to the $100(\delta/2)$ percentage point and $\chi^2_{1-\delta/2;n-1}$ relates to the $100[(1-\delta)/2]$ percentage point. This statement resulted from considering the nature of the distribution of

$$\frac{(n-1)s_x^2}{\sigma_x^2}$$

This can be shown to have a χ^2 distribution with $n-1$ degrees of freedom. Consequently,

$$P\left\{\chi^2_{1-\delta/2;n-1} \leqslant \frac{(n-1)s_x^2}{\sigma_x^2} \leqslant \chi^2_{\delta/2;n-1}\right\} = 1-\delta$$

Thus

$$P\left\{s_x\sqrt{\frac{(n-1)}{\chi^2_{\delta/2;n-1}}} \leqslant \sigma_x \leqslant s_x\sqrt{\frac{(n-1)}{\chi^2_{1-\delta/2;n-1}}}\right\} = 1-\delta \qquad (a)$$

The $100(1-\delta)\%$ upper one-sided confidence interval is

$$u = s_x\sqrt{\frac{n-1}{\chi^2_{1-\delta;n-1}}}$$

Example B

In the manufacture of a small rifle barrel, the processing capabilities for machining the firing chamber were studied. A sample of $n=25$ measurements were taken on internal diameters, from which a value $s_d = 0.00132$ in. was obtained for the standard deviation estimator. Assuming the measured dimension to be normally distributed, a 99% confidence on σ_d is given by equation (a).

From Appendix 8 and for $n-1=24$, the values for calculating the upper and lower confidence limits are

$$(l,u) = \left\{0.00132\sqrt{\frac{24}{45.558}}, 0.00132\sqrt{\frac{24}{9.886}}\right\} \text{ in.}$$

$$(l,u) = \{0.000958, 0.00206\} \text{ in.}$$

12.2 CONFIDENCE INTERVALS ON DIFFERENCES OF RANDOM VARIABLES

Since $P\{S-s>0\}$ is of central importance in probabilistic mechanical design, so also is the confidence interval of the difference between mean strength $\bar{S}$ and mean stress $\bar{s}$.

From the upper and lower confidence limits u and l of the confidence interval on $\mu_S - \mu_s$, it is a simple exercise to calculate the reliability bounds, R_u and R_l, associated with u and l in a design (see Section 12.3).

12.2.1 Confidence Interval on $\mu_S - \mu_s$, σ_S and σ_s Known

Let S be a normal-strength random variable with mean μ_S and standard deviation σ_S (known), and let s be a normal applied stress random variable, independent of S, with mean μ_s and standard deviation σ_s (known). Let $\bar{S}$ be the sample mean of n_S observations on S, and let $\bar{s}$ be the sample mean of n_s observations on s. The $100(1-\delta)$ confidence interval for $\mu_S - \mu_s$ is given by

$$(\bar{S} - \bar{s}) \pm k_{\delta/2} \sqrt{\frac{\sigma_S^2}{n_S} + \frac{\sigma_s^2}{n_s}}$$

where $\delta/2$ is the $100(\delta/2)$ percentage point of the normal distribution. This result is obtained by considering the distribution of [17, p. 220]

$$\frac{(\bar{S} - \bar{s}) - (\mu_S - \mu_s)}{\sqrt{\frac{\sigma_S^2}{n_S} + \frac{\sigma_s^2}{n_s}}}$$

which can be shown to be normally distributed $N(0,1)$. Thus

$$P\left\{-k_{\delta/2} \leq \frac{(\bar{S}-\bar{s})-(\mu_S-\mu_s)}{\sqrt{\frac{\sigma_S^2}{n_S}+\frac{\sigma_s^2}{n_s}}} \leq k_{\delta/2}\right\} = 1-\delta$$

which is equivalent to (see Figure 12.1)

$$P\left\{(\bar{S}-\bar{s}) - k_{\delta/2}\sqrt{\frac{\sigma_S^2}{n_S}+\frac{\sigma_s^2}{n_s}} \leq \mu_S - \mu_s \leq (\bar{S}-\bar{s}) + K_{\delta/2}\sqrt{\frac{\sigma_S^2}{n_S}+\frac{\sigma_s^2}{n_s}}\right\} = 1-\delta$$

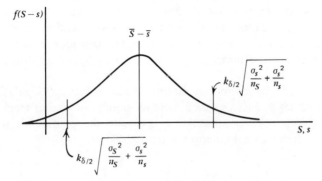

Figure 12.1 Confidence interval on M_S-M_s, with σ_S and σ_s known.

Example C

The upper and lower confidence limits on the difference between the means of strength $\bar{S}$ and applied stress $\bar{s}$ are needed. Assume the following data from the example in Section 9.3.1:

Strength	Applied Stress
$\bar{S}=27{,}000$ psi	$\bar{s}=18{,}400$ psi
$\sigma_S=3{,}200$ psi	$\sigma_s=1{,}500$ psi
$n_S=20$	$n_s=10$

A 95% confidence interval is desired; thus (from Appendix 1)

$$k_{\delta/2}=1.96$$

Calculations are as follows:

$$\bar{S}-\bar{s}=27{,}000-18{,}400=8600 \text{ psi}$$

$$\sqrt{\frac{\sigma_S^2}{n_S}+\frac{\sigma_s^2}{n_s}}=\sqrt{\frac{10{,}240{,}000}{20}+\frac{2{,}250{,}000}{10}}=858.5 \text{ psi}$$

and

$$k_{\delta/2}\cdot\sqrt{\frac{\sigma_S^2}{n_S}+\frac{\sigma_s^2}{n_s}}=1.96\,(858.5)=1682.7.$$

The 95% confidence interval is $\{6917 \leqslant \mu_S-\mu_s \leqslant 10{,}283\}$ psi.

552 Reliability Confidence Intervals

In the usual design problem situation, μ_S and μ_s are not known, σ_S and σ_s are not known and cannot be assumed equal, and n_S and n_s are probably not equal. The general problem of estimating the confidence interval for the difference between the means of allowable stress $\overline{S}$ and applied stress $\bar{s}$ are now examined.

Let $S_1, S_2, \ldots, S_n$ be a random sample of size n from a population that is $N(\mu_S, \sigma_S)$, and let $s_1, s_2, \ldots, s_m$ be a random sample of size m from a population that is $N(\mu_s, \sigma_s)$. It is desired to estimate the confidence interval for $\mu_S - \mu_s$ at $1 - \delta$. Applicable theorems are as follows:

Example B

1. Let $z = [(S - \mu_S)/\sigma_S]$; then $z \approx N(0, 1)$.
2. Let $z = [(\overline{S} - \mu_S)/\sigma_{\overline{S}}] \approx [(\overline{S} - \mu_S)/\sigma_{(S/\sqrt{n})}]$; then $z - N(0, 1)$.
3. If $\overline{S} - N(\mu_S, \sigma_{\overline{S}}^2)$ and $\overline{S} - \bar{s} \sim N(\mu_S - \mu_s, (\sigma_S^2/n) + (\sigma_s^2/m))$.
4. Let

$$z = \frac{(\overline{S} - \bar{s}) - (\mu_S - \mu_s)}{\sqrt{(\sigma_S^2/n) + (\sigma_s^2/m)}}$$

then $z \sim N(0, 1)$.

5. If $u \sim \chi_{\delta_1}^2$ and $v \sim \chi_{\delta_2}^2$ and if u and v are stochastically independent, then $w = u + v \sim \chi_{\delta_1 + \delta_2}^2$.
6.

$$\frac{n \hat{\sigma}_S^2}{\sigma_S^2} \sim \chi_{(n-1)}^2 \quad \text{and} \quad \frac{m \hat{\sigma}_s^2}{\sigma_s^2} \sim \chi_{m-1}^2,$$

where $\hat{\sigma}_S^2 = [\Sigma(S - \overline{S})^2/n]$ and $\hat{\sigma}_s^2 = [\Sigma(s - \bar{s})^2/m]$. Thus

$$\frac{n \hat{\sigma}_S^2}{\sigma_S^2} + \frac{m \hat{\sigma}_s^2}{\sigma_s^2} = V \approx \chi_{n+m-2}^2$$

7. Let $z \sim N(0, 1)$ and $V \sim \chi_\delta^2$, and let z and V be stochastically independent; then $T = z/\sqrt{(V/f)}$ has a student's t distribution with f degrees of freedom.
8. Let

$$z = \frac{(\overline{S} - \bar{s}) - (\mu_S - \mu_s)}{\sqrt{(\sigma_S^2/n) + (\sigma_s^2/m)}}; \quad V = \frac{n \hat{\sigma}_S^2}{\sigma_S^2} + \frac{m \hat{\sigma}_s^2}{\sigma_s^2}$$

then $T = z/\sqrt{(v/f)}$. Then

$$T = \frac{[(\bar{S}-\bar{s})-(\mu_S-\mu_s)]/\sqrt{(\sigma_S^2/n)+(\sigma_s^2/m)}}{\sqrt{[(ns_S^2/\sigma_S^2)+(ms_s^2/\sigma_s^2)]/(n+m-2)}}$$

has a student's t distribution with $n+m-2$ degrees of freedom. Also, if $\sigma_S^2 = \sigma_s^2$, the preceding expression reduces to

$$T = \frac{(\bar{S}-\bar{s})-(\mu_S-\mu_s)}{\sqrt{[(1/n)+(1/m)]\cdot[(ns_S^2+ms_s^2)/(n+m-2)]}}.$$

12.2.2 Confidence Interval for Difference between Mean Values of Two Normal Random Variables of Unknown but Equal Standard Deviations [15, p. 221]

Given the normally distributed random variable X with mean μ_x and standard deviation σ_x (unknown) and normally distributed random variable Y with mean μ_y and standard deviation σ_y (unknown). Also, let X and Y be statistically independent. The variable $\bar{x}$ is the sample mean based on n_x values, and $\bar{y}$ is the sample mean based on n_y values. The $100(1-\delta)\%$ confidence interval for $\mu_x - \mu_y$ is

$$(\bar{x}-\bar{y}) \pm (t_{\delta/2; n_x+n_y-2}) \cdot \sqrt{\frac{1}{n_x}+\frac{1}{n_y}} \cdot \sqrt{\frac{\sum(x_i-\bar{x})^2+\sum(y_i-\bar{y})^2}{n_x+n_y-2}}$$

in which $t_{\delta/2; n_x+n_y-2}$ is the $100(\delta/2)$ percentage point of a t distribution with n_x+n_y-2 degrees of freedom. The preceding result was developed from a consideration of the distribution:

$$\frac{(\bar{x}-\bar{y})-(\mu_x-\mu_y)}{\sqrt{(1/n_x)+(1/n_y)}\sqrt{[\sum(x_i-\bar{x})^2+\sum(y_i-\bar{y})^2]/(n_x+n_y-2)}}$$

This expression can be shown to have a t distribution with n_x+n_y-2 degrees of freedom. Thus

$$P\left\{-t_{\delta/2;\,n_x+n_y-2}\right.$$

$$\leq \frac{(\bar{x}-\bar{y})-(\mu_x-\mu_y)}{\sqrt{(1/n_x)+(1/n_y)}\sqrt{\left[\sum(x_i-x)^2+\sum(y_i-y)^2\right]/(n_x+n_y-2)}}$$

$$\left.\leq t_{\delta/2;\,n_x+n_y-2}\right\}=1-\delta$$

Thus in the notation of strength and stress:

$$P\left\{\bar{S}-\bar{s}-(t_{\delta/2;\,n_S+n_s-2})\cdot\sqrt{\frac{1}{n_S}+\frac{1}{n_s}}\sqrt{\frac{n_S\sigma_S^2+n_s\sigma_s^2}{n_S+n_s-2}}\right.$$

$$\left.\leq \mu_S-\mu_s \leq \bar{S}-\bar{s}+(t_{\delta/2;\,n_S+n_s-2})\cdot\sqrt{\frac{1}{n_S}+\frac{1}{n_s}}\sqrt{\frac{n_S\sigma_S^2+n_s\sigma_s^2}{n_S-n_s-2}}\right\}=1-\delta$$

which defines the $100(1-\delta)\%$ confidence interval. The $100(1-\delta)\%$ confidence limits are

$$l=\bar{S}-\bar{s}-(t_{\delta/2;\,n_S+n_s-2})\cdot\sqrt{\frac{1}{n_S}+\frac{1}{n_s}}\sqrt{\frac{n_S\sigma_S^2+n_s\sigma_s^2}{n_S+n_s-2}}$$

$$u=\bar{S}-\bar{s}\pm(t_{\delta/2;\,n_S+n_s-2})\cdot\sqrt{\frac{1}{n_S}+\frac{1}{n_s}}\sqrt{\frac{n_S\sigma_S^2+n_s\sigma_s^2}{n_S+n_s-2}}$$

Example D

Although it is unlikely that the standard deviations of allowable stress (strength) and applied stress would be equal, assume for this problem that $\sigma_S=\sigma_s=\sigma$. The following stress and strength statistics are assumed for illustration purposes:

Strength	Stress
$\bar{S}=49{,}000$ psi	$\bar{s}=38{,}000$ psi
$\sigma_S=2.800$ psi	$\sigma_s=3{,}400$ psi
$n_S=50$	$n_s=40$

Calculations are

$$\sum(S_i - \bar{S})^2 = 392 \cdot 10^6; \qquad \sum(s_i - \bar{s})^2 = 462.4 \cdot 10^6$$

$$\frac{\sum(S_i - \bar{S})^2 + \sum(s_i - \bar{s})^2}{n_S + n_s - 2} = 9.7092 \cdot 10^6$$

A 90% confidence interval for the difference of mean strength and mean stress is

$$\bar{S} - \bar{s} \pm (t_{\delta/2; n_x + n_y - 2}) \cdot \sqrt{\frac{1}{n_S} + \frac{1}{n_s}} \sqrt{\frac{(S-\bar{S})^2 + (s-\bar{s})^2}{n_S + n_s - 2}}$$

$$(l, u) = 49{,}000 - 38{,}000 \pm t_{0.05, 88} \sqrt{\frac{1}{50} + \frac{1}{40}} \sqrt{9.709 \cdot 10^6}$$

From Appendix 7 the value for $t_{0.05; 88} \approx 1.660$, and

$$(l, u) = 11{,}000 \pm 1.660 \, (0.210)(3120)$$

$$(l, u) = [9900, 12{,}100] \text{ psi}$$

(*Note:* Unknown $\sigma_S = \sigma_s = \sigma = 3200$ psi, assumed.)

Example E

Using the same data as in Example C, except for the assumption that the standard deviations are unknown but equal,

$$\bar{S} - \bar{s} = 27{,}000 - 18{,}400 = 8600 \text{ psi}$$

$$n_S + n_s - 2 = 20 + 10 - 2 = 28$$

$$\sqrt{\frac{1}{n_S} + \frac{1}{n_s}} = \sqrt{\frac{1}{20} + \frac{1}{10}} = 0.3880$$

$$\sqrt{\frac{20(3{,}200)^2 + 10(1{,}500)^2}{28}} = 2846$$

For the $100(1 - \delta)\% = 95\%$ confidence interval from Appendix 7:

$$t_{\delta/2; 28} = 2.048.$$

Then

$$l = 8600 - 2.048 \, (0.3870)(2846.) = 6{,}340 \text{ psi.}$$

$$u = 8600 + 2.048 \, (0.388)(2846.) = 10{,}860 \text{ psi.}$$

The 95% confidence interval is $\{6340 \leq \mu_S - \mu_s \leq 10{,}860\}$.

12.2.3 The General Case of Confidence

If σ_S^2 and σ_s^2 are not equal, the variances do not cancel out, and if this statistic is to be used to calculate a confidence interval, σ_S^2 and σ_s^2 must be known.

An approximate method does, however, exist for the case of unknown σ_S and σ_s based on the following statistic:

$$T = \frac{(\bar{S} - \bar{s}) - (\mu_S - \mu_s)}{\sqrt{[s_S^2/(n-1)] + [s_s^2/(m-1)]}}, \quad \text{(b)}$$

where T has an approximate student's t distribution with f degrees of freedom; f is calculated as follows [123]:

$$f \approx \frac{\left[s_S^2/(n-1)\right] + \left[s_s^2/(m-1)\right]^2}{\left\{[s_S^2/(n-1)]^2/n+1\right\} + \left\{[s_s^2/(m-1)]^2/m+1\right\}} - 2 \quad \text{(c)}$$

Example F

In this example it is assumed that s_S and σ_s are not known and not necessarily equal. Further, there is no assumptions regarding the equality of n_S and n_s.

Initially, the same data are used as in Example C in (then the sample sizes of n_S and n_s are modified and the effect on the confidence interval examined).

	Strength	Applied Stress
	$\bar{S} = 27{,}000$ psi	$\bar{s} = 18{,}400$ psi
	$s_S = 3200$ psi	$s_s = 1500$ psi
	$n_S = 20$	$n_s = 10$

A 95% confidence interval approximation is desired.

Part A. ($n_S = 20$ and $n_s = 10$):

First, the equivalent number of degrees of freedom f is calculated; f is utilized with the student's t-distribution tables (Appendix 7) to obtain the 100δ percentage points. Use of equation (c) yields

$$f \approx \frac{\left\{[(3200)^2/19] + [(1500)^2/9]\right\}^2}{\left\{[(3200)^2/19]^2/21\right\} + \left\{[(1500)^2/9]^2/11\right\}} - 2,$$

$$f = \frac{(538{,}947 + 250{,}000)^2}{13.832 \cdot 10^{10} + 5.681 \cdot 10^{10}} - 2 = 29.90$$

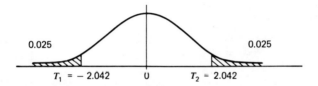

Figure 12.2 Ninety-five percent confidence interval.

Utilize the approximation $f' \simeq 30$. The 95% confidence interval for $t_{0.025;\,30} \approx 2.040$ from Appendix 7 is (see Figure 12.2)

$$P\left\{(\bar{S}-\bar{s})-2.040\sqrt{\frac{\sigma_S^2}{n_S-1}+\frac{\sigma_s^2}{n_s-1}}\right.$$

$$\left. \leqslant \mu_S - \mu_s \leqslant (\bar{S}-\bar{s})+2.040\sqrt{\frac{\sigma_S^2}{n_S-1}+\frac{\sigma_s^2}{n_s-1}}\right\}=0.95 \qquad (d)$$

Calculations are as follows:

$$\bar{S}-\bar{s} = 27{,}000 - 18{,}400 = 8600 \text{ psi}$$

$$\sqrt{\frac{\sigma_S^2}{n_S-1}+\frac{\sigma_s^2}{n_s-1}} = \sqrt{538{,}950 + 250{,}000} = 997$$

Substitution into equation (d) yields

$$P\{8600 - 2.040(996.7) \leqslant \mu_S - \mu_s \leqslant 8600 + 2.040(967)\} = 0.95$$

The 95% confidence interval is $\{l, u\} = \{6570, 10{,}630\}$ psi.

Part B. ($n_S = 50$ and $n_s = 50$):

The conditions of part B are the same as those in part A, except for sample sizes [from equation (c)]:

$$f = \frac{\left\{[(3200)^2/49]^2 + [(1500)^2/49]\right\}^2}{\left\{[(3200)^2/49]^2/51\right\} + \left\{[(1500)^2/49]^2/51\right\}} - 2 = 68.6$$

Utilize the approximation f' for the degree of freedom:

$$f' \approx 68$$

558 Reliability Confidence Intervals

The 95% confidence interval approximation is

$$\overline{S} - \overline{s} = 8600 \text{ psi}$$

$$\sqrt{\frac{\sigma_S^2}{n_S - 1} + \frac{\sigma_s^2}{n_s - 1}} = \sqrt{208{,}980 + 45{,}920} = 506.$$

The value of t for 95% confidence interval and $f' = 68$ is ≈ 1.995. Thus

$$P\{8600 - 1.995\,(506) \leqslant \mu_S - \mu_s \leqslant 8600 + 1.995\,(506)\} = 0.95,$$

and

$$\{l, u\} = \{7590; 9610\} \text{ psi}$$

12.3 CONFIDENCE INTERVAL ON RELIABILITY

In this section the reliability values R_l and R_u associated with the respective lower and upper confidence limits l and u on the mean value difference $\mu_S - \mu_s$ of a particular design are considered. Each case considered in Section 12.2, assuming a specific state of knowledge in a design problem, is discussed and exemplified in turn.

12.3.1 Confidence Interval on Reliability, σ_S and σ_s Known

In Section 12.2.1 confidence limits l and u on the difference $\mu_S - \mu_s$ were calculated. It was assumed that σ_S and σ_s were known.

Example G

In Example C, the upper and lower limits on a 95% confidence interval were calculated. The results were (see Figure 12.3):

$$l = 6920 \text{ psi}$$

$$\overline{S} - \overline{s} = 8600 \text{ psi}$$

$$u = 10{,}280 \text{ psi}$$

The standard deviation σ_{S-s} was

$$\sigma_{S-s} = \sqrt{\sigma_S^2 + \sigma_s^2} = \sqrt{(3200)^2 + (1500)^2} = 3530 \text{ psi}$$

From Section 9.3.2,

$$R = \int_0^\infty f(S-s)\,d(S-s)$$

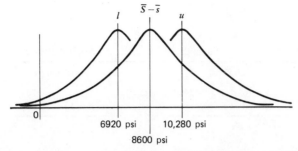

Figure 12.3 Upper and lower limits on a 95% interval on reliability.

Thus since $S-s$ in each case is normally distributed, the reliabilities corresponding to l, $\bar{S}-\bar{s}$, and u are determined by integrating the probability density functions defined by

$$N(6920, 3460)$$
$$N(8600, 3460)$$
$$N(10{,}280, 3460)$$

from 0 to ∞. The equivalent is to integrate the standardized normal probability density function from z_l, z, or z_u to ∞:

$$z_l = \frac{0-6920}{3{,}460} = -1.93; \quad R_l = 0.9732$$

$$z = \frac{0-8600}{3{,}460} = -2.48; \quad R = 0.9934$$

$$z_u = \frac{0-10{,}280}{3{,}460} = -3.07; \quad R_u = 0.9989$$

The reliability $R_l = 0.9732$, corresponding to the lower limit l, of the 95% confidence limit on $\mu_S - \mu_s$ is very likely not acceptable. The requirements for suitably reducing the confidence interval are discussed in the following sections. An alternative is to so adjust the target reliability that R_l takes on an acceptable value.

12.3.2 Confidence Interval on Reliability, $\sigma_S = \sigma_s$ but Unknown

In Section 12.2.2 confidence limits l and u on the difference $\mu_S - \mu_s$ were calculated. It was assumed that σ_S and σ_s were unknown but that σ_{S-s} was given. Note the data in Example H, which follows.

Example H

	Strength	Applied Stress
	$\bar{S} = 27{,}000$ psi	$\bar{s} = 18{,}400$ psi
	$\sigma = \sqrt{\sigma_S^2 + \sigma_s^2} = 3460$ (assumed)	
	$n_S = 20$	$n_s = 10$

The upper and lower confidence limits u and l on a 95% confidence interval were calculated. The results were

$$l = 6340 \text{ psi}$$

$$\bar{S} - \bar{s} = 8600 \text{ psi}$$

$$u = 10{,}860 \text{ psi}$$

From Section 9.3.2, equation (a):

$$R = \int_0^\infty f(S-s)\,d(S-s)$$

Thus since $S-s$ in each case is normally distributed, the reliabilities corresponding to l, $\bar{S}-\bar{s}$, and u are determined by integrating the probability density function defined by equation (a) from 0 to ∞:

$$N(6340, 3460)$$
$$N(8600, 3460)$$
$$N(10{,}860, 3460)$$

The equivalent is to integrate the standardized normal probability density function from z_l, z, and z_u:

$$z_l = \frac{0 - 6340}{3460} = -1.83; \quad R_l = 0.9664$$

$$z = \frac{0 - 8600}{3460} = -2.48; \quad R = 0.9934$$

$$z_u = \frac{0 - 10860}{3460} = -3.14; \quad R_u = 0.9992$$

to infinity in each case (by using Table 3.5).

12.3.3 Confidence Interval on Reliability, General Case

Example I

In Section 12.2.3 confidence limits l and u on the difference $\mu_S - \mu_s$ were estimated for the general case where σ_S and σ_s were unknown and not assumed equal. In Example F, part A it was assumed that $n_S = 20$ and $n_s = 10$. In part B it was assumed that $n_S = n_s = 50$.

The lower and upper limits on a 95% confidence interval were estimated. The results were (in Example F):

Part A	Part B
$l = 6{,}570$ psi	$l = 7{,}590$ psi
$\bar{S} - \bar{s} = 8{,}600$ psi	$\bar{S} - \bar{s} = 8{,}600$ psi
$u = 10{,}680$ psi	$u = 9{,}610$ psi

For both parts A and B, the standard deviation estimator s_{S-s} was

$$s_{S-s} = \sqrt{s_S^2 + s_s^2} = \sqrt{(3200)^2 + (1500)^2} = 3530 \text{ psi}$$

From Section 9.3.2, equation (a):

$$R = \int_0^\infty f(S-s)\,d(S-s)$$

As in the examples in Sections 12.3.1 and 12.3.2, reliability values are calculated by integrating the probability density functions defined by:

Part A	Part B
$N(6{,}570;\,3{,}530)$	$N(7{,}590;\,3{,}530)$
$N(8{,}600;\,3{,}530)$	$N(8{,}600;\,3{,}530)$
$N(10{,}630;\,3{,}530)$	$N(9{,}610;\,3{,}530)$

from 0 to ∞. The equivalent is to integrate the standardized normal probability density function from z_l, z, and z_u to ∞:

Part A ($n_S = 20;\, n_s = 10$)	Part B ($n_S = 50;\, n_s = 50$)
$z_l = \dfrac{0 - 6{,}570}{3530} = -1.86;$	$R_l = 0.9684$
$z = \dfrac{0 - 8{,}600}{3530} = -2.43;$	$R = 0.9925$
$z_u = \dfrac{0 - 10{,}630}{3530} = -3.01$	$R_u = 0.9987$
$z_l = \dfrac{0 - 7{,}590}{3530} = -2.15;$	$R_l = 0.9841$
$z = \dfrac{0 - 8{,}600}{3530} = -2.43;$	$R = 0.9925$
$z_u = \dfrac{0 - 0{,}610}{3530} = -2.72;$	$R_u = 0.9967$

It is clear that load stress cannot be measured directly but must be inferred from samples of other measured variables. Thus, samples of stress values of size n_s cannot be obtained. However, for certain cases of simple stress, geometric variables can be considered approximately deterministic. This permits considering the sample size n_p of loading to be utilized as the sample size n_s for load stress. Thus first approximations of the confidence intervals for reliability may be obtained.

APPENDIX 1

Table of the Standard Cumulative Normal Distribution

$$R = (2\pi)^{-1/2} \int_z^\infty e^{-z^2/2} dz.$$

z	.00	.01	.02	.03	.04	.05	.06	.07	.08	.09
0.0	.5000	.5040	.5080	.5120	.5160	.5199	.5239	.5279	.5319	.5359
0.1	.5398	.5438	.5478	.5517	.5557	.5596	.5636	.5675	.5714	.5753
0.2	.5793	.5832	.5871	.5910	.5948	.5987	.6026	.6064	.6103	.6141
0.3	.6179	.6217	.6255	.6293	.6331	.6368	.6406	.6443	.6480	.6517
0.4	.6554	.6591	.6628	.6664	.6700	.6736	.6772	.6808	.6844	.6879
0.5	.6915	.6950	.6985	.7019	.7054	.7088	.7123	.7157	.7190	.7224
0.6	.7257	.7291	.7324	.7357	.7389	.7422	.7454	.7486	.7517	.7549
0.7	.7580	.7611	.7642	.7673	.7703	.7734	.7764	.7794	.7823	.7852
0.8	.7881	.7910	.7939	.7967	.7995	.8023	.8051	.8078	.8106	.8133
0.9	.8159	.8186	.8212	.8238	.8264	.8289	.8315	.8340	.8365	.8389
1.0	.8413	.8438	.8461	.8485	.8508	.8531	.8554	.8577	.8599	.8621
1.1	.8643	.8665	.8686	.8708	.8729	.8749	.8770	.8790	.8810	.8830
1.2	.8849	.8869	.8888	.8907	.8925	.8944	.8962	.8980	.8997	.9^{0}147
1.3	.90320	.90490	.90658	.90824	.90988	.91149	.91309	.91466	.91621	.91774
1.4	.91924	.92073	.92220	.92364	.92507	.92647	.92785	.92922	.93056	.93189
1.5	.93319	.93448	.93574	.93699	.93822	.93943	.94062	.94179	.94295	.94408
1.6	.94520	.94630	.94738	.94845	.94950	.95053	.95154	.95254	.95352	.95449
1.7	.95543	.95637	.95728	.95818	.95907	.95994	.96080	.96164	.96246	.96327
1.8	.96407	.96485	.96562	.96638	.96712	.96784	.96856	.96926	.96995	.97062
1.9	.97128	.97193	.97257	.97320	.97381	.97441	.97500	.97558	.97615	.97670
2.0	.97725	.97778	.97831	.97882	.97932	.97982	.98030	.98077	.98124	.98169
2.1	.98214	.98257	.98300	.98341	.98382	.98422	.98461	.98500	.98537	.98574
2.2	.98610	.98645	.98679	.98713	.98745	.98778	.98809	.98840	.98870	.98899
2.3	.98928	.98956	.98983	.9^{2}0097	.9^{2}0358	.9^{2}0613	.9^{2}0863	.9^{2}1106	.9^{2}1344	.9^{2}1576
2.4	.9^{2}1802	.9^{2}2024	.9^{2}2240	.9^{2}2451	.9^{2}2656	.9^{2}2857	.9^{2}3053	.9^{2}3244	.9^{2}3431	.9^{2}3613
2.5	.9^{2}3790	.9^{2}3963	.9^{2}4132	.9^{2}4297	.9^{2}4457	.9^{2}4614	.9^{2}4766	.9^{2}4915	.9^{2}5060	.9^{2}5201
2.6	.9^{2}5339	.9^{2}5473	.9^{2}5604	.9^{2}5731	.9^{2}5855	.9^{2}5975	.9^{2}6093	.9^{2}6207	.9^{2}6319	.9^{2}6427
2.7	.9^{2}6533	.9^{2}6636	.9^{2}6736	.9^{2}6833	.9^{2}6928	.9^{2}7020	.9^{2}7110	.9^{2}7197	.9^{2}7282	.9^{2}7365
2.8	.9^{2}7445	.9^{2}7523	.9^{2}7599	.9^{2}7673	.9^{2}7744	.9^{2}7814	.9^{2}7882	.9^{2}7948	.9^{2}8012	.9^{2}8074
2.9	.9^{2}8134	.9^{2}8193	.9^{2}8250	.9^{2}8305	.9^{2}8359	.9^{2}8411	.9^{2}8462	.9^{2}8511	.9^{2}8559	.9^{2}8605
3.0	.9^{2}8650	.9^{2}8694	.9^{2}8736	.9^{2}8777	.9^{2}8817	.9^{2}8856	.9^{2}8893	.9^{2}8930	.9^{2}8965	.9^{2}8999
3.1	.9^{3}0324	.9^{3}0646	.9^{3}0957	.9^{3}1260	.9^{3}1553	.9^{3}1836	.9^{3}2112	.9^{3}2378	.9^{3}2636	.9^{3}2886
3.2	.9^{3}3129	.9^{3}3363	.9^{3}3590	.9^{3}3810	.9^{3}4024	.9^{3}4230	.9^{3}4429	.9^{3}4623	.9^{3}4810	.9^{3}4991

Appendix 1 *Continued*

$$R = (2\pi)^{-1/2} \int_z^\infty e^{-z^2/2} dz.$$

z	.00	.01	.02	.03	.04	.05	.06	.07	.08	.09
3.3	$.9^3 5166$	$.9^3 5335$	$.9^3 5499$	$.9^3 5658$	$.9^3 5811$	$.9^3 5959$	$.9^3 6103$	$.9^3 6242$	$.9^3 6376$	$.9^3 6505$
3.4	$.9^3 6631$	$.9^3 6752$	$.9^3 6869$	$.9^3 6982$	$.9^3 7091$	$.9^3 7197$	$.9^3 7299$	$.9^3 7398$	$.9^3 7493$	$.9^3 7585$
3.5	$.9^3 7674$	$.9^3 7759$	$.9^3 7842$	$.9^3 7922$	$.9^3 7999$	$.9^3 8074$	$.9^3 8146$	$.9^3 8215$	$.9^3 8282$	$.9^3 8347$
3.6	$.9^3 8409$	$.9^3 8469$	$.9^3 8527$	$.9^3 8583$	$.9^3 8637$	$.9^3 8689$	$.9^3 8739$	$.9^3 8787$	$.9^3 8834$	$.9^3 8879$
3.7	$.9^3 8922$	$.9^3 8964$	$.9^4 0039$	$.9^4 0426$	$.9^4 0799$	$.9^4 1158$	$.9^4 1504$	$.9^4 1838$	$.9^4 2159$	$.9^4 2468$
3.8	$.9^4 2765$	$.9^4 3052$	$.9^4 3327$	$.9^4 3593$	$.9^4 3848$	$.9^4 4094$	$.9^4 4331$	$.9^4 4558$	$.9^4 4777$	$.9^4 4988$
3.9	$.9^4 5190$	$.9^4 5385$	$.9^4 5573$	$.9^4 5753$	$.9^4 5926$	$.9^4 6092$	$.9^4 6253$	$.9^4 6406$	$.9^4 6554$	$.9^4 6696$
4.0	$.9^4 6833$	$.9^4 6964$	$.9^4 7090$	$.9^4 7211$	$.9^4 7327$	$.9^4 7439$	$.9^4 7546$	$.9^4 7649$	$.9^4 7748$	$.9^4 7843$
4.1	$.9^4 7934$	$.9^4 8022$	$.9^4 8106$	$.9^4 8186$	$.9^4 8263$	$.9^4 8338$	$.9^4 8409$	$.9^4 8477$	$.9^4 8542$	$.9^4 8605$
4.2	$.9^4 8665$	$.9^4 8723$	$.9^4 8778$	$.9^4 8832$	$.9^4 8882$	$.9^4 8931$	$.9^4 8978$	$.9^5 0226$	$.9^5 0655$	$.9^5 1066$
4.3	$.9^5 1460$	$.9^5 1837$	$.9^5 2199$	$.9^5 2545$	$.9^5 2876$	$.9^5 3193$	$.9^5 3497$	$.9^5 3788$	$.9^5 4066$	$.9^5 4332$
4.4	$.9^5 4587$	$.9^5 4831$	$.9^5 5065$	$.9^5 5288$	$.9^5 5502$	$.9^5 5706$	$.9^5 5902$	$.9^5 6089$	$.9^5 6268$	$.9^5 6439$
4.5	$.9^5 6602$	$.9^5 6759$	$.9^5 6908$	$.9^5 7051$	$.9^5 7187$	$.9^5 7318$	$.9^5 7442$	$.9^5 7561$	$.9^5 7675$	$.9^5 7784$
4.6	$.9^5 7888$	$.9^5 7987$	$.9^5 8081$	$.9^5 8172$	$.9^5 8258$	$.9^5 8340$	$.9^5 8419$	$.9^5 8494$	$.9^5 8566$	$.9^5 8634$
4.7	$.9^5 8699$	$.9^5 8761$	$.9^5 8821$	$.9^5 8877$	$.9^5 8931$	$.9^5 8983$	$.9^6 0320$	$.9^6 0789$	$.9^6 1235$	$.9^6 1661$
4.8	$.9^6 2067$	$.9^6 2453$	$.9^6 2822$	$.9^6 3173$	$.9^6 3508$	$.9^6 3827$	$.9^6 4131$	$.9^6 4420$	$.9^6 4696$	$.9^6 4958$
4.9	$.9^6 5208$	$.9^6 5446$	$.9^6 5673$	$.9^6 5889$	$.9^6 6094$	$.9^6 6289$	$.9^6 6475$	$.9^6 6652$	$.9^6 6821$	$.9^6 6981$

Source: A. Hald, *Statistical Tables and Formulas*, Wiley, New York 1952 (Table II).

APPENDIX 2

Tolerance Factors for Normal Distributions[a]

α n	γ = 0.75					γ = 0.90				
	0.25	0.10	0.05	0.01	0.001	0.25	0.10	0.05	0.01	0.001
2	4.498	6.301	7.414	9.531	11.920	11.407	15.978	18.800	24.167	30.227
3	2.501	3.538	4.187	5.431	6.844	4.132	5.847	6.919	8.974	11.309
4	2.035	2.892	3.431	4.471	5.657	2.932	4.166	4.943	6.440	8.149
5	1.825	2.599	3.088	4.033	5.117	2.454	3.494	4.152	5.423	6.879
6	1.704	2.429	2.889	3.779	4.802	2.196	3.131	3.723	4.870	6.188
7	1.624	2.318	2.757	3.611	4.593	2.034	2.902	3.452	4.521	5.750
8	1.568	2.238	2.663	3.491	4.444	1.921	2.743	3.264	4.278	5.446
9	1.525	2.178	2.593	3.400	4.330	1.839	2.626	3.125	4.098	5.220
10	1.492	2.131	2.537	3.328	4.241	1.775	2.535	3.018	3.959	5.046
11	1.465	2.093	2.493	3.271	4.169	1.724	2.463	2.933	3.849	4.906
12	1.443	2.062	2.456	3.223	4.110	1.683	2.404	2.863	3.758	4.792
13	1.425	2.036	2.424	3.183	4.059	1.648	2.355	2.805	3.682	4.697
14	1.409	2.013	2.398	3.148	4.016	1.619	2.314	2.756	3.618	4.615
15	1.395	1.994	2.375	3.118	3.979	1.594	2.278	2.713	3.562	4.545
16	1.383	1.977	2.355	3.092	3.946	1.572	2.246	2.676	3.514	4.484
17	1.372	1.962	2.337	3.069	3.917	1.552	2.219	2.643	3.471	4.430
18	1.363	1.948	2.321	3.048	3.891	1.535	2.194	2.614	3.433	4.382
19	1.355	1.936	2.307	3.030	3.867	1.520	2.172	2.588	3.399	4.339
20	1.347	1.925	2.294	3.013	3.846	1.506	2.152	2.564	3.368	4.300
21	1.340	1.915	2.282	2.998	3.827	1.493	2.135	2.543	3.340	4.264
22	1.334	1.906	2.271	2.984	3.809	1.482	2.118	2.524	3.315	4.232
23	1.328	1.898	2.261	2.971	3.793	1.471	2.103	2.506	3.292	4.203
24	1.322	1.891	2.252	2.959	3.778	1.462	2.089	2.489	3.270	4.176
25	1.317	1.883	2.244	2.948	3.764	1.453	2.077	2.474	3.251	4.151
26	1.313	1.877	2.236	2.938	3.751	1.444	2.065	2.460	3.232	4.127
27	1.309	1.871	2.229	2.929	3.740	1.437	2.054	2.447	3.215	4.106

Appendix 2 *Continued*

n	0.25	0.10	0.05	0.01	0.001	0.25	0.10	0.05	0.01	0.001
28	1.305	1.865	2.222	2.920	3.728	1.430	2.044	2.435	3.199	4.085
29	1.301	1.860	2.216	2.911	3.718	1.423	2.034	2.424	3.184	4.066
30	1.297	1.855	2.210	2.904	3.708	1.417	2.025	2.413	3.170	4.049
31	1.294	1.850	2.204	2.896	3.699	1.411	2.017	2.403	3.157	4.032
32	1.291	1.846	2.199	2.890	3.690	1.405	2.009	2.393	3.145	4.106
33	1.288	1.842	2.194	2.883	3.682	1.400	2.001	2.385	3.133	4.001
34	1.285	1.838	2.189	2.877	3.674	1.395	1.994	2.376	3.122	3.987
35	1.283	1.834	2.185	2.871	3.667	1.390	1.988	2.368	3.112	3.974
36	1.280	1.830	2.181	2.866	3.660	1.386	1.981	2.361	3.102	3.961
37	1.278	1.827	2.177	2.860	3.653	1.381	1.975	2.353	3.092	3.949
38	1.275	1.824	2.173	2.855	3.647	1.377	1.969	2.346	3.083	3.938
39	1.273	1.821	2.169	2.850	3.641	1.374	1.964	2.340	3.075	3.927
40	1.271	1.818	2.166	2.846	3.635	1.370	1.959	2.334	3.066	3.917
41	1.269	1.815	2.162	2.841	3.629	1.366	1.954	2.328	3.059	3.907
42	1.267	1.812	2.159	2.837	3.624	1.363	1.949	2.322	3.051	3.897
43	1.266	1.810	2.156	2.833	3.619	1.360	1.944	2.316	3.044	3.888
44	1.264	1.807	2.153	2.829	3.614	1.357	1.940	2.311	3.037	3.879
45	1.262	1.805	2.150	2.826	3.609	1.354	1.935	2.306	3.030	3.871
46	1.261	1.802	2.148	2.822	3.605	1.351	1.931	2.301	3.024	3.863
47	1.259	1.800	2.145	2.819	3.600	1.348	1.927	2.297	3.018	3.855
48	1.258	1.798	2.143	2.815	3.596	1.345	1.924	2.292	3.012	3.847
49	1.256	1.796	2.140	2.812	3.592	1.343	1.920	2.288	3.006	3.840
50	1.255	1.794	2.138	2.809	3.588	1.340	1.916	2.284	3.001	3.833

α / n	$\gamma = 0.95$					$\gamma = 0.99$				
	0.25	0.10	0.05	0.01	0.001	0.25	0.10	0.05	0.01	0.001
2	22.858	32.019	37.674	48.430	60.573	114.363	160.193	188.491	242.300	303.054
3	5.922	8.380	9.916	12.861	16.208	13.378	18.930	22.401	29.055	36.616
4	3.779	5.369	6.370	8.299	10.502	6.614	9.398	11.150	14.527	18.383
5	3.002	4.275	5.079	6.634	8.415	4.643	6.612	7.855	10.260	13.015
6	2.604	3.712	4.414	5.775	7.337	3.743	5.337	6.345	8.301	10.548
7	2.361	3.369	4.007	5.248	6.676	3.233	4.613	5.488	7.187	9.142
8	2.197	3.136	3.732	4.891	6.226	2.905	4.147	4.936	6.468	8.234
9	2.078	2.967	3.532	4.631	5.899	2.677	3.822	4.550	5.966	7.600
10	1.987	2.839	3.379	4.433	5.649	2.508	3.582	4.265	5.594	7.129
11	1.916	2.737	3.259	4.277	5.452	2.378	3.397	4.045	5.308	6.766
12	1.858	2.655	3.162	4.150	5.291	2.274	3.250	3.870	5.079	6.477
13	1.810	2.587	3.081	4.044	5.158	2.190	3.130	3.727	4.893	6.240
14	1.770	2.529	3.012	3.955	5.045	2.120	3.029	3.608	4.737	6.043
15	1.735	2.480	2.954	3.878	4.949	2.060	2.945	3.507	4.605	5.876

Appendix 2 *Continued*

α \ n	γ = 0.75					γ = 0.90				
	0.25	0.10	0.05	0.01	0.001	0.25	0.10	0.05	0.01	0.001
16	1.705	2.437	2.903	3.812	4.865	2.009	2.872	3.421	4.492	5.732
17	1.679	2.400	2.858	3.754	4.791	1.965	2.808	3.345	4.393	5.607
18	1.655	2.366	2.819	3.702	4.725	1.926	2.753	3.279	4.307	5.497
19	1.635	2.337	2.784	3.656	4.667	1.891	2.703	3.221	4.230	5.399
20	1.616	2.310	2.752	3.615	4.614	1.860	2.659	3.168	4.161	5.312
21	1.599	2.286	2.723	3.577	4.567	1.833	2.620	3.121	4.100	5.234
22	1.584	2.264	2.697	3.543	4.523	1.808	2.584	3.078	4.044	5.163
23	1.570	2.244	2.673	3.512	4.484	1.785	2.551	3.040	3.993	5.098
24	1.557	2.225	2.651	3.483	4.447	1.764	2.522	3.004	3.947	5.039
25	1.545	2.208	2.631	3.457	4.413	1.745	2.494	2.972	3.904	4.985
26	1.534	2.193	2.612	3.432	4.382	1.727	2.469	2.941	3.865	4.935
27	1.523	2.178	2.595	3.409	4.353	1.711	2.446	2.914	3.828	4.888
28	1.514	2.164	2.579	3.388	4.326	1.695	2.424	2.888	3.794	4.845
29	1.505	2.152	2.554	3.368	4.301	1.681	2.404	2.864	3.763	4.805
30	1.497	2.140	2.549	3.350	4.278	1.668	2.385	2.841	3.733	4.768
31	1.489	2.129	2.536	3.332	4.256	1.656	2.367	2.280	3.706	4.732
32	1.481	2.118	2.524	3.316	4.235	1.644	2.351	2.801	3.680	4.699
33	1.475	2.108	2.512	3.300	4.215	1.633	2.335	2.782	3.655	4.668
34	1.468	2.099	2.501	3.286	4.197	1.623	2.320	2.764	3.632	4.639
35	1.462	2.090	2.490	3.272	4.179	1.613	2.306	2.748	3.611	4.611
36	1.455	2.081	2.479	3.258	4.161	1.604	2.293	2.732	3.590	4.585
37	1.450	2.073	2.470	3.246	4.146	1.595	2.281	2.717	3.571	4.560
38	1.446	2.068	2.464	3.237	4.134	1.587	2.269	2.703	3.552	4.537
39	1.441	2.060	2.455	3.226	4.120	1.579	2.257	2.690	3.534	4.514
40	1.435	2.052	2.445	3.213	4.104	1.571	2.247	2.677	3.518	4.493
41	1.430	2.045	2.437	3.202	4.090	1.564	2.236	2.665	3.502	4.472
42	1.426	2.039	2.429	3.192	4.077	1.557	2.227	2.653	3.486	4.453
43	1.422	2.033	2.422	3.183	4.065	1.551	2.217	2.642	3.472	4.434
44	1.418	2.027	2.415	3.173	2.053	1.545	2.208	2.631	3.458	4.416
45	1.414	2.021	2.408	3.165	4.042	1.539	2.200	2.621	3.444	4.399
46	1.410	2.016	2.402	3.156	4.031	1.533	2.192	2.611	3.431	4.383
47	1.406	2.011	2.396	3.148	4.021	1.527	2.184	2.602	3.419	4.367
48	1.403	2.006	2.390	3.140	4.011	1.522	2.176	2.593	3.407	4.352
49	1.399	2.001	2.384	3.133	4.002	1.517	2.169	2.584	3.396	4.337
50	1.396	1.969	2.379	3.126	3.993	1.512	2.162	2.576	3.385	4.323

[a] Factors K such that the probability is γ that at least a proportion $1 - \alpha$ of the distribution will be included between $\bar{x} \pm K s$, where $\bar{x}$ and s are estimates of the mean and the standard deviation computed from a sample of n.

Source: C. Eisenhart, M. W. Hastay, W. A. Wallis, *Techniques of Statistical Analysis*, McGraw-Hill, New York, 1947, Chapter 2.

APPENDIX 3

One-sided Tolerance Limit Factors k for Normal Distribution, 0.95 Confidence, and n–1 Degrees of Freedom

n	.75000	.90000	.95000	.97500	.99000	.99900	.99990	.99999
2	11.763	20.581	26.260	31.257	37.094	49.276	59.304	68.010
3	3.806	6.155	7.656	8.986	10.553	13.857	16.598	18.986
4	2.618	4.162	5.144	6.015	7.042	9.214	11.019	12.593
5	2.150	3.407	4.203	4.909	5.741	7.502	8.966	10.243
6	1.895	3.006	3.708	4.329	5.062	6.612	7.901	9.025
7	1.732	2.755	3.399	3.970	4.642	6.063	7.244	8.275
8	1.618	2.582	3.187	3.723	4.354	5.688	6.796	7.763
9	1.532	2.454	3.031	3.542	4.143	5.413	6.469	7.390
10	1.465	2.355	2.911	3.402	3.981	5.203	6.219	7.105
11	1.411	2.275	2.815	3.292	3.852	5.036	6.020	6.878
12	1.366	2.210	2.736	3.201	3.747	4.900	5.858	6.694
13	1.328	2.155	2.671	3.125	3.659	4.787	5.723	6.540
14	1.296	2.109	2.614	3.060	3.585	4.690	5.609	6.409
15	1.268	2.068	2.566	3.005	3.520	4.607	5.510	6.297
16	1.243	2.033	2.524	2.956	3.464	4.535	5.424	6.199
17	1.220	2.002	2.486	2.913	3.414	4.471	5.348	6.113
18	1.201	1.974	2.453	2.875	3.370	4.415	5.281	6.037
19	1.183	1.949	2.423	2.841	3.331	4.364	5.221	5.968
20	1.166	1.926	2.396	2.810	3.295	4.318	5.167	5.906
21	1.152	1.905	2.371	2.781	3.263	4.277	5.118	5.850
22	1.138	1.886	2.349	2.756	3.233	4.239	5.073	5.799
23	1.125	1.869	2.328	2.732	3.206	4.204	5.031	5.752
24	1.114	1.853	2.309	2.710	3.181	4.172	4.994	5.709
25	1.103	1.838	2.292	2.690	3.158	4.142	4.959	5.670

Appendix 3 *Continued*

n	.75000	.90000	.95000	.97500	.99000	.99900	.99990	.99999
26	1.093	1.824	2.275	2.672	3.136	4.115	4.926	5.633
27	1.083	1.811	2.260	2.654	3.116	4.089	4.896	5.598
28	1.075	1.799	2.246	2.638	3.098	4.066	4.868	5.566
29	1.066	1.788	2.232	2.623	3.080	4.043	4.841	5.536
30	1.058	1.777	2.220	2.608	3.064	4.022	4.816	5.508
31	1.051	1.767	2.208	2.595	3.048	4.002	4.793	5.481
32	1.044	1.758	2.197	2.582	3.034	3.984	4.771	5.456
33	1.037	1.749	2.186	2.570	3.020	3.966	4.750	5.433
34	1.031	1.740	2.176	2.559	3.007	3.950	4.730	5.410
35	1.025	1.732	2.167	2.548	2.995	3.934	4.712	5.389
36	1.019	1.725	2.158	2.538	2.983	3.919	4.694	5.369
37	1.014	1.717	2.149	2.528	2.972	3.904	4.677	5.350
38	1.009	1.710	2.141	2.518	2.961	3.891	4.661	5.332
39	1.004	1.704	2.133	2.510	2.951	3.878	4.646	5.314
40	.999	1.697	2.125	2.501	2.941	3.865	4.631	5.298
41	.994	1.691	2.118	2.493	2.932	3.854	4.617	5.282
42	.990	1.685	2.111	2.485	2.923	3.842	4.603	5.266
43	.986	1.680	2.105	2.478	2.914	3.831	4.591	5.252
44	.982	1.674	2.098	2.470	2.906	3.821	4.578	5.238
45	.978	1.669	2.092	2.463	2.898	3.811	4.566	5.224
46	.974	1.664	2.086	2.457	2.890	3.801	4.555	5.211
47	.971	1.659	2.081	2.450	2.883	3.792	4.544	5.199
48	.967	1.654	2.075	2.444	2.876	3.783	4.533	5.187
49	.964	1.650	2.070	2.438	2.869	3.774	4.523	5.175
50	.960	1.646	2.065	2.432	2.862	3.766	4.513	5.164

Source: Ref. 66.

APPENDIX 4

Stress Conversion Table

ksi (1000 lb/in.²)	kg/mm²	ksi (1000 lb/in.²)	kg/mm²
½	0.3516	56	39.4
1	0.7031	58	40.8
2	1.406		
3	2.109	60	42.2
4	2.812	62	43.6
		64	45.0
5	3.52	66	46.4
6	4.22	68	47.8
7	4.92		
8	5.62	70	49.2
9	6.33	72	50.6
		74	52.0
10	7.03	76	53.4
11	7.73	78	54.8
12	8.44		
13	9.14	80	56.2
14	9.84	82	57.7
		84	59.1
15	10.55	86	60.5
16	11.25	88	61.9
17	11.95		
18	12.66	90	63.3
19	13.36	92	64.7
		94	66.1
20	14.06	96	67.5
21	14.77	98	68.9
22	15.47		
23	16.17	100	70.3
24	16.87	120	84.4
		140	98.4
25	17.58	160	112.5
26	18.28	180	126.6
27	18.98		
28	19.7	200	140.6

Appendix 4 *Continued*

ksi (1000 lb/in.^{2})	kg/mm^2	ksi (1000 lb/in.^{2})	kg/mm^2
29	20.4	220	154.7
		240	168.7
30	21.1	260	182.8
32	22.5	280	197.0
34	23.9		
36	25.3	300	211
38	26.7	320	225
		340	239
40	28.1	360	253
42	29.5	380	267
44	30.9		
46	32.3	400	281
48	33.7	420	295
		440	309
50	35.2	460	323
52	36.6	480	337
54	38.0	500	352

Source: Heywood [75].

APPENDIX 5

Statistics of Element of Sections

The following variables appear in this appendix:

A = Area of section
I = Moment of inertia about axis $I-I$
c = Distance from axis $I-I$ to remotest point of section
I/c = Section modulus
r_g = Radius of gyration

$$\mu_A = \bar{b}\bar{h} \qquad\qquad \mu_{(I/c)} = \frac{\bar{b}\bar{h}^2}{6}$$

$$\varsigma_A \approx \sqrt{\varsigma_h^2 \cdot \bar{b}^2 + \varsigma_b^2 \cdot \bar{h}^2}$$

$$\mu_c = 0.5\bar{h} \qquad\qquad \varsigma_{(I/c)} = \frac{\bar{h}}{6}\sqrt{4\bar{b}^2 \cdot \varsigma_h^2 + \varsigma_b^2 \cdot \bar{h}^2}$$

$$\varsigma_c = 0.5\varsigma_h$$

$$\mu_I = \frac{\bar{b}\bar{h}^3}{12} \qquad\qquad \mu_{r_g} = \frac{\bar{h}}{\sqrt{12}} = 0.289\bar{h}$$

$$\varsigma_I = \frac{\bar{h}^2}{12}\sqrt{9\bar{b}^2 \varsigma_h^2 + \bar{h}^2 \varsigma_b^2} \qquad\qquad \varsigma_{r_g} = 0.289\varsigma_h$$

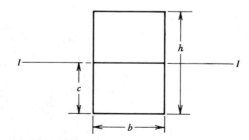

Figure A5.1 Rectangle: axis through center.

Statistics of Element of Sections

$\mu_A = \bar{b}\bar{h}$

$s_A = \sqrt{s_h^2 \cdot \bar{b}^2 + s_b^2 \cdot \bar{h}^2}$

$\mu_c = \bar{h}$

$s_c = s_h$

$\mu_I = \dfrac{\bar{b}\bar{h}^3}{3}$

$s_I = \dfrac{\bar{h}^2}{3}\sqrt{9\bar{b}^2 \cdot s_h^2 + \bar{h}^2 \cdot s_b^2}$

$\mu_{I/c} = \dfrac{\bar{b}\bar{h}^2}{3}$

$s_{I/c} = \dfrac{\bar{h}}{3}\sqrt{4\bar{b}^2 \cdot s_h^2 + \bar{h}^2 \cdot s_b^2}$

$\mu_{r_g} = \dfrac{\bar{h}}{\sqrt{3}} = 0.577\bar{h}$

$s_{r_g} = 0.577\, s_h$

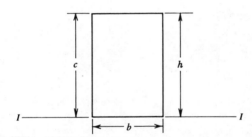

Figure A5.2 Rectangle: axis on base.

$\mu_A = \bar{b}_2\bar{h}_2 - \bar{b}_1\bar{h}_1$

$s_A = \sqrt{b_2^2 \cdot s_{h_2}^2 + h_2^2 \cdot s_{b_2}^2 + b_1^2 \cdot s_{h_1}^2 + h_1^2 \cdot s_{b_1}^2}$

$\mu_I = \dfrac{\bar{b}_2\bar{h}_2^3 - \bar{b}_1\bar{h}_1^3}{12}$

$s_I = \dfrac{1}{12}\sqrt{\bar{h}_2^6 s_{b_2}^2 + 9\bar{b}_2^2\bar{h}_2^4 s_{h_2}^2 + 9\bar{b}_1^2\bar{h}_1^4 s_{h_1}^2 + \bar{h}_1^6 s_{b_1}^2}$

$\mu_{I/c} = \dfrac{\bar{b}_2\bar{h}_2^3 - \bar{b}_1\bar{h}_1^3}{6\bar{h}_2}$

$s_{I/c} = \dfrac{1}{6}\sqrt{\bar{h}_2^4 s_{b_2}^2 + \dfrac{\bar{h}_1^6 s_{b_1}^2 + 9\bar{b}_1^2\bar{h}_1^4 s_{h_1}^2}{\bar{h}_2^2} + \left(\dfrac{2\bar{b}_2\bar{h}_2 + \bar{b}_1\bar{h}_1^3}{\bar{h}_2^2}\right)^2 s_{h_2}^2}$

$\bar{c} = \dfrac{\bar{h}_2}{2}$

$s_c = \dfrac{s_{h_2}}{2}$

Appendix 5

$$\mu_{r_g} = \sqrt{(\bar{b}_2\bar{h}_2^3 - \bar{b}_1\bar{h}_1^3)/12(\bar{b}_2\bar{h}_2 - \bar{b}_1\bar{h}_1)}$$

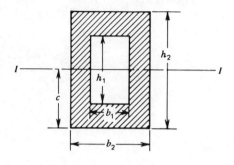

Figure A5.3 Hollow rectangle: axis through center.

$$\mu_A = \frac{\pi}{4}\bar{d}^2 \qquad \mu_{r_g} = \frac{\bar{d}}{4}$$

$${}^{\scriptscriptstyle\supset}\!A = \frac{\pi}{2}\bar{d}\,{}^{\scriptscriptstyle\supset}\!d \qquad {}^{\scriptscriptstyle\supset}\!{r_g} = \frac{{}^{\scriptscriptstyle\supset}\!d}{4}$$

$$\mu_c = \frac{\bar{d}}{z}$$

$${}^{\scriptscriptstyle\supset}\!c = \frac{{}^{\scriptscriptstyle\supset}\!d}{2}$$

$$\mu_I = \frac{\pi\bar{d}^4}{64} \qquad \mu_J = \frac{\pi\bar{d}^4}{32}$$

$${}^{\scriptscriptstyle\supset}\!I = \frac{\pi}{16}\bar{d}^3 \cdot {}^{\scriptscriptstyle\supset}\!d \qquad {}^{\scriptscriptstyle\supset}\!J = \frac{\pi\bar{d}^3}{8}\,{}^{\scriptscriptstyle\supset}\!d$$

$$\mu_{I/c} = \frac{\pi\bar{d}^3}{32}$$

$${}^{\scriptscriptstyle\supset}\!{I/c} = \frac{3\pi}{32}\bar{d}^2 \cdot {}^{\scriptscriptstyle\supset}\!d$$

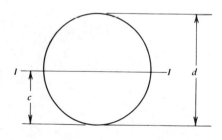

Figure A5.4 Circle: axis through center.

Statistics of Element of Sections 575

$$\mu_A = \frac{\pi}{4}(\bar{d}_2^2 - \bar{d}_1^2)$$

$$\mu_{I/c} = \frac{\pi}{32}\left[d_2^3 - \frac{\bar{d}_1^4}{\bar{d}_2}\right]$$

$$s_A = \frac{\pi}{2}\sqrt{\bar{d}_2^2 \cdot s_{d_2}^2 + \bar{d}_1^2 \cdot s_{d_1}^2}$$

$$s_{I/c} = \frac{\pi}{32}\left[\left\{3d_2^2 + \frac{\bar{d}_1^4}{\bar{d}_2^2}s_{d_2}^2 + \frac{16\bar{d}_1^6}{\bar{d}_2^2}s_{d_1}^2\right\}\right]^{1/2}$$

$$\mu_c = \frac{\bar{d}_2}{2}$$

$$\mu_{r_g} = \frac{(\bar{d}_2^2 + \bar{d}_1^2)^{1/2}}{4}$$

$$s_c = \frac{s_{d_2}}{2}$$

$$s_{r_g} = \frac{1}{4}\sqrt{(\bar{d}_2^2 s_{d_2}^2 + \bar{d}_1^2 s_{d_1}^2)/\bar{d}_2^2 + \bar{d}_1^2}$$

$$\mu_I = \frac{\pi}{64}(\bar{d}_2^4 - \bar{d}_1^4) \approx \pi \bar{r}^3 \bar{t}$$

$$s_I = \frac{\pi}{16}\sqrt{d_2^6 s_{d_2}^2 + d_1^6 s_{d_1}^2}$$

$$J = \frac{\pi(d_2^4 - d_1^4)}{32} \text{ in.}^4$$

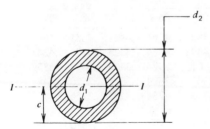

Figure A5.5

APPENDIX 6

Areas under the Normal Curve from K_α to ∞

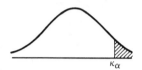

$$\int_{K_\alpha}^{\infty} \frac{1}{\sqrt{2\pi}} e^{-x^2/2} dx = \alpha$$

K_α	.00	.01	.02	.03	.04	.05	.06	.07	.08	.09
0.0	.5000	.4960	.4920	.4880	.4840	.4801	.4761	.4721	.4681	.4641
0.1	.4602	.4562	.4522	.4483	.4443	.4404	.4364	.4325	.4286	.4247
0.2	.4207	.4168	.4129	.4090	.4052	.4013	.3974	.3936	.3897	.3859
0.3	.3821	.3783	.3745	.3707	.3669	.3632	.3594	.3557	.3520	.3483
0.4	.3446	.3409	.3372	.3336	.3300	.3264	.3238	.3192	.3156	.3121
0.5	.3085	.3050	.3015	.2981	.2946	.2912	.2877	.2843	.2810	.2776
0.6	.2743	.2709	.2676	.2643	.2611	.2578	.2546	.2514	.2483	.2451
0.7	.2420	.2389	.2358	.2327	.2296	.2266	.2236	.2206	.2177	.2148
0.8	.2119	.2090	.2061	.2033	.2005	.1977	.1949	.1922	.1894	.1867
0.9	.1841	.1814	.1788	.1762	.1736	.1711	.1685	.1660	.1635	.1611
1.0	.1587	.1562	.1539	.1515	.1492	.1469	.1446	.1423	.1401	.1379
1.1	.1357	.1335	.1314	.1292	.1271	.1251	.1230	.1210	.1190	.1170
1.2	.1151	.1131	.1112	.1093	.1075	.1056	.1038	.1020	.1003	.0985
1.3	.0968	.0951	.0934	.0918	.0901	.0885	.0869	.0853	.0838	.0823
1.4	.0808	.0793	.0778	.0764	.0749	.0735	.0721	.0708	.0694	.0681
1.5	.0668	.0655	.0643	.0630	.0618	.0606	.0594	.0582	.0571	.0559
1.6	.0548	.0537	.0526	.0516	.0505	.0495	.0485	.0475	.0465	.0455
1.7	.0446	.0436	.0427	.0418	.0409	.0401	.0392	.0384	.0375	.0367
1.8	.0359	.0351	.0344	.0336	.0329	.0322	.0314	.0307	.0301	.0294
1.9	.0287	.0281	.0274	.0268	.0262	.0256	.0250	.0244	.0239	.0233

Appendix 6 *Continued*

$$\int_{K_\alpha}^{\infty} \frac{1}{\sqrt{2\pi}} e^{-x^2/2} dx = \alpha$$

K_α	.00	.01	.02	.03	.04	.05	.06	.07	.08	.09
2.0	.0228	.0222	.0217	.0212	.0207	.0202	.0197	.0192	.0188	.0183
2.1	.0179	.0174	.0170	.0166	.0162	.0158	.0154	.0150	.0146	.0143
2.2	.0139	.0136	.0132	.0129	.0125	.0122	.0119	.0116	.0113	.0110
2.3	.0107	.0104	.0102	.00990	.00964	.00939	.00914	.00889	.00866	.00842
2.4	.00820	.00798	.00776	.00755	.00734	.00714	.00695	.00676	.00657	.00639
2.5	.00621	.00604	.00587	.00570	.00554	.00539	.00523	.00508	.00494	.00480
2.6	.00466	.00453	.00440	.00427	.00415	.00402	.00391	.00379	.00368	.00357
2.7	.00347	.00336	.00326	.00317	.00307	.00298	.00289	.00280	.00272	.00264
2.8	.00256	.00248	.00240	.00233	.00226	.00219	.00212	.00205	.00199	.00193
2.9	.00187	.00181	.00175	.00169	.00164	.00159	.00154	.00149	.00144	.00139

K_α	.0	.1	.2	.3	.4	.5	.6	.7	.8	.9
3	.00135	$.0^3 968$	$.0^3 687$	$.0^3 483$	$.0^2 337$	$.0^3 233$	$.0^3 159$	$.0^3 108$	$.0^4 723$	$.0^4 481$
4	$.0^4 317$	$.0^4 207$	$.0^4 133$	$.0^5 854$	$.0^5 541$	$.0^5 340$	$.0^6 211$	$.0^6 130$	$.0^6 793$	$.0^6 479$
5	$.0^6 287$	$.0^6 170$	$.0^7 996$	$.0^7 579$	$.0^7 333$	$.0^7 190$	$.0^7 107$	$.0^8 599$	$.0^8 332$	$.0^8 182$
6	$.0^9 987$	$.0^9 530$	$.0^9 282$	$.0^9 149$	$.0^{10} 777$	$.0^{10} 402$	$.0^{10} 206$	$.0^{10} 104$	$.0^{11} 523$	$.0^{11} 260$

Source: From Frederick E. Croxton, *Elementary Statistics with Applications in Medicine*, Prentice-Hall, Englewood Cliffs, N.J., 1953, p. 323.

APPENDIX 7

Percentage Points of t Distribution [a]

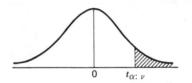

ν \ α	.40	.30	.20	.10	.050	.025	.010	.005	.001	.0005
1	0.325	0.727	1.376	3.078	6.314	12.71	31.82	63.66	318.3	636.6
2	0.289	0.617	1.061	1.886	2.920	4.303	6.965	9.925	22.33	31.60
3	0.277	0.584	0.978	1.638	2.353	3.182	4.541	5.841	10.22	12.94
4	0.271	0.569	0.941	1.533	2.132	2.776	3.747	4.604	7.173	8.610
5	0.267	0.559	0.920	1.476	2.015	2.571	3.365	4.032	5.893	6.859
6	0.265	0.553	0.906	1.440	1.943	2.447	3.143	3.707	5.208	5.959
7	0.263	0.549	0.896	1.415	1.895	2.365	2.998	3.499	4.785	5.405
8	0.262	0.546	0.889	1.397	1.860	2.306	2.896	3.355	4.501	5.041
9	0.261	0.543	0.883	1.383	1.833	2.262	2.821	3.250	4.297	4.781
10	0.260	0.542	0.879	1.372	1.812	2.228	2.764	3.169	4.144	4.587
11	0.260	0.540	0.876	1.363	1.796	2.201	2.718	3.106	4.025	4.437
12	0.259	0.539	0.873	1.356	1.782	2.179	2.681	3.055	3.930	4.318
13	0.259	0.538	0.870	1.350	1.771	2.160	2.650	3.012	3.852	4.221
14	0.258	0.537	0.868	1.345	1.761	2.145	2.624	2.977	3.787	4.140
15	0.258	0.536	0.866	1.341	1.753	2.131	2.602	2.947	3.733	4.073
16	0.258	0.535	0.865	1.337	1.746	2.120	2.583	2.921	3.686	4.015
17	0.257	0.534	0.863	1.333	1.740	2.110	2.567	2.898	3.646	3.965
18	0.257	0.534	0.862	1.330	1.734	2.101	2.552	2.878	3.611	3.922
19	0.257	0.533	0.861	1.328	1.729	2.093	2.539	2.861	3.579	3.883
20	0.257	0.533	0.860	1.325	1.725	2.086	2.528	2.845	3.552	2.850
21	0.257	0.532	0.859	1.323	1.721	2.080	2.518	2.831	3.527	3.819
22	0.256	0.532	0.858	1.321	1.717	2.074	2.508	2.819	3.505	3.792

Appendix 7 *Continued*

ν \ α	.40	.30	.20	.10	.050	.025	.010	.005	.001	.0005
23	0.256	0.532	0.858	1.319	1.714	2.069	2.500	2.807	3.485	3.767
24	0.256	0.531	0.857	1.318	1.711	2.064	2.492	2.797	3.467	3.745
25	0.256	0.531	0.856	1.316	1.708	2.060	2.485	2.787	3.450	3.725
26	0.256	0.531	0.856	1.315	1.706	2.056	2.479	2.779	3.435	3.707
27	0.256	0.531	0.855	1.314	1.703	2.052	2.473	2.771	3.421	3.690
28	0.256	0.530	0.855	1.313	1.701	2.048	2.467	2.763	3.408	3.674
29	0.256	0.530	0.854	1.311	1.699	2.045	2.462	2.756	3.396	3.659
30	0.256	0.530	0.854	1.310	1.697	2.042	2.457	2.750	3.385	3.646
40	0.255	0.529	0.851	1.303	1.684	2.021	2.423	2.704	3.307	3.551
50	0.255	0.528	0.849	1.298	1.676	2.009	2.403	2.678	3.262	3.495
60	0.254	0.527	0.848	1.296	1.671	2.000	2.390	2.660	3.232	3.460
80	0.254	0.527	0.846	1.292	1.664	1.990	2.374	2.639	3.195	3.415
100	0.254	0.526	0.845	1.290	1.660	1.984	2.365	2.626	3.174	3.389
200	0.254	0.525	0.843	1.286	1.653	1.972	2.345	2.601	3.131	3.339
500	0.253	0.525	0.842	1.283	1.648	1.965	2.334	2.586	3.106	3.310
∞	0.253	0.524	0.842	1.282	1.645	1.960	2.326	2.576	3.090	3.291

[a]Table of $t_{\alpha;\nu}$—the 100α percentage point of t distribution for ν degrees of freedom.
Source: This table is reproduced from A. Hald, *Statistical Tables and Formulas*, Wiley, New York, 1952.

APPENDIX 8

Cumulative χ^2 Distribution

$$F(u) = \int_0^u \frac{x^{(n-2)/2} e^{-x/2} dx}{2^{n/2}(n-2)/2} \ !$$

n \ F	.005	.010	.025	.050	.100	.250	.500	.750	.900	.950	.975	.990	.995
1	0.0^{x}393	0.0^{w}157	0.0^{w}982	0.0^{2}393	0.0158	0.102	0.455	1.32	2.71	3.84	5.02	6.63	7.88
2	0.0100	0.0201	0.0506	0.103	0.211	0.575	1.39	2.77	4.61	5.99	7.38	9.21	10.6
3	0.0717	0.115	0.216	0.352	0.584	1.21	2.37	4.11	6.25	7.81	9.35	11.3	12.8
4	0.207	0.297	0.484	0.711	1.06	1.92	3.36	5.39	7.78	9.49	11.1	13.3	14.9
5	0.412	0.554	0.831	1.15	1.61	2.67	4.35	6.63	9.24	11.1	12.8	15.1	16.7
6	0.676	0.872	1.24	1.64	2.20	3.45	5.35	7.84	10.6	12.6	14.4	16.8	18.5
7	0.989	1.24	1.69	2.17	2.83	4.25	6.35	9.04	12.0	14.1	16.0	18.5	20.3
8	1.34	1.65	2.18	2.73	3.49	5.07	7.34	10.2	13.4	15.5	17.5	20.1	22.0
9	1.73	2.09	2.70	3.33	4.17	5.90	8.34	11.4	14.7	16.9	19.0	21.7	23.6
10	2.16	2.56	3.25	3.94	4.87	6.74	9.34	12.5	16.0	18.3	20.5	23.2	25.2
11	2.60	3.05	3.82	4.57	5.58	7.58	10.3	13.7	17.3	19.7	21.9	24.7	26.8
12	3.07	3.57	4.40	5.23	6.30	8.44	11.3	14.8	18.5	21.0	23.3	26.2	28.3
13	3.57	4.11	5.01	5.89	7.04	9.30	12.3	16.0	19.8	22.4	24.7	27.7	29.8
14	4.07	4.66	5.63	6.57	7.79	10.2	13.3	17.1	21.1	23.7	26.1	29.1	31.3
15	4.60	5.23	6.26	7.26	8.55	11.0	14.3	18.2	22.3	25.0	27.5	30.6	32.8
16	5.14	5.81	6.91	7.96	9.31	11.9	15.3	19.4	23.5	26.3	28.8	32.0	34.3
17	5.70	6.41	7.56	8.67	10.1	12.8	16.3	20.5	24.8	27.6	30.2	33.4	35.7
18	6.26	7.01	8.23	9.39	10.9	13.7	17.3	21.6	26.0	28.9	31.5	34.8	37.2
19	6.84	7.63	8.91	10.1	11.7	14.6	18.3	22.7	27.2	30.1	32.9	36.2	38.6
20	7.43	8.26	9.59	10.9	12.4	15.5	19.3	23.8	28.4	31.4	34.2	37.6	40.0
21	8.03	8.90	10.3	11.6	13.2	16.3	20.3	24.9	29.6	32.7	35.5	38.9	41.4
22	8.64	9.54	11.0	12.3	14.0	17.2	21.3	26.0	30.8	33.9	36.8	40.3	42.8
23	9.26	10.2	11.7	13.1	14.8	18.1	22.3	27.1	32.0	35.2	38.1	41.6	44.2
24	9.89	10.9	12.4	13.8	15.7	19.0	23.3	28.2	33.2	36.4	39.4	43.0	45.6
25	10.5	11.5	13.1	14.6	16.5	19.9	24.3	29.3	34.4	37.7	40.6	44.3	46.9
26	11.2	12.2	13.8	15.4	17.3	20.8	25.3	30.4	35.6	38.9	41.9	45.6	48.3
27	11.8	12.9	14.6	16.2	18.1	21.7	26.3	31.5	36.7	40.1	43.2	47.0	49.6
28	12.5	13.6	15.3	16.9	18.9	22.7	27.3	32.6	37.9	41.3	44.5	48.3	51.0
29	13.1	14.3	16.0	17.7	19.8	23.6	28.3	33.7	39.1	42.6	45.7	49.6	52.3
30	13.8	15.0	16.8	18.5	20.6	24.5	29.3	34.8	40.3	43.8	47.0	50.9	53.7

Source: This table is abridged from Catherine M. Thompson, "Tables of Percentage Points of the Incomplete Beta Function and of the Chi-square Distribution," *Biometrika*, **32** (1941). Reprinted with permission.

APPENDIX 9

Relationships for Determining Theoretical Stress Concentration Factor

The relationships described in this appendix are for determining the theoretical stress concentration factors for various geometric configurations and loading situations. All the equations shown, through those for the geometry of a flat plate with a hole, have been derived by Heinz Neuber [61]. The equations for the filleted shaft loaded in bending are developed in Chapter 3 of this book. In addition, FORTRAN code equivalents for the algebraic equations are given. These equations are intended for use in program Monti, which is discussed in Chapter 7. The following change of variables was made in transforming the algebraic equations to the FORTRAN code:

$$B = VAR(1)/VAR(3)$$

$$C = SQRT(B)$$

$$\alpha_1 = A1$$

$$\alpha_2 = A2$$

$$K_{tk} = AKTK$$

$$K_{fk} = AKFK$$

$$K_t = VALUE$$

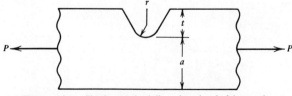

Figure A9.1 Single-notched flat plate loaded in tension.

$$\alpha_1 = \frac{2[(a/r)+1]\sqrt{(a/r)}}{[(a/r)+1]\tan^{-1}\sqrt{a/r} + \sqrt{a/r}}$$

$$\alpha_2 = \frac{4(a/r)\sqrt{(a/r)}}{3\left[\sqrt{(a/r)} + [(a/r)-1]\tan^{-1}\sqrt{(a/r)}\right]}$$

$$C = \frac{\alpha_1 - \sqrt{(a/r)+1}}{4/3\alpha_2 \sqrt{(a/r)+1} - 1}$$

$$K_{tk} = \frac{\alpha_1 - 2C}{1 - \dfrac{C}{\sqrt{(a/r)+1}}}$$

$$K_{fk} = 1 + 2\sqrt{\frac{t}{r}}$$

$$K_t = 1 + \frac{(K_{tk}-1)(K_{fk}-1)}{\sqrt{(K_{tk}-1)^2 + (K_{fk}-1)^2}}$$

The FORTRAN equivalent is as follows:
```
B=VAR(1)/VAR(3)
C=SQRT(B)
A1=(2.*(B+1.)*C)/((B+1.)*ATAN(C)+C)
A2=(4.*B*C)/(3.*(C+(B-1.)*ATAN(C)))
C=(A1-SQRT(B+1.))/((4./(3.*A2))*SQRT(B+1.)-1.)
AKTK=(A1-2.*C)/(1.-C/SQRT(B+1.))
AKFK=1.+2.*SQRT(VAR(2)/VAR(3))
VALUE=1.+((AKTK-1.)*(AKFK-1.))/SQRT((AKTK-1.)**2.+(AKFK-1.)**2.))
```

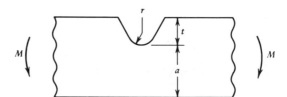

Figure A9.2 Single-notched flat plate loaded in bending.

$$\alpha_1 = \frac{2[(a/r)+1]\sqrt{a/r}}{[(a/r)+1]\tan^{-1}\sqrt{a/r} + \sqrt{a/r}}$$

$$\alpha_2 = \frac{4(a/r)\sqrt{a/r}}{3\left[\sqrt{a/r} + [(a/r)-1]\tan^{-1}\sqrt{a/r}\right]}$$

$$K_{tk} = \frac{2[(a/r)+1] - \alpha_1 \sqrt{(a/r)+1}}{(4/\alpha_2)[(a/r)+1] - 3\alpha_1} \qquad K_{fk} = 1 + 2\sqrt{\frac{t}{r}}$$

$$K_t = 1 + \frac{(K_{tk}-1)(K_{fk}-1)}{\sqrt{(K_{tk}-1)^2 + (K_{fk}-1)^2}}$$

The FORTRAN equivalent is as follows:
B = VAR(1)/VAR(3)
C = SQRT(B)
A1 = (2.*(B+1.)*C/((B+1.)*ATAN(C)+C)
A2 = (4.*B*C)/(3.*(C+(B−1.)*ATAN(C)))
AKTK = (2.*(B+1.)−A1*(SQRT(B+1.)))/((4./A2)*(B+1.)−(3.*A1))
AKFK = 1.+2.*(SQRT(VAR(2)/VAR(3))
VALUE = 1.+(((AKTK−1.)*(AKFK−1.))/SQRT((AKTK−1.)**2.+(AKFK−1.)**2.))

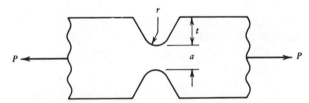

Figure A9.3 Double-notched flat plate loaded in tension.

$$K_{tk} = \frac{2[(a/r)+1]\sqrt{a/r}}{[(a/r)+1]\tan^{-1}\sqrt{a/r} + \sqrt{a/r}} \qquad K_{fk} = 1 + 2\sqrt{\frac{t}{r}}$$

$$K_t = 1 + \frac{(K_{tk}-1)(K_{fk}-1)}{\sqrt{(K_{tk}-1)^2 + (K_{fk}-1)^2}}$$

The FORTRAN equivalent is as follows:
B = VAR(1)/VAR(3)
C = SQRT(B)
AKTK = (2.*(B+1.)*C)/((B+1.)*ATAN(C)+C)
AKFK = 1.+2.*SQRT(VAR(2)/VAR(3))
VALUE = 1.+(((AKTK−1.)*(AKFK−1.))/(SQRT((AKTK−1.)**2.+(AKFK−1.)**2.)))

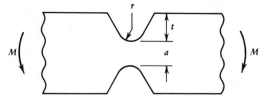

Figure A9.4 Double-notched flat plate loaded in bending.

$$K_{tk} = \frac{4(a/r) + \sqrt{a/r}}{3\left[\sqrt{a/r} + [(a/r)-1]\tan^{-1}\sqrt{a/r}\right]} \qquad K_{fk} = 1 + 2\sqrt{\frac{a}{r}}$$

$$K_t = 1 + \frac{(K_{tk}-1)(K_{fk}-1)}{\sqrt{(K_{tk}-1)^2 + (K_{fk}-1)^2}}$$

The FORTRAN equivalent is as follows:
B = VAR(1)/VAR(3)
C = SQRT(B)
AKTK = (4.*B*C)/(3.*(C+(B−1.)*ATAN(C)))
AKFK = 1.+2.*SQRT(VAR(2)/VAR(3))
VALUE = 1.+(((AKTK−1.)*(AKFK−1.))/(SQRT((AKTK−1.)**2.+(AKFK−1.)**2.)))

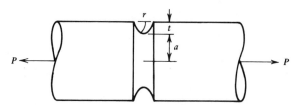

Figure A9.5 Circumferentially notched shaft loaded in tension.

$$\frac{1}{m} = (\text{Poisson ratio}) \qquad N = \frac{a}{r} + \frac{2}{m}\sqrt{\frac{a}{r}+1} + 2$$

$$K_{tk} = \frac{1}{N}\left[\frac{a}{r}\sqrt{\frac{a}{r}+1} + \left(0.5+\frac{1}{m}\right)\frac{a}{r} + \left(1+\frac{1}{m}\right)\left(\sqrt{\frac{a}{r}+1}+1\right)\right]$$

$$K_{fk} = 1+2\sqrt{\frac{t}{r}} \qquad K_t = 1+\frac{(K_{tk}-1)(K_{fk}-1)}{\sqrt{(K_{tk}-1)^2 + (K_{fk}-1)^2}}$$

The FORTRAN equivalent is as follows:
```
B=VAR(1)/VAR(3)
D=SQRT(B+1)
AM=(Poisson ratio)
AN=B+2.*AM*D+2
AKTK=(1./AN)*(B*D+(.5+AM)*B+(1.+AM)*(D+1.))
AKFK=1.+2.*SQRT(VAR(2)/VAR(3))
VALUE=1.+(((AKTK−1.)*(AKFK−1.))/SQRT((AKTK−
1.)**2.+(AKFK−1.)**2.))
```

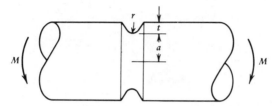

Figure A9.6 Circumferentially notched shaft loaded in bending.

$$\frac{1}{m} = (\text{Poisson ratio})$$

$$N = 3\left(\frac{a}{r}+1\right)+\left(1+\frac{4}{m}\right)\sqrt{\frac{a}{r}+1} + \frac{1+(1/m)}{1+\sqrt{(a/r)+1}}$$

$$K_{tk} = \frac{3}{4N}\left(\sqrt{\frac{a}{r}+1}+1\right)\left[3\frac{a}{r}-\left(1-\frac{2}{m}\right)\sqrt{\frac{a}{r}+1}+4+\frac{1}{m}\right]$$

$$K_{fk} = 1+2\sqrt{\frac{t}{r}} \qquad K_t = 1+\frac{(K_{tk}-1)(K_{fk}-1)}{\sqrt{(K_{tk}-1)^2+(K_{fk}-1)^2}}$$

The FORTRAN equivalent is as follows:
```
B=VAR(1)/VAR(3)
D=SQRT(B+1)
AM=(Poisson ratio)
AN=3.*(B+1)+(1.+4.*AM)*D+(1.+AM)/(1.+D)
AKTK=(3./(4.*AN))*(D+1.)*(3.*B−(1.−2.*AM)*D+4.+AM)
AKFK=1.+2.*SQRT(VAR(2)/VAR(3))
VALUE=1.+(((AKTK−1.)*(AKFK−1.))/SQRT((AKTK−
1.)**2.+(AKFK−1.)**2.))
```

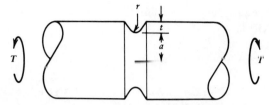

Figure A9.7 Circumferentially notched shaft loaded in torsion.

$$K_{tk} = \frac{3\left[1+\sqrt{(a/r)+1}\right]^2}{4\left[1+2\sqrt{(a/r)+1}\right]} \qquad K_{fk} = 1+\sqrt{\frac{t}{r}}$$

$$K_t = 1 + \frac{(K_{tk}-1)(K_{fk}-1)}{\sqrt{(K_{tk}-1)^2+(K_{fk}-1)^2}}$$

The FORTRAN equivalent is as follows:
B = VAR(1)/VAR(3)
D = SQRT(B+1)
AKTK = (3.*(1.+D)**2.)/(4.*(1.+2.*D))
AKFK = 1.+SQRT(VAR(2)/VAR(3))
VALUE = 1.+(((AKTK−1.)*(AKFK−1.))/SQRT((AKTK−1.)**2.+(AKFK−1.)**2.))

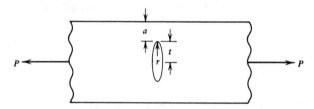

Figure A9.8 Flat plate with a hole loaded in tension.

$$\alpha_1 = \frac{2(a/r+1)\sqrt{a/r}}{[(a/r)+1]\tan^{-1}\sqrt{a/r}+\sqrt{a/r}}$$

$$\alpha_2 = \frac{4(a/r)\sqrt{a/r}}{3\left\{[(a/r)-1]\tan^{-1}\sqrt{a/r}+\sqrt{a/r}\right\}}$$

$$\alpha_3 = \frac{2[(a/r)+1]-\alpha_1\sqrt{a/r+1}}{(4/\alpha_2)[(a/r)+1]-3\alpha_1}$$

$$K_{tk} = \alpha_1 + 0.8\alpha_3 \qquad\qquad K_{fk} = 1 + 2\sqrt{\frac{t}{r}}$$

$$K_t = 1 + \frac{(K_{tk}-1)(K_{fk}-1)}{\sqrt{(K_{tk}-1)^2 + (K_{fk}-1)^2}}$$

The FORTRAN equivalent is as follows:
B = VAR(1)/VAR(3)
C = SQRT(B)
A1 = (2.*(B+1.)*C)/((B+1.)*ATAN(C)+C)
A2 = (4.*B*C)/(3.*((B−1.)*ATAN(C)+C))
A3 = (2.*(B+1.)−A1*SQRT(B+1.))/((4./A2)*(B+1.)−(3.*A1))
AKTK = A1+.8*A3
AKFK = 1.+2.*SQRT(VAR(2)/VAR(3))
VALUE = 1.+(((AKTK−1.)*(AKFK−1.))/SQRT((AKTK−1.)**2.+(AKFK−1.)**2.))

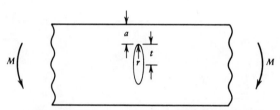

Figure A9.9 Flat plate with a hole loaded in bending.

$$\alpha_1 = \frac{2[(a/r)+1]\sqrt{a/r}}{[(a/r)+1]\tan^{-1}\sqrt{a/r} + \sqrt{a/r}}$$

$$\alpha_2 = \frac{4(a/r)\sqrt{a/r}}{3\{[(a/r)-1]\tan^{-1}\sqrt{a/r}\ \sqrt{a/r}\}}$$

$$\alpha_3 = \frac{2[(a/r)+1]-\alpha_1\sqrt{a/r+1}}{(a/\alpha_2)[(a/r)+1]-3\alpha_1}$$

$$K_{tk} = \alpha_1 + 0.8\alpha_3$$

$$K_{fk} = 1 + \sqrt{\frac{t}{r}} \qquad\qquad K_t = 1 + \frac{(K_{tk}-1)(K_{fk}-1)}{\sqrt{(K_{tk}-1)^2 + (K_{fk}-1)^2}}$$

The FORTRAN equivalent is as follows:
B = VAR(1)/VAR(3)
C + SQRT(B)
A1 = (2.*(B+1.)*C)/((B+1.)*ATAN(C)+C)

A2=(4.*B*C)/(3.*((B−1.)*ATAN(C)+C))
A3=(2.*(B+1.)−A1*(SQRT(B+1.)))/((VAR(1)/A2)*(B+1.)−(3.*A1))
AKTK=A1+0.8*A3
AKFK=1.+SQRT(VAR(2)/VAR(3))
VALUE=1.+(((AKTK−1.)*(AKFK−1.))/SQRT((AKTK−1.)**2.+(AKFK−1.)**2.))

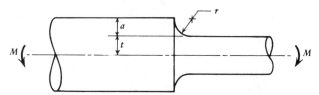

Figure A9.10 Filleted shaft loaded in bending.

$$K_{tk_1} = \frac{4\frac{a}{r} + \sqrt{\frac{a}{r}}}{3\left[\sqrt{\frac{a}{r}} + \left(\frac{a}{r} - 1\right)\tan^{-1}\sqrt{\frac{a}{r}}\right]}$$

$$K_{fk_1} = 1 + 2\sqrt{\frac{a}{r}}$$

$$K_{t_{2g}} = 1 + \frac{(K_{tk_1} - 1)(K_{fk_1} - 1)}{\sqrt{(K_{tk_1} - 1)^2 + (K_{fk_1} - 1)^2}}$$

$$\frac{1}{m} = (\text{Poisson ratio})$$

$$N = 3\left(\frac{a}{r} + 1\right) + \left(1 + \frac{4}{m}\right)\sqrt{\frac{a}{r} + 1} + \frac{1 + (1/m)}{1 + \sqrt{(a/r) + 1}}$$

$$K_{tk_2} = \frac{3}{4N}\left(\sqrt{\frac{a}{r} + 1} + 1\right)\left[3\frac{a}{r} - \left(1 - \frac{2}{m}\right)\sqrt{\frac{a}{r} + 1} + 4 + \frac{1}{m}\right]$$

$$K_{fk_2} = 1 + 2\sqrt{\frac{t}{r}}$$

$$K_{t_{3g}} = 1 + \frac{(K_{tk_2} - 1)(K_{fk_2} - 1)}{\sqrt{(K_{tk_2} - 1)^2 + (K_{fk_2} - 1)^2}}$$

$$K_{t_{2f}} = 1 + \tanh\frac{[(a + t/a) - 1]^{1/4}}{1 - (r/2a)} \frac{0.13 + 0.65[1 - (r/2a)]^4}{(r/2a)^{1/3}}$$

$$K_{t_{3f}} = 1(K_{t_{2f}} - 1)\frac{(K_{t_{3g}} - 1)}{(K_{t_{2g}} - 1)}$$

The FORTRAN equivalent is as follows:

```
B = VAR(1)/VAR(3)              E = (VAR(1) + VAR(2))/VAR(1)
C = SQRT(B)                    F = VAR(3)/(2.*VAR(1))
D = SQRT(B + 1)                EX = 1./3.
AM = .3
AKTK1 = (4.*B*C)/(3.*(C + (B − 1.)*ATAN(C)))
AKFK1 = 1. + 2.*SQRT(VAR(1)/VAR(3))
AKT2G = 1. + (((AKTK1 − 1.)*(AKFK1 − 1.))/SQRT((AKTK1 − 1.)**2.
                                              + (AKFK − 1.)**2.))
AN = 3.*(B + 1.) + (1. + 4.*AM)*D + (1. + AM)/(1. + D)
AKTK2 = (3./(4.*AN))*(D + 1.)*(3.*B − (1. − 2.*AM)*D + 4. + AM)
AKFK2 = 1. + 2.*SQRT(VAR(2)/VAR(3))
AKT3G = 1. + (((AKTK2 − 1.)*(AKFK − 1.))/SQRT((AKTK2 − 1.)**2.
                                              + (AKFK2 − 1.)**2.))
AKT2F = 1. + TANH((E − 1.)**.25/(1. − F))*((.13 + .65*(1. − F)**4.)
                                                         /(F**EX))
VALUE = 1. + (AKT2F − 1.)*((AKT3G − 1.)/(AKT2G − 1))
```

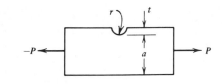

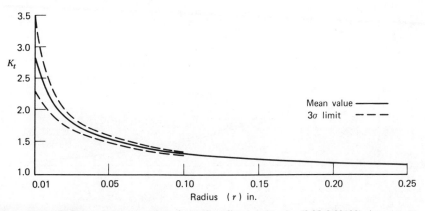

Figure A9.11 Values of K_t as a function of radius r; $(\bar{a}, s_a) = (0.25, 0.00133)$ in., $(t, s_t) = (0.25, 0.00333)$ in., $\bar{r}$ (varies), $s_r = 0.00333$ in.; (a) for a single-notched flat plate loaded in tension, (b) for a single-notched flat plate loaded in bending, (c) for a double-notched flat plate loaded in tension, (d) for a double-notched flat plate loaded in bending, (e) for a hole in a flat bar loaded in bending, (f) for a hole in a flat plate loaded in tension, (g) for a circumferentially notched shaft loaded in tension, (h) for a circumferentially notched shaft loaded in bending, (i) for a circumferentially notched shaft loaded in torsion, (j) for a filleted shaft loaded in bending.

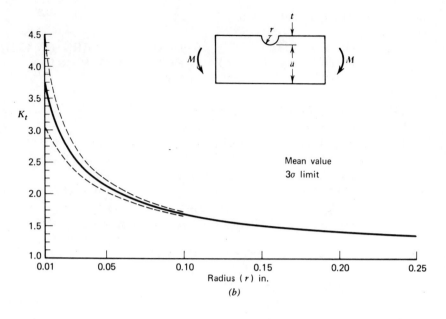

(b)

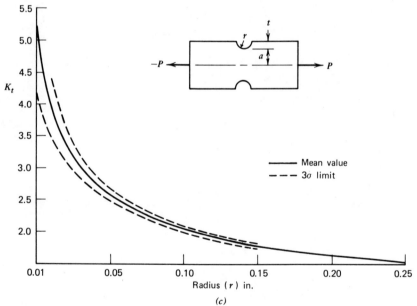

(c)

Figure A9.11 *Continued*

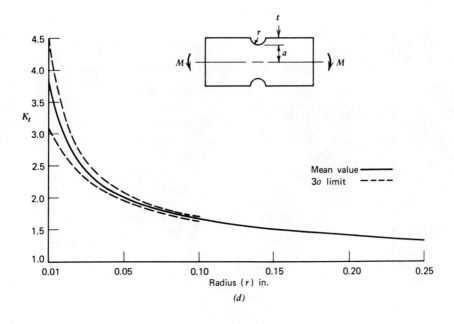

(d)

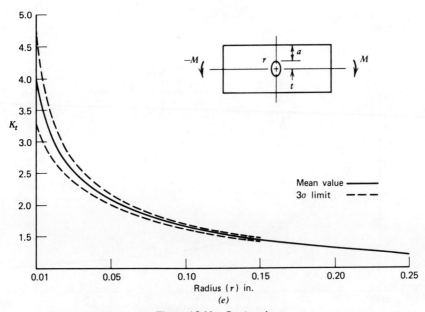

(e)

Figure A9.11 Continued

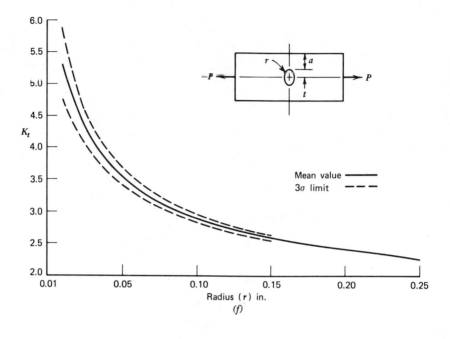

(f)

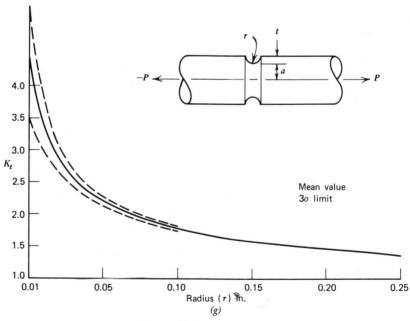

(g)

Figure A9.11 *Continued*

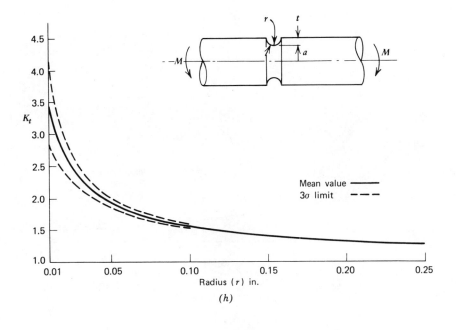

(h)

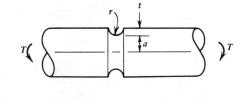

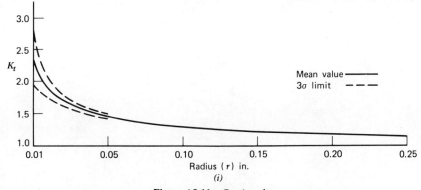

(i)

Figure A9.11 *Continued*

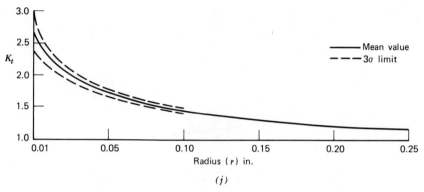

(j)

Figure A9.11 *Continued*

APPENDIX 10.A

Static Strength of Materials

Material	Reference	Remarks	Condition	Ultimate Tensile, S_u (ksi)			Tensile Yield, S_y (ksi)		
				Mean	s	Sample	Mean	s	Sample
2014 (AMS 4135)	54	Aluminum Alloys	Forging	70.0	1.89	36	63.0	2.23	35
2014-T651	127	Plate, L-T	0.50–1.50 in.	85.0	2.76	20	75.0	1.75	20
		Plate, T-L	0.50–1.50 in.	83.0	1.34	20	72.0	2.07	19
2014-T651	127	Plate L-T	Ambient temp.	69.0	1.82	20	63.0	1.35	20
		Plate T-L	Ambient temp.	69.0	1.42	20	61.0	1.65	20
2024-0	68	Clad sheet L-T	0.023–0.067	28.1	1.73	69	13.0	0.89	69
2024-T3	54	Sheet	Bare	67.0	0.93	248			
		Sheet	Clad	50.6	2.94	61			
2024-T4		Clad sheet	0.014–0.022 T-L	60.1	2.53	62	36.1	1.90	6
			0.023–0.028 L-T	66.4	2.42	153	41.4	2.14	153
			T-L	60.9	2.27	73	36.6	1.91	73
	68		0.029–0.035 L-T	64.9	2.07	257	40.1	2.02	257
			T-L	61.6	2.17	116	37.3	1.91	116
			0.068–0.076 L-T	65.8	1.59	64	41.5	2.46	64
2024-T6	68	Clad sheet	0.029–0.035 L-T	63.6	2.51	299	50.1	2.85	299
			0.058–0.067	65.9	2.56	194	51.8	1.82	194
2024-T81	68	Clad sheet	0.029–0.057 L-T	65.9	2.34	18	59.7	3.25	18
			0.068–0.140 L-T	69.9	1.70	15	64.4	2.41	15
2024-T86	68	Clad sheet	0.036–0.045 L-T	71.0	1.48	57	66.9	2.20	57
			0.086–0.096 L-T	75.2	1.53	69	70.5	2.84	69
2219-T87	127	Plate	0.50–1.5 L-T	67.0	1.64	17	54.0	1.53	17
			T-L	67.0	1.85	17	54.0	2.01	17
6061-T4	68	Sheet	0.032–0.125	36.6	1.69	1461	20.1	2.30	1461
6061-T6	68	Sheet	0.032–0.125	45.6	1.91	1648	41.7	2.89	1648
7039-T6	127	Plate—320°F	0.50–1.50 L-T	78.9	0.56	8	65.9	0.94	8
			T-L	80.6	0.65	7	65.3	0.41	7
7039-T6	127	Plate—ambient	0.50–1.50 L-T	62.5	0.86	8	54.9	1.25	8
			T-L	63.7	0.78	7	54.1	0.72	7
7039-T63		Forgings	−320°F	90.8	1.40	85	74.1	1.40	83
			Ambient temp.	65.3	2.16	85	58.4	1.4	83

69			200°F	59.8	2.18	64	53.6	2.27	63
			300°F	43.4	1.40	64	38.7	1.58	63
	7075-T6	Bare sheet		80.5	3.99	908	70.2	3.12	873
54	7075-T6	Clad sheet	0.025 L-T	75.5	2.10	250	63.7	1.98	250
			T-V	73.5	1.84	50	62.6	1.87	50
68			0.040 L-T	76.8	2.04	517	65.1	2.19	517
			T-L	75.6	2.16	248	64.3	1.97	2
			0.125 L-T	76.2	1.08	300	65.3	1.52	300
54	7075-T6 (AMS 4139)		Forging L-T	83.3	3.53	102	73.8	4.16	102
69	7075-T73	Hand forging	Ambient L-T	72.2	2.18	65	60.5	2.32	62
			T-L	71.4	2.19	68	59.0	2.40	68
67	7079-T6	Die forging		76.6	2.06	73	67.4	2.34	73
			11.5-in. section	74.1	1.18	15	62.9	1.50	15
			4.0-in. section	78.0	1.92	19	67.4	2.76	19
67	142-T77	Sand castings	Heat Treated	27.6	2.07	107			
67	355-T6	Sand cast spec.	Heat treated and aged	35.5	2.11	70	25.6	2.08	122
67	355-T71	Sand cast spec.	Heat treated and aged	34.8	1.68	83			
54	356-T6	Permanent mold casting		47.9	1.79	39	40.6	1.46	39
54		Sand castings		37.3	2.66	73	26.7	2.84	73
	355-T6	Sand castings		35.7	1.50	14	26.0	1.85	14

Magnesium Alloys

66	AZ 31 B-0	Sheet L-T	0.016–0.060	40.9	3.81	—	20.2	0.95	—
66	AZ 31 B-0	Sheet ambient	0.061–0.249	40.9	3.81	—	23.9	3.81	—
66	AZ 31-B-H24	Sheet and plate	0.016–0.249 L-T	41.2	0.95	—	31.2	0.95	—
54	AZ 91-T4	Castings	QQ-M-56	38.5	2.26	30	14.3	0.67	30
54	AZ 91-T6	Castings		38.5	1.87	76	18.8	0.98	76
66	HK 31 A-0	Sheet and plate	0.016–0.249	34.4	1.90	—	20.2	0.95	—
69	AZ 318-H24	Sheet and plate	0.016–0.249	41.78	0.97	14800	31.54	1.34	13580

Titanium Alloys

68	Titanium 99%	Commerically pure	T-L	101.9	6.98	1363	85.1	9.03	1354
			L-T	101.9	6.98	1363	85.1	9.03	1354
68	Ti-6Al-4V	Sheet and bar	Annealed 75°F	135.5	6.7	2542	130.6	7.2	2611
7	Ti-6Al-4V		Temperature 572°F	99.1	4.49	462	81.0	5.25	556
68	Ti-6Al-4V	Sheet and bar	Forgings	143.0	4.98	89	135.3	4.75	89

Appendix 10.A *Continued*

Material	Reference	Remarks	Condition	Ultimate Tensile, S_u (ksi)			Tensile Yield, S_y (ksi)		
				Mean	s	Sample	Mean	s	Sample
Titanium Alloys									
Ti-6Al-4V	68	Sheet	Heat treated and aged	175.4	7.91	603	163.7	9.03	603
Ti-6Al-4V	68	Sheet	Solution treated + aged T-L	180.6	11.17	318	164.7	10.96	318
			+ L-T	172.1	8.83	336	154.0	7.71	336
Ti-6Al-2.5SN	68	Sheet	Annealed	127.4	6.79	1640	120.3	7.03	1640
Ti-4Al-3MD-1V	68	Sheet	Solution treated + aged T-L	201.	8.91	1438	173.8	10.63	1438
			+ L-T	200.0	7.91	1426	172.1	10.37	1426
Ti-5Al-2.5SN	69	Ring forging	Annealed—320°F	190.0	2.95	16	175.0	5.87	7
			Ambient	120.0	3.68	34	111.0	4.35	34
Ti-5Al-2.5SN	69	Forging	Radial—320°F	185.0	4.77	8	170.0	7.35	8
			Ambient	117.0	3.09	20	107.0	4.34	22
Ti-5Al-2.5SN	69	Pancake forging	−423°F	210.6	3.52	22	190.6	4.5	32
			−320°F	188.2	3.28	11	174.6	4.21	11
Ti-5Al-2.5SN	67	Ring forging	−320°F	187.5	2.44	31	172.8	4.02	32
			Ambient	118.8	1.45	54	108.9	1.83	53
Ti-4Al-3No-1V	67	Sheet T-L	Solution treated + aged	200.9	8.91	1438	174.5	10.63	1438
		L-T		200.1	7.91	1426	172.4	10.37	1426
Ti-2.5Al-16V	68	Sheet T-L	Solution treated + aged	180.2	8.34	745	167.6	9.02	745
		L-T		176.3	7.85	755	162.3	7.85	755
Ti-6Al-4V 2SN	66	Sheet, Strip, Plate	Annealed ≤ 0.187	150.6	8.7	113	142.9	11.5	112
Ti-5Al-2.8Cr 1.2 Fe	7	Sheet and Plate	Solution treated + aged	187.1	14.9	377	163.6	8.6	355
Carbon Steels									
C1006	29	Sheet	Hot rolled	48.3	0.52	5	35.7	0.80	5
C1018	68	Round bars	Cold drawn	87.6	5.74	50	78.4	5.90	50
C1020	65	Structural steel	Hot rolled	63.7	2.63	1011			
C1035	68	Round bars	Hot rolled, 1–9-in. dia.	86.2	3.92	913	49.5	5.36	899
	68	Bars	Cold drawn	92.1	6.1				
C1038	68	Bolts	3/8-in. dia. × 1.5	133.4	3.38	75			

Material		Description					
C1045	68	Round bars	Cold drawn 3/4–5/4-in. dia.	117.7	7.13	30	95.5	6.59	25
1117	68	Round bars	Cold drawn, 1-in. dia.	83.1	5.25	145	81.4	4.71	135
1137	68	Round bars	Cold drawn, 1-in. dia.	106.5	6.15	145	98.1	4.24	150
ASTM A201	68	Plate	Hot rolled 1/4–2 in.-dia.				37.3	1.43	26
ASTM A212	68	Plate, grade A	Hot rolled, $\frac{1}{4}$–2 in.				44.2	2.18	34
ASTM A212	68	Plate, grade B	Hot rolled, $\frac{1}{4}$–2 in.				45.4	1.89	33
ASTM A285	68	Plate, grade A	Hot rolled, $\frac{1}{4}$–2 in.				32.6	1.46	21
Low-carbon steel	68	Sheet, commercial	Hot rolled, 0.0598 in.	49.1	1.81	135	32.6	2.14	140
Low-carbon steel	68	Sheet, drawing quality	Hot rolled, 0.075 in.	47.4	1.26	205	34.0	2.25	215
Low-carbon steel	68	Sheet, drawing quality	Cold rolled	44.7	1.12	140	25.6	1.87	140
Low-carbon 0.17C, 0.75Mn	68	Casting	As cast	66.7	2.01	200	34.8	0.84	200
Low-carbon 0.17C, 0.75Mn	68	Casting	Normalized and tempered	66.7	2.01	200	34.8	0.84	200
Low-carbon 0.17C, 0.75Mn	68	Casting	Normalized	81.0	3.55	53	52.6	3.54	53
Low-carbon 0.37%, 0.75Mn	68	Casting	Normalized and tempered	100.8	4.80	513	63.1	3.23	513
Medium-carbon 0.25 & 0.45C	68	Casting		99.1	4.71	513	55.3	3.52	513

Low-alloy Steels

Material		Description					
5Cr-Mo-V Alloy		5 Cr–1.5 Mo–0.4V– 0.35C	$\frac{3}{4}$-in.-dia. bar aircraft steel	296.5	1.84	25	240.8	3.10	25
	67	1 Cr–1.00 Mo	Casting normalized and tempered	92.1	4.46	25	65.6	5.35	25
Nickel steel	68	0.3C–1.5 Mn–0.3 Mo	Quenched and tempered	145.8	5.34	50	131.4	5.50	50
Ni-Cr-Mo Alloy	67	0.2C–2.25 Ni Casting	Normalized and tempered	83.2	2.35	200	55.4	1.54	200
Ni-Cr-Mo Alloy	67	0.24C–2.0 Ni–0.8 Cr 0.35 Mo	Normalized and quenched casting	129.2	5.48	22	112.7	7.96	22
	67	0.28C–2Mi–0.8Cr– 0.35 Mo	Normalized, quenched, and tempered	161.1	4.97	44	148.1	4.98	44
High-strength structural steel	68	0.2C–1.25Mn–0.025P 0.25Si–0.25 Cu	0.075–0.25 in.	73.7	5.05	301			
			0.375–1.00 in.	77.5	5.30	437			

Appendix 10.A *Continued*

Material	Reference	Remarks	Condition	Ultimate Tensile, S_u (ksi)			Tensile Yield, S_y (ksi)		
				Mean	s	Sample	Mean	s	Sample
		Low-alloy Steels							
High-strength structural steel	68	0.22C–1.35Mn–0.025P 0.25Si–0.25 Cu	0.50–1.50 in.	76.9	2.06	109	49.6	3.69	113
			As rolled, 0.134–4.0 in.	79.1	4.65	214	51.0	4.63	234
High-strength structural steel	68	0.17C–1.1Mn–0.12P– 0.08Si–0.3Cu–0.05V	Hot rolled, 0.075–0.25 in.	69.8	4.93	312	50.2	3.79	324
			Hot rolled, 0.188–0.625 in.	74.4	3.39	212	50.3	3.04	2.3
High-strength structural steel	68	0.1C–0.75Mn–0.09P–0.1Si– <1.5 in. 1.15Cu–0.6Ni–0.13Mo– 0.2Al		75.4	3.18	306	55.9	2.87	3.69
AISI 4130		Bars, forgings, sheet, plate, and tubing	Ambient	51.8	1.1		57.8	1.1	
			300°F	47.8	1.1		56.8	1.1	
	127		Normalized 500°F	44.8	1.1		54.8	1.1	
			700°F	43.6	2.2		51.8	1.1	
			900°F	37.6	2.2		45.8	1.1	
AISI 4130	4	Rod, $\frac{3}{8}$-in. dia.	Cold drawn and annealed	104.6	1.93	34			
AISI 4340	4	Rod, $\frac{3}{8}$-in. dia.	Cold drawn and annealed	116.4	1.30	33			
AISI 4340	4	Wire, 0.0625-in. dia. 0.0937-in. dia.	Cold drawn, annealed, and straightened	113.0	1.60	50			
				103.4	2.30	50			
AISI 4140 & 4340	127	Bars, forgings, sheet, plate, tubing	Heat treated 140 @ room temp.						
			Ambient	129.3	3.2		89.6	2.2	
			300°F	118.3	3.2		88.3	2.2	
			500°F	109.3	3.2		83.3	2.2	
			700°F	98.3	3.2		70.8	2.2	
			900°F	71.6	3.2		57.6	2.2	
AISI 4140 & 4340	127	Bars, forgings, sheet, plate, and tubing	Heat treated to 160 @ room temp.						
			Ambient	156.1	4.3		104.3	3.2	
			300°F	139.1	4.3		100.3	3.2	
			500°F	124.3	3.2		94.3	3.2	

Material		Form	Composition	Size/Condition	Tensile strength (ksi)	% El		Yield strength (ksi)	% El	BHN
AISI 4140 & 4340	127	Bars, forgings, sheet, plate, and tubing		Heat treated to 180 @ room temp.						
				Ambient	178.9	5.4		117.3	3.2	
				300°F	157.1	4.3		112.3	3.2	
				500°F	143.1	4.3		105.3	3.2	
				700°F	124.3	3.2		89.6	2.2	
				900°F	87.6	2.2		64.6	2.2	
ASTM A7	127	Plate (heavy)		Ambient temp.	64.8	2.93	141	39.8	4.15	141
ASTM A7	54	Structural shapes		Angles, channels, I Beams	65.6	2.43	166	42.7	4.86	166
High-strength structural steel	68	Structural Steel	0.22C–1.35Mn–0.025P–0.25Si–0.25Cu Steel #2							
				0.50–1.50 in.	76.9	2.06	109	49.6	3.69	113
				0.1875–1.50 in.	80.1	4.80	113	49.4	4.43	113
		Plate		As rolled, 0.134–4.00 in.	79.1	4.65	214	51.0	4.63	234
		Coil		Hot rolled, 0.075–0.250 in.	69.8	4.93	312	50.2	3.79	324
High-strength structural steel	68	Plate 0.188–0.625 in.	0.17C–1.1Mn–0.12P–0.08Si–0.3Cu–0.05V	Hot rolled	74.4	3.39	212	50.3	3.04	213
High-strength structural steel	68		0.1C–0.75Mn–0.09P–0.1Si–1.15Cu–0.6Ni–0.13Mo–0.2Al	< 1.5 in.	75.4	3.18	306	55.9	2.87	309
ASTM A7 Structural steel	54	Rolled WF beams		18 WF 50				35.9	1.98	6
				18 WF 55				31.5	1.22	6
				18 WF 60				32.8	1.55	10
				21 WF 62				32.1	1.08	4
				18 WF 96				29.9	1.04	24
ASTM A7	54	Plate, $\frac{7}{16}$ in.						34.8	1.02	24
ASTM A7	54	Structural shapes		Angles, channels, and I-beams	65.6	2.43	166	42.6	4.86	166
Intermediate-grade structural steel	24	$\frac{3}{8}$–1.25-in. Bars (reinforcing)		No. 3–10 Bars	47.7	5.95	171			
				No. 5 Bars	49.9	3.64	20			

Stainless Steels

Material		Form		Condition						
Type 202	67	Strip, austenitic		Cold rolled, 0.017–0.022 in.	99.7	2.71	25	49.9	1.32	25
Type 301	67	Strip and sheet, austenitic		Annealed, 0.038–0.195 in.	105.0	5.68	100	46.8	4.70	100

Appendix 10.A *Continued*

Material	Source	Remarks	Condition	Ultimate Tensile, S_u (ksi)			Tensile Yield, S_y (ksi)		
				Mean	s	Sample	Mean	s	Sample
		Stainless Steel							
Type 301	67	Strip austenitic	Cold-rolled, br. finish	191.2	5.82	17	166.8	9.37	14
Type 301	67	Sheet and strip austenitic	Cold-worked, 20–28%	161.4	9.05	182	118.4	5.55	182
Type 302	67	Sheet 0.003	Annealed	90.4	3.01	45	42.4	3.53	45
Type 304	68	Round bars	Annealed, 0.50–4.625 in.	85.0	4.14	45	37.9	3.76	35
Type 304	68	Tubing, austenitic	Annealed, 0.125–1.00-in. OD, 0.155–0.188-in. wall	89.1	4.22	204	40.4	4.50	204
Type 310	68	Sheet, austenitic	Annealed, 0.040–0.100 in.	84.8	4.23	45			
Type 316	68	Sheet, austenitic	Annealed, 0.040–0.120	84.0	2.92	41			
Type 321	68	Sheet, austenitic	0.030–0.070 in.	86.8	4.10	80			
Type 310	69	Bar 0.75-in. dia.	Annealed @ 2150°F	182.0	5.0	4	99.5	1.34	4
			–423°F	156.0	3.9	3	70.0	1.41	3
			–320°F	85.0	3.94	6	32.0	2.13	6
			Ambient						
		Sheet 0.062 in.	Annealed @ 2150°F Ambient	84.8	2.98	7	40.3	1.56	7
Type 310	69	Pancake forging	Annealed @ 1900°F	82.46	1.11	108	42.04	2.98	108
Type 347	68	Tubing, austenitic	Annealed, 0.125–1.00-in. OD 0.15–0.188-in. wall	93.8	3.47	149	46.4	7.45	36
Type 403	68	Bar, ferritic	Hot rolled, 0.750–2.125 in.	109.7	4.48	549	81.2	6.86	549
Type 410	68	Bar, ferritic	Annealed, 0.125–1.00-in. OD 0.015–0.188-in. wall	105.3	3.09	204	78.5	3.91	168
Type 410	68	Tubing, ferritic	–100°F	72.4	3.96	88			
Type 347	69	Forging	70°F	118.0	3.34	23	34.5	1.15	187
			600°F	81.0	1.45	18	29.8	1.15	187
			1200°F	56.3	1.33	24	21.3	1.15	187
			–302°F	43.6	0.61	22	18.7	1.15	187
Type 347	69	Sheet	75°F Longitudinal				54.5	2.63	33
			800°F				37.3	2.16	36
							25.7	1.80	33

Material	Ref	Form	Condition						
Type 430	67	Sheet, 0.001 in.		72.6	2.08	54	42.4	2.42	54
Type 446	67	Tubing, ferritic	Annealed, 0.125–1.00-in. OD 0.015–0.188-in. wall						
17-7 PH	68	Sheet	0.016–0.062 in.	81.8	2.41	60			
17-7 PH	68	Sheet	0.005–0.010 in.	198.8	9.51	106	189.4	11.47	103
AM-350	68	Sheet	Annealed, 0.025–0.125 in.	199.0	12.61	101	193.6	13.47	102
AM-355	68	Bar stock		152.8	6.70	194	63.4	4.04	188
				186.0	9.18	44	171.6	8.47	36

High-strength Steels and Super Alloys

Material	Ref	Form	Condition						
Inconel 718	69	Welded sheet	Solution treated @ 1950°F	173.0	2.39	15	141.2	2.34	15
718 Alloy	69	Forging, pancake	Annealed and aged @ 1950°F						
			−423°F	259.0	4.97	35	198.0	7.48	33
			−320°F	240.0	4.97	22	186.0	7.48	22
			Ambient	186.0	4.97	13	163.0	7.48	13
718 Alloy	69	Forgings	Annealed, −320°F	244.5	1.55	8	197.1	1.68	4
A-286 Alloy	67	Bar	Treated @ 1650°F, quenched	111.0	12.1	30	92.1	8.16	30
A-286 Alloy	69	Forgings, bolts	Treated @ 1650°F						
			Ambient	153.0	2.87	8	102.2	2.25	8
			Treated and cold worked	202.0	4.3	12	180.4	5.6	12
Hasteloy C	65	Investment cast.	1500°F, Short time S_u	55.0	3.15	219			
Rene 41	54		Ambient	193.4	8.7	803			
Thelaloy	68	As cast	Ambient, 1500°F	75.4	5.27	250			
				54.4	3.39	250			
Udimet	67	Nickel-based alloy	Rolled bar 1200°F	178.3	2.56	80			
Hastelloy X	69	Plate	−320°F	135.2	2.01	4	71.2	1.77	4

Heat-resistant Alloys

Material	Ref	Form	Condition						
SAE 70309 (HH Alloy)	67	26% Cr–12% Ni	As cast	83.2	4.73	80	45.8	3.97	80
SAE 70310 (HK Alloy)	67	26% Cr–20% Ni	As cast	72.4	5.12	183	43.5	2.90	366
S-816	67	42% Co–20% Cr–20% Ni–etc.	1350°F	91.0	3.56	24			
HS-21		62% Co–27% Cr–etc.	As cast, 1500°F	61.0	4.46	302			

Appendix 10.A Continued

Material	Reference	Remarks	Condition	Ultimate Tensile, S_u (ksi)			Tensile Yield, S_y (ksi)		
				Mean	s	Sample	Mean	s	Sample
HS-31 (X-40)	67	55% Co–25% Cr– 10% Ni–8% V	As cast, 1500°F	62.7	4.55	30			
Thetalloy	68	50% Ni–25% Cr– 12.5% Co–etc.	As cast, Ambient 1500°F	75.4 54.4	5.27 3.39	250 250			
Hastelloy C	68	55% Ni–17% Mo– 16% Cr–etc.	Invest. casting 1500°F	55.0	3.15	299			
Udinet 500	67	50% Ni–19% Cr 19% Co–etc.	Rolled bar, 1200°F	178.3	2.56	80			
M-252	67	54% Ni–19% Cr–10% Co–10% Mo–etc.	Rolled bar, 1350°F	153.4	5.84	89			
A-286	67	15% Cr–26% Ni– 1.25 Mo–etc.	Bar, heat treated and quenched	111.0	12.1	30	92.1	8.16	30
		Corrosion-resistant Steel							
CA-15	68	Corrosion-resistant alloy	Casting	107.0	5.64	23	80.7	4.31	23
CF-8	68	Corrosion-resistant alloy	Casting	77.4	4.8	292	37.8	4.20	292
CK-20	68	Corrosion-resistant alloy	Casting	76.8	5.48	138	38.2	6.18	138
CN-7M	68	Corrosion-resistant alloy	Casting	70.4	3.96	37	32.9	3.26	37
		Cast Iron							
Malleable Iron Grade, 32510	68	Ferritic Grade 32510	Foundry A Foundry B	53.4 56.0	2.68 1.41	434 785	34.9 39.1	1.47 1.22	434 785
Malleable, iron grade 35018	68	Ferritic	Grade 35018	53.3	1.59	159	38.5	1.42	159
Malleable iron	68	Pearlitic 217–233 BHN	Spheroidized	93.9	4.33	172	61.8	2.48	172
Malleable	68	Pearlitic	Spheroidized	89.1	3.61	161	60.1	2.19	161

Nodular iron grade 60-45-10	67	As cast	Grade 60-45-10	73.2	5.37	412	54.5	4.65	412
Nodular iron	68	Nodular iron	Grade 60-45-15	62.2	1.55	385	49.0	4.20	385
Nodular iron	68	Nodular iron	Grade 100-70-04	122.2	7.65	55	79.3	4.51	55
Nodular iron	68	As cast	Auto crankshafts	99.3	6.14	125	74.9	4.62	125

Source: For additional materials design data, see Refs. 7, 13, 54, 67, 68, 69, and the Material Properties Data Center, Traverse City, Michigan.

APPENDIX 10.B

Dynamic Strength of Materials

Materials behavior in this appendix [78, 79] is modeled by (1) mean values and standard deviation $\mu = \bar{S}_f$ and $\sigma = \,_{>}S_f$ and (2) Weibull parameters x_0, b, and θ. Since the statistics $\bar{S}_f$ and $\,_{>}S_f$ are usually employed in engineering applications the transformation [equation (8.6a)] is utilized:

$$S_f = x_0 + (\theta - x_0)\Gamma\frac{1+1}{b}$$

$$C_{S_f} = \frac{\,_{>}S_f}{\bar{S}_f} = \frac{\bar{S}_f - x_0}{\bar{S}_f} b^{-0.926}, \quad \text{for} \quad 0.70 < b < 10.0$$

Figure 8.30 provides values of $\Gamma[(1+1)/b]$ as a function of b.

AISI 1045 Steel—S_u = 105, 120 ksi, Rotary Bending

HT[a]	S_u	Spec.	x_0 (ksi)			b			θ (ksi)		
		Cycles	10^4	10^5	10^6	10^4	10^5	10^6	10^4	10^5	10^6
1.	105	No–N	79.0	67.0	56.7	2.6	2.75	2.85	86.2	73.0	61.65

[a] Heat treatment:
   1. WQ for 1520°F tempered at 1210°F, S_u = 105 ksi and S_y = 82 ksi.

AMS 5727 Steel—S_u = 120 ksi, S_y = 30 ksi, Axial Load, Completely Reversed

T (°F)	K_t		x_0 (ksi)			b			θ (ksi)		
		Cycles	10^4	10^5	10^6	10^4	10^5	10^6	10^4	10^5	10^6
80	1.0		68	61	55	2.5	2.75	2.8	74	67	61
1200	1.0		51	47	44	2.85	2.85	3.0	58	54	50
80	2.4		59	42	30	2.75	2.95	2.98	66	47	34
1200	2.4		29	24	20	2.21	2.3	2.32	36	31	26

D_{6AC} Steel—$S_u = 270$ ksi, $S_y = 231$ ksi, Axial Load, Completely Reversed

T (°F)	K_t	X_0 (ksi)			b			θ (ksi)		
	Cycles	10^4	10^5	10^6	10^4	10^5	10^6	10^4	10^5	10^6
80	1.0	160	90	60	2.8	2.9	3.0	191	106	60
450	1.0	145	115	90	3.15	3.3	4.6	162	129	102
550	1.0	125	100	78	3.7	3.8	4.0	161	125	98
80	3.0	52	40	30	3.3	3.4	3.8	82	66	53
450	3.0	73	41	34	2.1	2.54	3.4	81	46	40
550	3.0	63	44	35	2.75	3.0	3.4	70	51	40

M10 Tool Steel—$S_u = 330$ ksi, Rotary Bending

HT	K_t	Spec.	X_0 (ksi)			b			θ (ksi)		
	Cycles		10^4	10^5	10^6	10^4	10^5	10^6	10^4	10^5	10^6
2.	1.0	No–N	152	133	117	2.67	2.7	2.73	185.7	163.7	144
3.	1.0	No–N	127	119	111	1.89	1.95	2.0	163.5	153.2	143.5

Heat treatment:
  2. Preheat 1450°F $\frac{1}{2}$ hr, harden 2150°F 5 min, OQ until black, AC, temp. 1100°F 2 hr, AC retemp. 1100°F 2 hr, AC, after finishing op. Nitrided 975°F 48 hr.
  3. Same as step 2 but instead of nitriding, stress relieve at 1000°F in protective atmosphere FC.

A-286 Stainless Steel—$S_u = 90$ ksi, $S_y = 46$ ksi, Axial Load, Completely Reversed

T (°F)	K_t	X_0 (ksi)			b			θ (ksi)		
		10^4	10^5	10^6	10^4	10^5	10^6	10^4	10^5	10^6
80	1.0	40	31	24	1.84	2.1	2.2	54	43	34
−320	1.0	61	58	54	2.3	2.34	2.45	78	74	70
−423	1.0	73	67	62	2.12	2.3	2.33	84	78	72
80	3.5	30	22		2.7	2.9		36	27	
−320	3.5	35	25	17	1.58	1.58	1.8	56	39	27
−423	3.5	30	12	5	2.15	2.25	2.37	86	36	15

347 Stainless Steel—$S_u = 92$ ksi, $S_y = 46$ ksi, Axial Load

K_t	S_m	X_0 (ksi)				b				θ (ksi)			
Life	Cycles	10^3	10^4	10^5	10^6	10^3	10^4	10^5	10^6	10^3	10^4	10^5	10^6
1.0	25–35			28	26			1.47	1.55			32	30
2.0	25–35	54	35	23		3.9	5.2	5.6		58	38	30	
4.0	25–35	32	22	15		2.9	2.7	2.25		36	25	17	
1.0	38–45	38	36	35		4.3	4.7	4.8		43	41	39	
2.0	38–45	42	34	28		4.9	5.5	5.8		47	39	32	

PH 15-7 MO Stainless Steel—$S_u = 201$ ksi, $S_y = 196$ ksi, Axial Load, Completely Reversed

T (°F)	K_t	X_0 (ksi)			b			θ (ksi)		
		Cycles 10^4	10^5	10^6	10^4	10^5	10^6	10^4	10^5	10^6
80	1.0	145	104	75	1.68	1.73	1.78	169	119	86
500	1.0	120	73		2.8	2.85		129	79	
80	4.0	42	31	23	2.35	2.4	2.43	56	41	30
500	4.0	36	32	29	2.55	2.58	2.7	41	37	32

17-7PH—$S_u = 205$ ksi, $S_y = 195$ ksi, Axial Load, Completely Reversed

K_t	Spec.	X_0 (ksi)			b			θ (ksi)		
		Cycles 10^4	10^5	10^6	10^4	10^5	10^6	10^4	10^5	10^6
1	No–N	90.0	82.0	63.0	8.3	6.0	6.0	153.6	126.6	88.6

AISI 2340 Steel—$S_u = 116$–122 ksi, $S_y = 76$–96 ksi, Rotary Bending

HT	Spec.	S_y	S_u	X_0 (ksi)			b			θ (ksi)		
				Cycles 10^4	10^5	10^6	10^4	10^5	10^6	10^4	10^5	10^6
4.	No–N	96	116	84.0	74.0	61.0	4.3	3.4	4.1	101.5	85.4	72.0
5.	No–N	79	119	81.0	71.0	62.0	2.8	2.8	2.8	88.7	77.6	67.8
6.	No–N	76	122	87.0	75.0	64.0	4.4	4.4	4.9	94.4	81.55	70.5

Heat treatment:
    4. Oil quenched from 1450°F, tempered at 1200°F.
    5. Air blast quenched from 1450°F, tempered at 700°F.
    6. Air-blast quenched from 1450°F, no tempering.

AISI 3140 Steel—$S_u = 108, 109$ ksi, $S_y = 87, 75$ ksi, Rotary Bending

HT	Spec.	X_0 (ksi)				b				θ (ksi)			
		Cycles 10^3	10^4	10^5	10^6	10^3	10^4	10^5	10^6	10^3	10^4	10^5	10^6
7.	No–N	89	74	66	57	3.7	5.2	5	5.5	100.4	87.7	76.7	67.2

Heat treatment:
    7. OQ from 1520°F, tempered at 1300°F, $S_u = 108$ ksi, $S_y = 87$ ksi.

4140 Steel—$S_u = 135$ ksi, $S_y = 122$ ksi, Rotary Bending

	X_0 (ksi)				b				θ (ksi)			
Cycles	10^4	10^5	10^6	10^7	10^4	10^5	10^6	10^7	10^4	10^5	10^6	10^7
Unnotched	74	68	63	58	2.4	2.5	2.7	2.8	90.8	83.5	76.8	70.4

4340 Steel—$S_u = 206$–280 ksi, Rotary Bending

HT	Spec.	S_u	X_0 (ksi)				b				θ (ksi)			
		Cycles	10^4	10^5	10^6	10^7	10^4	10^5	10^6	10^7	10^4	10^5	10^6	10^7
8.	8	No–N	246	75	64	57	49	3.05	3.3	3.5	3.8	128.5	114	101.3	90
9.	8	No–N	222	77	65	55	45	2.75	2.85	3.0	3.2	113.5	94.5	79.4	66.2

For heat treatment:
  8. Normalize 1550°F, OQ, temper at 400°F AC.
  9. Normalize 1550°F, quench, temper at 775°F AC.

4340 Steel—$S_u = 144$–290 ksi, Rotary Bending

HT	K_t	S_u	X_0 (ksi)			b			θ (ksi)		
		Cycles	10^4	10^5	10^6	10^4	10^5	10^6	10^4	10^5	10^6
10.	1.0	158	100	72	53	2.55	2.8	3.0	124.8	93.1	67.9
11.	1.0	171	122	100	84	2.4	2.7	3.15	140.3	117.2	99.2
12.	1.0	275	100	85	75	2.2	2.35	2.5	138.9	119.5	102
13.	1.0	290	133	123	113	2.55	2.7	3.1	165.9	153.5	142.9

Heat treatment:
  10. Normalize at 1600°F, 2 hr, AC; aust. 1500°F, 2 hr, OQ, temper at 1150°F, 4 hr, AC.
  11. Normalize at 1600°F, 1 hr.
  12. Austenitize at 1550°F, salt bath 20 min, OQ to 120–150°F, temper at 400°F, 4 hr melt practice—electric furnace.
  13. Austenitize at 1550°F, salt bath 20 min, OQ to 120–150°F, temper at 400°F, 4 hr melt practice—vacuum furnace.

THERMOLD J—$S_u = 294$ ksi, Axial Load, Completely Reversed

K_t	Spec.	X_0 (ksi)					b					θ (ksi)				
		Cycles	10^3	10^4	10^5	10^6	10^7	10^3	10^4	10^5	10^6	10^7	10^3	10^4	10^5	10^6	10^7
1.0	No–N				100	102	90			4.4	2.4	2.1			143	124.3	108.3
2.0	V–N		120	85	62	59		5.2	5.9	5.9	6		139.7	100.3	72.7	69.1	
3.0	V–N		98	73	54			5.2	5.5	5.4			112.2	82.8	61.8		

Fe–5.5 Mo–2.5 Cr–0.5C—$S_u = 314$ ksi, $S_y = 267$ ksi, Rotary Bending

K_t	Spec.	X_0 (ksi)				b				θ (ksi)			
		Cycles	10^4	10^5	10^6	10^7	10^4	10^5	10^6	10^7	10^4	10^5	10^6	10^7
1.0	No–N	160	150	140	130	3.8	3.6	3.6	3.6	195.8	182	169.1	156.5
2.6	V–N	72.60	69.0	66.0	63.0	5.0	4.8	4.8	4.8	82.9	79.3	75.6	72.1

321 Stainless Steel—$S_u = 38$ ksi, $S_y = 38$ ksi, Axial Load, Completely Reversed

K_t	T (°F)	X_0 (ksi)				b				θ (ksi)			
	Cycles	10^3	10^4	10^5	10^6	10^3	10^4	10^5	10^6	10^3	10^4	10^5	10^6
1.0	80	44.0	39.6	32.4	27.9	1.5	1.4	1.4	1.4	47.15	42.5	34.8	29.9

Multiment N-155 Stainless Steel—$S_u = 114$–126 ksi, $S_y = 60$–73 ksi, Rotary, Plate Bending and Axial Loading, Completely Reversed

Surface Finish (RMS)	T of L	Surface Prep.	S_u	X_0 (ksi)			b			θ (ksi)		
			Cycles	10^5	10^6	10^7	10^5	10^6	10^7	10^5	10^6	10^7
4		M.P.	119	52	48	45	4	4.2	4.8	61.9	57.8	53.8
22		M.P.	119	52	46	41	2.6	2.8	3.05	60.5	54.5	48.8
75		M.P.	119	59	55	51	2.8	2.95	3.2	64.9	60.8	57

1100 Aluminum—$S_u = 10$ ksi, $S_y = 9$ ksi

Test Conditions						Fatigue Strength Distributions and Parameters	
Heat treat.	Surf. fin.	Freq.	Temp.	K_t	Other	Dist.	Param.	Life in Cycles*
								1×10^3 1×10^4 5×10^4 1×10^5 5×10^5 1×10^6 5×10^6 1×10^7 1×10^8

Rotary-beam Bending—miscellaneous

Milled #600 Gr.	80	1.0		Weibull	b	3.470 3.460 3.460 3.460 3.460 3.460 3.460 3.460^a
					θ	7.318 5.416 4.757 3.521 3.092 2.289 2.010 1.307
					x_0	3.077 2.284 2.006 1.484 1.304 0.9653 0.8479 0.5512

2014 Aluminum—$S_u = 68$–78 ksi, $S_y = 60$–71 ksi

Test Conditions						Fatigue Strength Distributions and Parameters	
Heat treat.	Surf. fin.	Freq.	Temp.	K_t	Other	Dist.	Param.	Life in Cycles
								1×10^3 1×10^4 5×10^4 1×10^5 5×10^5 1×10^6 5×10^6 1×10^6 1×10^7 1×10^8

Rotary-beam Bending—Miscellaneous

T-6 Wrought		80	1.0	Weibull	b	1.921^a 1.781 1.682 1.636 1.770 1.522 1.642 1.689 1.824^a
					θ	75.93 63.92 56.68 53.81 47.75 45.31 40.20 38.18 32.17
					x_0	68.38 57.85 51.46 48.94 43.23 41.34 36.55 34.66 29.07

2024 Aluminum (Sheet or Plate)—$S_u = 61$–81 ksi, $S_y = 45$–61 ksi

Test Conditions						Fatigue Strength Distributions and Parameters	
Heat treat.	Surf. fin.	Freq.	Temp.	K_t	Other	Dist.	Param.	Life in Cycles
								1×10^3 1×10^4 5×10^4 1×10^5 5×10^5 1×10^6 5×10^6 1×10^6 1×10^7 1×10^8

Plate Bending (Completely Reversed)–Miscellaneous

T-4	m. p.	30	80	1.0	SP-8	Normal	σ	3.123^a 2.376^a 2.112^a 1.607 1.428 1.087 0.996 0.6536^a
							μ	57.64 43.85 38.98 29.66 26.36 20.06 17.83 12.06

2024 Aluminum (Bars)—$S_u = 61$–81 ksi, $S_y = 45$–61 ksi

Test Conditions							Fatigue Strength Distributions and Parameters								
Heat treat.	Surf. fin.	Freq.	Temp.	K_t	Other	Dist.	Param.	\multicolumn{9}{c}{Life in Cycles}
								1×10^3	1×10^4	5×10^4	1×10^5	5×10^5	1×10^6	5×10^6	1×10^7	1×10^8

Rotary-beam Bending

Heat treat.	Surf. fin.	Freq.	Temp.	K_t	Other	Dist.	Param.	1×10^3	1×10^4	5×10^4	1×10^5	5×10^5	1×10^6	5×10^6	1×10^7	1×10^8
T-4	Ground	25	80	3.9	Cold rd.	Normal	σ	4.684^a	3.815	3.306	3.108	2.693	2.532	2.194	2.026^a	1.680^a
							μ	25.46	20.74	17.97	16.89	14.63	13.76	11.92	11.21	9.133

2024 Aluminum (Bars)—$S_u = 61$–81 ksi, $S_y = 45$–61 ksi

Axial (Completely Reversed)

Heat treat.	Surf. fin.	Freq.	Temp.	K_t	Other	Dist.	Param.	1×10^3	1×10^4	5×10^4	1×10^5	5×10^5	1×10^6	5×10^6	1×10^7	1×10^8
T-4			80	1.0		Normal	σ	3.818^a	3.020	2.564	2.389	2.028	1.890	1.604	1.495^a	1.183
							μ	62.30	49.28	41.84	38.99	33.10	30.84	26.18	24.40	19.30

2026 Aluminum—$S_u = 77.5$ ksi, $S_y = 69.5$ ksi

Axial (Completely Reversed)

Heat treat.	Surf. fin.	Freq.	Temp.	K_t	Other	Dist.	Param.	1×10^3	1×10^4	5×10^4	1×10^5	5×10^5	1×10^6	5×10^6	1×10^7	1×10^8
Fully ht. trt.			80	1.0		Normal	σ	2.485^a	2.140^a	1.928^a	1.843	1.660	1.587	1.429	1.366	1.177^a
							μ	49.12	42.30	38.10	36.42	32.81	31.36	28.25	27.01	23.26

*Note. 1×10^3 indicates one thousand (10^3) cycles of life. $1\times10^4 = 10{,}000$ cycles, $5\times10^4 =$ fifty thousand cycles of life, etc. $1\times10^8 =$ one hundred million cycles of life. The above indications of cycle life will be used in the remainder of Appendix 10B.

2219 Aluminum—$S_u = 59$–67 ksi, $S_y = 46$–54 ksi

Axial (Completely Reversed)

Heat treat.	Surf. fin.	Freq.	Temp.	K_t	Other	Dist.	Param.	1×10^3	1×10^4	5×10^4	1×10^5	5×10^5	1×10^6	5×10^6	1×10^7	1×10^8
T-62	m. p.		80	3.5		Normal	σ	2.014^a	1.268	0.9173	0.7979	0.5773	0.5201	0.3633	0.3160^a	0.1988^a
							μ	28.56	17.97	13.00	11.31	8.183	7.118	5.150	4.494	2.819

5083 Aluminum—$S_u = 50$ ksi, $S_y = 34.4$ ksi

Axial (Completely Reversed)—Miscellaneous

Heat treat.	Surf. fin.	Freq.	Temp.	K_t	Other	Dist.	Param.	1×10^3	1×10^4	5×10^4	1×10^5	5×10^5	1×10^6	5×10^6	1×10^7	1×10^8
H-113			80	1.0		Weibull	b		1.440	1.669	1.490^a	1.069^a	1.355^a	$.9732^a$	1.669^a	
							θ		44.44	40.78	33.33	30.52	24.98	22.83	17.26	
							x_0		42.78	39.06	32.05	29.58	24.09	22.21	16.48	

5086 Aluminum—$S_u = 45$ ksi, $S_y = 30$ ksi

Test Conditions							Fatigue Strength Distributions and Parameters								
Heat treat.	Surf. fin	Freq.	Temp.	K_t	Other	Dist.	Param.	\multicolumn{8}{c}{Life in Cycles}
								1×10^3	1×10^4	5×10^4	1×10^5	5×10^5	1×10^6	5×10^6	1×10^7	1×10^8

Rotary-beam Bending—Miscellaneous

Heat treat.	Surf. fin	Freq.	Temp.	K_t	Other	Dist.	Param.									
H-34	m. p.	33	80	1.0		Normal	σ	2.372^a	1.990	1.845	1.547	1.434	1.203	1.215^a	0.667^a	
							μ	37.05	31.06	28.79	24.15	22.39	18.78	17.41	13.53	

7001 Aluminum—$S_u = 100$ ksi, $S_y = 90$ ksi

Axial (Completely Reversed)

Heat treat.	Surf. fin	Freq.	Temp.	K_t	Other	Dist.	Param.									
T-75 Forged		30	80	1.0	Longitudinal	Normal	σ	4.707^a	3.681	3.099	2.878	2.424	2.251	1.895	1.760^a	1.376^a
							μ	60.65	47.43	39.94	37.09	31.23	29.00	24.42	22.68	17.73

7075 Aluminum—$S_u = 71$–87 ksi, $S_y = 67$–79 ksi

Rotary-beam Bending

Heat treat.	Surf. fin	Freq.	Temp.	K_t	Other	Dist.	Param.									
0° F5 24 hrs.		20	80	1.0		Normal	σ	7.583^a	5.836^a	4.859	4.491	3.740	3.456	2.878	2.660	2.047
							μ	77.75	59.70	49.71	45.94	38.26	35.35	29.44	27.21	20.94

7079 Aluminum—$S_u = 70$–79 ksi, $S_y = 63$–66 ksi

Rotary-beam Bending—Miscellaneous

Heat treat.	Surf. fin	Freq.	Temp.	K_t	Other	Dist.	Param.									
T-652	m. p. 10 RMS		80	1.0	Forged	Weibull	b	3.113^a	3.040	3.059	3.218	3.130	3.096	3.230	3.204^a	3.132^a
							θ	63.26	48.75	40.83	37.58	31.32	28.96	24.14	22.32	17.20
							x_0	44.73	34.71	28.88	26.30	22.11	20.51	16.87	15.64	12.14

7079 Aluminum—$S_u = 70$–79 ksi, $S_y = 63$–66 ksi

Axial (Completely Reversed)—Miscellaneous

Heat treat.	Surf. fin	Freq.	Temp.	K_t	Other	Dist.	Param.									
T-6	m. p. 10 RMS		80	1.0	Drawn	Normal	σ	6.466^a	5.555	4.995	4.772	4.291	4.099	3.686	3.521	3.025^a
							μ	43.50	37.37	33.60	32.10	28.87	27.58	24.80	23.69	20.35

2.5 Al Magnesium

Test Conditions						Fatigue Strength Distributions and Parameters									
Heat treat.	Surf. fin.	Freq.	Temp.	K_t	Other	Dist.	Param.	\multicolumn{8}{c	}{Life in Cycles}
								1×10^3	1×10^4	5×10^4	1×10^5	5×10^5	1×10^6	5×10^6	1×10^7	1×10^8

Rotary-beam Bending

Heat treat.	Surf. fin.	Freq.	Temp.	K_t	Other	Dist.	Param.	Values
			80	1.0		Normal	σ	2.273^a 1.892^a 1.664^a 1.575 1.385 1.311 1.153 1.091 0.9086^a
							μ	13.35 11.11 9.778 9.252 8.139 7.702 6.775 6.412 5.337

AZ61A Magnesium—$S_u = 45$–46 ksi, $S_y = 29$–34 ksi

Rotary-beam Bending

Heat treat.	Surf. fin.	Freq.	Temp.	K_t	Other	Dist.	Param.	Values
m.p.	25		80	1.0		Weibull	b	1.735^a 1.832 1.789 1.718 1.832 1.818^a 1.712^a 1.832^a 1.832^a
							θ	42.57 31.82 25.95 23.76 19.39 17.76 14.48 13.27 9.920
							x_0	35.77 26.57 21.73 19.99 16.20 14.85 12.19 11.08 8.285

AZ61Z Magnesium—$S_u = 45$–46 ksi, $S_y = 29$–34 ksi

Plate Bending (Completely Reversed)

Heat treat.	Surf. fin.	Freq.	Temp.	K_t	Other	Dist.	Param.	Values
HT-16		29	80	1.0		Normal	σ	4.008^a 3.395^a 3.023^a 2.876 2.561 2.436 2.169 2.063 1.747^a
							μ	23.11 19.57 17.43 16.58 14.76 14.04 12.50 11.89 10.07

Hastelloy (RL35)—$S_u = 64$ ksi

Axial (Completely Reversed)

Heat treat.	Surf. fin.	Freq.	Temp.	K_t	Other	Dist.	Param.	Values
HT-19		60	1200	1.0		Normal	σ	3.540^a 3.222 3.018 2.933 2.747 2.670 2.501 2.431 2.213^a
							μ	56.79 51.70 48.42 47.07 44.08 42.85 40.13 39.01 35.51

Incoloy 901 (AMS-5560A)—$S_u = 182$ ksi, $S_y = 116$ ksi

Axial (Completely Reversed)

Heat treat.	Surf. fin.	Freq.	Temp.	K_t	Other	Dist.	Param.	Values
450°F		30	600	1.0	Bored	Normal	σ	5.726^a 4.169^a 3.339 3.035 2.431^a 2.209^a 1.770^a 1.608^a 1.171^a
							μ	123.1 89.67 71.83 65.28 52.29 47.52 38.07 34.60 25.19

Inconel-X (AMS-5667) — S_u = 186–225 ksi, S_y = 146–194 ksi

Test Conditions							Fatigue Strength Distributions and Parameters								
Heat treat.	Surf. fin.	Freq.	Temp.	K_t	Other	Dist.	Param.	\multicolumn{9}{c}{Life in Cycles}
								1×10^3	1×10^4	5×10^4	1×10^5	5×10^5	1×10^6	5×10^6	1×10^7	1×10^8

Rotary-beam Bending

| HT-22 | | 183 | 80 | 1.0 | Vacuum | Normal | σ | 15.06^a | 11.21^a | 9.124^a | 8.349^a | 6.794 | 6.217 | 5.058 | 4.629 | 3.446^a |
| | | | | | | | μ | 179.1 | 133.4 | 108.5 | 99.32 | 80.82 | 73.95 | 60.17 | 55.06 | 41.00 |

Inconel-X (AMS-5667) — S_u = 186–225 ksi, S_y = 146–194 ksi

Axial (Completely Reversed) — Miscellaneous

Heat treat.	Surf. fin.	Freq.	Temp.	K_t	Other	Dist.	Param.	Life in Cycles
			80	1.0		Normal	σ	6.772^a 4.787 3.775 3.408 2.688 2.427^a 1.914^a 1.728^a 1.230^a
							μ	148.9 106.0 83.63 75.51 59.55 53.77 42.41 38.29 27.26

FS-27 Nickel-base Cermet — S_u = 75 ksi

Plate Bending (Completely Reversed)

Heat treat.	Surf. fin.	Freq.	Temp.	K_t	Other	Dist.	Param.	Life in Cycles
			1800	1.0		Weibull	b	1.593^a 1.374 1.404 1.406 1.632 1.589 1.493 1.454 1.664^a
							θ	36.83 30.92 27.39 25.99 23.05 21.87 19.36 18.37 15.45
							x_0	31.98 27.11 23.97 22.75 19.91 18.94 16.86 16.03 13.32

Hastelloy C — S_u = 120 ksi, S_y = 65 ksi

Axial (Completely Reversed)

Heat treat.	Surf. fin.	Freq.	Temp.	K_t	Other	Dist.	Param.	Life in Cycles
			80	1.0		Weibull	b	1.855 1.855 1.855 1.855 1.909 1.882 1.885^a $1.88t^a$
							θ	79.26 69.67 63.66 61.24 55.97 53.83 49.19 47.31
							x_0	67.52 59.35 54.23 52.17 47.53 45.79 41.90 40.31

Inconel 718 — S_u = 121–225 ksi, S_y = 75–189 ksi

Axial (Completely Reversed)

Heat treat.	Surf. fin.	Freq.	Temp.	K_t	Other	Dist.	Param.	Life in Cycles
		50	1400	1.0		Normal	σ	3.024^a 2.913 2.839 2.807 2.735 2.705 2.635 2.606 2.511^a
							μ	48.36 46.60 45.40 44.90 43.75 43.26 42.15 41.68 40.16

Inconel 718—$S_u = 121–225$ ksi, $S_y = 75–189$ ksi

Test Conditions								Fatigue Strength Distributions and Parameters									
Heat treat.	Surf. fin.	Freq.	Temp.	K_t	Other	Dist.	Param.	Life in Cycles									
								1×10^3	1×10^4	5×10^4	1×10^5	5×10^5	1×10^6	5×10^6	1×10^7	1×10^8	
								Axial (Completely Reversed)									
		50	80	3.0		Weibull	b	1.088^a	$.7352^a$	1.053^a	$.5474^a$	0.7763	0.8313	0.9320	0.9679^a	1.069^a	
							θ	37.78	32.75	29.89	28.53	25.94	24.91	22.66	21.75	18.99	
							x_0	34.00	30.00	26.95	26.22	23.73	22.73	20.57	19.71	17.10	

Waspalloy—$S_u = 185$ ksi, $S_y = 125$ ksi

Test Conditions								Fatigue Strength Distributions and Parameters									
Heat treat.	Surf. fin.	Freq.	Temp.	K_t	Other	Dist.	Param.	Life in Cycles									
								1×10^3	1×10^4	5×10^4	1×10^5	5×10^5	1×10^6	5×10^6	1×10^6	1×10^7	1×10^8
								Rotary-beam Bending									
HT-25			80	1.0		Normal	σ	9.185^a	7.641^a	6.719^a	6.356^a	5.589	5.288	4.650	4.399	3.659^a	
							μ	128.6	107.0	94.11	89.04	78.29	74.07	65.13	61.62	51.26	

Rene-41 (AMS-5713)—$S_u = 182–197$ ksi, $S_y = 131–150$ ksi

Test Conditions								Fatigue Strength Distributions and Parameters									
Heat treat.	Surf. fin.	Freq.	Temp.	K_t	Other	Dist.	Param.	Life in Cycles									
								1×10^3	1×10^4	5×10^4	1×10^5	5×10^5	1×10^6	5×10^6	1×10^6	1×10^7	1×10^8
								Axial (Completely Reversed)									
HT-33			80	1.0		Normal	σ	4.529^a	3.991^a	3.618	3.468	3.144	3.014	2.732^a	2.619^a	2.276^a	
							μ	136.6	118.7	107.6	103.2	93.59	89.72	81.34	77.97	67.77	

Udimet-500—$S_u = 182–188$ ksi, $S_y = 111–147$ ksi

Test Conditions								Fatigue Strength Distributions and Parameters									
Heat treat.	Surf. fin.	Freq.	Temp.	K_t	Other	Dist.	Param.	Life in Cycles									
								1×10^3	1×10^4	5×10^4	1×10^5	5×10^5	1×10^6	5×10^6	1×10^6	1×10^7	1×10^8
								Rotary-beam Bending—Miscellaneous									
HT-34		183	80	1.0		Normal	σ	9.187^a	6.621^a	5.266^a	4.771^a	3.795	3.439	2.735	2.478	1.786^a	
							μ	170.7	123.0	97.89	88.70	70.55	63.92	50.84	46.07	33.20	

Commercially Pure Titanium (Ti-A55)—$S_u = 70–120$ ksi, $S_y = 57–65$ ksi

Test Conditions								Fatigue Strength Distributions and Parameters									
Heat treat.	Surf. fin.	Freq.	Temp.	K_t	Other	Dist.	Param.	Life in Cycles									
								1×10^3	1×10^4	5×10^4	1×10^5	5×10^5	1×10^6	5×10^6	1×10^6	1×10^7	1×10^8
								Rotary-beam Bending									
		30	80	1.0		Normal	σ	4.714^a	4.073	3.677	3.519	3.177	3.041	2.745^a	2.627^a		
							μ	93.03	80.38	72.58	69.46	62.71	60.01	54.188	51.85		

Ti–4 Al–3 Mo–1 V — $S_u = 119$–201 ksi, $S_y = 84$–180 ksi

Test Conditions							Fatigue Strength Distributions and Parameters								
Heat treat.	Surf. fin.	Freq.	Temp.	K_t	Other	Dist.	Param.	\multicolumn{9}{c}{Life in Cycles}
								1×10^3	1×10^4	5×10^4	1×10^5	5×10^5	1×10^6	5×10^6	1×10^7	1×10^8
\multicolumn{8}{l}{Axial (Completely Reversed)}									
HT-43	Ground	30	400	1.0	0.063 in. Thick	Normal	σ	19.51^a	13.49	10.42	9.328	7.207	6.449	4.982	4.458^a	
							μ	129.7	89.67	69.28	61.99	47.90	42.86	33.11	29.63	

Ti–4 Al–3 Mo–1 — $S_u = 119$–201 ksi, $S_y = 84$–180 ksi

Test Conditions							Fatigue Strength Distributions and Parameters								
Heat treat.	Surf. fin.	Freq.	Temp.	K_t	Other	Dist.	Param.	\multicolumn{9}{c}{Life in Cycles}
								1×10^3	1×10^4	5×10^4	1×10^5	5×10^5	1×10^6	5×10^6	1×10^6	1×10^7	1×10^8
\multicolumn{8}{l}{Axial (Completely Reversed)}									
HT-43		30	600	1.0	0.020 in. Thick	Normal	σ	19.88^a	15.46	12.97	12.02	10.08	9.353	7.845	7.273^a	
							μ	102.5	79.74	66.88	62.01	52.01	48.22	40.45	37.50	

Ti–4 Al–3 Mo–1 — $S_u = 119$–201 ksi, $S_y = 84$–180 ksi

Test Conditions							Fatigue Strength Distributions and Parameters								
Heat treat.	Surf. fin.	Freq.	Temp.	K_t	Other	Dist.	Param.	\multicolumn{9}{c}{Life in Cycles}
								1×10^3	1×10^4	5×10^4	1×10^5	5×10^5	1×10^6	5×10^6	1×10^6	1×10^7	1×10^8
\multicolumn{8}{l}{Axial (Completely Reversed)}									
		30	600	1.0	0.063 Thick	Normal	σ	27.89^a	18.65	14.08	12.48	9.425	8.350	6.305	5.586^a	
							μ	123.7	82.75	62.48	55.36	41.80	37.03	27.98	24.77	

Ti–5–Al-2.5 Sn-0.07 N$_2$ — $S_u = 134$ ksi, $S_y = 126$ ksi

Test Conditions							Fatigue Strength Distributions and Parameters								
Heat treat.	Surf. fin.	Freq.	Temp.	Other	Dist.	Param.	\multicolumn{9}{c}{Life in Cycles}
							1×10^3	1×10^4	5×10^4	1×10^5	5×10^5	1×10^6	5×10^6	1×10^6	1×10^7	1×10^8
\multicolumn{7}{l}{Rotary-beam Bending}									
HT-44	cal & polish	80		1.0		Normal	σ	7.116^a	6.479^a	6.068	5.898	5.524	5.370	5.029	4.889	
							μ	102.1	92.97	87.07	84.65	79.27	77.07	72.17	70.16	

Ti–6 Al–4 V — $S_u = 132$–170 ksi, $S_y = 128$–160 ksi

Test Conditions							Fatigue Strength Distributions and Parameters								
heat treat.	Surf. fin.	Freq.	Temp.	K_t	Other	Dist.	Param.	\multicolumn{9}{c}{Life in Cycles}
								1×10^3	1×10^4	5×10^4	1×10^5	5×10^5	1×10^6	5×10^6	1×10^6	1×10^7	1×10^8
\multicolumn{8}{l}{Rotary-beam Bending}									
HT-46		134	80	1.0		Normal	σ	5.989^a	5.754	5.594	5.527	5.374	5.310	5.163	5.101	
							μ	107.5	103.3	100.4	99.25	96.50	95.35	92.71	91.59	

Ti–8 Al–1 Mo–1 V—$S_u = 147$ ksi, $S_y = 133$ ksi

Test Conditions							Fatigue Strength Distributions and Parameters								
Heat treat.	Surf. fin.	Freq.	Temp.	K_t	Other	Dist.	Param.	\multicolumn{8}{l}{Life in Cycles}
								1×10^3	1×10^4	5×10^4	1×10^5	5×10^5	1×10^6	5×10^6	1×10^7	1×10^8
							\multicolumn{9}{l}{Axial (Completely Reversed)}
HT-53		20	80	10		Normal	σ	5.955^a	4.949	4.570	3.798	3.507	2.9.4^a	2.691^a	
							μ	119.1	99.02	91.43	75.98	70.16	58.31	53.84	

Ti–16 V–2.5 Sn—$S_u = 107\text{–}186$ ksi, $S_y = 78\text{–}173$ ksi

Test Conditions							Fatigue Strength Distributions and Parameters								
Heat treat.	Surf. fin.	Freq.	Temp.	K_t	Other	Dist.	Param.	\multicolumn{8}{l}{Life in Cycles}
								1×10^3	1×10^4	5×10^4	1×10^5	5×10^5	1×10^6	5×10^6	1×10^7	1×10^8
							\multicolumn{9}{l}{Axial (Completely Reversed)}
							σ	19.37	14.95	12.47	11.54	9.629	8.907	7.432	6.875^a
							μ	113.0	87.24	72.79	67.33	56.18	51.97	43.36	40.11

^{a}Extrapolated values.

References

[1] J. L. Shigley, *Mechanical Engineering Design*, 3rd ed., McGraw-Hill, New York, 1977.
[2] E. V. Krick, *An Introduction to Engineering and Engineering Design*, Wiley, New York, 1967.
[3] *Fatigue Design Handbook*, Society of Automotive Engineers, Two Pennsylvania Plaza, New York, N.Y., 1968.
[4] D. Kececioglu and E. Haugen, *Interactions Among the Various Phenomena Involved in the Design of Dynamic and Rotary Machinery and Their Effects on Reliability*, 2nd technical report to the ONR, Washington, D. C., July 1969.
[5] H. R. Jaeckel and S. R. Swanson, *Random Loads Spectrum Test to Determine Durability of Structural Components of Automotive Vehicles*, Ford Motor Company, Dearborn, Michigan, and MTS Systems Corporation, Minneapolis, Minnesota, 1970.
[6] *A Statistical Summary of Mechanical-Property Data For Titanium Alloys*, Defense Metals Information Center, Battelle Memorial Institute, Columbus 1, Ohio, February 1961.
[7] *Airplane Strength and Rigidity Reliability Requirements, Repeated Loads and Fatigue*, Mil-A-8866 (ASG), May 1960, Military Specifications.
[8] J. W. Brewer, A. L. Austin, and M. L. Brewer, *Introduction to Socio-Technological Systems Analysis*, Tech Report, University of California, Department of Mechanical Engineering, Davis, California, 1970.
[9] E. B. Haugen, *Probabilistic Approaches to Design*, Wiley, New York, 1968.
[10] G. E. Ingram, "A Basic Approach For Structural Reliability," Eleventh National Symposium on Reliability and Quality Control, August 1964, Arinc Research Corp.
[11] H. D. Brunk, *An Introduction to Mathematical Statistics*, Ginn, New York, 1960.
[12] J. M. H. Olmsted, *Real Variables*, Appleton-Century-Crofts, New York, 1959.
[13] G. J. Hahn and S. S. Shapiro, *Statistical Models in Engineering*, Wiley, New York, 1967.
[14] G. W. Snedecor, *Statistical Methods*, (5th ed.,) Iowa State University Press, Ames, 1962.
[15] A. H. Bowker and G. J. Lieberman, *Engineering Statistics*, Prentice Hall, Englewood Cliffs, N. J., 1959.
[16] E. B. Haugen, "Statistical Algebra for Engineering Applications," 35th Session of the International Statistical Institute, Belgrade, Yugoslavia, September 1965.
[17] G. Heckman and D. Dietrich, "The Probabilistic Pythagorean Theorem," thesis, The University of Arizona, Tucson, 1972.
[18] L. A. Bryan and E. B. Haugen, "Numerical Method for Evaluating the Variance of a Complicated Function," National Annual Meeting, American Statistical Association, Montreal, Canada, August 1972.
[19] F. Scheid, *Theory and Problems of Numerical Analysis*, McGraw-Hill, New York, 1968.
[20] E. Knierim and E. Haugen, *Algebra of Lognormal Random Variables*, thesis, The University of Arizona, Tucson, 1971.
[21] J. Aitchison and J. A. C. Brown, *The Lognormal Distribution*, Cambridge University Press, London, 1969.
[22] E. Haugen and S. Eckert, *Reliability Error Due to Assumed Normality of Stress and Strength in Mechanical Systems*, 129th Annual Conference of The American Statistical Association, New York, August 1969.

[23] A. Hald, *Statistical Theory with Engineering Applications*, Wiley, New York, 1952.
[24] J. R. Benjamin and C. Allin Cornell, *Probability, Statistics and Decision for Civil Engineers*, McGraw-Hill, New York, 1970.
[25] J. B. Scarborough, *Numerical Mathematical Analysis*, John Hopkins University Press, Baltimore, 1962.
[26] A. F. Madayag, Ed., *Metal Fatigue*, Wiley, New York, 1969.
[27] R. C. Juvinall, *Stress, Strain, and Strength*, McGraw-Hill, New York, 1967.
[28] J. Taylor, *Manual on Aircraft Loads*, Pergamon, New York, 1965.
[29] S. R. Swanson, "Random Load Fatigue Testing: A State of the Art Survey," *Materials Research and Standards*, Vol. 8, No. 4, ASTM, April 1968.
[30] M. Bratt, G. Reethof, and G. Weber, *A Model for Time Varying and Interfering Stress Strength Probability Density Distributions with Consideration for Failure Incidence and Property Degradation*, Large Jet Engine Department, General Electric Company, 1967.
[31] J. E. Hayes, *Structural Design Criteria for Boost Vehicles by Statistical Methods*, SID-64J-290, North American Aviation Inc., Space Division, March 4, 1965.
[32] S. L. Bussa, N. J. Sheth, and S. R. Swanson, "Development of a Random Load Life Prediction Model," *Materials Research and Standards*, ASTM, March 1972.
[33] K. D. Wood, *Aerospace Vehicle Design*, Vol. I, Johnson Publishing Company, Boulder, Colo., 1968.
[34] E. Parzen, *Stochastic Processes*, Holden Day, San Francisco, 1962.
[35] E. G. Kirkpatrick, *Quality Control for Managers and Engineers*, Wiley, New York, 1970.
[36] G. H. Lee, *Experimental Stress Analysis*, Wiley, New York, 1950.
[37] J. Dally and W. Riley, *Experimental Stress Analysis*, McGraw-Hill, New York, 1965.
[38] J. S. Bendat, *The Application of Statistics to the Flight Vehicle Vibration Problem*, Ramo Woolridge Corporation, ASD Technical report No. 61-123, December 1961.
[39] *Handbook of 400 Series Reliability Procedures*, Nerva Program, Contract SNP-1, Aerojet General Corporation, Sacramento, California, 1970.
[40] *Tables of Normal Probability Functions*, National Bureau of Standards, Applied Mathematics Series 23, U.S. Government Printing Office, Washington, D. C., 1953.
[41] S. R. Swanson, "A Survey of Fatigue Testing Using Stationary and Non-Stationary Random Vibration," 84th Meeting of the Acoustical Society of America, November 28, 1972.
[42] R. M. Olsen, *Engineering Fluid Mechanics*, International Textbook Company, Scranton, Pa., 1966.
[43] H. J. Grover, S. A. Gordon, and L. R. Jackson, *The Fatigue of Metals and Structures*, NAVWEPS 00-25-534, Battelle Memorial Institute, 1 June 1967.
[44] G. Sines and J. L. Waisman, *Metal Fatigue*, McGraw-Hill, New York, 1959.
[45] S. H. Crandall and W. D. Mark, *Random Vibration in Mechanical Systems*, Academic, New York, 1963.
[46] J. L. Morrison and G. Z. Libertiny, *Notes on Fatigue*, Tech Report, University of Bristol, England, 1967.
[47] D. Kececioglu and E. Haugen, *First Technical Report*, Office of Naval Research, Contract N00014-67-A-0209-0002, April 1968.
[48] H. O. Fuchs et al., *Shortcuts in Cumulative Damage Analysis*, SAE, Automobile Engineering Meeting, Detroit, Mich., May 1973.
[49] N. F. Spotts, *Design of Machine Elements*, 4th ed., Prentice Hall, Englewood Cliffs, N. J., 1971.
[50] D. W. Dudley, *Gear Handbook*, McGraw-Hill, New York, 1967.
[51] J. Datsko, *Material Properties and Manufacturing Process*, Wiley, New York, 1966.
[52] S. D. Volkov, *Statistical Strength Theory*, Gordon and Breach, New York, 1962.
[53] S. Timoshenko, *Strength of Materials*, Part I, D. Van Nostrand, New York, 1948.
[54] E. Haugen and D. Mutti, *Statistical Metals Manual*, The University of Arizona, Tucson, Arizona, 1973 (unpublished communication).

References

[55] D. Karnapp and F. C. Valls, "Structural Reliability Predictions Using Finite Element Programs," *Annals of Reliability and Maintainability*, 1971. Tech Report

[56] *Kent's Mechanical Engineers Handbook*, 12th ed., Wiley, New York, 1967.

[57] R. A. Anderson, *Fundamentals of Vibration*, Macmillan, New York, 1967.

[58] *Steel Construction Manual*, 5th ed., AISC, New York, 1973.

[59] R. E. Petersen, *Stress Concentration Design Factors*, Wiley, New York, 1974.

[60] *Theory of Elasticity*, S. Timoshenko and J. Goodier, 3rd ed., McGraw-Hill, New York, 1970.

[61] Heinz Neuber, *Theory of Notch Stresses*, J. W. Edwards Publishers, Ann Arbor, Mich., 1946.

[62] J. E. Wells, *Stress Concentration in Probabilistic Design*, Master's report, The University of Arizona, June 1972.

[63] P. Kuhn and H. Hadrath, *An Engineering Method for Estimating Notch Size Effect in Fatigue Tests on Steel*, NACA, technical note No. 2805, October 1952.

[64] E. B. Haugen and P. H. Wirsching, "Probabilistic Fatigue Design Alternative to Miner's Cumulative Damage Rule," 1973 Annual Reliability and Maintainability Symposium Proceeding, Philadelphia, January 23–25, 1973.

[65] V. M. Faires, *Design of Machine Elements*, Macmillan, New York, 1965.

[66] *Metallic Materials and Elements for Flight Vehicle Structures*, Mil-Handbook 5B, Department of Defense, Washington, D. C., September 1, 1971.

[67] *The Metals Handbook*, 8th ed., American Society of Metals, Metals Park, Ohio, 1961.

[68] C. R. Mischke, *Some Tentative Weibullian Descriptions of the Properties of Steels, Aluminums, and Titaniums*, Engineering Research Institute, Iowa State University, Ames, 1971.

[69] *Manual of Materials Data Release Memorandum*, Nerva Program, SNP-1, Aerojet General Corporation, Sacramento, Calif., 1971.

[70] J. Rumpf and J. Fisher, Calibration of A325 Bolts, *Journal of Structural Division of ASCE*, **89**, (ST6) (December 1963).

[71] Damage Tolerance Design Handbook, compiled by Battelle Columbus Laboratories, September 1973. National Technical Information Service, U.S. Department of Commerce, Springfield, Va. 22151.

[72] H. O'Neill, *Hardness Measurement of Metal and Alloys*, Chapman and Hall, London, 1967.

[73] *Achievement of High Fatigue Resistance in Metals and Alloys*, ASTM, special technical publication No. 467, p. 190, 1970.

[74] E. B. Haugen, "Statistical Definition of Combined Stress Fatigue Strength of Low Alloy Steels," International Conference on Mechanical Behavior of Materials, Kyoto, Japan, 1971.

[75] R. B. Heywood, *Designing Against Fatigue*, distributed in the United States by Barnes and Noble; printed in Great Britain by Jarrold and Sons, London, 1962.

[76] W. J. Harris, *Metallic Fatigue*, Pergamon, London, 1961.

[77] R. Ikegami, *The Fatigue of Aluminum Alloys Subjected to Random Loadings*, University of California, Berkeley, Ph.D. thesis, 1969.

[78] C. Lipson, N. J. Sheth, and R. Disney, *Reliability Prediction–Mechanical Stress/Strength Interference (Ferrous)*, TR RADC-TR-66-710, March 1967, AD 813574.

[79] C. Lipson, N. J. Sheth, R. Disney, and M. Alton, *Reliability Prediction–Mechanical Stress/Strength Interference, (Nonferrous)*, TR RADC-TR-68-403, February 1969, AD 856021.

[80] A. M. Freudenthal, *Safety, Safety Factors, and Reliability of Mechanical Systems*, Columbia University, New York.

[81] C. Lipson and N. J. Sheth, *Statistical Design and Analysis of Experiments*, McGraw-Hill, New York, 1973.

[82] J. C. Whittaker, *Development of Titanium and Steel Fatigue Variability Model for Application of Reliability Analysis Approach to Aircraft Structures*, The Boeing Company, Technical Report AFML-TR-72-236, 1965.

References

[83] R. Smith, M. Hirshberg, and S. S. Manson, *Fatigue Behavior of Materials Under Strain Cycling in Low and Intermediate Life*, NASA technical note No. D-1574.

[84] E. Haugen and J. Kritz, "A Redefinition of Endurance Life Design Strength Criteria By Statistical Methods," *Proceedings of the Air Force Conference on Fatigue and Fracture of Aircraft Structures and Materials*, Miami, Florida, September 1970.

[85] Charles R. Mischke, *Experiencing System and Community Level Design*: Step I—*Organization of the Computer*, Tech Report, Iowa State University, Ames, June 1973.

[86] S. L. Bussa, *Fatigue Life of a Low Carbon Steel Notched Specimen Under Stochastic Conditions*, Advance Testing Laboratory, Ford Motor Company, 1967.

[87] A. Ang and W. Tang, *Probability Concepts in Engineering Planning and Design*, Wiley, New York, 1975.

[88] B. M. Hillberry, *Fatigue of 2024-T3 Aluminum Alloy Due to Broad-Band and Narrow-Band Random Loading*, Iowa State University of Science and Technology, Ph.D. dissertation, 1967, University of Microfilms, Ann Arbor, Michigan.

[89] G. J. Peterson, *Design Refinement and Experimental Research on Wire Fatigue Machines*, Master's report, Department of Aerospace and Mechanical Engineering, The University of Arizona, Tucson, 1969.

[90] *Saturn Launch Vehicle Reliability Study*, Arinc Research Corporation, publication no. 141-2-199, NASA, December 1960.

[91] R. J. Roark, *Just How Safe Are You?*, Product Engineering, August 16, 1965.

[92] C. Allin Cornell, "Bayesian Statistical Decision Theory and Reliability Based Design," International Conference on Safety and Reliability of Engineering Structures, Washington, D. C., April 1969.

[93] B. R. Ellingwood, and A. H. S. Ang, *A Probabilistic Study of Safety Criteria For Design*, Civil Engineering Studies, Structural Research Series No. 387, University of Illinois, Urbana, June 1972.

[94] H. E. Davis, G. Troxell, and C. Wiskocil *The Testing and Inspection of Engineering Materials* (*Local Buckling of Round Tubes*), McGraw-Hill, New York, 1955.

[95] *Error in Assumption of Normality*, Nerva Program Procedure #R101-NRP-601, Aerojet General Company, Sacramento, Calif., 1972.

[96] *Computer Applications of Statistical Methods in Engineering Design*, Chasis Engineering Design Office, Ford Motor Company, Dearborn, Michigan, January 1971.

[97] D. Kececioglu and E. Haugen, *Fatigue Reliability Design Data for Dynamic and Rotary Machines*, ASME, United Engineering Center, 345 E. 47th Street, New York, 70-Av/SpT-36.

[98] *Unified and American Screw Threads*, ASA B1, published by the ASME, 1-1960.

[99] Channcey Starr, *An Overview of the Problem of Public Safety*, National Academy of Engineering, 1970.

[100] Joseph Marin, *Mechanical Behavior of Engineering Materials*, Prentice Hall, Englewood Cliffs, N. J., 1962.

[101] Charles Mischke, "A Rationale for Mechanical Design to a Reliability Specification," *Proceedings of the First Technology Transfer Conference of ASME*, 1974.

[102] W. L. Collins and J. O. Smith, "Effects of Temperature on Fatigue Limit of Gray Iron," *Proceedings of ASTM*, **41**, 797 (1941).

[103] M. R. Spiegel, *Theory and Problems of Statistics*, Schanms Outline p. 123, 1961.

[104] Carl C. Osgood, *A Basic Course in Fracture Mechanics*, Machine Design, Cleveland, Ohio, July 22, 1971.

[105] G. Wadsworth and J. Bryan, *Introduction to Probability and Random Varibles*, McGraw-Hill, New York, 1960.

[106] *AFBMA Standards*, Section 9, Anti-Friction Bearing Mfg. Association, New York, 1959.

[107] Tedric A. Harris, "Predicting Bearing Reliability," *Machine Design*, **35** (1), 129–132 (January 3, 1963).

[108] P. H. Wirsching and E. B. Haugen, "Probabilistic Design for Random Fatigue Loads," *Journal of Engineering Mechanics* (in press).

[109] P. H. Wirsching and E. B. Haugen, "A General Statistical Model for Random Fatigue," *Journal of Engineering Materials and Technology*.
[110] James Miller, "Low Cycle Fatigue Under Biaxial Strain Controlled Conditions," *JMLSA*, 7 (3), 307–314 (September 1972).
[111] R. L. Maxwell, *Kinematics and Dynamics of Machinery*, Prentice Hall, Englewood Cliffs, N. J., 1964.
[112] Leo Kovalevsky, *Statistical Analysis of Statically Indeterminate Structures*, STR-163, North American Rockwell Company, Downey, Calif., 1966.
[113] Ralph I. Stevens, *Linear Elastic Fracture Mechanics and Its Application to Fatigue*, Tech Report, The University of Iowa, Ames, 1970.
[114] K. E. Hofer, "Equations for Fracture Mechanics," *Machine Design* (February 1, 1969).
[115] H. Liebowitz, Ed., *Fracture, An Advanced Treatise*, Vols. IV and V, Academic, New York, 1960.
[116] Daniel Daves, *Application of Probabilistic Design Procedures to Fracture Mechanics*, M.S. thesis, The University of Arizona, Tucson, July 1973.
[117] "Mechanical Behavior of Materials (Fracture Mechanics)," *Proceedings of the International Conference on Mechanical Behavior of Materials*, The Society of Materials Science, Kyoto, Japan, August 15–20, 1971.
[118] T. W. Crooker, "The Roles of Fracture Mechanics in Fatigue Design," ASME paper No. 76-DE-5, April 1976.
[119] L. Cooper and D. Steinberg, *Introduction to a Method of Optimization*, Saunders, Philadelphia, 1970.
[120] R. C. Johnson, *Optimum Design of Machine Elements*, Wiley, New York, 1961.
[121] Charles R. Mischke, *An Introduction to Computer-Aided Design*, Prentice-Hall, Englewood Cliffs, N. J., 1968.
[122] W. F. Brown and J. E. Srawley, *Plane Strain Crack Toughness Testing of High Strength Metallic Materials*, ASTM Special Technical Publication No. 410, 1966.
[123] Mary G. Natrella, *Experimental Statistics*, National Bureau of Standards Handbook 91, U.S. Printing Office, Washington, D. C., August 1, 1963.
[124] C. Allin Cornell, Structural Safety Specifications Based on Second Moment Reliability Analysis, MIT Cambridge, Mass., 1969.
[125] S. S. Manson, *Thermal Stress and Low Cycle Fatigue*, McGraw-Hill, New York, 1966.
[126] E. L. Grant, *Statistical Quality Control*, 3rd ed., McGraw-Hill, New York, 1964.
[127] *Structures Reliability Report, Methods and Data Design Manual*, Vol. 1, Martin Marietta Corporation, Baltimore, December 1961.

Index

Note: Numbers in parentheses are reference numbers and indicate that the author's work is referred to although his name is not mentioned in the text.

Accuracy, cost of, 150
Analysis of stress, 166
Ang and Ellingwood, 345
Approximate calculations, accuracy, 77

Ball bearing stresses, 219
Bayes procedure, 87
Bayes's theorem in design, 430
Beam deflection, concentrated load, 231
Biaxial stress static, 195
Bowker, A., 456(15), 553(15)

Camp Meidell, inequality, 76
Castigliano's Theorm, 247
Central limit theorem, 82, 88
Chebysheff's inequality, 76
Closure, 58
Column, hinged and flat ended, 261
   random behavior, 251
Confidence, 557
  in design, 367
  interval, 546, 547
  limits, 546
Constant life diagram, 339
Contact stress equations, 217
Convergence, 59
Cooper, L., 515(119)
Cornell, A., 365(92)
Cormier, D., 361
Correlated random variables, 18
Correlated R. V.'s: products, 47
  quotients, 53
  sums, 41
Correlation, 101
Correlation-coefficients, 18
Coupling formula, 365

Crandall and Mark, 319(45), 433(45)
Critical damping, 266
  load, 253
  unit load, 256
Cylinder, thin walled, 452

Data reduction, 82
Deflection: of beams, 228
  model, 225
  random variables, 222
  with superposition, 233
Design, 1
  classical versus probabilistic, 13
  criterion, 11
  deterministic approach, 12
  development, 7
  economic considerations, 15
  effort, 2
  equations, 520
  factors, 456
  finite life, 419, 472
  for endurance, 399
  for failure, 385
  for non-normal stress, 386
  mechanical, 1
  preliminary, 11
  procedure, 368
  reliability, 15
  theoretical validity, 14
  theoritical k_t, 282
  theory, 9
Dietrich, D., 550(17)
Differences: of correlated R. V.'s, 44
  of independent R. V.'s, 43
  of mean and variance, 29
Dimensional combinations, 141

Dimensional magnitude, 129
Distortion energy theory, 446
Distribution: exponential, 18
  gamma, 18
  lognormal, 16
  normal, 18
  Rayleigh, 22
  two-parameter, 16
  two-parameter Weibull, 18
Dynamic strength, 318
Dynamic stress: models, 198
  multi-axial, 200

Elastic limit, 294
Endurance limit, 321
  versus finite life, 477
  torsional, 333
Ergodic random process, 126
Euler's formula, 253
Euler's models, 257
Expectation, algebra of, 25, 34
Expected value, 22

Failure modes, gear tooth, 210
  probability, 362
Failure theories, 438
Failure theory comparison, 448
Faires, V. M., 295(65), 315(65), 318(65)
Fatigue Design Handbook, 308(3), 331(3), 461(3)
Fatigue failure, 467
  research, 5
  strength, 321
  stresses, random, 196
Finite element analysis, 5
Flow diagram, Monte Carlo, 158
Fluctuating stress, 334
Force random variable, 94
Fracture mechanics, 5
Fracture toughness, 309
  threshold, 310
Frequency ratio, 267
Function, moment generation, 35

Gear tooth: design, 424
  terminology, 210
General reliability model, 360
Geometric random variables, 128
Gerber fatigue envelope, 337
Goodman diagram, 336
Goodness of fit test, 86
Graphical estimates, 87

Hahn, G., 323, 387, 433
Hardness: versus mechanical properties, 315
  Rockwell, 4
Harris, W. J., 319(76)
Haugen, E., 306, 325, 330, 397, 398, 399
Heckmann, C., 550
Heywood, R. B., 319(75), 333(75), 334(75), 454(75)

Ikegami, R., 319(77)
Independent and correlated loads, 103
Indeterminate problems, 236
Instability, 251
Invariance of stress sums, 175

Johnson, R. C., 520(120)
Juvinall, R. C., 334(27), 538(27), 438(27), 454(27), 459(27)

Kececioglu, D., 306, 325(47), 330(47), 361, 397(47), 398(47), 399(47)
K_t by Monet Carlo methods, 284
K_t statistics, 283

Law of large numbers, 130
Laws of combination, 57
Lewis dynamic stress model, 220
Libertiny, G., 335
Lieberman, G., 456, 553
Limit equations, 521
Limiting tolerances, gears, 148
Limit load, 80
Limits of engineering variables, 76
Lipson, C. and Sheth, 323, 340
Lipson, C., Sheth, and Disney, 456(78), 321(78), 329(78), 456(79), 462(78)
Loading, 79
Load measurement, 82
  modifying parameters, 116
Load record ensembles, 122
Loads: constant amplitude, 104
  dynamics, 108
  Histogram, 83
  narrow band random, 111
Lognormal density function, 69
Lognormal-random variables, 68
Low cycle fatigue, 327

Madayag, A. F., 323(26)
Manson, S. S., Smith, and Hirshberg, 330(87)
Mathematics, supporting, 16

Index 625

Maximum deflection, 232
Maximum normal stress theory, 438
Maximum shear stress theory, 443
Mean square, 27(n5)
Mean value, 16, 60
Mechanical optimization, 522
Metal products, tolerances, 135
Metals Handbook, 291(67), 316(67)
Mischke, C., 291(68), 391(68), 427(68), 457(68), 459(68), 460(68), 515(85), 522(121)
Models, engineering variables, 7
Modulus of elasticity, 191
Moment generating function, bivariate normal, 38
Moment of inertia, 378
Moment statistics, 377
Monte Carlo-concept, 74
Monte Carlo-simulation, 155
Monte Carlo-simulation of K, 237
Morrison, J., 335

Narrow band fatigue data, 334
Natural frequency, 268
Natural tolerance limits, 132
Nonconventional applications, 381
Nonlinear combinations, 154
Normal distributions: algebra of, 39
  as model, 59
Normally distributed stress, 168
Normal-stress and strength, 363
Normal-random stresses, sum, 169
Notch sensitivity, 462
Number, properties of, 17
Numbers, random, 75

O'Neill, H., 314(72)
Opposing failure modes, 540
Optimization, 5, 515
  problems, 516
  with stress gradient, 524
  with zero stress gradient, 532
Optimum design, 519
Osgood, C., 468(104)
Overload, 100

Parabolic Gerber envelopes, 339
Parzen, E., 350(34)
Photoelastic methods, 280
Point estimate, 546
Poisson ratio, 194
Pricing, Warranties and Inventory control, 434

Probabilistic-approach, 357
Probabilistic-Pythagorean theorem, 63
Probability considerations, 18
Product of random variables, 45
  and inverse, 51
Program Monti-flow chart, 286

Quadratics-statistics of, 56
Quality of elasticity, 191
Quotient, mean and variance, 32

Random load-fatigue data, 343
Random-life prediction, 321(32)
Random processes, 119
Random variables, 16
  continuous, 24
Rayleigh distribution, 124
Reliability, 3, 361
  assurance, 428
  calculation, 365
  confidence intervals, 546
  performance measure, 367
Resonant responses, random, 198
Rockwell C hardness, 317
Root mean square (RMS), 27
Rupture, modulus of, 306

Safety factors, 358
Sample size, 75
Sensitivity analysis, 67
Sensitivity of R, 375
Sensitivity to k_f, 476
Shapiro, S., 323, 387, 433
Shear stress, torsion and bending, 66
Shigley, 1, 298, 314, 315, 326, 336, 466
Significant root in design, 374
Significant strength, 454
Significant stress, 437, 455
Simpson's rule, 397
Sines, G., 314, 324, 455
Size effect, 458
Slenderness ratio, 252
S-N curve, 322
S-N evenlope estimation, 326
Spotts, 366, 400, 461
Springs in parallel, 268
Springs in series, 268
Spring rate, k, 225
Spur gear-stresses, 216
Spur gear-tooth systems, 213
Staircase Test Method, 339
Starr, C., 437
Static ultimate strength, 7

Stationary random process, 125
Statistical-models, 336
Statistical-tolerance limits, 134
Statistics: of arbitrarily functions, 60
  endurance limit, 326
  functions of one R. V., 25
  functions of two R. V.'s, 28
  product, 30
  τ, 224
Steinberg, D., 515
Strain energy, 238
  general expression, 244
  in tension, 238
Stress:
  bending, 182
  constant amplitude, 148, 469
  narrow band random, 198
  random equivalent, 203
  random principal, 176
  random variable, 162
  shearing, 185
  static, 163
  static biaxial, 172
  and strain, 189
  torsional, 178
Stress concentration:
  envelopes, 287
  factor, 277
  and static loads, 281
Stress confidence limits, 169
Stress-strain curves, 192
Stress statistics with finite elements, 204
Students t distribution, 558
Sum of correlated normal R.V.'s, 41
Sum of independent normal R.V.'s, 41

Sum of random variables, 28
Superposition of R.V.'s, 187
Surface finish, 149
Swanson, S. S., 321(29), 340(29), 343(29), 348(29)

Taylor series, 61
Temperature: Factor, 460
  transition, 308
Tensile strength, 295
Time to first failure, 433
Tolerance limits, 132
Torsional formula, 222
Torsional-shearing stresses, 181
Transformation of parameters, 70

Ultimate load, 80
Ultimate tensile strength, 5
Unit strain, 189

Variance, 16, 22
  of random functions, 61
Variation, coefficient of, 56
Vibration, 262
  and damping, 265
  records, 263
  steady state, 266

Waisman, J., 314, 324, 455
Wirsching, P., 327
Work, and energy, 239

Yield, torsional, 305
Yield point, 295
Yield strength, 295